PARALLELOGRAM

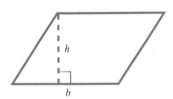

Area: $A = bh$

TRAPEZOID

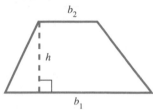

Area: $A = \dfrac{1}{2}h(b_1 + b_2)$

RECTANGULAR SOLID

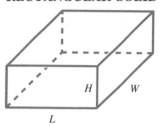

Volume: $V = LWH$

CIRCLE

Area: $A = \pi r^2$
Circumference: $C = \pi D$

SPHERE

Volume: $V = \dfrac{4}{3}\pi r^3$
Surface area: $S = 4\pi r^2$

RIGHT CIRCULAR CYLINDER

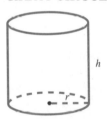

Volume: $V = \pi r^2 h$
Lateral surface area: $S = 2\pi rh$

CONE

Volume: $V = \dfrac{1}{3}\pi r^2 h$
Surface area: $S = \pi r \sqrt{r^2 + h^2}$

Other Formulas

Distance: $D = RT$ R = rate, T = time
Percent: $p = br$ p = percentage, b = base, r = rate
Temperature: $F = \dfrac{9}{5}C + 32$ $C = \dfrac{5}{9}(F - 32)$
Simple interest: $I = Prt$ P = principal, r = rate or percent, t = time in years
Amount: $A = P + Prt = P(1 + rt)$
Sale price: $S = L = rL$ S = sale price, L = list price, r = rate of discount

Elementary Algebra

ANNOTATED INSTRUCTOR'S EDITION

Elementary Algebra

MARK DUGOPOLSKI

Southeastern Louisiana University

ADDISON-WESLEY PUBLISHING COMPANY

Reading, Massachusetts • Menlo Park, California • New York
Don Mills, Ontario • Wokingham, England • Amsterdam • Bonn
Sydney • Singapore • Tokyo • Madrid • San Juan

Sponsoring Editor	Charles B. Glaser
Developmental Editor	Stephanie Botvin
Managing Editor	Karen M. Guardino
Production Supervisor	Marion E. Howe
Copy Editor	Barbara Willette
Proofreader	Laura Michaels
Text Designer	Geri Davis, Quadrata, Inc.
Design Consultant	Meredith Nightingale
Layout Artist	Lorraine Hodsdon
Cover Designer	Marshall Henrichs
Art Consultant	Loretta M. Bailey
Illustrator	Tech-Graphics
Photo Researcher	Susan Van Etten
Manufacturing Manager	Roy Logan
Marketing Manager	Melissa Acuña
Production Services Manager	Herbert Nolan
Compositor, Color Separator	York Graphic Services, Inc.
Printer	R.R. Donnelley & Sons

Library of Congress Cataloging-in-Publication Data

Dugopolski, Mark.
    Elementary algebra/by Mark Dugopolski.—Annotated instructor's edition.
      p. cm.
    Includes index.
    ISBN 0-201-52118-0
    1. Algebra. I. Title.
    QA152.2.D84 1991b
    512.9—dc20   90-39348
                                              CIP

ABCDEFGHIJ-DO-94321

To the Instructor

This text is designed for a one term course in elementary algebra, with an emphasis on the basics in the textual material as well as in the exercises. A complete development for each topic is provided and no prior training in algebra is required. Even though we assume a fundamental knowledge of arithmetic, the basic operations of arithmetic are reviewed as needed.

My primary goal in writing this book was to write one that students can read, understand, and enjoy, while gaining confidence in their ability to use mathematics in their everyday lives. Toward this end, I have endeavored to design a comprehensive, yet flexible, presentation of elementary algebra that will satisfy the individual needs of instructors and a variety of students. I have taken what I refer to as a "common-ground" approach whereby, whenever possible, the text reminds students of ideas they are already familiar with, and then builds on those ideas. This approach allows the instructor to meet the students on a common ground, before advancing to a new level.

Key Features

A complete pedagogical system that is designed to motivate the student and make mathematics more accessible is detailed in the following pages.

Ancillary Materials

A complete ancillary package for the instructor is also described in detail at the end of this section.

CHAPTER **4**

Factoring

A glider-pilot is participating in a contest in which he must attempt to drop a sandbag onto a target from a moving glider. How does he know when to release the sandbag? He certainly cannot wait until he is directly above the target. The sandbag has the same forward motion as the glider, and if released directly above the target, it will surely miss. There are many factors to be taken into account in solving this problem. The two most important are the speed of the glider and the altitude of the glider.

If the sandbag takes 4 seconds to fall to the ground, then the sandbag must be released 4 seconds before the glider is above the target. The sandbag continues the same forward motion as the glider, and it will hit the target when the glider is directly over the target. The time it takes an object to fall to the ground from a given altitude can be found by solving an equation involving a polynomial. In Exercise 41 of Section 4.6, we will determine how long it takes for a sandbag to fall from an altitude of 784 feet.

Chapter Opener
Each chapter begins with a realistic problem situation. After the student has learned the necessary skills, the problem is presented in the exercise set. See pages 143 and 171.

39. The sum of two numbers is 11, and their product is 30. Find the numbers. 5 and 6

40. Molly's age is twice Anita's. If the sum of the squares of their two ages is 80, then what are their ages? 4 and 8

41. If an object is dropped from a height of s_0 feet, then its altitude S after t seconds is given by the formula $S = -16t^2 + s_0$. If a sandbag is dropped by a glider-pilot from a height of 784 feet, then how long does it take for the sandbag to reach the ground? (On the ground, $S = 0$.) 7 seconds

42. If the glider-pilot of Exercise 41 throws his sandbag downward from an altitude of 128 feet with an initial velocity of 32 feet per second, then its altitude after t seconds is given by the formula $S = -16t^2 - 32t + 128$. How long does it take for the sandbag to reach the earth? 2 seconds

43. One end of a glass prism is in the shape of a triangle that has a height that is 1 inch longer than twice the base. If the area of the triangle is 39 square inches, then how long are the base and height? 6 inches, 13 inches

45. Last year Otto's garden was square in shape. This year he plans to make it smaller by shortening one side by 5 feet and the other by 8 feet. If the area of the smaller garden will be 180 square feet, then what was the size of Otto's garden last year? 20 feet by 20 feet

46. Rosita's Christmas present from Carlos is in a box that has a width that is 3 inches shorter than the height. The length of the base is 5 inches longer than the height. If the area of the base is 84 square inches, then what is the height of the package? 9 inches

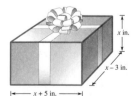

x in.

$x - 3$ in.

$x + 5$ in.

Figure for Exercise 46

47. Imelda and Gordon have designed a new kite. While Imelda is flying the kite, Gordon _____ directly below the kite. The kite is de____ ____ger than

2.2 More Linear Equations

IN THIS SECTION:

- Identities
- Conditional Equations
- Inconsistent Equations
- Equations Involving Fractions
- Equations Involving Decimals
- Mental Algebra

In This Section
Every section of the text begins with a list of the topic headings within that section. This aids the student in reading the section and in studying for tests and quizzes. See page 67.

In this section we will solve more equations of the type that we solved in Section 2.1. However, some equations in this section have infinitely many solutions, and some have no solution.

Identities

Have students write some identities of their own, where both sides are not exactly identical; e.g., $x + 3 = 3 + x$.

It is easy to find equations that are satisfied by any real number that we choose as a replacement for the variable. For example, the equations

$$x \div 2 = \frac{1}{2}x, \qquad x + x = 2x, \qquad \text{and} \qquad x + 1 = x + 1$$

Marginal Notes for Instructors
Suggestions and comments to the instructor appear in the text margin of the Annotated Instructor's Edition. These comments give tips on teaching the material, including alternative examples and explanations, and explain how the current material connects to later chapters. These notes are printed in blue so the instructor can quickly distinguish them from the textual material. See pages 67 and 68.

Review why division by 0 is undefined. How would you explain to your child why division by 0 is undefined?

are satisfied by all real numbers. The equation

$$\frac{5}{x} = \frac{5}{x}$$

is satisfied by any real number except 0, because division by 0 is undefined.

Boxed Definitions and Rules
Important definitions and rules are boxed for quick reference. See page 68.

DEFINITION
Identity

> An equation that is satisfied by every number for which both sides of the equation are defined is called an **identity.**

Emphasize that an identity does not have to be true for every real number.

We cannot recognize that the equation in the next example is an identity until we have simplified each side.

Encoding and Decoding

Algebra is useful because it can be used to solve problems. Since problems are generally communicated verbally, we must be able to translate verbal expressions into algebraic expressions (**encode**) and translate algebraic expressions into verbal expressions (**decode**). Consider the following examples of verbal expressions and their corresponding algebraic expressions.

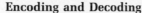

Verbal Expressions and Corresponding Algebraic Expressions

Verbal Expression	Algebraic Expression
The **sum** of $5x$ and 3	$5x + 3$
The **product** of 5 and $x + 3$	$5(x + 3)$
The **sum** of 8 and $x/3$	$8 + \dfrac{x}{3}$
The **quotient** of $8 + x$ and 3	$\dfrac{8 + x}{3}$
The **difference** of 3 and x^2	$3 - x^2$
The **square** of $3 - x$	$(3 - x)^2$

Encoding Phrases

Many verbal phrases occur repeatedly in applications. The following box contains a list of some frequently occurring verbal phrases and their equivalent algebraic expressions.

Encoding Verbal Phrases

Verbal Phrase	Algebraic Expression
Addition	
The sum of a number and 8	$x + 8$
Five is added to a number	$x + 5$
Two more than a number	$x + 2$
A number increased by 3	$x + 3$
Subtraction	
Four is subtracted from a number	$x - 4$
Three less than a number	$x - 3$
The difference between 7 and a number	$7 - x$
Some number decreased by 2	$x - 2$
Multiplication	
The product of 5 and a number	$5x$
Twice a number	$2x$
One half of a number	$\dfrac{1}{2}x$
Five percent of a number	$.05x$
Division	
The ratio of a number and 6	$x/6$
The quotient of 5 and a number	$5/x$
Three divided by some number	$3/x$

I emphasize that we want to solve simple equations mentally, because later on we must solve simple equations as part of more complicated problems such as finding intercepts.

Mental Algebra

It is very important to develop the skill of solving equations in a systematic way, writing down every step as we have been doing. It is also important to be able to solve simple equations quickly. In Chapter 1 we learned to solve simple equations mentally by decoding. The method of solving equations that we have been using in this chapter can also be done mentally if the solution involves only one or two steps. For example, to solve

$$x + 5 = 0,$$

we can subtract 5 from each side mentally and write

$$x = -5.$$

To solve the equation

$$3x - 7 = 0,$$

we add 7 to each side mentally and then divide each side by 3 to get

$$x = \frac{7}{3}.$$

EXAMPLE 6 Solve each equation mentally.

a) $x - 3 = 0$ b) $2x + 5 = 0$ c) $3x = \frac{1}{2}$

Mental Math
Once a skill is learned, reminders within the text point out to the student where steps can be combined or performed mentally. This frees the student from writing down each step and adds to the student's confidence in solving problems. See pages 71–72.

Solution

a) Add 3 to each side to get $x = 3$. The solution is 3.
b) Subtract 5 from each side and then divide by 2 to get $x = -5/2$. The solution is $-5/2$.
c) Multiply each side by 1/3. Since

$$\frac{1}{3} \cdot \frac{1}{2} = \frac{1}{6}$$

we get $x = 1/6$. The solution is 1/6. ◀

Math At Work
A Math At Work box shows students the mathematics of that chapter in the context of everyday life. Applications are taken from a variety of fields to maximize student interest. The exercise in this box is optional, but it may be used to promote classroom discussion. See page 77.

Math at Work

Food chemists and manufacturers want to know: How hot is hot?

The spiciness of hot peppers could not be measured systematically until W. L. Scoville invented a method in 1912. Using the Scoville method, a panel of tasters sipped measured concentrations of hot pepper diluted with sugar water and determined how long it took for the pepper to stop burning. The longer the time, the higher the pepper's rating on what is now called the Scoville scale.

As spicy foods cooked with hot peppers have become more common in the United States, scientists and food manufacturers have searched for a more accurate way to measure spiciness. Using a method of chemical analysis called chromatography, food chemists can now measure the amount of the powerful chemical called capsaicin that gives peppers their heat.

Today, most food manufacturers use chromatography rather than the Scoville method to measure the hotness of pepper products. Advocates claim that chromatography is more efficient and creates more uniform results than the Scoville method. However, critics point out that chromatography ignores the other chemicals in the pepper that contribute to the overall burning sensation.

A food chemist has determined through experimentation that the relationship between Scoville units (SU) and chromatography measurements (in parts per million) is given by the formula

$$\text{ppm} = \frac{\text{SU}}{15}.$$

How many parts per million of capsaicin are contained in the following foods?

Chili con carne	20 Scoville units	1.3 ppm
Chile pepper, dry ground	975 Scoville units	65 ppm
Jalapeno pepper, fresh and green	1700 Scoville units	111.3 ppm
Pure capsaicin	15,000,000 Scoville units	1,000,000 ppm

General Strategy for All Word Problems

The steps to follow in providing a complete solution to a word problem can be stated as follows.

STRATEGY
Solving Word Problems

1. Read the problem.
2. If possible, draw a diagram to illustrate the problem.
3. Choose a variable and write down what it represents.
4. Represent any other unknowns in terms of that variable.
5. Write an equation that fits the situation.
6. Solve the equation.
7. Be sure that your solution answers the question posed in the original problem.
8. Check your answer by using it to solve the original problem (not the equation).

Strategy Boxes
Strategy boxes teach students how to approach new problems using skills they have already learned. The steps are consistently applied throughout the text. This gives students confidence in solving problems, especially word problems. See pages 89 and 163.

The Factoring Strategy

The following strategy is a summary of the ideas that we use to factor a polynomial completely.

STRATEGY
Factoring Polynomials Completely

1. If there are any common factors, factor them out first.
2. When factoring a binomial, check to see whether it is the difference of two squares. Remember that a sum of two squares does not factor.
3. When factoring a trinomial, check to see whether it is a perfect square trinomial.
4. When factoring a trinomial that is not a perfect square, use the ac method.
5. If the polynomial has four terms, try factoring by grouping.

We will use the factoring strategy in the next example.

EXAMPLE 1 Factor each polynomial completely:

a) $2a^2b - 24ab + 72b$
b) $3x^3 + 6x^2 - 75x - 150$

Solution

a) $2a^2b - 24ab + 72b = 2b(a^2 - 12a + 36)$ **First factor out the GCF $2b$.**
   $\qquad\qquad\qquad\quad = 2b(a - 6)^2$ **Factor the perfect square trinomial.**
b) $3x^3 + 6x^2 - 75x - 150 = 3[x^3 + 2x^2 - 25x - 50]$ **First factor out the GCF 3.**
   $\qquad\qquad\qquad\qquad\quad = 3[x^2(x + 2) - 25(x + 2)]$ **Factor by grouping.**
   $\qquad\qquad\qquad\qquad\quad = 3(x^2 - 25)(x + 2)$ **Factor out $x + 2$.**
   $\qquad\qquad\qquad\qquad\quad = 3(x + 5)(x - 5)(x + 2)$ **Factor the difference of two squares.** ◄

Student Annotations
Carefully worded notes guide the student through successive steps of an example. See page 164.

Warm-ups

Warm-ups
Prior to each exercise set is a set of 10 true-false statements. These statements review the concepts developed in the section and many point out common student errors. They can be used in class to stimulate discussion or by the student to study for exams. The Warm-ups do not require any written work and can be answered easily through understanding of the material. See pages 79–80.

True or false?

1. If we solve $D = R \cdot T$ for T, we get $T \cdot R = D$. F
2. If we solve $a - b = 3a - m$ for a, we get $a = 3a - m + b$. F
3. Solving $A = L \cdot W$ for L, we get $L = W/A$. F
4. Solving $D = R \cdot T$ for R, we get $R = d/t$. F
5. The perimeter of a rectangle is found by multiplying its length and width. F
6. The volume of a rectangular box is found by multiplying its length, width, and height. T
7. The length of an NFL football field, excluding the end zones, is 99 yards. F

8. Solving $y - x = 5$ for y gives us $y = x + 5$. T
9. If $x = -1$ and $y = -3x + 6$, then $y = 3$. F
10. The perimeter of a rectangle is a measurement of the total distance around the outside edge. T

2.3 EXERCISES

Keyed and Non-keyed Exercises
Each exercise set begins with exercises specifically keyed to the examples. These are followed by exercises that are not strictly linked to the examples. Keyed exercises build confidence; non-keyed exercises are designed to test the student's ability to synthesize ideas. See page 80.

Solve each formula for the specified variable. See Examples 1 and 2.

1. $D = R \cdot T$ for R $R = D/T$
2. $A = L \cdot W$ for W $W = A/L$
3. $I = Prt$ for P $P = \dfrac{I}{rt}$
4. $I = Prt$ for t $t = \dfrac{I}{Pr}$
5. $F = \dfrac{9}{5}C + 32$ for C $C = \dfrac{5}{9}(F - 32)$
6. $C = 2\pi r$ for r $r = \dfrac{C}{2\pi}$
7. $A = \dfrac{1}{2}bh$ for h $h = 2A/b$
8. $A = \dfrac{1}{2}bh$ for b $b = 2A/h$
9. $P = 2L + 2W$ for L $L = \dfrac{P - 2W}{2}$
10. $P = 2L + 2W$ for W $W = \dfrac{P - 2L}{2}$
11. $C = \pi D$ for π $\pi = C/D$
12. $F = ma$ for a $a = F/m$
13. $A = \dfrac{1}{2}(a + b)$ for a $a = 2A - b$
14. $A = \dfrac{1}{2}(a + b)$ for b $b = 2A - a$
15. $S = P + Prt$ for r $r = \dfrac{S - P}{Pt}$
16. $S = P + Prt$ for t $t = \dfrac{S - P}{Pr}$
17. $A = \dfrac{1}{2}h(a + b)$ for a $a = \dfrac{2A - bh}{h}$
18. $A = \dfrac{1}{2}h(a + b)$ for b $b = \dfrac{2A - ah}{h}$

Solve each equation for x. See Example 3.

19. $5x + a = 3x + b$ $x = \dfrac{b - a}{2}$
20. $2x - x = 4x + b - 5b$ $x = \dfrac{c + 5b}{5}$
21. $4(a + x) - 3(x - a) = 0$ $x = -7a$
22. $-2(x - b) - (5a - x) = a + b$ $x = b - 6a$
23. $3x - 2(a - 3) = 4x$ $x = 12 - a$
24. $2(x - 3w) + 3(x + w) = 0$ $x = 3w/5$
25. $3x + 2ab = 4x - 5ab$ $x = 7ab$
26. $x - a = -x + a + 4b$ $x = a + 2b$

Solve for y. See Examples 4 and 5.

27. $x + y = -9$ $y = -x - 9$
28. $3x + y = -5$ $y = -3x - 5$
29. $x + y - 6 = 0$ $y = -x + 6$
30. $4x + y - 2 = 0$ $y = -4x + 2$
31. $2x - y = 2$ $y = 2x - 2$
32. $x - y = -3$ $y = x + 3$
33. $3x - y + 4 = 0$ $y = 3x + 4$
34. $-2x - y + 5 = 0$ $y = -2x + 5$
35. $y - x = 7$ $y = x + 7$
36. $y - x + 3 = 0$ $y = x - 3$
37. $x + 2y = 4$ $y = -\dfrac{1}{2}x + 2$
38. $3x + 2y = 6$ $y = -\dfrac{3}{2}x + 3$
39. $2x - 2y = 1$ $y = x - \dfrac{1}{2}$
40. $3x - 2y = -6$ $y = \dfrac{3}{2}x + 3$
41. $y + 2 = 3(x - 4)$ $y = 3x - 14$
42. $y - 1 = \dfrac{1}{2}(x - 2)$ $y = \dfrac{1}{2}x$
43. $\dfrac{1}{2}x - \dfrac{1}{3}y = -2$ $y = \dfrac{3}{2}x + 6$
44. $\dfrac{x}{2} + \dfrac{y}{4} = \dfrac{1}{2}$ $y = -2x + 2$
45. $y - 3 = -3(x - 1)$ $y = -3x + 6$
46. $y + 9 = -1(x - 1)$ $y = -x - 8$
47. $y - 5 = -6(x + 4)$ $y = -6x - 19$
48. $y - 4 = -\dfrac{2}{3}(x - 9)$ $y = -\dfrac{2}{3}x + 10$
49. $y - 7 = -\dfrac{1}{2}(x - 4)$ $y = -\dfrac{1}{2}x + 9$
50. $\dfrac{2}{3}x - \dfrac{1}{2}y = \dfrac{1}{6}$ $y = \dfrac{4}{3}x - \dfrac{1}{3}$

Calculator Exercises

Many exercise sets include exercises that can be done most easily using a calculator. These exercises are optional and are marked by an icon. Calculator exercises provide the student with the necessary practice to become proficient with a scientific calculator. See page 281.

Perform the following computations with the aid of a scientific calculator. Write answers in scientific notation. Round the decimal part to three decimal places.

45. $(6.3 \times 10^6)(1.45 \times 10^{-4})$ 9.135×10^2

46. $(8.35 \times 10^9)(4.5 \times 10^3)$ 3.758×10^{13}

47. $(5.36 \times 10^{-4}) + (3.55 \times 10^{-5})$ 5.715×10^{-4}

48. $(8.79 \times 10^8) + (6.48 \times 10^9)$ 7.359×10^9

49. $(3.56 \times 10^{85})(4.43 \times 10^{96})$ 1.577×10^{182}

50. $(4.36 \times 10^{55})(7.7 \times 10^{88})$ 3.357×10^{144}

51. $(8 \times 10^{99}) + (3 \times 10^{99})$ 1.1×10^{100}

52. $(8 \times 10^{-99}) + (9 \times 10^{-99})$ 1.7×10^{-98}

53. $\dfrac{(3.5 \times 10^5)(4.3 \times 10^{-6})}{3.4 \times 10^{-8}}$ 4.426×10^7

54. $\dfrac{(3.5 \times 10^{-8})(4.4 \times 10^{-4})}{2.43 \times 10^{45}}$ 6.337×10^{-57}

55. The distance from the earth to the sun is 93 million miles. Express this distance in feet. (1 mile = 5280 feet.) 4.910×10^{11} feet

56. The speed of light is 9.83569×10^8 feet per second. How long does it take light to get from the sun to the earth?

57. How long does it take a spacecraft traveling at 2×10^{35} miles per hour (warp factor 4) to travel 93 million miles. 4.65×10^{-28} hours

58. If the radius of a very small circle is 2.35×10^{-8} centimeters, then what is the circle's area?

59. If the circumference of a circle is 5.68×10^9 feet, then what is its radius? 9.040×10^8 feet

60. If the diameter of a circle is 1.3×10^{-12} meters, then what is its radius? 6.5×10^{-13} meters

56. 499.2 seconds or 8.3 minutes **58.** 1.735×10^{-15} square centimeters

Geometry

Numerous geometric problems throughout the text review and reinforce basic geometric relationships. Illustrations highlight the geometry within the problem situations. A summary of common geometric formulas and figures can be found inside the front cover. See page 310.

7.4 EXERCISES

Find the exact solution to each problem. See Example 1.

1. The length of a rectangle is 2 meters longer than the width. If the area is 10 square meters, then what are the length and width? $L = 1 + \sqrt{11}$ meters, $W = -1 + \sqrt{11}$ meters

2. One leg of a right triangle is 4 centimeters longer than the other leg. If the area of this triangle is 8 square centimeters, then what are the lengths of the legs?

2. $-2 + 2\sqrt{5}$ centimeters, $2 + 2\sqrt{5}$ centimeters

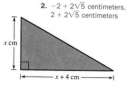

Figure for Exercise 2

3. If the diagonal of a square is 8 feet long, then what is the length of the side of the square? $4\sqrt{2}$ feet

4. If one side of a rectangle is 2 meters shorter than the other side and the diagonal is 10 meters long, then what are the dimensions of the rectangle? 6 meters by 8 meters

5. The base of a parallelogram is 6 inches longer than its height. If the area of the parallelogram is 10 square inches, then what are the base and height?

5. Base: $3 + \sqrt{19}$ inches, height: $-3 + \sqrt{19}$ inches

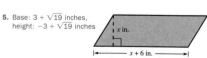

Figure for Exercise 5

Answers for Instructors

Answers to both even and odd numerical exercises are printed directly in the text of the Annotated Instructor's Edition (answers requiring graphs appear in a separate section following this section). These answers are printed in blue so they can be distinguished from the exercises. See page 310.

39. The sum of two numbers is 11, and their product is 30. Find the numbers. 5 and 6

40. Molly's age is twice Anita's. If the sum of the squares of their two ages is 80, then what are their ages? 4 and 8

41. If an object is dropped from a height of s_0 feet, then its altitude S after t seconds is given by the formula $S = -16t^2 + s_0$. If a sandbag is dropped by a glider-pilot from a height of 784 feet, then how long does it take for the sandbag to reach the ground? (On the ground, $S = 0$.) 7 seconds

42. If the glider-pilot of Exercise 41 throws his sandbag downward from an altitude of 128 feet with an initial velocity of 32 feet per second, then its altitude after t seconds is given by the formula $S = -16t^2 - 32t + 128$. How long does it take for the sandbag to reach the earth? 2 seconds

43. One end of a glass prism is in the shape of a triangle that has a height that is 1 inch longer than twice the base. If the area of the triangle is 39 square inches, then how long are the base and height? 6 inches, 13 inches

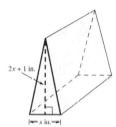

$2x + 1$ in.

x in.

Figure for Exercise 43

44. The radius of a circle is 1 meter longer than the radius of another circle. If their areas differ by 5π square meters, then what is the radius of each? 2 meters and 3 meters

45. Last year Otto's garden was square in shape. This year he plans to make it smaller by shortening one side by 5 feet and the other by 8 feet. If the area of the smaller garden will be 180 square feet, then what was the size of Otto's garden last year? 20 feet by 20 feet

46. Rosita's Christmas present from Carlos is in a box that has a width that is 3 inches shorter than the height. The length of the base is 5 inches longer than the height. If the area of the base is 84 square inches, then what is the height of the package? 9 inches

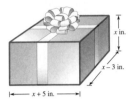

x in.

$x - 3$ in.

$x + 5$ in.

Figure for Exercise 46

47. Imelda and Gordon have designed a new kite. While Imelda is flying the kite, Gordon is standing directly below the kite. The kite is designed so that its altitude is always 20 feet larger than the distance between Imelda and Gordon. What is the altitude of the kite when it is 100 feet from Imelda? 80 feet

48. A car is traveling on a road that is perpendicular to a railroad track. When the car is 30 meters from the crossing, the car's new collision detector warns the driver that there is a train 50 meters from the car and heading toward the same crossing. How far is the train from the crossing?

49. Virginia is buying carpet for two square rooms. One room is 3 yards wider than the other. If she needs 45 square yards of carpet, then what are the dimensions of each room? 3 yards by 3 yards and 6 yards by 6 yards

50. Ahmed has one-half of a treasure map, which indicates that the treasure is buried in the desert $2x + 6$ paces from Castle Rock. Vanessa has the other half, which indicates that to find the treasure one must get to Castle Rock, walk x paces to the north, and then $2x + 4$ paces to the east. If they pool their information to save a lot of digging, then what is the value of x?

Applications
Teaching students to solve applied problems is one of the main goals of this text. Word problems occur in over 50% of the exercise sets. Photographs and drawings illustrate many of the applications and help the student relate to the situation. See page 171.

Wrap-up

Chapter Summary
Each chapter ends with a summary that illustrates each important idea of the chapter and gives an example. The chapter summaries are beneficial to the student in reviewing the chapter and in studying for exams. See page 139.

CHAPTER 3

SUMMARY

Concepts		Examples
Term of a polynomial	A number or the product of a number and one or more variables	$5x^3$, $-4x$, 7
Polynomial	A single term or a finite sum of terms	$2x^5 - 4x^2 + 5$
Add or subtract polynomials	Add or subtract the like terms.	$(x + 1) + (x - 4) = 2x - 3$ $(x^2 - 3x) - (4x^2 - x) = -3x^2 - 2x$
Product rule	$x^m \cdot x^n = x^{m+n}$	$a \cdot a^3 = a^4$ $3x^6 \cdot 4x^9 = 12x^{15}$
Multiply polynomials	Multiply each term of one polynomial by every term of the other polynomial, then combine like terms.	$x^2 + 2x + 5$ $\underline{\quad\quad x - 1}$ $-x^2 - 2x - 5$ $\underline{x^3 + 2x^2 + 5x\quad}$

REVIEW EXERCISES

3.1 *Perform the indicated operations.*

1. $(2w - 6) + (3w + 4)$ $5w - 2$

2. $(1 - 3x) + (4x - 6)$ $x - 5$

3. $(x^2 - 2x - 5) - (x^2 + 4x - 9)$ $-6x + 4$

4. $(3 - 5x - x^2) - (x^2 - 7x + 8)$ $-2x^2 + 2x - 5$

5. $(5 - 3x + x^2) + (x^2 - 4x - 9)$ $2x^2 - 7x - 4$

6. $(-2x^2 + 3x - 4) + (x^2 - 7x + 2)$ $-x^2 - 4x - 2$

7. $(4 - 3x - x^2) - (x^2 - 6x + 5)$ $-2x^2 + 3x - 1$

8. $(x^3 - x) - (x^2 + 5)$ $x^3 - x^2 - x - 5$

Chapter Review
The chapter review consists of sets of exercises keyed to the sections of the chapter, followed by miscellaneous exercises that are designed to test the student's ability to synthesize the various concepts. See page 140.

Chapter Test
Following the Chapter Review, the Chapter Test gives the student a way to check his or her readiness for a test on that chapter. In order to help the students to be independent of the examples, the chapter test is not keyed to the text sections. See page 141.

CHAPTER 3 TEST

Perform the indicated operations.

1. $7x^3 + 4x^2 + 2x - 11$ **2.** $-x^2 - 9x + 2$

1. $(7x^3 - x^2 - 6) + (5x^2 + 2x - 5)$

2. $(x^2 - 3x - 5) - (2x^2 + 6x - 7)$

3. $-5x^3 \cdot 7x^5$ $-35x^8$

4. $3x^3 \cdot 3x^3$ $9x^6$

5. $3x^3 + 3x^3$ $6x^3$

6. $3x^3 \div 3x^3$ 1

7. $-4a^6b^5 \div (-2a^5b)$ $2ab^4$

8. $\dfrac{-6a^7b^6}{-2a^3b^4}$ $3a^4b^2$

9. $\dfrac{6y^3 - 9y^2}{-3y}$ $-2y^2 + 3y$

10. $(x^2 - 5x + 3)(x - 2)$

11. $(x^3 - 2x^2 - 4x + 3) \div (x - 3)$

12. $3x^2(5x^3 - 7x^2 + 4x - 1)$

13. $(x - 2) \div (2 - x)$ -1

10. $x^3 - 7x^2 + 13x - 6$ **11.** $x^2 + x - 1$ **12.** $15x^5 - 21x^4 + 12x^3 - 3x^2$

Tying It All Together

CHAPTERS 1–3

Simplify each expression.

1. $-16 \div (-2)$ 8

2. $(-2)^3 - 1$ -9

3. $(-5)^2 - 3(-5) + 1$ 41

4. $2^{10} \cdot 2^{15}$ 2^{25}

5. $2^{15} \div 2^{10}$ 2^5

6. $2^{10} - 2^5$ 992

7. $3^2 \cdot 4^2$ 144

8. $(172 - 85) \div (85 - 172)$ -1

9. $(5 + 3)^2$ 64

10. $5^2 + 3^2$ 34

11. $(30 - 1)(30 + 1)$ 899

12. $(30 + 1)^2$ 961

Perform the indicated operations.

13. $(x + 3)(x + 5)$ $x^2 + 8x + 15$

14. $(x^2 + 8x + 15) \div (x + 5)$ $x + 3$

15. $x + 3(x + 5)$ $4x + 15$

16. $(x^2 + 8x + 15)(x + 5)$ $x^3 + 13x^2 + 55x + 75$

17. $-5t^3v \cdot 3t^2v^6$ $-15t^5v^7$

18. $(-10t^3v^2) \div (-2t^2v)$ $5tv$

19. $(-6y^3 + 8y^2) \div (-2y^2)$ $3y - 4$

20. $(y^2 - 3y - 9) - (-3y^2 + 2y - 6)$ $4y^2 - 5y - 3$

Tying it All Together
Beginning in Chapter 2, Tying it All Together presents exercises that help the student to synthesize material from all of the preceding chapters. There may also be exercises that will help in understanding the upcoming chapters. See page 142.

Ancillary Materials

Annotated Instructor's Edition

In addition to the material presented in the student text, the AIE includes marginal notes for instructors; answers to all numerical exercises in the text; and answers to all graphing exercises in the instructor's answer section, which follows this preface.

Instructor's Solutions Manual

This supplement includes detailed solutions to every exercise in the text.

Student's Solutions Manual

Includes detailed solutions to the odd-numbered exercises

Student's Study Guide

Includes diagnostic tests for each chapter, as well as detailed solutions and explanations for each test question.

COMPUTERIZED TESTING

AWTest (Apple)

Algorithmic-based testing system keyed to text: multiple choice, open-ended, and true/false questions.

OmniTest (IBM)

A powerful new testing system developed exclusively for Addison-Wesley. It is an algorithm-driven system that allows the user to create up to 99 perfectly parallel forms of any test effortlessly, and allows the user to add his or her own test items and edit existing items with an easy to use, on-screen "What-You-See-Is-What-You-Get" text editor.

Mac Test (Macintosh)

Test Item bank containing 10 questions for each objective drawn from the text. Questions are in multiple choice format.

A demonstration package, including examples of each type of testing package, is available upon request.

Printed Test Bank

A printed test bank is available that contains three alternate forms of tests for each chapter, as well as cumulative review tests and final exams.

Video Tapes

An extensive set of video tapes is available free upon adoption. These videos cover every topic presented in the text. Particular attention has been paid to working examples and giving complete explanation of the steps involved.

TUTORIAL SOFTWARE

IMPACT: An Interactive Mathematics Tutorial (IBM & Mac)

Tutorial correlated to each section of the text. This package will generate exercises similar to those in the text exercise set. If the student gets an exercise wrong, he or she can see either an example or a solution. The program will lead students interactively through step-by-step solutions to the exercise so that they can easily pinpoint specific trouble areas if they make an error.

The Math Lab (IBM & Apple II)

Students pick the topic area, level of difficulty, and number of problems. If they get a wrong answer, the program will prompt them with the first step of the solution. This program keeps detailed records of students' results, which can be stored on the disk.

Professor Weissman's Software (IBM)

Students pick the topic area and level of difficulty. The program generates and gives step-by-step solutions if students get a wrong answer. Problems increase in difficulty as students are successful. This program also keeps records of students' scores.

Algebra Problem Solver (IBM)

Students choose a topic and can either request an exercise from the computer or enter their own homework exercise. The computer will show them a step-by-step solution to the problem.

24-HOUR HINTS BY PHONE

The Math Hotline

This innovative new idea in teaching and learning packages was created by Mark Dugopolski specifically with the struggling student in mind. Suppose one of your students is at home attempting to complete an assignment and just doesn't know how to start solving a problem. Now, simply by calling the Math Hotline (504) 542-0658 [(800) 874-4896 September 1990 through June 1991] from a touch-tone phone and responding to a few questions with the keypad, the student can receive a hint for any odd-numbered exercise. This means thousands of hints are as close as the phone, 24-hours a day, 365 days a year.

Instructor's Answer Section

CHAPTER 1

Section 1.2

2.
   5 6 7 8 9...

4.
   −2 −1 0 1 2

6.
   ...−6 −5 −4 −3 −2

8.
   1 2 3 4 5 6

10.
   0 1 2

CHAPTER 2

Section 2.6

16.
   −11 −10 −9 −8 −7 −6 −5

18.
   2 3 4 5 6 7 8

20.
   −4 −3 −2 −1 0 1 2

22.
   −9 −8 −7 −6 −5 −4 −3

24.
   0 1 2 3 4 5 6

26.
   −5 −4 −3 −2 −1 0 1

28.
   200 240 280

30.
   −3.4
   −7 −6 −5 −4 −3 −2

32.
   0 1 2 3 4 5

34.
   −5 −4 −3 −2 −1 0 1

36. (number line, open circle at −2, filled circle at 2; scale −3 −2 −1 0 1 2 3)

38. (number line, filled circle at 0, open circle at 6; scale 0 1 2 3 4 5 6)

Section 2.7

2. (number line, scale −3 −2 −1 0 1 2 3)

4. (number line, open circle at 4; scale 0 1 2 3 4 5 6)

6. (number line, open circle at −2; scale −4 −3 −2 −1 0 1 2)

8. (number line, filled circle at −5; scale −9 −8 −7 −6 −5 −4 −3)

10. (number line, filled circle at 6; scale 2 3 4 5 6 7 8)

12. (number line, open circle at 5; scale 2 3 4 5 6 7 8)

14. (number line, open circle at 13; scale 9 10 11 12 13 14 15)

16. (number line, filled circle at −3; scale −5 −4 −3 −2 −1 0 1)

18. (number line, open circle at −4; scale −8 −7 −6 −5 −4 −3 −2)

20. (number line, open circle at 18; scale 16 17 18 19 20 21 22)

22. (number line, open circle at 3.91; scale 2 3 4 5 6 7)

24. (number line, open circle at 7 and open circle at 10; scale 6 7 8 9 10 11 12)

26. (number line, open circle at $-\frac{7}{3}$ and open circle at 1; scale −3 −2 −1 0 1 2 3)

28. (number line, open circle at −1 and open circle at 4; scale −1 0 1 2 3 4)

30. (number line, open circle at −1 and filled circle at 1; scale −3 −2 −1 0 1 2 3)

32. (number line, open circle at $-\frac{15}{2}$ and filled circle at $\frac{3}{2}$)

34. (number line, filled circle at −5 and filled circle at 1; scale −5 −4 −3 −2 −1 0 1)

36. (number line, open circle at −2.69 and open circle at −.91)

Chapter 2 Review Exercises

106. (number line, open circle at −11; scale −14 −13 −12 −11 −10 −9 −8)

108. (number line, open circle at 10; scale 8 9 10 11 12 13 14)

110. (number line, filled circle at −15; scale −17 −16 −15 −14 −13 −12 −11)

112. (number line, open circle at −1 and filled circle at $\frac{1}{2}$; scale −3 −2 −1 0 1 2 3)

114. (number line, open circle at 0 and open circle at 8; scale 0 2 4 6 8 10)

CHAPTER 8

Section 8.1

26.

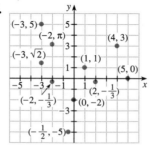

28.

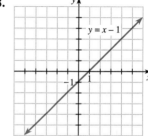

30.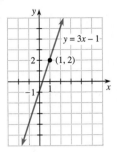

32.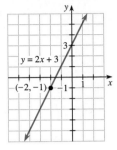

34.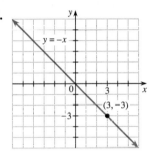

36.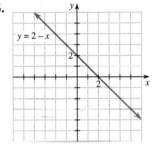

38.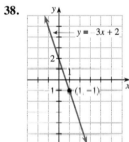

40.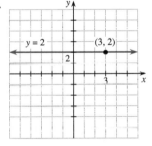

42.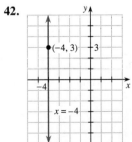

44.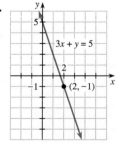

46.

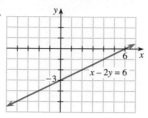

48.

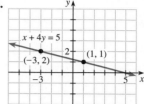

50.

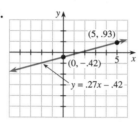

64.

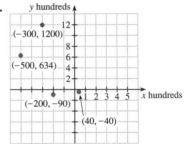

72.

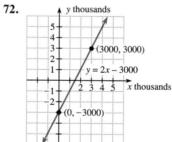

74.

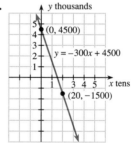

76.

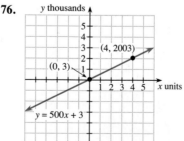

Section 8.3

32.

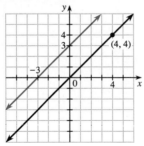

34.

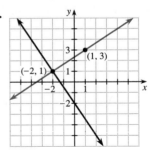

36.

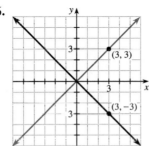

38.

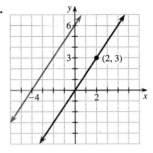

40.

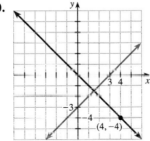

Section 8.4

42.

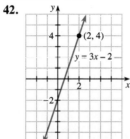

44.

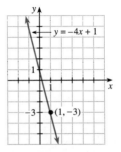

46.

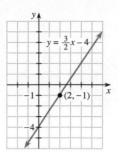

48.

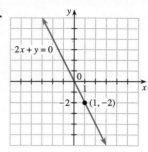

50.

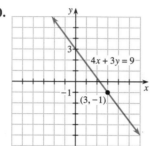

52.

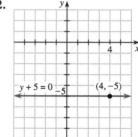

54.

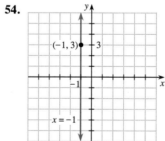

56.

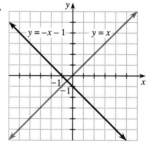

58.

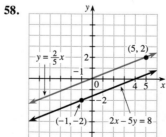

60.

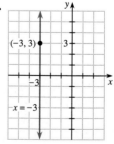

62.

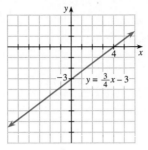

Section 8.5

22.

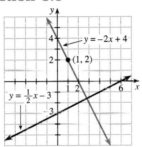

24.

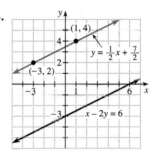

Section 8.6

10.

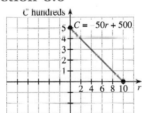

12.

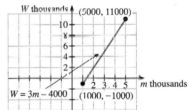

14.

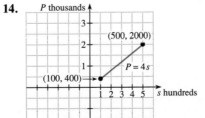

16.

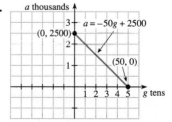

18.

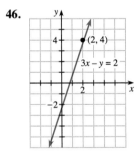

V = −4000a + 30000

Chapter 8 Review Exercises

14.

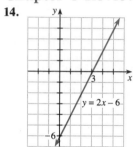

y = 2x − 6

16.

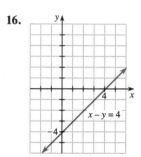

x − y = 4

44.

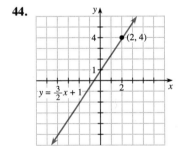

$y = \frac{3}{2}x + 1$

46.

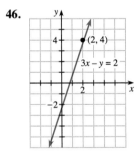

3x − y = 2

48.

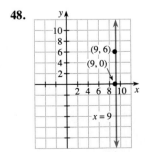

x = 9

68.

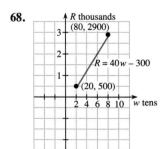

R = 40w − 300

70.

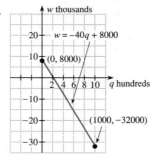

CHAPTER 9

Section 9.4

8.

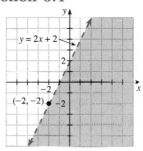

10.

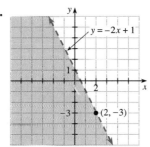

12.

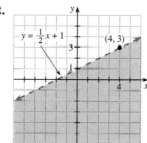

14.

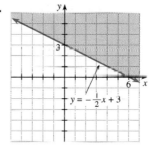

16.

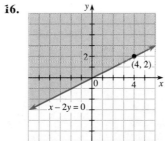

18.

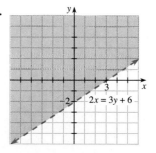

20.

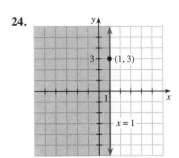

22.

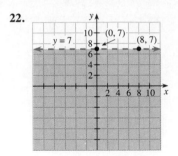

24.

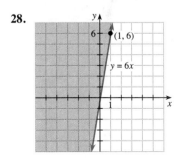

26.

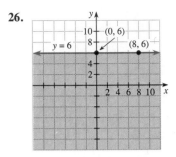

28.

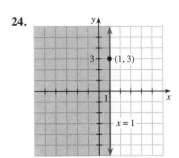

30.

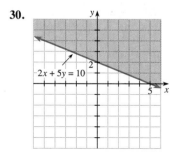

32.

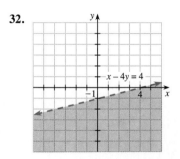

34.

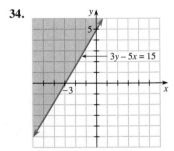

36.

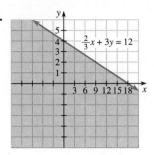

$-\frac{2}{3}x + 3y = 12$

38.

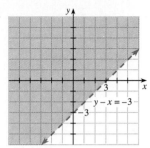

$y - x = -3$

40.

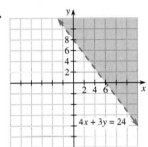

$4x + 3y = 24$

42.

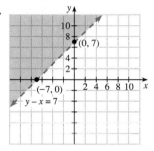

$(0, 7)$

$(-7, 0)$

$y - x = 7$

Section 9.5

8.

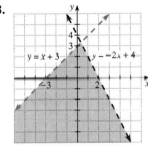

$y = x + 3$

$y = -2x + 4$

10.

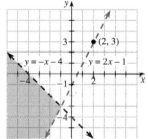

$(2, 3)$

$y = -x - 4$ $y = 2x - 1$

12.

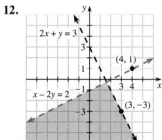

$2x + y = 3$

$(4, 1)$

$x - 2y = 2$

$(3, -3)$

14.

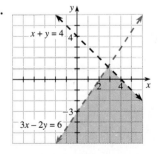

$x + y = 4$

$3x - 2y = 6$

16.
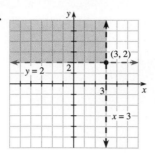

18.
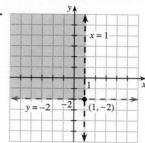

20.
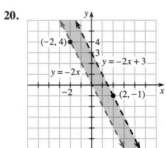

22.
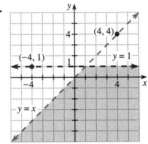

24.
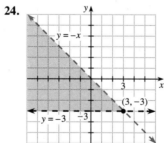

26.
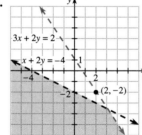

28.
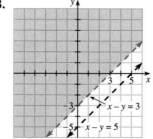

30.
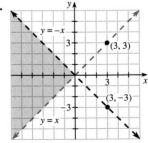

32.
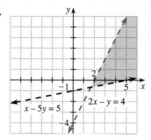

34.
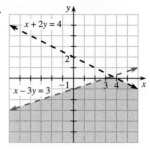

36.
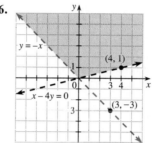

Chapter 9 Review Exercises

22.
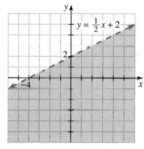

24.
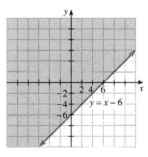

26.
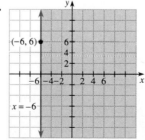

28.
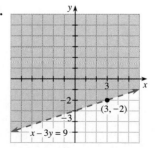

30.

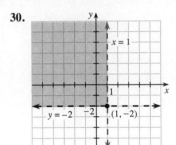

32.

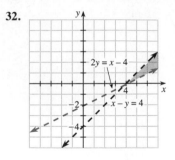

34.

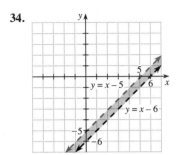

36.

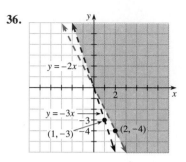

Elementary Algebra

Elementary Algebra

MARK DUGOPOLSKI

Southeastern Louisiana University

ADDISON-WESLEY PUBLISHING COMPANY

Reading, Massachusetts • Menlo Park, California • New York
Don Mills, Ontario • Wokingham, England • Amsterdam • Bonn
Sydney • Singapore • Tokyo • Madrid • San Juan

Sponsoring Editor	Charles B. Glaser
Developmental Editor	Stephanie Botvin
Managing Editor	Karen M. Guardino
Production Supervisor	Marion E. Howe
Copy Editor	Barbara Willette
Proofreader	Laura Michaels
Text Designer	Geri Davis, Quadrata, Inc.
Design Consultant	Meredith Nightingale
Layout Artist	Lorraine Hodsdon
Cover Designer	Marshall Henrichs
Art Consultant	Loretta M. Bailey
Illustrator	Tech-Graphics
Photo Researcher	Susan Van Etten
Manufacturing Manager	Roy Logan
Marketing Manager	Melissa Acuña
Production Services Manager	Herbert Nolan
Compositor, Color Separator	York Graphic Services, Inc.
Printer	R.R. Donnelley & Sons

Library of Congress Cataloging-in-Publication Data

Dugopolski, Mark.
    Elementary algebra/Mark Dugopolski.
        p.   cm.
    Includes index.
    ISBN 0-201-50839-7
    1. Algebra.   I. Title.
    QA152.2.D83   1991
    512.9—dc20   90-39346
                                                        CIP

To my parents,
Walter and Anne Dugopolski

Preface

This text is designed for a one term course in elementary algebra, with an emphasis on the basics in the textual material as well as in the exercises. A complete development for each topic is provided and no prior training in algebra is required. Even though we assume a fundamental knowledge of arithmetic, the basic operations of arithmetic are reviewed as needed. The unifying theme of the text is first the development of the skills necessary for solving equations and inequalities and then application of those skills to solving applied problems.

My primary goal in writing this book was to write one that students can read, understand, and enjoy, while gaining confidence in their ability to use mathematics. Toward this end, I have endeavored to design a comprehensive, yet flexible, presentation of elementary algebra that will satisfy the individual needs of instructors and a variety of students. I have taken what I refer to as a "common-ground" approach. Whenever possible, the text reminds students of ideas they are already familiar with and then builds on those ideas. This approach allows the instructor to meet the students on a common ground, before advancing to a new level.

Key Features

▶ Every chapter begins with a **Chapter Opener** that features an application of an idea developed in that chapter. This gives the student a concrete idea of what will be accomplished in that chapter.

▶ Every section begins with a list of topics that tells what is **In This Section.** The sections are divided into subsections and the subsection titles correspond to the topics from the In-This-Section list.

▶ Important ideas are set apart in boxes for quick reference. These boxes are used for **definitions, rules, summaries, and strategies.**

▶ Simple **fractions and decimals** are used throughout the text. This feature helps to reinforce the basic arithmetic skills that are necessary for success in algebra.

▶ One of the main goals of the text is the development of the skills necessary for solving equations and inequalities and then application of those skills to solving **applied problems.** For this reason, word problems occur wherever possible. This text contains over 450 word problems and they occur in over 50% of the exercise sets.

▶ **Inequalities** are used in applications as well as equations. This helps students to see the need for solving inequalities. There are word problems involving both simple and compound linear inequalities.

▶ There are numerous **geometric word problems** throughout the text, which are designed to review basic geometric facts and figures. A summary of common geometric facts can be found inside the front cover.

▶ Each exercise set is preceded by **Warm-ups,** a set of 10 simple statements that are to be marked either true or false. These exercises are designed to bridge the gap between the lecture and the exercise sets and to stimulate discussion of the concepts presented in the section. Many of the false statements point out examples of common student errors. The answers to all of these exercises are given in the answer section.

▶ The exercise sets throughout the text contain **keyed and non-keyed exercises.** The exercises at the end of each section follow the same order as the textual material and they are keyed to the examples. This organization allows instructors the flexibility of being able to cover only part of a section and easily see which exercises are appropriate to assign. The keyed exercises always give the student a place to start and build confidence. Wherever appropriate, non-keyed exercises follow the keyed exercises and are designed to bring all of the ideas together. Answers to all odd-numbered exercises are in the answer section.

▶ **Calculator exercises** are included whenever appropriate. These exercises may be omitted by instructors who do not require calculators in their courses. For those classes where calculators are required, these exercises will provide good periodic practice with the calculator.

▶ **Mental Exercises** occur at the end of many exercise sets and in the chapter review exercises. These exercises consist mainly of simple equations that can be solved in one or two steps, or simple expressions that can be simplified mentally. These exercises are meant to be done (perhaps more than once) only after the written exercises. Mental exercises point out to all students what many of them would discover on their own.

▶ Each chapter contains a **Math at Work** feature that illustrates how an idea from that chapter occurs in a real-life situation. The applications are drawn from a variety of fields to maximize student interest. Each Math at Work contains an optional exercise pertaining to the application presented.

▶ Each chapter ends with a four-part **Wrap-up.**

The **Chapter Summary** is a summary of the important concepts from the chapter along with brief illustrative examples.

The **Review Exercises** are intended to provide a review of each section of the chapter. The Review Exercises contain exercises that are keyed to the sections of the chapter, as well as miscellaneous exercises. Answers to the odd-numbered Review Exercises are in the answer section.

The **Chapter Test** is designed to help the student test his or her readiness for a chapter test. To help the student to be independent of the examples and sections, the Chapter Test has no keyed exercises. All answers for each Chapter Test are given in the answer section.

Tying It All Together exercises are designed to help students fit the current ideas in with ideas from the previous chapters. They may also contain exercises that will help in understanding the upcoming chapter. These exercises could be worked after a chapter test, but prior to starting the next chapter. All answers for these exercises are given in the answer section.

Ancillary Materials

▶ An **Annotated Instructor's Edition** is available to instructors. In addition to the material presented in the student edition, the AIE contains marginal notes to the instructor, in-text answers to all numerical exercises, and answers to all even-numbered graphing exercises in the instructor's answer section, which follows the To the Instructor introductory material.

▶ The **Instructor's Solution Manual** contains a detailed solution to each exercise in the text.

▶ The **Student's Solution Manual** contains detailed solutions to all of the odd-numbered exercises as well as solutions to all exercises from the Warm-ups, Chapter Tests, and Tying It All Together sections.

▶ The **Student's Study Guide** includes diagnostic tests for each chapter as well as detailed solutions and explanations for each test question.

▶ There are several options available for **Computerized Testing.**

AWTest (Apple) is an algorithm-based testing system keyed to the text. AW-Test will generate multiple choice, open-ended, and true-false questions.

Omni Test (IBM) is a powerful new testing system developed exclusively for Addison-Wesley. It is an algorithm-driven system that allows the user to create up to 99 perfectly parallel forms of any test effortlessly. It also allows the user to add his or her own test items and edit existing items with an easy to use, on screen "What-You-See-Is-What-You-Get" text editor.

Mac Test (Macintosh) is a test item bank containing ten questions for each objective drawn from the text. Questions are in a multiple choice format.

▶ The **Printed Test Bank** contains three alternate forms of tests for each chapter, as well as cumulative review tests and final exams.

▶ **Tutorial Software** is available in several forms.

IMPACT: An Interactive Mathematics Tutorial (IBM & Mac) is a tutorial package correlated to each section of the text. This package will generate exercises similar to those in the text exercise sets. If the student gets an exercise wrong, he or she can see either an example or a solution. The program will lead students interactively through step-by-step solutions to the exercises so that they can easily pinpoint specific trouble areas if they make an error.

Students using **The Math Lab** (IBM & Apple II) pick the topic area, level of difficulty, and number of problems. If they get a wrong answer, the program will prompt them with the first step of the solution. This program keeps detailed records of students' results, which can be stored on the disk.

In **Professor Weissman's Software** (IBM), students pick the topic area and level of difficulty. The program generates problems and gives step-by-step solutions if they get a wrong answer. Problems increase in difficulty as students are successful. This program also keeps records of students' scores.

For the **Algebra Problem Solver** (IBM), students choose a topic and can either request an exercise from the computer or enter their own homework exercise. The computer will show them a step-by-step solution to the problem.

▶ An extensive set of **video tapes** is available free upon adoption. These videos cover every topic presented in the text. Particular attention has been paid to working examples and giving complete explanations of the steps involved.

▶ **The Math Hotline** is a new idea in teaching and learning packages created especially for this text. Suppose a student is at home studying and can't get a problem started. Now, simply by calling the Math Hotline (504-542-0658) [800-874-4896 September 1990 through June 1991] from a touch-tone phone and responding to a few questions with the keypad, a student can receive a hint for any odd-numbered exercise. This means thousands of hints are as close as the phone. The Math Hotline will provide hints 24-hours a day, 365 days a year.

Acknowledgments

I would like to express my appreciation to the reviewers whose comments and suggestions were invaluable in writing this book:

Donley A. Chandler
El Paso County Community College

Sally I. Copeland
Johnson County Community College

Arthur P. Dull
Diablo Valley College

Carol J. Flakus
Lower Columbia College

Margaret Greene
Florida Community College

Donald K. Hostetler
Mesa Community College

Deborah E. Harvey Kell
Mercer Community College

Bettye Knox
San Antonio College

Donna Krichiver
Johnson County Community College

Richard Marshall
Eastern Michigan University

Myrna Mitchell
Pima County Community College

I thank Beth Gray, Southeastern Louisiana University, for working all of the problems, and David Busekist, Southeastern Louisiana University, for class-testing the manuscript. I thank James Morgan, Southeastern Louisiana University, and Wayne Andrepont, University of Southwestern Louisiana, for error checking.

I thank the staff at Addison-Wesley for all of their help and encouragement throughout this project, especially Chuck Glaser, Betsy Burr, Stuart Johnson, and Stephanie Botvin.

I also want to express my sincere appreciation to my wife, Cheryl, and my daughters, Sarah and Alisha, for their patience and support. I could not have completed the project without them.

Hammond, Louisiana   M.D.

To the
Student

This text was written to help you learn algebra. I have made the explanations as clear as possible and I have provided plenty of examples to illustrate the ideas. Even though I assume that you have a fundamental knowledge of arithmetic, the basic operations of arithmetic are reviewed as needed. This text was written with you in mind. You should be aware of the many features that were designed into this text to make the text easy to use.

The exercises follow the same order as the textual material and they are keyed to the examples in the text. This feature is to help you get started. Don't try to understand an entire section before beginning the exercises. Get started by doing exercises. If you don't understand how to do an exercise, then refer to the example that corresponds to that exercise. If you miss class and you need to work on your own, read the section in pieces. Read until you get through an example, then work the corresponding exercises.

To learn algebra you must do algebra and doing algebra means working problems. I have written an abundance of problems for this text. I have tried to avoid overly complicated exercises to give you plenty of practice with the basics. The answers to all odd-numbered exercises are given in the back of the book. These answers should be consulted only after you have worked a problem and checked your work.

To help you bring each chapter together, a four-part Wrap-up is included at the end of each chapter. In each Wrap-up you will find a Chapter Summary that lists the key ideas in each chapter along with illustrative examples. You will also find Review Exercises that are keyed to the sections of the chapter, along with miscellaneous exercises for the chapter. A Chapter Test is included for each chapter to let you check your progress. Complete answers for chapter tests are given in the back of the book. The exercises in the Chapter Tests are not keyed to the section from which they come. After each Chapter Test (starting with Chapter 2) I have included some exercises called Tying It All Together. These exercises are designed to help you fit current ideas in with ideas from previous chapters. Sometimes these exercises include problems that will be helpful to review before starting the next chapter. The Tying It All Together exercises could be worked after an in-class test but prior to the next class meeting.

If you need additional help with the exercises, there is a Student's Solution Manual available that includes complete solutions to those problems that have answers in the book. There is also a Student's Study Guide available to help you to prepare for tests. I have also prepared a computerized help line that you can call from a touch tone phone and get a brief message on how to work a problem. Help is available 24 hours a day for those exercises that have answers in the answer section of the book. The number is 504-542-0658 [800-874-4896 September 1990 through June 1991]. Relax and do your homework faithfully, and I am sure that you will be successful in algebra.

Mark Dugopolski

Contents

6 Powers and Roots   234

7 Quadratic Equations   289

8 Linear Equations in Two Variables   325

From Arithmetic to Algebra

Is the price of a hamburger, a Coke, and French fries the same as the price of French fries, a hamburger, and a Coke? For an oil change and a lube job, is the price the same regardless of which is done first? Are these isolated instances, or do we have something more general here? Of course, the answer to the first two questions is yes. These examples are certainly not isolated instances. They illustrate some of the properties of addition that we will study in Section 1.7 of this chapter. We use the properties of arithmetic every day without being aware of them. However, in algebra we must understand the properties of the operations of arithmetic, as well as how to perform the operations.

1.1 Fractions

IN THIS SECTION:

- **Equivalent Fractions**
- **Multiplying Fractions**
- **Dividing Fractions**
- **Adding and Subtracting Fractions**
- **Fractions, Decimals, and Percents**

The objective of Chapter 1 is to provide a smooth transition from arithmetic to algebra. You may wish to omit certain sections of this chapter if your students are well prepared for the course.

Algebra is an extension of arithmetic. Many of the ideas we encounter in algebra are based on the basic operations with fractions. In this section we review the operations with fractions that we will use in algebra. There are variations in the methods taught to perform the basic operations with fractions. The methods used in this section are the methods that will be most beneficial to us in algebra. So try to perform the operations as shown in the examples of this section.

Equivalent Fractions

The number above the fraction bar of a fraction is called the **numerator,** and the number below the fraction bar is called the **denominator.**

$$\frac{3}{7} \begin{array}{l}\longleftarrow \text{ Numerator}\\ \longleftarrow \text{ Denominator}\end{array}$$

Every fraction can be written in infinitely many equivalent forms. For example,

$$\frac{2}{3} = \frac{4}{6} = \frac{6}{9} = \frac{8}{12} = \frac{10}{15} = \cdots \quad \text{The 3 dots mean ``and so on.''}$$

Each fraction equivalent to 2/3 is obtained from 2/3 by multiplying the numerator and denominator of 2/3 by the same number. We are **building up** the fraction. For example,

$$\frac{2}{3} = \frac{2 \cdot 5}{3 \cdot 5} = \frac{10}{15}. \quad \text{The raised dot indicates multiplication.}$$

We say that 2/3 is in **lowest terms.**

If we write 10 as $2 \cdot 5$, then we have **factored** 10. The numbers 2 and 5 are **factors** of 10. When we convert an equivalent fraction to 2/3, we are **reducing it to lowest terms.** To reduce 10/15 to lowest terms, we factor 10 as $2 \cdot 5$ and factor

15 as $3 \cdot 5$, then **divide out** the common factor 5:

$$\frac{10}{15} = \frac{2 \cdot \cancel{5}}{3 \cdot \cancel{5}} = \frac{2}{3} \qquad \textbf{\textit{Note:}} \begin{array}{l} \mathbf{10 \div 5 = 2} \\ \mathbf{15 \div 5 = 3} \end{array}$$

We can state a general rule for reducing fractions. When we state rules, we use letters to represent numbers. A letter used to represent some numbers is called a **variable.** Variables are used extensively in algebra.

RULE
Reducing Fractions

If $a \neq 0$ and $c \neq 0$, then

$$\frac{a \cdot b}{a \cdot c} = \frac{b}{c}.$$

EXAMPLE 1   Reduce each fraction to lowest terms.

a) $\dfrac{15}{24}$ b) $\dfrac{30}{42}$

Solution   For each fraction, factor the numerator and denominator and then divide by the common factor:

a) $\dfrac{15}{24} = \dfrac{\cancel{3} \cdot 5}{\cancel{3} \cdot 8} = \dfrac{5}{8}$ b) $\dfrac{30}{42} = \dfrac{5 \cdot \cancel{6}}{7 \cdot \cancel{6}} = \dfrac{5}{7}$ ◀

To add fractions, it may be necessary to write a fraction as an equivalent fraction with a larger denominator. This is referred to as **building up the denominator.** We build up the denominator by multiplying the numerator and denominator by the same number.

RULE
Building Up Fractions

If $a \neq 0$ and $c \neq 0$, then

$$\frac{b}{c} = \frac{a \cdot b}{a \cdot c}.$$

EXAMPLE 2   Build up each fraction so that it is equivalent to a fraction with the indicated denominator.

a) $\dfrac{3}{4} = \dfrac{?}{28}$ b) $\dfrac{5}{3} = \dfrac{?}{30}$

Solution

a) Because $4 \cdot 7 = 28$, we multiply the numerator and denominator by 7:

$$\frac{3}{4} = \frac{3 \cdot 7}{4 \cdot 7} = \frac{21}{28}$$

b) Because $3 \cdot 10 = 30$, we multiply the numerator and denominator by 10:

$$\frac{5}{3} = \frac{5 \cdot 10}{3 \cdot 10} = \frac{50}{30}$$ ◄

STRATEGY
Obtaining Equivalent Fractions

Remember that equivalent fractions can be obtained by

1. **multiplying** the numerator and denominator by the same nonzero number or
2. **dividing** the numerator and denominator by the same nonzero number.

Multiplying Fractions

To multiply two fractions, we multiply their numerators and multiply their denominators. For example,

$$\frac{2}{3} \cdot \frac{5}{8} = \frac{10}{24} = \frac{\cancel{2} \cdot 5}{\cancel{2} \cdot 12} = \frac{5}{12}.$$

In general, we have the following definition.

DEFINITION
Multiplication of Fractions

If $b \neq 0$ and $d \neq 0$, then

$$\frac{a}{b} \cdot \frac{c}{d} = \frac{ac}{bd}.$$

In the next example of multiplication, reducing is done before multiplying rather than after.

EXAMPLE 3   Find the indicated products.

a) $\dfrac{1}{3} \cdot \dfrac{3}{4}$
  b) $\dfrac{4}{5} \cdot \dfrac{15}{22}$

Solution

a) $\dfrac{1}{3} \cdot \dfrac{3}{4} = \dfrac{1}{\cancel{3}} \cdot \dfrac{\cancel{3}}{4} = \dfrac{1}{4}$

b) Factor the numerators and denominators, then divide out the common factors before multiplying.

$$\frac{4}{5} \cdot \frac{15}{22} = \frac{2 \cdot 2}{5} \cdot \frac{3 \cdot 5}{2 \cdot 11} = \frac{6}{11}$$ ◀

Dividing Fractions

To divide two fractions we invert the divisor and multiply.

EXAMPLE 4   Find the indicated quotients.

a) $\dfrac{1}{3} \div \dfrac{7}{6}$ **b)** $\dfrac{2}{3} \div 5$

Math at Work

What is the difference between cooking and baking? A chef might answer that cooking is an art, but baking is a science.

A good cook can be creative in devising new flavors. In fact, new cuisines often evolve from blending the traditional ingredients of one cuisine with the preparation style of another. The creole cuisine of New Orleans, for example, evolved by combining native American ingredients with the preparation style of French immigrants.

Baking, on the other hand, requires a scientific understanding of how ingredients react together chemically. A cake that is prepared with an incorrect ratio of dry ingredients to wet ingredients will be either too moist or too dry; without the right proportion of yeast, a loaf of bread will not rise.

A professional baker often needs to take a basic recipe and mass-produce it. Using the recipe listed below, which yields one loaf of pumpernickel bread, create a recipe to produce 48 loaves.

Pumpernickel Bread

$\frac{3}{4}$ cup warm water	36	$1\frac{1}{2}$ tablespoons butter	72
$1\frac{1}{2}$ tablespoons dry yeast	72	1 teaspoon salt	48
$\frac{1}{2}$ teaspoon sugar	4	1 tablespoon caraway seeds	48
$\frac{3}{4}$ cup skim milk	36	$1\frac{1}{2}$ cups rye flour	72
$\frac{1}{6}$ cup molasses	8	$\frac{3}{4}$ cup whole wheat flour	36
1 tablespoon honey	48	$1\frac{1}{2}$ cups all-purpose flour	72
1 tablespoon brown sugar	48	$\frac{1}{16}$ cup cornmeal	3

Solution   In each case we invert the divisor (the number on the right) and multiply.

a) $\dfrac{1}{3} \div \dfrac{7}{6} = \dfrac{1}{3} \cdot \dfrac{6}{7} = \dfrac{1}{\cancel{3}} \cdot \dfrac{2 \cdot \cancel{3}}{7} = \dfrac{2}{7}$ **b)** $\dfrac{2}{3} \div 5 = \dfrac{2}{3} \cdot \dfrac{1}{5} = \dfrac{2}{15}$ ◄

We can state the definition for dividing fractions as follows.

DEFINITION
Division of Fractions

> If $b \neq 0$, $c \neq 0$, and $d \neq 0$, then
> $$\frac{a}{b} \div \frac{c}{d} = \frac{a}{b} \cdot \frac{d}{c}.$$

Adding and Subtracting Fractions

Fractions with identical denominators are added or subtracted by adding or subtracting their numerators. For example,

$$\frac{1}{7} + \frac{2}{7} = \frac{3}{7}$$

and

$$\frac{7}{10} - \frac{3}{10} = \frac{4}{10} = \frac{\cancel{2} \cdot 2}{\cancel{2} \cdot 5} = \frac{2}{5}.$$

The definition for adding and subtracting fractions is stated as follows.

DEFINITION
Addition and Subtraction of Fractions

> If $b \neq 0$, then
> $$\frac{a}{b} + \frac{c}{b} = \frac{a+c}{b} \quad \text{and} \quad \frac{a}{b} - \frac{c}{b} = \frac{a-c}{b}.$$

If the fractions have different denominators, we must convert them to equivalent fractions with the same denominator and then add or subtract. For example, to add the fractions 1/2 and 1/3, we build up each denominator to get a denominator of 6. The denominator 6 is the smallest number that is a multiple of both 2 and 3. For this reason, 6 is called the **least common denominator (LCD).**

Emphasize the method. We will use the same method for adding rational expressions.

EXAMPLE 5   Perform the indicated operations.

a) $\dfrac{1}{2} + \dfrac{1}{3}$ **b)** $\dfrac{1}{3} - \dfrac{1}{12}$

Solution

a)

$$\frac{1}{2} + \frac{1}{3} = \frac{1 \cdot 3}{2 \cdot 3} + \frac{1 \cdot 2}{3 \cdot 2}$$

$$= \frac{3}{6} + \frac{2}{6} \qquad \text{Build each denominator to a denominator of 6.}$$

$$= \frac{5}{6} \qquad \text{Then add.}$$

b) The LCD for the denominators 3 and 12 is 12. In this case we change the denominator of only one of the fractions:

$$\frac{1}{3} - \frac{1}{12} = \frac{1 \cdot 4}{3 \cdot 4} - \frac{1}{12}$$

$$= \frac{4}{12} - \frac{1}{12} \qquad \text{Multiply numerator and denominator of 1/3 by 4.}$$

$$= \frac{3}{12} \qquad \text{Then subtract.}$$

$$= \frac{1}{4} \qquad \text{Reduce to lowest terms.} \qquad \blacktriangleleft$$

Fractions, Decimals, and Percents

In the decimal number system, fractions with a denominator of 10, 100, 1000, etc., are written as decimal numbers. For example,

$$\frac{3}{10} = .3, \qquad \frac{25}{100} = .25, \qquad \text{and} \qquad \frac{5}{1000} = .005.$$

Fractions with a denominator of 100 are often written as percents. Think of the percent symbol (%) as representing the denominator of 100. For example,

$$\frac{25}{100} = 25\%, \qquad \frac{5}{100} = 5\%, \qquad \text{and} \qquad \frac{300}{100} = 300\%.$$

The next example will illustrate further how to convert from any one of the forms (fraction, decimal, percent) to the others.

EXAMPLE 6 Convert each given fraction, decimal, or percent into its other two forms.

a) $\frac{1}{5}$ **b)** 6%   **c)** .1

Solution

a)
$$\frac{1}{5} = \frac{1 \cdot 20}{5 \cdot 20} = \frac{20}{100} = .20 = 20\%$$

$$\text{So } \frac{1}{5} = .2 = 20\%.$$

b)
$$6\% = \frac{6}{100} = .06 \qquad \frac{6}{100} = \frac{2 \cdot 3}{2 \cdot 50} = \frac{3}{50}$$

$$\text{So } 6\% = .06 = \frac{3}{50}.$$

c)
$$.1 = \frac{1}{10} = \frac{1 \cdot 10}{10 \cdot 10} = \frac{10}{100} = 10\%$$

$$\text{So } .1 = \frac{1}{10} = 10\%. \qquad \blacktriangleleft$$

Warm-ups

Ask students for a correct version of each false statement. Answer all 10 questions before looking for correct answers in back of text.

True or false?

1. Every fraction is equal to infinitely many equivalent fractions. T

2. The fraction $\frac{8}{12}$ is equivalent to the fraction $\frac{4}{6}$. T

3. The fraction $\frac{8}{12}$ reduced to lowest terms is $\frac{4}{6}$. F

4. $\frac{1}{2} \cdot \frac{2}{3} = \frac{1}{3}$ T   5. $\frac{1}{2} \cdot \frac{3}{5} = \frac{3}{10}$ T   6. $\frac{1}{2} \cdot \frac{6}{5} = \frac{6}{10}$ T

7. $\frac{1}{2} \div 3 = \frac{1}{6}$ T   8. $5 \div \frac{1}{2} = 10$ T   9. $\frac{1}{2} + \frac{1}{4} = \frac{2}{6}$ F

10. $2 - \frac{1}{2} = \frac{3}{2}$ T

1.1 EXERCISES

Reduce each fraction to lowest terms. See Example 1.

1. $\frac{3}{6}$ $\frac{1}{2}$ 2. $\frac{2}{10}$ $\frac{1}{5}$ 3. $\frac{12}{18}$ $\frac{2}{3}$ 4. $\frac{30}{40}$ $\frac{3}{4}$ 5. $\frac{15}{5}$ 3

6. $\frac{39}{13}$ 3   7. $\frac{50}{100}$ $\frac{1}{2}$ 8. $\frac{5}{1000}$ $\frac{1}{200}$ 9. $\frac{200}{100}$ 2   10. $\frac{125}{100}$ $\frac{5}{4}$

Convert each fraction or whole number into an equivalent fraction with the indicated denominator. See Example 2.

11. $\dfrac{3}{4} = \dfrac{?}{8}$ $\dfrac{6}{8}$

12. $\dfrac{5}{7} = \dfrac{?}{21}$ $\dfrac{15}{21}$

13. $\dfrac{8}{3} = \dfrac{?}{12}$ $\dfrac{32}{12}$

14. $\dfrac{7}{2} = \dfrac{?}{8}$ $\dfrac{28}{8}$

15. $5 = \dfrac{?}{2}$ $\dfrac{10}{2}$

16. $9 = \dfrac{?}{3}$ $\dfrac{27}{3}$

17. $\dfrac{3}{4} = \dfrac{?}{100}$ $\dfrac{75}{100}$

18. $\dfrac{1}{2} = \dfrac{?}{100}$ $\dfrac{50}{100}$

19. $\dfrac{3}{10} = \dfrac{?}{100}$ $\dfrac{30}{100}$

20. $\dfrac{2}{5} = \dfrac{?}{100}$ $\dfrac{40}{100}$

Find the indicated products. See Example 3.

21. $\dfrac{2}{3} \cdot \dfrac{5}{9}$ $\dfrac{10}{27}$

22. $\dfrac{1}{8} \cdot \dfrac{1}{8}$ $\dfrac{1}{64}$

23. $\dfrac{1}{3} \cdot 15$ 5

24. $\dfrac{1}{4} \cdot 16$ 4

25. $\dfrac{3}{4} \cdot \dfrac{14}{15}$ $\dfrac{7}{10}$

26. $\dfrac{5}{8} \cdot \dfrac{12}{35}$ $\dfrac{3}{14}$

27. $\dfrac{2}{5} \cdot 300$ 120

28. $\dfrac{3}{10} \cdot \dfrac{20}{21}$ $\dfrac{2}{7}$

29. $\dfrac{1}{2} \cdot \dfrac{6}{5}$ $\dfrac{3}{5}$

30. $\dfrac{1}{2} \cdot \dfrac{3}{5}$ $\dfrac{3}{10}$

Find each quotient. See Example 4.

31. $\dfrac{3}{4} \div \dfrac{1}{4}$ 3

32. $\dfrac{2}{3} \div \dfrac{1}{2}$ $\dfrac{4}{3}$

33. $\dfrac{1}{3} \div 5$ $\dfrac{1}{15}$

34. $\dfrac{3}{5} \div 3$ $\dfrac{1}{5}$

35. $5 \div \dfrac{1}{4}$ 20

36. $8 \div \dfrac{2}{3}$ 12

37. $\dfrac{6}{10} \div \dfrac{1}{2}$ $\dfrac{6}{5}$

38. $\dfrac{1}{3} \div \dfrac{1}{3}$ 1

Find each sum or difference. See Example 5.

39. $\dfrac{1}{3} + \dfrac{1}{4}$ $\dfrac{7}{12}$

40. $\dfrac{1}{2} + \dfrac{3}{5}$ $\dfrac{11}{10}$

41. $\dfrac{3}{4} - \dfrac{2}{3}$ $\dfrac{1}{12}$

42. $\dfrac{4}{5} - \dfrac{3}{4}$ $\dfrac{1}{20}$

43. $\dfrac{1}{6} + \dfrac{5}{8}$ $\dfrac{19}{24}$

44. $\dfrac{3}{4} + \dfrac{1}{6}$ $\dfrac{11}{12}$

45. $\dfrac{1}{4} + \dfrac{1}{4}$ $\dfrac{1}{2}$

46. $\dfrac{1}{10} - \dfrac{1}{10}$ 0

47. $\dfrac{1}{2} - \dfrac{1}{4}$ $\dfrac{1}{4}$

48. $\dfrac{1}{3} + \dfrac{1}{6}$ $\dfrac{1}{2}$

Convert each given fraction, decimal, or percent into its other two forms. See Example 6.

49. $\dfrac{3}{5}$ $.6, 60\%$

50. $\dfrac{19}{20}$ $.95, 95\%$

51. $.08$ $\dfrac{2}{25}, 8\%$

52. $.4$ $\dfrac{2}{5}, 40\%$

53. $.01$ $\dfrac{1}{100}, 1\%$

54. $.005$ $\dfrac{1}{200}, .5\%$

55. 9% $\dfrac{9}{100}, .09$

56. 60% $\dfrac{3}{5}, .6$

57. 2% $\dfrac{1}{50}, .02$

58. 120% $\dfrac{6}{5}, 1.2$

Perform the indicated operations.

59. $\dfrac{6}{10} + \dfrac{2}{5}$ 1

60. $\dfrac{5}{6} - \dfrac{1}{6}$ $\dfrac{2}{3}$

61. $\dfrac{1}{2} \cdot \dfrac{1}{2} \cdot \dfrac{1}{2}$ $\dfrac{1}{8}$

62. $\dfrac{3}{8} \div \dfrac{1}{8}$ 3

63. $\dfrac{1}{2} + \dfrac{1}{3} + \dfrac{1}{4}$ $\dfrac{13}{12}$

64. $\dfrac{1}{2} + \dfrac{1}{3} - \dfrac{1}{6}$ $\dfrac{2}{3}$

65. $\dfrac{7}{8} \div \dfrac{3}{14}$ $\dfrac{49}{12}$

66. $\dfrac{2}{3} \cdot \dfrac{2}{3} \cdot \dfrac{2}{3}$ $\dfrac{8}{27}$

67. $\dfrac{3}{4} - \dfrac{3}{8}$ $\dfrac{3}{8}$

68. $\dfrac{5}{16} \cdot \dfrac{3}{10}$ $\dfrac{3}{32}$

69. $\dfrac{5}{8} + \dfrac{1}{6}$ $\dfrac{19}{24}$

70. $\dfrac{2}{3} - \dfrac{1}{6}$ $\dfrac{1}{2}$

1.2 The Real Numbers

IN THIS SECTION:

- The Integers
- The Rational Numbers
- The Number Line
- The Real Numbers
- Absolute Value

In arithmetic we use only positive numbers, but in algebra we use negative numbers also. The numbers that we use in algebra are called the real numbers. We start the discussion of the real numbers with some simpler sets of numbers.

The Integers

The most fundamental collection or **set** of numbers is the set of **counting numbers.** Of course, these are the numbers that we use for counting. These numbers are also called the **natural numbers.** The set of natural numbers is written in symbols as follows.

The Natural Numbers

$$\{1, 2, 3, \ldots\}$$

Braces, { }, are used to indicate a set of numbers. The three dots after 1, 2, and 3 mean that the pattern continues; the dots are read as "and so on." There are infinitely many natural numbers.

The natural numbers, together with the number 0, are called the **whole numbers.** The set of whole numbers is written as follows.

The Whole Numbers

$$\{0, 1, 2, 3, \ldots\}$$

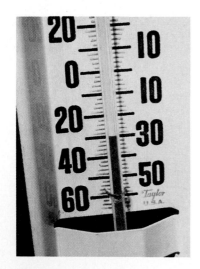

A negative number is used to represent a loss or a debt. A debt of $10 can be expressed by the negative number -10 (negative ten). When a thermometer reads 30 degrees below zero on a Fahrenheit scale, we say that the temperature is $-30°F$.

The whole numbers together with the negatives of the counting numbers form the set of **integers.**

The Integers

$$\{. . . , -3, -2, -1, 0, 1, 2, 3, . . .\}$$

The Rational Numbers

The fractions that we studied in Section 1.1 were all positive numbers. The set of rational numbers includes both positive and negative fractions. A **rational number** is any number that can be expressed as a ratio (or quotient) of two integers. We cannot list the rational numbers as easily as we listed the numbers in other sets above. The set of rational numbers is written in symbols using **set-builder notation** as follows.

The Rational Numbers

$$\{a/b \mid a \text{ and } b \text{ are integers, with } b \neq 0\}$$

The set of ———⌐   ⌐— such that

We read this notation as "the set of numbers of the form a/b such that a and b are integers, with b not equal to 0."

Examples of rational numbers are

$$3/1, \quad 5/4, \quad -7/10, \quad 0/6, \quad 5/1, \quad -77/3, \quad \text{and} \quad -3/-6.$$

Note that these examples of rational numbers are not all in simplest form. For instance, $3/1 = 3$ and $0/6 = 0$. From these examples we see that the rational numbers include the integers.

The Number Line

The **number line** is a diagram that helps us to visualize numbers and their relationships to each other. To construct a number line, we draw a straight line and label any convenient point with the number 0. Now we choose any convenient length and use it to locate other points. Points to the right of 0 correspond to the positive integers, and points to the left of 0 correspond to the negative integers. The number line is shown in Fig. 1.1.

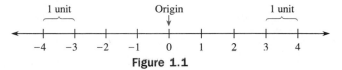

Figure 1.1

The numbers corresponding to the points on the line are called the **coordinates** of the points. The distance between two consecutive integers is called a **unit** and is the same for any two consecutive integers. The point with coordinate 0 is called the **origin.** The numbers on the number line increase in size from left to right. *When we compare the size of any two numbers, the larger number lies to the right of the smaller on the number line.* See Fig. 1.1.

The set of integers is illustrated or graphed as in Fig. 1.2. The three dots to the right and left on the number line indicate that the numbers go on indefinitely in both directions.

Figure 1.2

EXAMPLE 1   List the numbers described and graph the numbers on a number line.

a) The whole numbers less than 4
b) The integers between 3 and 9
c) The integers greater than −3

Solution

a) The whole numbers less than 4 are 0, 1, 2, and 3. These numbers are shown in Fig. 1.3.

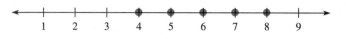

Figure 1.3

b) The integers between 3 and 9 are 4, 5, 6, 7, and 8. Note that 3 and 9 are not considered to be *between* 3 and 9. The graph is shown in Fig. 1.4.

Figure 1.4

c) The integers greater than −3 are −2, −1, 0, 1, and so on. To indicate the continuing pattern, we use 3 dots on the graph as shown in Fig. 1.5.

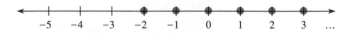

Figure 1.5   ◄

The Real Numbers

For every rational number there is a point on the number line. For example, the number 1/2 corresponds to a point halfway between 0 and 1 on the number line, and $-5/4$ corresponds to a point one and one-quarter units to the left of 0, as shown in Fig. 1.6.

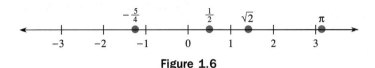

Figure 1.6

It is helpful to make a few rational guesses for $\sqrt{2}$ and check them by multiplying. For example;
$$(1.5)^2 = 2.25,$$
$$(1.4)^2 = 1.96,$$
$$(1.42)^2 = 2.0164,$$
etc.

The set of numbers that corresponds to *all* points on a number line is called the set of **real numbers.** There are points on the number line that do not correspond to rational numbers. Those real numbers that are not rational are called **irrational.** It can be shown that numbers such as $\sqrt{2}$ (the square root of 2) and π (Greek letter pi) are irrational. The number $\sqrt{2}$ is a number that can be multiplied by itself to obtain 2 ($\sqrt{2} \cdot \sqrt{2} = 2$). The number π is the ratio of the circumference and diameter of any circle. Irrational numbers are not as easy to represent as rational numbers. That is why we use symbols such as $\sqrt{2}$, $\sqrt{3}$, and π for irrational numbers. When we perform computations with irrational numbers, we use rational approximations for them. For example, $\sqrt{2}$ is approximately equal to 1.414, and π is approximately 3.14. Note that not all square roots are irrational. For example, $\sqrt{9} = 3$, since $3 \cdot 3 = 9$. We will deal with irrational numbers in greater depth when we study roots in Chapter 6.

Figure 1.7 summarizes the sets of numbers that make up the real numbers and shows the relationships between them.

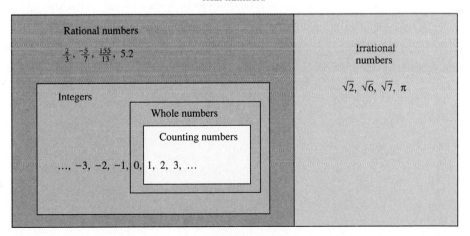

Figure 1.7

EXAMPLE 2   Determine whether each statement is true or false.

a) Every rational number is an integer.
b) Every counting number is an integer.
c) Every irrational number is a real number.

Solution

a) False because 1/2 is a rational number that is not an integer.
b) True
c) True   ◄

Absolute Value

The concept of absolute value will be used to define the basic operations with real numbers in Section 1.3. The absolute value of a number is determined by using the number line. The real numbers are the coordinates of the points on the number line. However, to simplify matters, we generally refer to the points as numbers. The numbers 5 and -5 are both five units away from 0 on the number line. So the absolute value of each of these numbers is 5. See Fig. 1.8.

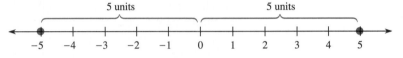

Figure 1.8

A number's distance from 0 on the number line is called the **absolute value** of the number. We write $|a|$ for "the absolute value of a." Therefore $|5| = 5$ and $|-5| = 5$. The distance from 0 to the number 1/4 is 1/4 unit, so $|1/4| = 1/4$. The distance from 0 to the number -6.23 is 6.23 units, so $|-6.23| = 6.23$.

EXAMPLE 3   Determine the values of the following.

a) $|3|$ **b)** $|-3|$ **c)** $|0|$

d) $\left|\dfrac{2}{3}\right|$ **e)** $|-.39|$

Solution

a) $|3| = 3$, because 3 is 3 units away from 0.
b) $|-3| = 3$, because -3 is 3 units away from 0.
c) $|0| = 0$, because 0 is 0 units away from 0.
d) $\left|\dfrac{2}{3}\right| = \dfrac{2}{3}$
e) $|-.39| = .39$ ◄

Note that $|a|$ represents distance, and distance is never negative. *For any number a, $|a|$ is greater than or equal to zero.*

Two numbers that are located on opposite sides of zero and have the same absolute value are called **opposites** of each other. The opposite of zero is zero. The numbers 5 and -5 are opposites of one another. We say that the opposite of 5 is -5 and the opposite of -5 is 5. The symbol "$-$" is used to indicate "opposite" as well as "negative." When the negative sign is used before a number, it should be read as "negative." When it is used in front of parentheses or a variable, it should be read as "opposite." For example,

$$-(5) = -5 \text{ means "the opposite of 5 is negative 5,"}$$

and

$$-(-5) = 5 \text{ means "the opposite of negative 5 is 5."}$$

In general, $-a$ means "the opposite of a." If a is positive, $-a$ is negative. If a is negative, $-a$ is positive. Opposites have the following property.

PROPERTY
Opposite of an Opposite

For any real number a,　　$-(-a) = a$.

Remember that we have defined $|a|$ to be the distance between 0 and a on the number line. Using opposites, we can give a symbolic definition of absolute value.

DEFINITION
Absolute Value

$$|a| = \begin{cases} a & \text{if } a \text{ is positive or zero} \\ -a & \text{if } a \text{ is negative} \end{cases}$$

According to this definition, the absolute value of a nonnegative number is that number. Using this definition, we write

$$|8| = 8$$

since 8 is positive. The second line of the definition says that the absolute value of a negative number is the opposite of that number. For example, the absolute value of -8 is the opposite of -8:

$$|-8| = -(-8) = 8$$

Warm-ups

True or false?
1. The natural numbers and the counting numbers are the same. T
2. The number 8,134,562,877,565 is a counting number. T
3. Zero is a counting number. F

4. Zero is not a rational number. F
5. The opposite of negative 3 is positive 3. T
6. The absolute value of 4 is −4. F
7. −(−9) = 9 T
8. −(−b) = b for any number b. T
9. Negative six is greater than negative three. F
10. Negative five is between four and six. F

1.2 EXERCISES

List the numbers described and graph them on a number line. See Example 1.   Graphs in Answers sections.

1. The counting numbers smaller than 6   1, 2, 3, 4, 5
2. The natural numbers larger than 4   5, 6, 7, 8, 9, . . .
3. The whole numbers smaller than 5   0, 1, 2, 3, 4
4. The integers between −3 and 3   −2, −1, 0, 1, 2
5. The whole numbers between −5 and 5   0, 1, 2, 3, 4
6. The integers smaller than −1   −2, −3, −4, −5, . . .
7. The counting numbers larger than −4   1, 2, 3, 4, 5, . . .
8. The natural numbers between −5 and 7   1, 2, 3, 4, 5, 6
9. The integers larger than 1/2   1, 2, 3, 4, 5, . . .
10. The whole numbers smaller than 7/3   0, 1, 2

Determine whether each statement is true or false. See Example 2.

11. Every integer is a rational number. T
12. Every counting number is a whole number. T
13. Zero is a counting number. F
14. Every whole number is a counting number. F
15. The ratio of the circumference and diameter of a circle is an irrational number. T
16. Every rational number can be expressed as a ratio of integers. T
17. Every whole number can be expressed as a ratio of integers. T
18. Some of the rational numbers are integers. T
19. Some of the integers are natural numbers. T
20. There are infinitely many rational numbers. T
21. Zero is an irrational number. F
22. Every irrational number is a real number. T

Determine the values of the following. See Example 3.

23. $|-6|$ 6
24. $|4|$ 4
25. $|0|$ 0
26. $|7|$ 7
27. $|-7|$ 7
28. $|-45|$ 45
29. $|-30|$ 30
30. $|0|$ 0
31. $|-9|$ 9
32. $|-2|$ 2
33. $\left|\dfrac{3}{4}\right|$ $\dfrac{3}{4}$
34. $\left|-\dfrac{1}{2}\right|$ $\dfrac{1}{2}$
35. $|-5.09|$ 5.09
36. $|.00987|$ .00987

True or false?

37. The integer −16 is smaller than 9. T
38. The integer −12 is larger than −7. F
39. The rational number −5/2 is smaller than −9/4. T
40. The rational number 5/8 is larger than 6/7. F
41. The absolute value of −3 is larger than 2. T
42. The absolute value of −6 is larger than 0. T
43. The absolute value of −4 is smaller than 3. F
44. The absolute value of 5 is smaller than −4. F
45. The absolute value of −5 is larger than the absolute value of −9. F

46. The absolute value of -12 is larger than the absolute value of 8.   T

47. Of the two numbers -16 and 9, -16 has the larger absolute value.   T

48. Of the two numbers -12 and -7, -7 has the larger absolute value.   F

49. If we add the absolute values of -3 and -5, we get 8.   T

50. If we multiply the absolute values of -2 and 5, we get 10.   T

51. The absolute value of any negative number is greater than 0.   T

52. The absolute value of any positive number is less than 0.   F

1.3 Addition and Subtraction of Real Numbers

IN THIS SECTION:
- Addition of Two Negative Numbers
- Addition of Numbers with Opposite Signs
- Subtraction of Signed Numbers

In arithmetic we add and subtract only positive numbers. In Section 1.2 we introduced the concept of absolute value of a number. Now we will use absolute value to extend the operations of addition and subtraction to the real numbers. We will work only with rational numbers in this chapter. We will learn to perform operations with irrational numbers in Chapter 6.

Addition of Two Negative Numbers

Students generally find money easier to understand than the number-line approach to addition and subtraction.

A good way to understand positive and negative numbers is to think of the *positive numbers as assets* and the *negative numbers as debts*. For this illustration we can think of assets simply as cash. Think of debts as unpaid bills such as the electric bill or the phone bill. If you have debts of $7 and $8, then your total debt is $15. This can be expressed symbolically as

$$(-7) \quad + \quad (-8) \quad = \quad -15.$$

$7 debt ⌐   plus   ⌐$8 debt   ⌐$15 debt

We think of this addition as adding the absolute values of -7 and -8 $(7 + 8 = 15)$ and then putting a negative sign on that result to get -15. This illustrates the following rule.

**RULE
Sum of Two Negative
Numbers**

> To find the sum of two negative numbers, add their absolute values and then put a negative sign on the result.

EXAMPLE 1   Perform the indicated operations.

a) $(-12) + (-9)$ b) $(-3.5) + (-6.28)$ c) $\left(-\dfrac{1}{2}\right) + \left(-\dfrac{1}{4}\right)$

Solution

a) The absolute values of -12 and -9 are 12 and 9, and their sum is 21. So $(-12) + (-9) = -21$.

b) Add the absolute values and put a negative sign on the result. Remember to line up the decimal points when adding decimal numbers: $(-3.5) + (-6.28) = -9.78$.

c) $\left(-\dfrac{1}{2}\right) + \left(-\dfrac{1}{4}\right) = \left(-\dfrac{2}{4}\right) + \left(-\dfrac{1}{4}\right) = -\dfrac{3}{4}$ ◀

Addition of Numbers with Opposite Signs

If you have a debt of $5 and have only $5 in cash, then your debts equal your assets (in absolute value), and your **net worth** is $0. Net worth is the total of debts and assets. Symbolically,

$$-5 + 5 = 0.$$

$5 debt ⎤   ⎤   ⎤ Net worth
$5 cash ⎦

For any number a, a and its opposite $-a$ have a sum of zero. For this reason, a and $-a$ are called **additive inverses** of one another. Note that the words "negative," "opposite," and "additive inverse" are often used interchangeably.

Additive Inverse Property

> For any number a, $a + (-a) = 0$ and $(-a) + a = 0$.

To understand the sum of a positive and a negative number, consider the following situation. If you have a debt of $6 and you have $10 in cash, you may have $10 in hand, but your net worth is only $4. Your assets exceed your debts (in absolute value), and you have a positive net worth. In symbols,

$$-6 + 10 = 4.$$

Note that to get 4, we actually subtract 6 from 10.

If you have a debt of $7 but have only $5 in cash, then your debts exceed your assets (in absolute value). You have a negative net worth of $-\$2$. In symbols,

$$-7 + 5 = -2.$$

Note that to get the 2 in the answer, we subtract 5 from 7.

Note that the additive inverse property takes care of the addition of two numbers with opposite signs and the *same* absolute value.

As you can see from these examples, the sum of a positive number and a negative number (with different absolute values) may be either positive or negative. These examples help us to understand the rule for adding numbers with opposite signs and different absolute values.

RULE
Sum of Two Numbers with Opposite Signs (and Different Absolute Values)

If a and b have opposite signs, the sum of a and b is found by subtracting the absolute values of the numbers.

1. The answer is positive if the number with the larger absolute value is positive.
2. The answer is negative if the number with the larger absolute value is negative.

EXAMPLE 2 Evaluate.

a) $-5 + 13$ b) $6 + (-7)$ c) $7 + (-7)$

d) $\left(-\dfrac{1}{3}\right) + \left(\dfrac{1}{2}\right)$ e) $-6.4 + 2.1$ f) $-5 + .09$

Solution

a) The absolute values of -5 and 13 are 5 and 13. Subtract them to get 8. Since the number with the larger absolute value is 13 and it is positive, the result is positive 8: $-5 + 13 = 8$.

b) The absolute values of 6 and -7 are 6 and 7. Subtract them to get 1. Since -7 has the larger absolute value, the result is negative: $6 + (-7) = -1$.

c) Since 7 and -7 are additive inverses, $7 + (-7) = 0$.

d) $\left(-\dfrac{1}{3}\right) + \left(\dfrac{1}{2}\right) = \left(-\dfrac{2}{6}\right) + \left(\dfrac{3}{6}\right) = \dfrac{1}{6}$

e) Line up the decimal points and subtract 2.1 from 6.4. Since 6.4 is larger than 2.1 and 6.4 has a negative sign, the sign of the answer is negative: $-6.4 + 2.1 = -4.3$.

f) Line up the decimal points and subtract .09 from 5.00. Since 5.00 is larger than .09, and 5.00 has the negative sign, the sign of the answer is negative. $-5 + .09 = -4.91$. ◄

Subtraction of Signed Numbers

Point out that even with whole numbers we learn addition before subtraction.

Each subtraction problem with signed numbers is solved by doing an equivalent addition problem. So before attempting subtraction of signed numbers, be sure that you understand addition of signed numbers.

Now think of subtraction as removing debts or assets and addition as receiving debts or assets. If you have $10 in cash and $3 is taken from you, your resulting net worth is the same as if you have $10 cash and a phone bill for $3 arrives in the mail. In symbols,

$$10 - 3 = 10 + (-3).$$

Remove —⌐ ↑   ↑ ⌐— Debt
    Cash ⌐   ⌐— Receive

Suppose you have $15 but owe a friend $5. Your net worth is only $10. If the debt of $5 is cancelled or forgiven, your net worth will go up to $15, the same as if you received $5 in cash. In symbols,

$$10 - (-5) = 10 + 5.$$

Remove —⌐ ↑   ↑ ↑ ⌐— Cash
    Debt ⌐   ⌐— Receive

Removing a debt is equivalent to receiving cash.

Notice that each subtraction problem is equivalent to an addition problem in which we add the opposite of what we were going to subtract. These examples illustrate the definition of subtraction.

DEFINITION
Subtraction of Real Numbers

For any numbers a and b,

$$a - b = a + (-b).$$

EXAMPLE 3   Perform each subtraction.

a) $5 - 3$ **b)** $-5 - 3$ **c)** $-5 - (-3)$

d) $\dfrac{1}{2} - \left(-\dfrac{1}{4}\right)$ **e)** $-3.6 - (-5)$ **f)** $.02 - 8$

Solution   To do *any* subtraction, we can change it to *addition of the opposite*. Note that in part (a) it is certainly not necessary to do so.

Subtract⌐   ⌐Positive 3   Add⌐   ⌐Negative 3
a) $5 \quad - \quad 3 \quad = \quad 5 \quad + \quad (-3) \quad = \quad 2$

b) $-5 - 3 = -5 + (-3) = -8$

Subtract⌐   ⌐Negative 3   Add⌐   ⌐Positive 3
c) $-5 \quad - \quad (-3) \quad = \quad -5 \quad + \quad 3 \quad = \quad -2$

d) $\dfrac{1}{2} - \left(-\dfrac{1}{4}\right) = \dfrac{1}{2} + \dfrac{1}{4} = \dfrac{3}{4}$

e) $-3.6 - (-5) = -3.6 + 5 = 1.4$

f) $.02 - 8 = .02 + (-8) = -7.98$ ◄

Warm-ups

For T–F questions involving variables, encourage students to use some numbers in place of the variables.

True or false?

1. The additive inverse of -3 is 0.   F
2. If b is a negative number, then $-b$ is a positive number.   T
3. $-9 + 8 = -1$ T
4. $(-2) + (-4) = -6$ T
5. The sum of a positive number and a negative number is a negative number.   F
6. The result of a subtracted from b is the same as b plus the opposite of a.   T
7. If a and b are negative numbers, then $a - b$ is a negative number.   F
8. $0 - 7 = -7$ T
9. $5 - (-2) = 3$ F
10. $-5 - (-2) = -7$ F

1.3 EXERCISES

Perform each operation. See Example 1.

1. $(-3) + (-10)$ -13
2. $(-81) + (-19)$ -100
3. $-.25 + (-.9)$ -1.15
4. $-.8 + (-2.35)$ -3.15
5. $\left(-\dfrac{1}{3}\right) + \left(-\dfrac{1}{6}\right)$ $-\dfrac{1}{2}$
6. $\left(-\dfrac{2}{3}\right) + \left(-\dfrac{1}{3}\right)$ -1

Evaluate. See Example 2.

7. $-8 + 8$ 0
8. $20 + (-20)$ 0
9. $-7 + 9$ 2
10. $10 + (-30)$ -20
11. $7 + (-13)$ -6
12. $-8 + 20$ 12
13. $8.6 + (-3)$ 5.6
14. $-9.5 + 12$ 2.5
15. $\dfrac{1}{4} + \left(-\dfrac{1}{2}\right)$ $-\dfrac{1}{4}$
16. $\dfrac{3}{4} + 2$ $\dfrac{5}{4}$

Fill in the parentheses to make each statement correct. See Example 3.

17. $8 - 2 = 8 + (?)$ -2
18. $3.5 - 1.2 = 3.5 + (?)$ -1.2
19. $4 - 12 = 4 + (?)$ -12
20. $\dfrac{1}{2} - \dfrac{5}{6} = \dfrac{1}{2} + \left(?\right)$ $-\dfrac{5}{6}$
21. $-3 - (-8) = -3 + (?)$ 8
22. $-9 - (-2.3) = -9 + (?)$ 2.3
23. $8.3 - (-1.5) = 8.3 + (?)$ 1.5
24. $10 - (-6) = 10 + (?)$ 6

Perform each subtraction. See Example 3.

25. $6 - 10$ -4
26. $3 - 19$ -16
27. $-3 - 7$ -10
28. $-3 - 12$ -15
29. $5 - (-6)$ 11
30. $5 - (-9)$ 14
31. $-6 - 5$ -11
32. $-3 - 6$ -9

33. $\dfrac{1}{4} - \dfrac{1}{2}$ $-\dfrac{1}{4}$ **34.** $\dfrac{2}{5} - \dfrac{2}{3}$ $-\dfrac{4}{15}$ **35.** $\dfrac{1}{2} - \left(-\dfrac{1}{4}\right)$ $\dfrac{3}{4}$ **36.** $\dfrac{2}{3} - \left(-\dfrac{1}{6}\right)$ $\dfrac{5}{6}$

37. $10 - 3$ 7 **38.** $13 - 3$ 10 **39.** $1 - .07$ .93 **40.** $.03 - 1$ $-.97$

41. $7.3 - (-2)$ 9.3 **42.** $-5.1 - .15$ -5.25 **43.** $-.03 - 5$ -5.03 **44.** $.7 - (-.3)$ 1

Perform the indicated operations.

45. $-5 + 8$ 3 **46.** $-6 + 10$ 4 **47.** $-6 + (-3)$ -9 **48.** $(-13) + (-12)$ -25

49. $-8 - 4$ -12 **50.** $4 - (-15)$ 19 **51.** $6 - (-17)$ 23 **52.** $-9 - 13$ -22

53. $(-12) + (-15)$ -27 **54.** $-12 + 12$ 0 **55.** $13 + (-20)$ -7 **56.** $5 + (-19)$ -14

57. $-12 - 9$ -21 **58.** $-9 - (-7)$ -2 **59.** $-6 - 6$ -12 **60.** $-9 - 8$ -17

61. $-16 + .03$ -15.97 **62.** $.59 + (-3.4)$ -2.81 **63.** $.08 - 3$ -2.92 **64.** $1.8 - 9$ -7.2

65. $-3.7 + (-.03)$ -3.73 **66.** $.9 + (-1)$ $-.1$

67. $\dfrac{3}{4} + \left(-\dfrac{3}{5}\right)$ $\dfrac{3}{20}$ **68.** $-\dfrac{2}{3} + \dfrac{1}{6}$ $-\dfrac{1}{2}$ **69.** $-\dfrac{1}{2} - \left(-\dfrac{3}{4}\right)$ $\dfrac{1}{4}$ **70.** $-\dfrac{1}{8} - \left(-\dfrac{1}{8}\right)$ 0

71. $45.87 + (-49.36)$ -3.49 **72.** $-.357 + (-3.465)$ -3.822 **73.** $.6578 + (-1)$ $-.3422$

74. $-2.347 + (-3.5)$ -5.847 **75.** $-3.45 - 45.39$ -48.84 **76.** $9.8 - 9.974$ $-.174$

77. $-5.79 - 3.06$ -8.85 **78.** $0 - (-4.537)$ 4.537

1.4 Multiplication and Division of Real Numbers

IN THIS SECTION:

- **Multiplication of Real Numbers**
- **Division of Real Numbers**
- **Division by Zero**

In this section we will complete the study of the four basic operations with real numbers.

Multiplication of Real Numbers

The result of multiplying two numbers is referred to as the **product** of the numbers. The numbers multiplied are referred to as **factors.** In algebra we use a raised dot, "·", between the factors to indicate multiplication, or we place symbols next to one another to indicate multiplication. Thus ab and $a \cdot b$ are both referred to as the product of a and b. When multiplying numbers, we may enclose them in parentheses to make the meaning clear. To write 5 times 3, we may write it as $5 \cdot 3$, $5(3)$, $(5)3$, or $(5)(3)$. In multiplying a number and a variable, no sign is used between them. Thus $5x$ is used to represent the product of 5 and x.

Multiplication is just a short way to do repeated additions. Adding threes together five times gives

$$3 + 3 + 3 + 3 + 3 = 15.$$

So we have the multiplication fact $5 \cdot 3 = 15$. Adding together five negative threes gives

$$(-3) + (-3) + (-3) + (-3) + (-3) = -15.$$

So we should have $5(-3) = -15$. We can think of $5(-3) = -15$ as saying that taking on 5 debts of \$3 each is equivalent to a debt of \$15. Losing 5 debts of \$3 each is equivalent to gaining \$15, so we should have $(-5)(-3) = 15$.

These examples illustrate the rules for multiplying signed numbers.

RULES
Product of Signed Numbers

1. The product of two numbers with the same sign is a positive number, equal to the product of their absolute values.
2. The product of two numbers with opposite signs is a negative number, found by multiplying their absolute values and then putting a negative sign on the result.

For example, to multiply -2 and -3, we multiply their absolute values: $(2 \cdot 3 = 6)$. Thus $(-2)(-3) = 6$. To multiply -4 and 7, we multiply their absolute values: $(4 \cdot 7 = 28)$ and then put a negative sign on the result. Thus $-4 \cdot 7 = -28$. This definition is simply remembered as

Same sign $\longleftrightarrow$ positive result,
Opposite signs $\longleftrightarrow$ negative result.

EXAMPLE 1 Evaluate the following products.

a) $(-2)(-3)$ b) $(3)(6)$ c) $(-5)(10)$

d) $\left(\dfrac{1}{3}\right)\left(-\dfrac{1}{2}\right)$ e) $(-.02)(.08)$ f) $(-300)(-.06)$

Solution

a) First find the product of the absolute values:

$$|-2| \cdot |-3| = 2 \cdot 3 = 6$$

Because -2 and -3 have the same sign, we get $(-2)(-3) = 6$.

b) $(3)(6) = 18$ **Same sign, positive result**

c) $(-5)(10) = -50$ **Opposite signs, negative result**

d) $\left(\dfrac{1}{3}\right)\left(-\dfrac{1}{2}\right) = -\dfrac{1}{6}$

e) When multiplying decimals, we total the number of decimal places in the factors to get the number of decimal places in the product. Thus $(-.02)(.08) = -.0016$.

f) $(-300)(-.06) = 18$ ◀

Division of Real Numbers

We say that $10 \div 5 = 2$ because $2 \cdot 5 = 10$. This illustrates how division can be defined in terms of multiplication.

DEFINITION
Division of Real Numbers

If a, b, and c are any numbers with $b \neq 0$, then

$$a \div b = c \qquad \text{provided that} \qquad c \cdot b = a.$$

The number c in the above definition is called the **quotient** of a and b. We also refer to both $a \div b$ and a/b as the quotient of a and b.

Using the definition of division, we get

$$10 \div (-2) = -5 \qquad \text{because } (-5)(-2) = 10,$$
$$-10 \div 2 = -5 \qquad \text{because } (-5)(2) = -10,$$
$$-10 \div (-2) = 5 \qquad \text{because } (5)(-2) = -10.$$

From these examples we see that the rules for dividing signed numbers are similar to those for multiplying signed numbers.

RULES
Division of Signed Numbers

1. If a and b ($b \neq 0$) have the same sign, the quotient of a divided by b is the same as the quotient of their absolute values.
2. To find the quotient of a and b ($b \neq 0$) when they have opposite signs, find the quotient of their absolute values and put a negative sign on that result.

This rule can be simply remembered as

Same sign ⟷ positive result,
Opposite signs ⟷ negative result.

EXAMPLE 2   Evaluate.

a) $(-8) \div (-4)$ b) $(-8) \div 8$ c) $8 \div (-4)$

d) $(-4) \div \dfrac{1}{3}$ e) $(-2.5) \div (.05)$

Solution

a) $(-8) \div (-4) = 2$ Same sign, positive result

b) $(-8) \div 8 = -1$ Opposite signs, negative result

c) $8 \div (-4) = -2$

d) $(-4) \div \dfrac{1}{3} = (-4) \cdot \dfrac{3}{1} = -12$

e) $(-2.5) \div (.05) = -50$ ◄

We use the same rules for division when division is indicated by a fraction bar. For example,

$$\frac{-9}{3} = -3, \qquad \frac{9}{-3} = -3, \qquad \frac{-1}{2} = \frac{1}{-2} = -\frac{1}{2}, \qquad \text{and} \qquad \frac{-4}{-2} = 2.$$

Note that if one negative sign appears in a fraction, the fraction has the same value whether the negative sign is in the numerator, in the denominator, or in front of the fraction. If the numerator and denominator of a fraction are both negative, then the fraction has a positive value.

Division by Zero

Division by zero is certainly contrary to the intuitive idea of division. For example, we can divide \$8 among 4 people to get \$2 per person, and we can divide \$0 among 4 people to get \$0 per person, but it doesn't make sense to divide \$8 among 0 people. How many dollars per person would we get?

Now consider how division by zero would fit the mathematical definition of division given above. If we write $10 \div 0 = c$, we need to find a number c such that $c \cdot 0 = 10$. This is impossible. If we write $0 \div 0 = c$, we need to find a number c such that $c \cdot 0 = 0$. In fact, $c \cdot 0 = 0$ is true for any value of c. Having $0 \div 0$ equal to any number would be confusing in doing computations. Thus $a \div b$ is only defined for $b \neq 0$. Quotients such as

$$8 \div 0, \qquad 0 \div 0, \qquad \frac{8}{0}, \qquad \text{and} \qquad \frac{0}{0}$$

are said to be **undefined**.

Warm-ups

True or false?

1. The product of 7 and y is written as $7y$.   T

2. The product of -2 and 5 is 10.   F

3. $\left(-\dfrac{1}{2}\right)\left(-\dfrac{1}{2}\right) = \dfrac{1}{4}$ T

4. $(-.2)(.2) = -.4$ F

5. The quotient of x and 3 can be written as $x \div 3$ or $x/3$. T

6. $(-9) \div (-3) = 3$ T

7. $6 \div (-2) = -3$ T

8. $\left(-\dfrac{1}{2}\right) \div \left(-\dfrac{1}{2}\right) = 1$ T

9. $\dfrac{0}{0} = 0$ F

10. $0 \div 6$ is undefined. F

1.4 EXERCISES

Evaluate. See Example 1.

1. $(-3)(9)$ -27
2. $(6)(-4)$ -24
3. $(.5)(-.6)$ $-.3$
4. $(-.3)(.3)$ $-.09$
5. $(-12)(-12)$ 144
6. $(-11)(-11)$ 121
7. $\left(-\dfrac{3}{4}\right)\left(\dfrac{4}{9}\right)$ $-\dfrac{1}{3}$
8. $\left(-\dfrac{2}{3}\right)\left(-\dfrac{6}{7}\right)$ $\dfrac{4}{7}$

Evaluate. See Example 2.

9. $8 \div (-8)$ -1
10. $(-90) \div (-30)$ 3
11. $(40) \div (-.5)$ -80
12. $(3) \div (.1)$ 30
13. $(-6) \div (-2)$ 3
14. $(-20) \div (-40)$ $\dfrac{1}{2}$
15. $(.5) \div (-2)$ $-.25$
16. $(-.75) \div (-.5)$ 1.5
17. $\left(-\dfrac{2}{3}\right) \div \left(-\dfrac{4}{5}\right)$ $\dfrac{5}{6}$
18. $\left(-\dfrac{1}{3}\right) \div \left(\dfrac{4}{9}\right)$ $-\dfrac{3}{4}$

Perform the indicated operations.

19. $(25)(-4)$ -100
20. $(5)(-4)$ -20
21. $(-3)(-9)$ 27
22. $(-51) \div (-3)$ 17
23. $-9 \div 3$ -3
24. $86 \div (-2)$ -43
25. $20 \div (-5)$ -4
26. $(-8)(-6)$ 48
27. $(-6)(5)$ -30
28. $(-18) \div 3$ -6
29. $(-57) \div (-3)$ 19
30. $(-30)(4)$ -120
31. $(.6)(-.3)$ $-.18$
32. $(-.2)(-.5)$ $.1$
33. $(-.03)(-10)$ $.3$
34. $(.05)(-1.5)$ $-.075$
35. $(-.6) \div (.1)$ -6
36. $8 \div (-.5)$ -16
37. $(-.6) \div (-.4)$ 1.5
38. $(-63) \div (-.9)$ 70
39. $(.45)(-365)$ -164.25
40. $8.5 \div (-.15)$ -56.667
41. $(-52) \div (-.034)$
42. $(-4.8)(5.6)$ -26.88

41. 1529.41

Perform the indicated operations.

43. $(-4)(-4)$ 16
44. $-4 - 4$ -8
45. $-4 + (-4)$ -8
46. $-4 \div (-4)$ 1
47. $-4 + 4$ 0
48. $-4 \cdot 4$ -16
49. $-4 - (-4)$ 0
50. $0 \div (-4)$ 0
51. $.1 - 4$ -3.9
52. $(.1)(-4)$ $-.4$
53. $(-4) \div (.1)$ -40
54. $-.1 - 4$ -4.1
55. $(-.1)(-4)$ $.4$
56. $-.1 + 4$ 3.9
57. $|-.4|$ $.4$
58. $|.4|$ $.4$
59. $\dfrac{-6}{3}$ -2
60. $\dfrac{2}{-3}$ $-\dfrac{2}{3}$
61. $\dfrac{3}{-4}$ $-\dfrac{3}{4}$
62. $\dfrac{-12}{-3}$ 4

63. $-\dfrac{1}{5} + \dfrac{1}{6}$ $-\dfrac{1}{30}$

64. $-\dfrac{3}{5} - \dfrac{1}{4}$ $-\dfrac{17}{20}$

65. $\left(-\dfrac{3}{4}\right)\left(\dfrac{2}{15}\right)$ $-\dfrac{1}{10}$

66. $-1 \div \left(-\dfrac{1}{4}\right)$ 4

67. $\dfrac{45.37}{6}$ 7.5617

68. $(-345) \div (28)$ -12.321

69. $(-4.3)(-4.5)$ 19.35

70. $\dfrac{-12.34}{-3}$ 4.113

1.5   Arithmetic Expressions

IN THIS SECTION:

- Exponential Expressions
- Arithmetic Expressions

In Sections 1.2 and 1.3 we learned how to perform operations with a pair of real numbers to obtain a third real number. In this section we will learn to evaluate expressions involving several numbers and operations.

Exponential Expressions

We use the notation of exponents to simplify the writing of repeated multiplication. For example,

$$2 \cdot 2 \cdot 2 = 2^3 \quad \text{and} \quad 5 \cdot 5 = 5^2.$$

DEFINITION
Exponents

> For any counting number n,
>
> $$a^n = \underbrace{a \cdot a \cdot a \cdot \ldots \cdot a.}_{n \text{ factors of } a}$$
>
> We call a the **base**, n the **exponent**, and a^n an **exponential expression.**

We read a^n as "a to the nth power." For 3^5 and 10^6 we would say "3 to the fifth power" and "10 to the sixth power." We can also use the words "squared" and "cubed" for the second and third powers. For example, x^2 and 2^3 would be read as "x squared" and "2 cubed," respectively.

EXAMPLE 1   Write each product in exponential notation and each exponential expression as a product without exponents.

a) $6 \cdot 6 \cdot 6 \cdot 6 \cdot 6$

b) $(-2)(-2)(-2)(-2)$

c) $t \cdot t \cdot t$

d) 3^2

e) y^6

Solution

a) $6 \cdot 6 \cdot 6 \cdot 6 \cdot 6 = 6^5$

b) $(-2)(-2)(-2)(-2) = (-2)^4$

Note that parentheses are essential in the expression $(-2)^4$, because $(-2)^4$ and -2^4 have different values. The reason for this will be discussed following Example 4 of this section.

c) $t \cdot t \cdot t = t^3$

d) $3^2 = 3 \cdot 3$

e) $y^6 = y \cdot y \cdot y \cdot y \cdot y \cdot y$ ◄

EXAMPLE 2   Evaluate the following exponential expressions.

a) 3^3 **b)** $(-2)^3$

c) $(-10)^4$ **d)** $(.5)^2$

Solution

a) $3^3 = 3 \cdot 3 \cdot 3 = 9 \cdot 3 = 27$

b) $(-2)^3 = (-2)(-2)(-2) = 4(-2) = -8$

c) $(-10)^4 = (-10)(-10)(-10)(-10) = 10,000$

d) $(.5)^2 = (.5)(.5) = .25$ ◄

Arithmetic Expressions

Ask students to give a nonmeaningful combination of numbers and operations.

The result of writing numbers in meaningful combination with the ordinary operations of arithmetic is called an **arithmetic expression** or simply an **expression.** Consider the expressions

$$(3 + 2) \cdot 5, \qquad 3 + (2 \cdot 5), \qquad \text{and} \qquad 3 + 2 \cdot 5.$$

The parentheses are used as **grouping symbols** and indicate which operation to perform first. Absolute value symbols and fraction bars are also used as grouping symbols. We evaluate the first two expressions as follows:

$$(3 + 2) \cdot 5 = 5 \cdot 5 = 25$$
$$3 + (2 \cdot 5) = 3 + 10 = 13$$

To simplify the writing of expressions, parentheses are often omitted, as in the expression $3 + 2 \cdot 5$. When no parentheses are present, we agree to perform **multiplication before addition.** Thus

$$3 + 2 \cdot 5 = 3 + 10 = 13.$$

Each of the three expressions above could be read as "3 plus 2 times 5," when reading from left to right. Note that this does not accurately describe the expressions, because these expressions do not all have the same value. In Section 1.6 we will learn how to read an expression in a way that accurately describes the expression.

To evaluate expressions consistently, we follow an accepted **order of operations.** When no grouping symbols are present, we agree to perform operations in the following order:

Order of Operations

1. Evaluate each exponential expression (in order from left to right).

2. Perform multiplication and division (in order from left to right).

3. Perform addition and subtraction (in order from left to right).

When parentheses or absolute value symbols are involved, we first evaluate expressions within each set of parentheses or absolute value symbols using the order of operations. When an expression involves a fraction bar, the numerator and denominator are each treated as if they are in parentheses. The use of the order of operations is further illustrated in the following examples.

EXAMPLE 3 Use the order of operations to evaluate each expression.

a) $2^3 \cdot 3^2$ b) $2 \cdot 5 - 3 \cdot 4 + 4^2$ c) $2 \cdot 3 \cdot 4 - 3^3 + \dfrac{8}{2}$

Solution

a) $2^3 \cdot 3^2 = 8 \cdot 9$ Evaluate exponential expressions before multiplying.
$= 72$

b) $2 \cdot 5 - 3 \cdot 4 + 4^2 = 2 \cdot 5 - 3 \cdot 4 + 16$ Exponential expressions first.
$\qquad\qquad\qquad\quad = 10 - 12 + 16$ Multiplication second.
$\qquad\qquad\qquad\quad = 14$ Then addition and subtraction from left to right.

c) $2 \cdot 3 \cdot 4 - 3^3 + \dfrac{8}{2} = 2 \cdot 3 \cdot 4 - 27 + \dfrac{8}{2}$ Exponential expressions first.
$\qquad\qquad\qquad\qquad = 24 - 27 + 4$ Multiplication and division second.
$\qquad\qquad\qquad\qquad = 1$ Addition and subtraction from left to right.   ◀

The next example involves grouping symbols.

EXAMPLE 4 Use the order of operations to evaluate each expression.

a) $3 - 2(7 - 2^3)$ b) $\dfrac{9 - 5 + 4}{5^2 - 3 \cdot 7}$ c) $3 - |7 - 3 \cdot 4|$

Solution

a) $3 - 2(7 - 2^3) = 3 - 2(7 - 8)$
$$= 3 - 2(-1) \quad \text{Evaluate within parentheses first.}$$
$$= 3 - (-2) \quad \text{Multiply.}$$
$$= 5 \quad \text{Subtract.}$$

b) $\dfrac{9 - 5 + 4}{5^2 - 3 \cdot 7} = \dfrac{8}{25 - 21} = \dfrac{8}{4} = 2$ **Numerator and denominator are treated as if in parentheses.**

c) $3 - |7 - 3 \cdot 4| = 3 - |7 - 12|$
$$= 3 - |-5| \quad \text{Evaluate within the absolute value symbols first.}$$
$$= 3 - 5 \quad \text{Evaluate the absolute value.}$$
$$= -2 \quad \text{Subtract.} \qquad \blacktriangleleft$$

This idea is essential for evaluating algebraic expressions like $3 - x^4$ in the next section.

Be careful when evaluating expressions like $(-2)^4$ and -2^4. The parentheses in $(-2)^4$ make it clear how to evaluate it:

$$(-2)^4 = (-2)(-2)(-2)(-2) = 16$$

The expression -2^4 is understood to be the same as the expression $-1 \cdot 2^4$. Thus

$$-2^4 = -1 \cdot 2^4 = -1 \cdot 16 = -16.$$

In evaluating -2^4 we evaluate the exponential expression first and then take the opposite.

EXAMPLE 5   Evaluate each expression.

a) -6^2 **b)** $(-6)^2$ **c)** $-(5 - 8)^2$

Solution

a) $-6^2 = -36$ **Square first, then take the opposite.**

b) $(-6)^2 = (-6)(-6) = 36$

c) The negative sign in front of the parentheses will be used last:
$$-(5 - 8)^2 = -(-3)^2 \quad \text{Evaluate within parentheses first.}$$
$$= -9 \quad \text{Square } -3 \text{ to get 9, then take the opposite of 9 to get } -9. \qquad \blacktriangleleft$$

In the next example, grouping symbols occur within grouping symbols. We evaluate within the innermost grouping symbol first. Brackets, [ ], may be used as parentheses when grouping occurs within grouping.

EXAMPLE 6   Evaluate each expression.

a) $6 - 4[5 - (7 - 9)]$ **b)** $-2|3 - (9 - 5)| - |-3|$

Solution

a) $6 - 4[5 - (7 - 9)] = 6 - 4[5 - (-2)]$ **Innermost parentheses first.**
$$= 6 - 4[7] \quad \text{Next evaluate within the brackets.}$$
$$= 6 - 28 \quad \text{Multiply.}$$
$$= -22 \quad \text{Subtract.}$$

b) $-2|3 - (9 - 5)| - |-3| = -2|3 - 4| - |-3|$

$\qquad\qquad\qquad\quad = -2|-1| - |-3|$ **Evaluate within the first absolute value.**

$\qquad\qquad\qquad\quad = -2 \cdot 1 - 3$ **Evaluate absolute values.**

$\qquad\qquad\qquad\quad = -2 - 3$ **Multiply.**

$\qquad\qquad\qquad\quad = -5$ **Subtract.** ◀

Warm-ups

True or false?

1. $(-3)^2 = -6$ F
2. $5 - 3 \cdot 2 = 4$ F
3. $(5 - 3)2 = 4$ T
4. $|5 - 6| = |5| - |6|$ F
5. $5 + 6 \cdot 2 = (5 + 6) \cdot 2$ F
6. $(2 + 3)^2 = 2^2 + 3^2$ F
7. $5 - 3^3 = 8$ F
8. $(5 - 3)^3 = 8$ T
9. $6 - \dfrac{6}{2} = \dfrac{0}{2}$ F
10. $\dfrac{6 - 6}{2} = 0$ T

1.5 EXERCISES

Write each product as an exponential expression and each exponential expression as a product without exponents. See Example 1.

1. $4 \cdot 4 \cdot 4$ 4^3
2. $(-5)(-5)(-5)(-5)$ $(-5)^4$
3. $(-y)(-y)(-y)$ $(-y)^3$
4. $t \cdot t$ t^2
5. 5^3 $5 \cdot 5 \cdot 5$
6. $(-3)^4$ $(-3)(-3)(-3)(-3)$
7. $(-a)^5$ $(-a)(-a)(-a)(-a)(-a)$
8. b^2 $b \cdot b$
9. $(-1)(-1)$ $(-1)^2$
10. $1 \cdot 1 \cdot 1 \cdot 1$ 1^4

Evaluate each exponential expression. See Example 2.

11. 3^4 81
12. 5^2 25
13. 5^3 125
14. 2^5 32
15. 1^5 1
16. 1^6 1
17. $(-1)^6$ 1
18. $(-1)^3$ -1
19. $(-2)^7$ -128
20. $(-2)^8$ 256
21. $(12)^2$ 144
22. $(-11)^2$ 121
23. 10^2 100
24. 10^3 1000

Evaluate each expression. See Example 3.

25. $-3 + 4 \cdot 6$ 21
26. $-3 \cdot 2 + 4 \cdot 5$ 14
27. $-3 \cdot 2^3$ -24
28. $-3 + 2^3$ 5
29. $2 - 3^2 + 4 \cdot 0$ -7
30. $3^2 \cdot 2^2$ 36
31. $5 \cdot 10^2$ 500
32. $3 - 5(-1)^3$ 8
33. $-3 - \dfrac{6}{2}$ -6
34. $36 \div 3^2$ 4
35. $9^2 \div 3^3$ 3
36. $3 - 2^3$ -5

Evaluate each expression. See Example 4.

37. $(-3 + 4)6$ 6
38. $-3 \cdot (2 + 4) \cdot 5$ -90
39. $(-3 \cdot 2)^3$ -216
40. $(-3 + 2)^3$ -1
41. $2 - 5(3 - 4 \cdot 2)$ 27
42. $(3 - 7)(4 - 6 \cdot 2)$ 32
43. $3 - |5 - 6|$ 2
44. $(3^2 - 5)(3 \cdot 2 - 8)$ -8

45. $|4 - 6 \cdot 3| + |9|$ 23   **46.** $3 - |6 - 7 \cdot 3|$ −12   **47.** $\dfrac{3 - 4 \cdot 6}{7 - 10}$ 7   **48.** $\dfrac{6 - (-8)}{-3 - (-1)}$ −7

49. $\dfrac{7 - 9}{9 - 7}$ −1   **50.** $\dfrac{3^2 - 2 \cdot 4}{-30 + 2 \cdot 4^2}$ $\dfrac{1}{2}$

Evaluate each expression. See Example 5.

51. -8^2 −64   **52.** $3 - 7^2$ −46   **53.** $4 - (-8)^2$ −60

54. $-(7 - 10)^3$ 27   **55.** $-(6 - 9)^4$ −81   **56.** $-2^2 - 3^2$ −13

Evaluate each expression. See Example 6.

57. $[3 - (2 - 4)][3 + (2 - 4)]$ 5   **58.** $9 + 3[5 - (3 - 6)^2]$ −3

59. $|3 - (7 - 3)|$ 1   **60.** $2 + 3 \cdot |4 - (7^2 - 6^2)|$ 29

61. $6^2 - [(2 + 3)^2 - 10]$ 21   **62.** $4 - 5 \cdot |3 - (3^2 - 7)|$ −1

63. $3 + 4[9 - 6(2 - 5)]$ 111   **64.** $3[(2 - 3)^2 + (6 - 4)^2]$ 15

Evaluate each expression.

65. $(-2 + 5)4$ 12   **66.** $(-3 - 7)3$ −30   **67.** $1 + 2^3$ 9

68. $(1 + 2)^3$ 27   **69.** $(-2)^2 - 4(-1)(3)$ 16   **70.** $(-2)^2 - 4(-2)(-3)$ −20

71. $4^2 - 4(1)(-3)$ 28   **72.** $3^2 - 4(-2)(3)$ 33   **73.** $(-11)^2 - 4(5)(0)$ 121

74. $(-12)^2 - 4(3)(0)$ 144   **75.** $5^2 - 4(2)(1)$ 17   **76.** $6^2 - 4(3)(5)$ −24

77. $[3 + 2(-4)]^2$ 25   **78.** $[6 - 2(-3)]^2$ 144   **79.** $-|-1|$ −1

80. $|1 - 7|$ 6   **81.** $\dfrac{4 - (-4)}{-2 - 2}$ −2   **82.** $\dfrac{3 - (-7)}{3 - 5}$ −5

83. $3(-1)^2 - 5(-1) + 4$ 12   **84.** $-2(1)^2 - 5(1) - 6$ −13   **85.** $5 - 2^2$ 1

86. $5 + (-2)^2$ 9   **87.** $-6 \cdot |9 - 6|$ −18   **88.** $8 - 3|5 - 9 + 1|$ −1

89. $1 - 5|5 - (9 + 1)|$ −24   **90.** $|6 - 3 \cdot 7| + |7 - (5 - 2)|$ 19

1.6 Algebraic Expressions

IN THIS SECTION:
- Identifying Algebraic Expressions
- Encoding and Decoding
- Evaluating Algebraic Expressions
- Solving Equations

In Section 1.5 we studied arithmetic expressions. In this section we will study expressions that are more general. The expressions of this section involve variables.

Identifying Algebraic Expressions

Since variables (or letters) are used to represent numbers, we can use variables in arithmetic expressions. The result of combining numbers and variables with the ordinary operations of arithmetic (in some meaningful way) is called an **algebraic expression** or simply an **expression.** For example,

$$x + 2, \qquad \pi r^2, \qquad b^2 - 4ac, \qquad \text{and} \qquad \frac{a - b}{c - d}$$

are algebraic expressions. The expression $x + 2$ is referred to as a **sum,** since the only operation in the expression is addition. The expression $a - b$ is referred to as a **difference,** as is the expression $4 - 9$. *If there is more than one operation in an expression, the expression is identified by the last operation to be performed.*

EXAMPLE 1   Identify each expression as either a sum, difference, product, quotient, or square.

a) $3(x + 2)$

b) $b^2 - 4ac$

c) $\dfrac{a - b}{c - d}$

d) $(a - b)^2$

Solution

a) The parentheses in $3(x + 2)$ indicate that we add before we multiply. So this expression is a product.

b) By the order of operations the last operation to perform in $b^2 - 4ac$ is subtraction. So this expression is a difference.

c) The last operation to perform in this expression is division. This expression is a quotient.

d) The last operation to perform is squaring, so this expression is a square.   ◄

Encoding and Decoding

Algebra is useful because it can be used to solve problems. Since problems are generally communicated verbally, we must be able to translate verbal expressions into algebraic expressions (**encode**) and translate algebraic expressions into verbal expressions (**decode**). Consider the following examples of verbal expressions and their corresponding algebraic expressions.

We emphasize the names of expressions now so that phrases like "the difference of two squares" make sense when we discuss factoring.

Verbal Expressions and Corresponding Algebraic Expressions

Verbal Expression	*Algebraic Expression*
The **sum** of $5x$ and 3	$5x + 3$
The **product** of 5 and $x + 3$	$5(x + 3)$
The **sum** of 8 and $x/3$	$8 + \dfrac{x}{3}$

The **quotient** of $8 + x$ and 3	$\dfrac{8 + x}{3}$
The **difference** of 3 and x^2	$3 - x^2$
The **square** of $3 - x$	$(3 - x)^2$

Because of the order of operations, reading from left to right does not always accurately describe an expression. For example, $5x + 3$ and $5(x + 3)$ can both be read as "5 times x plus 3" when read from left to right. *If we refer to an expression as a sum, difference, product, quotient, or square, then the meaning is clear because we know the last operation to be performed.* We will study translating verbal expressions into algebraic expressions again in Section 2.4.

EXAMPLE 2   Decode (write in words) each algebraic expression. Use the word sum, difference, product, quotient, or square.

a) $3/x$ **b)** $2y + 1$
c) $3x - 2$ **d)** $(a - b)(a + b)$

Solution

a) The quotient of 3 and x **b)** The sum of $2y$ and 1
c) The difference of $3x$ and 2   **d)** The product of $a - b$ and $a + b$

◄

EXAMPLE 3   Encode (write in symbols) each of the following verbal expressions.

a) The quotient of $a - b$ and 3   **b)** The difference of x^2 and y^2
c) The product of π and r^2 **d)** The square of $x - y$

Solution

a) $\dfrac{a - b}{3}$ or $(a - b) \div 3$ **b)** $x^2 - y^2$

c) πr^2 **d)** $(x - y)^2$ ◄

Evaluating Algebraic Expressions

The value of an algebraic expression depends on the values given to the variables. For example, the value of $x - 2y$ when $x = -2$ and $y = -3$ is found by replacing x and y by -2 and -3, respectively:

$$x - 2y = -2 - 2(-3) = -2 - (-6) = 4$$

If $x = 1$ and $y = 2$, the value of $x - 2y$ is found by replacing x by 1 and y by 2, respectively:

$$x - 2y = 1 - 2(2) = 1 - 4 = -3$$

Note that we use the order of operations in evaluating an algebraic expression.

EXAMPLE 4 If $a = 3$, $b = -2$, and $c = -4$, evaluate each expression.

a) $a^2 + 2ab + b^2$

b) $(a - b)(a + b)$

c) $b^2 - 4ac$

d) $\dfrac{a - b}{c - b}$

Solution

Note that the expressions that we are evaluating here are the same expressions that we will use later in the text.

a) $a^2 + 2ab + b^2 = 3^2 + 2(3)(-2) + (-2)^2$ **Replace.**

$\qquad\qquad\qquad = 9 + (-12) + 4$ **Evaluate.**

$\qquad\qquad\qquad = 1$

b) $(a - b)(a + b) = [3 - (-2)][3 + (-2)]$ **Replace.**

$\qquad\qquad\qquad = (5)(1)$ **Evaluate within parentheses.**

$\qquad\qquad\qquad = 5$ **Multiply.**

c) $b^2 - 4ac = (-2)^2 - 4(3)(-4)$ **Replace.**

$\qquad\qquad\quad = 4 - (-48)$ **Square -2 and perform the multiplication before**

$\qquad\qquad\quad = 52$ **subtracting.**

d) $\dfrac{a - b}{c - b} = \dfrac{3 - (-2)}{-4 - (-2)} = \dfrac{5}{-2} = -\dfrac{5}{2}$ ◄

Solving Equations

An **equation** is a statement of equality of two algebraic expressions. For example,

$$x + 3 = 9, \qquad 2x + 5 = 13, \qquad \text{and} \qquad \frac{x}{2} - 4 = 1$$

are equations. A number is a **solution** or **root** of an equation if we get a true statement when we replace the variable by the number. For example, 6 is a solution to $x + 3 = 9$, since $6 + 3 = 9$ is true. We say that 6 **satisfies** or **solves** the equation $x + 3 = 9$. The number 5 does not satisfy the equation $x + 3 = 9$, since $5 + 3 = 9$ is false. We have **solved** an equation when we have found all roots to the equation. Since 6 is the only solution to $x + 3 = 9$, we have solved the equation.

EXAMPLE 5 Determine whether the given number satisfies the equation following it.

a) 6, $3x - 7 = 9$

b) -3, $\dfrac{2x - 4}{5} = -2$

c) -5, $-x - 2 = 3$

Solution

a) Replace x by 6 in the equation $3x - 7 = 9$:

$$3(6) - 7 = 9$$
$$18 - 7 = 9$$
$$11 = 9 \qquad \text{False.}$$

The number 6 does not satisfy the equation $3x - 7 = 9$.

b) $\dfrac{2(-3) - 4}{5} = -2$

$\dfrac{-10}{5} = -2$ **True.**

The number -3 is the solution to this equation.

c) Replace x by -5 in $-x - 2 = 3$:

$$-(-5) - 2 = 3$$
$$5 - 2 = 3 \qquad \textbf{True.}$$

The number -5 satisfies this equation.   ◄

Many equations can be solved by decoding, *simply reading the equation*. Reading equations now will help us to write equations in Chapter 2. For example, the equation $3x = 21$ is read as

the product of 3 and x is 21.

Since the product of 3 and 7 is 21, x must be 7. The solution to the equation is the number 7.

The equation $2x + 5 = 13$ is decoded as

the sum of $2x$ and 5 is 13.

Since the sum of 8 and 5 is 13, $2x$ must have a value of 8. Now $2x$ is the product of 2 and x. Since the product of 2 and 4 is 8, x must be 4. So 4 is the solution to the equation. This solution is correct, since $2(4) + 5 = 13$ is correct.

EXAMPLE 6   Solve each equation by decoding and check your answer.

a) $\dfrac{x}{2} - 4 = 1$ **b)** $3(x + 4) = 18$

c) $-x + 7 = 10$

Solution

a) $\dfrac{x}{2} - 4 = 1$

$\dfrac{x}{2} = 5$ **Decode: The difference of $x/2$ and 4 is 1. Since $5 - 4 = 1$, $x/2$ must be 5.**

$x = 10$ **Decode: The quotient of x and 2 is 5. Since $10/2 = 5$, x must be 10.**

So 10 is the solution to the equation. To check this answer, replace x by 10 in the original equation:

$$\dfrac{10}{2} - 4 = 1$$

Since this equation is correct, we can be sure that 10 is the solution to the equation.

b) $3(x + 4) = 18$

$x + 4 = 6$ **Decode: The product of 3 and $x + 4$ is 18. Since $3 \cdot 6 = 18$, $x + 4$ must be 6.**

$x = 2$ **Decode: The sum of x and 4 is 6. Since $2 + 4 = 6$, x must be 2.**

Check: $3(2 + 4) = 18$. The solution to the equation is 2.

c) $-x + 7 = 10$

$-x = 3$ **Decode: The sum of $-x$ and 7 is 10. Since $3 + 7 = 10$, $-x$ must be 3.**

$x = -3$ **Decode: The opposite of x is 3. Since the opposite of -3 is 3, x must be -3.**

Check: $-(-3) + 7 = 10$. The solution to the equation is -3.   ◀

Complicated equations involving several operations, negative numbers, or fractions would be difficult to solve by decoding. We will solve only simple equations by decoding. We study decoding so that we may better understand what a root of an equation is and to help us understand the methods for solving more complicated equations presented in Section 2.1. Solving equations by decoding is actually a mental process. After solving the exercises as done in Example 6, you should try to solve them without writing anything except the answer.

Warm-ups

True or false?

1. The expression $2x + 3y$ is referred to as a sum.   T
2. The expression $5(y - 9)$ is a difference.   F
3. The expression $2(x + 3y)$ is a product.   T
4. The expression $\dfrac{x}{2} + \dfrac{y}{3}$ is a quotient.   F
5. The expression $(a - b)(a + b)$ is a product of a sum and a difference.   T
6. If x is -2, then the value of $2x + 4$ is 8.   F
7. If $a = -3$, then $a^3 - 5 = 22$.   F
8. The solution to the equation $2x - 3 = 13$ is 5.   F
9. "The product of $x + 3$ and $x + 5$ is 35" is encoded as $(x + 3)(x + 5) = 35$.   T
10. The expression $2(x + 7)$ is decoded as "the sum of 2 times x plus 7."   F

1.6 EXERCISES

Identify each expression as a sum, difference, product, quotient, square, or cube. See Example 1.

1. $a^3 - 1$ Difference

2. $a(a - 1)$ Product

3. $(a - 1)^3$ Cube

4. $3x$ Product

5. $3x + 5y$ Sum

6. $\dfrac{a - b}{b - a}$ Quotient

7. $\dfrac{a}{b} - \dfrac{b}{a}$ Difference

8. $(a - b)^2$ Square

9. $x^2 + y^2$ Sum

10. $a - \dfrac{a}{2}$ Difference

Use the term sum, difference, product, quotient, square, or cube to decode (write in words) each expression. See Example 2.

11. $x^2 - a^2$ The difference of x^2 and a^2

12. $a^3 + b^3$ The sum of a^3 and b^3

13. $(x - a)^2$ The square of $x - a$

14. $(a + b)^3$ The cube of $a + b$

15. $\dfrac{x - 4}{2}$ The quotient of $x - 4$ and 2

16. $2(x - 3)$ The product of 2 and $x - 3$

17. $\dfrac{x}{2} - 4$ The difference of $\dfrac{x}{2}$ and 4

18. $2x - 3$ The difference of $2x$ and 3

19. $(ab)^3$ The cube of ab.

20. a^3b^3 The product of a^3 and b^3

Encode (write in symbols) each verbal expression. See Example 3.

21. The sum of $2x$ and $3x$ $2x + 3x$

22. The product of $2x$ and $3x$ $(2x)(3x)$

23. The difference of 2 and $3x$ $2 - 3x$

24. The quotient of 2 and $x + 3$ $\dfrac{2}{x + 3}$

25. The square of $a + b$ $(a + b)^2$

26. The difference of a^3 and b^3 $a^3 - b^3$

27. The product of $x + 2$ and $x + 3$ $(x + 2)(x + 3)$

28. The cube of x x^3

29. The quotient of $x - 7$ and $7 - x$ $\dfrac{x - 7}{7 - x}$

30. The product of -3 and $x - 1$ $-3(x - 1)$

Evaluate each expression if $a = -1$, $b = 2$, and $c = -3$. See Example 4.

31. $-(a - b)$ 3

32. $b - a$ 3

33. $-a + 7$ 8

34. $-(c - a) + b$ 4

35. $b^2 - 4ac$ -8

36. $a^2 - 4bc$ 25

37. $\dfrac{a - c}{a - b}$ $-\dfrac{2}{3}$

38. $\dfrac{b - c}{b + a}$ 5

39. $(a - b)(a + b)$ -3

40. $(a - c)(a + c)$ -8

41. $(a - b)(a + c)$ 12

42. $(a + c)(a + b)$ -4

43. $c^2 - 2c + 1$ 16

44. $b^2 - 2b + 4$ 4

45. $a^3 - b^3$ -9

46. $b^3 - c^3$ 35

47. $\dfrac{2}{a} + \dfrac{6}{b} - \dfrac{9}{c}$ 4

48. $\dfrac{c}{a} + \dfrac{6}{c} - \dfrac{b}{a}$ 3

49. $|a|$ 1

50. $|a| \div a$ -1

51. $-|a|$ -1

52. $|c|$ 3

53. $|-c|$ 3

54. $|c| \div c$ -1

55. $|a - b|$ 3

56. $|b + c|$ 1

Determine whether the given number satisfies the equation following it. See Example 5.

57. 2, $3x + 7 = 13$ Yes

58. 4, $3x - 7 = x + 1$ Yes

59. -2, $-x + 4 = 6$ Yes

60. -1, $-3x + 7 = 10$ Yes

61. 5, $3x - 7 = 2x + 1$ No

62. 3, $-2(x - 1) = 2 - 2x$ Yes

63. -8, $x - 9 = -(9 - x)$ Yes

64. 1, $x^2 + 3x - 4 = 0$ Yes

65. -1, $x^2 + 5x + 4 = 0$ Yes

66. 8, $\dfrac{x}{x - 8} = 4$ No

67. -7, $-x - 4 = 3$ Yes

68. -9, $-x + 3 = 12$ Yes

Solve each equation by decoding. See Example 6.

69. $x + 9 = 17$ 8

70. $x - 7 = 7$ 14

71. $4x = 20$ 5

72. $12x = 144$ 12

73. $\dfrac{1}{2}x = 5$ 10

74. $\dfrac{x}{3} = 7$ 21

75. $5x = 0$ 0

76. $\dfrac{x}{199} = 0$ 0

77. $9 - x = 9$ 0

78. $x + 13 = 13$ 0

79. $2x + 1 = 21$ 10

80. $8 - 3x = 2$ 2

81. $3(x - 6) = 30$ 16

82. $-3(x + 2) = -15$ 3

83. $9 = 6x - 3$ 2

84. $7x - 3 = 11$ 2

85. $-x - 9 = 4$ −13

86. $-x + 6 = 8$ −2

87. $\dfrac{x + 7}{2} = 6$ 5

88. $\dfrac{8 - x}{3} = 2$ 2

89. $4 = \dfrac{x}{2} - 3$ 14

90. $4 - \dfrac{x}{3} = 0$ 12

91. $\dfrac{12}{x + 1} = 3$ 3

92. $\dfrac{20}{3x + 1} = 2$ 3

Find the value of $b^2 - 4ac$ for each of the following choices of a, b, and c.

93. $a = -1$, $b = -3$, $c = 2$ 17

94. $a = 1$, $b = 5$, $c = 2$ 17

95. $a = 2$, $b = 3$, $c = -4$ 41

96. $a = -2$, $b = -3$, $c = -1$ 1

97. $a = 4$, $b = 1$, $c = 3$ −47

98. $a = 3$, $b = 1$, $c = 0$ 1

99. $a = 1$, $b = -3$, $c = 0$ 9

100. $a = 1$, $b = -3$, $c = -5$ 29

101. $a = .5$, $b = 8$, $c = 2$ 60

102. $a = .2$, $b = 3$, $c = 5$ 5

103. $a = 4.2$, $b = 6.7$, $c = 1.8$ 14.65

104. $a = -3.5$, $b = 9.1$, $c = 3.6$ 133.21

105. $a = -1.2$, $b = 3.2$, $c = 5.6$ 37.12

106. $a = 2.4$, $b = -8.5$, $c = -5.8$ 127.93

1.7 Properties of the Real Numbers

IN THIS SECTION:

- The Commutative Properties
- The Associative Properties
- The Distributive Properties
- Identity Properties
- Inverse Properties
- Multiplication Property of Zero

Everyone knows that the price of a hamburger plus the price of a Coke is the same as the price of a Coke plus the price of a hamburger. But does everyone know that this illustrates the commutative property of addition? The properties of the real numbers are commonly used by anyone who performs the operations of arithmetic. In algebra we must have a thorough understanding of these properties.

The Commutative Properties

We get the same result whether we evaluate $3 + 5$ or $5 + 3$. This illustrates the **commutative property of addition.** The fact that $4 \cdot 6$ and $6 \cdot 4$ are the same illustrates the **commutative property of multiplication.**

Commutative Properties

> For any real numbers a and b,
>
> $$a + b = b + a$$
>
> and
>
> $$a \cdot b = b \cdot a.$$

EXAMPLE 1   Use the commutative property of addition to rewrite each expression.

a) $2 + (-10)$ b) $8 + x^2$ c) $2y - 4x$

Solution

a) $2 + (-10) = -10 + 2$ b) $8 + x^2 = x^2 + 8$

c) $2y - 4x = 2y + (-4x) = -4x + 2y$ ◄

EXAMPLE 2   Use the commutative property of multiplication to rewrite each expression.

a) $n \cdot 3$ b) $(x + 2) \cdot 3$ c) $5 - yx$

Solution

a) $n \cdot 3 = 3n$ b) $(x + 2) \cdot 3 = 3(x + 2)$

c) $5 - yx = 5 - xy$ ◄

Addition and multiplication are commutative operations, but what about subtraction and division? Since $5 - 3 = 2$ and $3 - 5 = -2$, subtraction is *not* commutative. To see that division is not commutative, try dividing \$8 among 4 people and \$4 among 8 people.

The Associative Properties

Consider the computation of $2 + 3 + 6$. Addition is performed on only two numbers at a time. We get the same result regardless of which of the following ways we perform the addition:

$$(2 + 3) + 6 = 5 + 6 = 11$$
$$2 + (3 + 6) = 2 + 9 = 11$$

This property of addition is called the **associative property of addition.** The associative property allows us to leave out the parentheses when writing $2 + 3 + 6$. The commutative and associative properties of addition are the reason that a hamburger, a Coke, and French fries cost the same as French fries, a hamburger, and a Coke.

We also have an **associative property of multiplication.** Consider the following:

$$(2 \cdot 3) \cdot 4 = 6 \cdot 4 = 24$$
$$2 \cdot (3 \cdot 4) = 2 \cdot 12 = 24$$

We get the same result for either order of multiplication. The associative property allows us to write $2 \cdot 3 \cdot 4$ without parentheses.

Associative Properties

For any real numbers a, b, and c,

$$(a + b) + c = a + (b + c)$$

and

$$(ab)c = a(bc).$$

EXAMPLE 3 Use the commutative and associative properties of multiplication and exponential notation to rewrite each product.

a) $(3x)(x)$ **b)** $(xy)(5yx)$

Solution

a) $(3x)(x) = 3(x \cdot x) = 3x^2$

b) The commutative and associative properties of multiplication allow us to rearrange the multiplication in any order. We generally write numbers before variables and variables in alphabetical order.

$$(xy)(5yx) = 5xxyy = 5x^2y^2$$

Note that it is not necessary to write the middle step in this process. When rewriting expressions such as these, you should write only the answer.   ◀

Consider the expression

$$3 - 9 + 7 - 5 - 8 + 4 - 13.$$

According to the accepted order of operations, we could evaluate this by computing from left to right. However, using the definition of subtraction, we can rewrite this expression as addition:

$$3 + (-9) + 7 + (-5) + (-8) + 4 + (-13)$$

The commutative and associative properties of addition allow us to add these in any order we choose. It is usually faster to add the positive numbers, add the negative numbers, and then combine those two totals:

$$3 + 7 + 4 + (-9) + (-5) + (-8) + (-13) = 14 + (-35) = -21$$

Note that by performing the operations in this manner, we must subtract only once. There is no need to rewrite this expression as we have done here. We can sum the positive numbers and the negative numbers from the original expression and then combine their totals.

EXAMPLE 4   Evaluate.

a) $3 - 7 + 9 - 5$ **b)** $4 - 5 - 9 + 6 - 2 + 4 - 8$

Solution

a) $3 - 7 + 9 - 5 \quad = \quad 12 + (-12) \quad = \quad 0$

   Sum of positive numbers ⌐   ⌐ Sum of negative numbers

b) $4 - 5 - 9 + 6 - 2 + 4 - 8 = 14 + (-24) = -10$ ◄

It is certainly not essential to evaluate the expressions of Example 4 as shown. We get the same answer by adding and subtracting from left to right. However, in algebra just getting the answer is not always the most important point. Learning to do something in more than one way often helps us to understand algebra.

Note that subtraction is not an associative operation. For example,

$$(8 - 4) - 3 \neq 8 - (4 - 3)$$

since $(8 - 4) - 3 = 1$ and $8 - (4 - 3) = 7$. Division is also not associative. For example,

$$(16 \div 4) \div 2 \neq 16 \div (4 \div 2)$$

since $(16 \div 4) \div 2 = 2$ and $16 \div (4 \div 2) = 8$.

The Distributive Properties

Consider the evaluation of the following expressions:

$$3(4 + 5) = 3 \cdot 9 = 27$$
$$3 \cdot 4 + 3 \cdot 5 = 12 + 15 = 27$$

Since both expressions have the same value, we can write

$$3(4 + 5) = \underline{3} \cdot 4 + \underline{3} \cdot 5.$$

Notice that the multiplication by 3 is **distributed over** the addition. Consider the following expressions involving multiplication and subtraction:

$$5(6 - 4) = 5 \cdot 2 = 10$$
$$5 \cdot 6 - 5 \cdot 4 = 30 - 20 = 10$$

For a real example of the distributive property, count the people sitting in 4 rows of 3 people each followed by 5 rows of 3 people each:

xxx
xxx
xxx
xxx

xxx
xxx
xxx
xxx
xxx

Get the total number of people in two different ways.

Since both expressions have the same value, we can write

$$\underline{5}(6 - 4) = \underline{5} \cdot 6 - \underline{5} \cdot 4.$$

Notice that the multiplication by 5 is distributed over each number in the parentheses. These examples illustrate the distributive properties.

Distributive Properties

> For any real numbers a, b, and c,
>
> $$a(b + c) = ab + ac$$
>
> and
>
> $$a(b - c) = ab - ac.$$

If we start with $4(x + 3)$ and write

$$\underline{4}(x + 3) = \underline{4} \cdot x + \underline{4} \cdot 3 = 4x + 12,$$

we are using the distributive property to remove the parentheses. We are multiplying 4 and $x + 3$ to get $4x + 12$.
     If we start with $3x - 15$ and write

$$3x - 15 = \underline{3} \cdot x - \underline{3} \cdot 5 = \underline{3}(x - 5),$$

we are using the distributive property to **factor** $3x - 15$. We are factoring out the common factor 3.

EXAMPLE 5   Use the distributive property to rewrite each expression. Either factor out the common factor or multiply, whichever is appropriate.

a) $7x - 21$ **b)** $a(3 - b)$
c) $5a + 5$ **d)** $-3(x - 2)$

Solution

a) Factor out 7: $7x - 21 = 7(x - 3)$
b) Multiply: $a(3 - b) = 3a - ab$ *Note:* $a \cdot 3 = 3a.$
c) Factor out 5: $5a + 5 = 5(a + 1)$
d) Multiply: $-3(x - 2) = -3x - (-3)(2)$ **Distributive property.**
         $= -3x - (-6)$ **$(-3)(2) = -6.$**
         $= -3x + 6$ **Simplify.**   ◀

Identity Properties

The numbers 0 and 1 have special properties. Multiplication of a number by 1 does not change the number, and addition of 0 to a number does not change the number. That is why 1 is called the **multiplicative identity** and 0 is called the **additive identity.**

Identity Properties

For any real number a,

$$a \cdot 1 = 1 \cdot a = a$$

and

$$a + 0 = 0 + a = a.$$

Inverse Properties

The idea of additive inverses was introduced in Section 1.3. Every real number a has an **additive inverse** or **opposite,** $-a$, such that $a + (-a) = 0$. Every nonzero real number a also has a **multiplicative inverse** or **reciprocal,** written $1/a$, such that $a(1/a) = 1$. Note that the sum of additive inverses is the additive identity and the product of multiplicative inverses is the multiplicative identity.

Inverse Properties

For any real number a there is a number $-a$ such that

$$a + (-a) = 0.$$

For any nonzero real number a there is a number $1/a$ such that

$$a \cdot \left(\frac{1}{a}\right) = 1.$$

For rational numbers the multiplicative inverse is easy to find. For example, the multiplicative inverse of 2/3 is 3/2 because

$$\frac{2}{3} \cdot \frac{3}{2} = \frac{6}{6} = 1.$$

EXAMPLE 6 Find the multiplicative inverse of each of the following numbers.

a) 5   b) .3
c) $-3/4$ d) 1.7

Solution

Technically $\dfrac{1}{.3}$ is the inverse of .3. However, it is not obvious what number $\dfrac{1}{.3}$ represents.

a) The multiplicative inverse of 5 is 1/5, since $5 \cdot \dfrac{1}{5} = 1$.

b) To find the reciprocal of .3, we write .3 as a ratio of integers:

$$.3 = \frac{3}{10}$$

The multiplicative inverse of .3 is 10/3, because

$$\frac{3}{10} \cdot \frac{10}{3} = 1.$$

c) The reciprocal of $-3/4$ is $-4/3$, because

$$\left(-\frac{3}{4}\right)\left(-\frac{4}{3}\right) = 1.$$

d) First convert 1.7 to a ratio of integers:

$$1.7 = 1\frac{7}{10} = \frac{17}{10}$$

The multiplicative inverse is 10/17.   ◀

Multiplication Property of Zero

Zero has a property that no other number has. Multiplication involving zero always results in zero.

Multiplication Property of Zero

> For any real number a, $0 \cdot a = 0$ and $a \cdot 0 = 0$.

EXAMPLE 7   Identify the property that justifies each equality.

a) $5 \cdot 7 = 7 \cdot 5$ b) $(4)(1/4) = 1$ c) $1 \cdot 864 = 864$
d) $6 + (5 + x) = (6 + 5) + x$ e) $3x + 5x = (3 + 5)x$
f) $6 + (x + 5) = 6 + (5 + x)$ g) $\pi x^2 + \pi y^2 = \pi(x^2 + y^2)$
h) $325 + 0 = 325$ i) $-3 + 3 = 0$
j) $455 \cdot 0 = 0$ k) $7(a - 1) = 7a - 7$

Solution

a) Commutative   b) Inverse   c) Identity
d) Associative   e) Distributive   f) Commutative
g) Distributive   h) Identity   i) Inverse
j) Multiplication property of 0   k) Distributive   ◀

Warm-ups

True or false?

1. Multiplication is a commutative operation. T
2. $24 \div (4 \div 2) = (24 \div 4) \div 2$ F

3. $1 \div 2 = 2 \div 1$ F
4. $6 - 5 = -5 + 6$ T
5. $9 - (4 - 3) = (9 - 4) - 3$ F
6. $5x + 5 = 5(x + 1)$ for any value of x. T
7. The multiplicative inverse of .02 is 50. T
8. $-3(x - 2) = -3x + 6$ for any value of x. T
9. $3x + 2x = (3 + 2)x$ for any value of x. T
10. The additive inverse of 0 is 0. T

1.7 EXERCISES

Use the commutative property of addition to rewrite each expression. See Example 1.

1. $9 + r$ $r + 9$
2. $2L + 2W$ $2W + 2L$
3. $3(2 + x)$ $3(x + 2)$
4. $P(1 + rt)$ $P(rt + 1)$
5. $4 - 5x$ $-5x + 4$
6. $b - 2a$ $-2a + b$

Use the commutative property of multiplication to rewrite each expression. See Example 2.

7. $x \cdot 6$ $6x$
8. $y^2 \cdot (-9)$ $-9y^2$
9. $(x - 4)(-2)$ $-2(x - 4)$
10. $a(b + c)$ $(b + c)a$
11. $4 - y \cdot 8$ $4 - 8y$
12. $z \cdot 9 - 2$ $9z - 2$

Use the commutative and associative properties of multiplication and exponential notation to rewrite each product. See Example 3.

13. $(4w)(w)$ $4w^2$
14. $(y)(2y)$ $2y^2$
15. $3a(ba)$ $3a^2b$
16. $(x)(9x)(xz)$ $9x^3z$
17. $(x \cdot x)(7x)$ $7x^3$
18. $y(y \cdot 5)(wy)$ $5wy^3$

Evaluate by finding first the sum of the positive numbers and then the sum of the negative numbers. See Example 4.

19. $8 - 4 + 3 - 10$ -3
20. $-3 + 5 - 12 + 10$ 0
21. $8 - 10 + 7 - 8 - 7$ -10
22. $6 - 11 + 7 - 9 + 13 - 2$ 4
23. $-4 - 11 + 7 - 8 + 15 - 20$ -21
24. $-8 + 13 - 9 - 15 + 7 - 22 + 5$ -29
25. $-3.2 + 2.4 - 2.8 + 5.8 - 1.6$ $.6$
26. $5.4 - 5.1 + 6.6 - 2.3 + 9.1$ 13.7
27. $3.26 - 13.41 + 5.1 - 12.35 - 5$ -22.4
28. $5.89 - 6.1 + 8.58 - 6.06 - 2.34$ $-.03$

Use the distributive property to either multiply or factor, whichever is appropriate. See Example 5.

29. $3(x - 5)$ $3x - 15$
30. $4(b - 1)$ $4b - 4$
31. $2m + 12$ $2(m + 6)$
32. $3y + 6$ $3(y + 2)$
33. $a(2 + t)$ $2a + at$
34. $b(a + w)$ $ab + bw$
35. $-3(w - 6)$ $-3w + 18$
36. $-3(m - 5)$ $-3m + 15$
37. $-4(5 - y)$ $-20 + 4y$
38. $-3(6 - p)$ $-18 + 3p$
39. $4x - 4$ $4(x - 1)$
40. $6y + 6$ $6(y + 1)$
41. $-1(a - 7)$ $-a + 7$
42. $-1(c - 8)$ $-c + 8$
43. $-1(t + 4)$ $-t - 4$
44. $-1(x + 7)$ $-x - 7$
45. $4y - 16$ $4(y - 4)$
46. $5x + 15$ $5(x + 3)$
47. $4a + 8$ $4(a + 2)$
48. $7a - 35$ $7(a - 5)$

Give the multiplicative inverse (reciprocal) of each of the following. See Example 6.

49. $\dfrac{1}{2}$ 2   **50.** $\dfrac{1}{3}$ 3   **51.** -5 $-\dfrac{1}{5}$ **52.** -6 $-\dfrac{1}{6}$ **53.** 7 $\dfrac{1}{7}$ **54.** 8 $\dfrac{1}{8}$

55. .25 4   **56.** .75 $\dfrac{4}{3}$ **57.** .9 $\dfrac{10}{9}$ **58.** $-.8$ $-\dfrac{5}{4}$ **59.** 2.5 $\dfrac{2}{5}$ **60.** 3.5 $\dfrac{2}{7}$

61. 1 1   **62.** -1 -1

Name the property that justifies each statement. See Example 7.

63. $3 \cdot x = x \cdot 3$ Commutative

64. $x + 5 = 5 + x$ Commutative

65. $2(x - 3) = 2x - 6$ Distributive

66. $a(bc) = (ab)c$ Associative

67. $-3(xy) = (-3x)y$ Associative

68. $3(x + 1) = 3x + 3$ Distributive

69. $4 + (-4) = 0$ Inverse

70. $1.3 + 9 = 9 + 1.3$ Commutative

71. $x^2 \cdot 5 = 5x^2$ Commutative

72. $0 \cdot \pi = 0$ Multiplication property of 0

73. $1 \cdot 3y = 3y$ Identity

74. $(.1)(10) = 1$ Inverse

75. $2a + 5a = (2 + 5)a$ Distributive

76. $3 + 0 = 3$ Identity

77. $-7 + 7 = 0$ Inverse

78. $1 \cdot b = b$ Identity

79. $(2346)0 = 0$ Multiplication property of 0

80. $4x + 4 = 4(x + 1)$ Distributive

81. $ay + y = y(a + 1)$ Distributive

82. $ab + bc = b(a + c)$ Distributive

Complete each statement, using the property named.

83. $a + y = $ _____ , commutative $y + a$

84. $6x + 6 = $ _____ , distributive $6(x + 1)$

85. $5(aw) = $ _____ , associative $(5a)w$

86. $x + 3 = $ _____ , commutative $3 + x$

87. $\dfrac{1}{2}x + \dfrac{1}{2} = $ _____ , distributive $\dfrac{1}{2}(x + 1)$

88. $-3(x - 7) = $ _____ , distributive $-3x + 21$

89. $6x + 15 = $ _____ , distributive $3(2x + 5)$

90. $(x + 6) + 1 = $ _____ , associative $x + (6 + 1)$

91. $4(.25) = $ _____ , inverse property 1

92. $-1(5 - y) = $ _____ , distributive $-5 + y$

93. $0 = 96($ _____ $)$, multiplication property of zero 0

94. $3 \cdot ($ _____ $) = 3$, identity property 1

95. $.33($ _____ $) = 1$, inverse property $\dfrac{100}{33}$

96. $-8(1) = $ _____ , identity property -8

1.8 Using the Properties

IN THIS SECTION:

- Using the Properties in Computation
- Like Terms
- Combining Like Terms
- Product and Quotients
- Removing Parentheses

The properties of the real numbers can be helpful when we are doing computations. In this section we will see how the properties can be applied in arithmetic and algebra.

Using the Properties in Computation

Consider the product of 26 and 200:

$$
\begin{array}{r}
200 \\
\times\ \ 26 \\
\hline
1200 \\
400 \\
\hline
5200
\end{array}
$$

Using the associative property, we can write

$$(26)(200) = (26)(2 \cdot 100) = (26 \cdot 2)(100) = 52 \cdot 100 = 5200.$$

To find this product mentally, first multiply 26 by 2 to get 52, then multiply 52 by 100 to get 5200.

EXAMPLE 1 Make use of the appropriate property to perform these computations mentally.

a) $347 + 35 + 65$ **b)** $3 \cdot 435 \cdot \dfrac{1}{3}$ **c)** $6 \cdot 28 + 4 \cdot 28$

Solution

a) To perform this addition mentally, the associative property can be applied as follows:

$$347 + (35 + 65) = 347 + 100 = 447$$

b) Use the commutative and associative properties and mentally rearrange this product:

$$3 \cdot 435 \cdot \frac{1}{3} = 435\left(3 \cdot \frac{1}{3}\right) \qquad \textbf{Commutative and associative properties.}$$

$$= 435 \cdot 1 \qquad \textbf{The inverse property.}$$

$$= 435 \qquad \textbf{The identity property.}$$

c) We can use the distributive property to rewrite this computation and do it mentally:

$$6 \cdot 28 + 4 \cdot 28 = (6 + 4)28 = 10 \cdot 28 = 280 \qquad \blacktriangleleft$$

Like Terms

An expression containing a number or the product of a number and one or more variables is called a **term.** For example,

$$-3, \quad 5x, \quad -3x^2y, \quad a, \quad \text{and} \quad -abc$$

are terms. The number preceding the variables in a term is called the **coefficient.** In the term $5x$ the coefficient of x is 5. In the term $-3x^2y$ the coefficient of x^2y is -3. In the term a the coefficient of a is 1 because $a = 1 \cdot a$. In the term $-abc$ the coefficient of abc is -1 because $-abc = -1 \cdot abc$. If two terms contain the same variables with the same exponents, they are called **like terms.** For example, $3x^2$ and $-5x^2$ are like terms.

Combining Like Terms

Using the distributive property on an expression involving the sum of like terms allows us to combine the like terms. For example,

$$3x + 5x = (3 + 5)x \qquad \text{The distributive property.}$$

$$= 8x \qquad \text{Add the coefficients.}$$

The distributive property says that $a(b + c) = ab + ac$ for any values of a, b, and c. In using the distributive property above, x could be any number. Thus $3x + 5x = 8x$ no matter what number is used for x.

We can also use the distributive property on a difference of like terms to combine them. For example,

$$-5xy - (-4xy) = [-5 - (-4)]xy \qquad \text{The distributive property.}$$

$$= -1xy \qquad -5 - (-4) = -1.$$

$$= -xy \qquad \text{Multiplying by } -1 \text{ is the}$$
$$\text{same as taking the opposite.}$$

Of course, we do not want to write out all of the steps shown here every time we combine like terms. *We can combine like terms as easily as we can add or subtract their coefficients.*

EXAMPLE 2 Combine the like terms.

a) $-3a + (-7a)$ b) $w + 2w$ c) $-2x + 5x$

d) $7xy - (-12xy)$ e) $2x^2 + 4x^2$

Solution

a) $-3a + (-7a) = -10a$

b) $w + 2w = 1w + 2w = 3w$

c) $-2x + 5x = 3x$

d) $7xy - (-12xy) = 19xy$

e) $2x^2 + 4x^2 = 6x^2$ ◄

Expressions such as

$$2 + 5x, \quad 3xy + 5x, \quad 3w + 5a, \quad \text{and} \quad 3x^2 + 5x$$

do not involve like terms. The terms in these expressions cannot be combined.

Products and Quotients

We can use the associative property of multiplication to simplify the product of two terms. For example,

$$3(5x) = (3 \cdot 5)x \qquad \text{The associative property.}$$
$$= (15)x \qquad \text{Multiply.}$$
$$= 15x \qquad \text{Remove unnecessary parentheses.}$$

Notice the difference between $3(5x)$ and $3(5 + x)$. In $3(5 + x)$ the distributive property is used to distribute the 3 over the addition. The 5 and the x are both multiplied by the 3:

$$3(5 + x) = 15 + 3x$$

In the next example we use the fact that dividing by 2 is the same as multiplying by $1/2$:

$$2\left(\frac{x}{2}\right) = 2\left(\frac{1}{2} \cdot x\right) \qquad \text{Multiplying by 1/2 is the same as dividing by 2.}$$
$$= \left(2 \cdot \frac{1}{2}\right)x \qquad \text{The associative property.}$$
$$= 1 \cdot x \qquad 2 \cdot \frac{1}{2} = 1.$$
$$= x \qquad \text{The multiplicative identity is 1.}$$

To simplify the product $(4x)(6x)$, we use both the commutative and associative properties to rewrite it:

$$(4x)(6x) = 4 \cdot 6 \cdot x \cdot x = 24x^2$$

We have shown how the properties are used to simplify products. However, in practice we do not write out any steps for these problems. We must be able to write only the answer.

EXAMPLE 3 Find the following products.

a) $(-3)(4x)$ **b)** $(-4a)(-7a)$

c) $(-3a)\left(\dfrac{b}{3}\right)$ **d)** $6 \cdot \dfrac{x}{2}$

Solution

a) $-12x$ **b)** $28a^2$

c) $-ab$ **d)** $3x$ ◄

EXAMPLE 4 Simplify each quotient.

a) $\dfrac{10x}{5}$ **b)** $\dfrac{4x + 8}{2}$

Solution

a) Since dividing by 5 is equivalent to multiplying by 1/5, we can use the associative property to write

$$\frac{10x}{5} = \frac{1}{5}(10x) = \left(\frac{1}{5} \cdot 10\right)x = (2)x = 2x.$$

All that we have really done is divide 10 by 5 to get 2.

b) Since dividing by 2 is equivalent to multiplying by 1/2, we can use the distributive property:

$$\frac{4x + 8}{2} = \frac{1}{2}(4x + 8) = 2x + 4 \qquad \text{Note that both 4 and 8 are divided by 2.} \blacktriangleleft$$

Removing Parentheses

Multiplying a number by -1 merely changes the sign of the number. For example,

$$(-1)(7) = -7 \qquad \text{and} \qquad (-1)(-8) = 8.$$

Thus we can say that -1 *times* a number is the *opposite* of the number. In symbols we write

$$(-1)x = -x \qquad \text{or} \qquad -1(y + 5) = -(y + 5).$$

When a negative sign appears in front of a sum or a difference, we think of it as multiplication by -1 and use the distributive property. For example,

$$-(w + 4) = -1(w + 4) = (-1)w + (-1)4 = -w + (-4) = -w - 4,$$
$$-(x - 3) = -1(x - 3) = (-1)x - (-1)3 = -x - (-3) = -x + 3.$$

Note that a negative sign in front of a set of parentheses is distributed over each term in the parentheses, changing the sign of each term.

EXAMPLE 5 Simplify each expression.

a) $5 - (x + 3)$ **b)** $3x - 6 - (2x - 4)$ **c)** $4x - (-x + 2)$

Solution

a) $5 - (x + 3) = 5 - x - 3$ Change the sign of each term in parentheses.
           $= 5 - 3 - x$ Commutative property.
           $= 2 - x$ Combine like terms.

b) $3x - 6 - (2x - 4) = 3x - 6 - 2x + 4$ Remove parentheses.
               $= 3x - 2x - 6 + 4$ Commutative property.
               $= x - 2$ Combine like terms.

c) $4x - (-x + 2) = 4x + x - 2$ Remove parentheses.
           $= 5x - 2$ Combine like terms. $\blacktriangleleft$

The commutative and associative properties of addition allow us to rearrange the terms so that we may combine the like terms. However, it is not necessary to actually write down the rearrangement. We can identify the like terms and combine them without rearranging.

EXAMPLE 6 Simplify the algebraic expressions.

a) $(-2x + 3) + (5x - 7)$ **b)** $-3x + 6x + 5(4 - 2x)$
c) $-2x(3x - 7) - (x - 6)$ **d)** $x - .02(x + 500)$

Solution

a) $(-2x + 3) + (5x - 7) = 3x - 4$ **Combine like terms.**
b) $-3x + 6x + 5(4 - 2x) = -3x + 6x + 20 - 10x$ **Distributive property.**
$= -7x + 20$ **Combine like terms.**
c) $-2x(3x - 7) - (x - 6) = -6x^2 + 14x - x + 6$ **Distributive property.**
$= -6x^2 + 13x + 6$ **Combine like terms.**
d) $x - .02(x + 500) = 1x - .02x - 10$ $(-.02)(500) = -10$.
$= .98x - 10$ $1 - .02 = .98$.   ◀

Warm-ups

True or false?

A statement involving variables should be marked true only if it is true for all values of the variable.

1. $3(x + 6) = 3x + 18$ T   **2.** $-3x + 9 = -3(x + 9)$ F
3. $-1(x - 4) = -x + 4$ T   **4.** $3a + 4a = 7a$ T
5. $(3a)(4a) = 12a$ F   **6.** $3(5 \cdot 2) = 15 \cdot 6$ F
7. $x + x = x^2$ F   **8.** $x \cdot x = 2x$ F
9. $3 + 2x = 5x$ F   **10.** $-(5x - 2) = -5x + 2$ T

1.8 EXERCISES

Perform each computation mentally, making use of the appropriate properties. See Example 1.

1. $35(200)$ 7000   **2.** $15(300)$ 4500   **3.** $\frac{4}{3}(.75)$ 1   **4.** $5(.2)$ 1

5. $(256 + 78) + 22$ 356   **6.** $12 + (88 + 376)$ 476   **7.** $35 \cdot 3 + 35 \cdot 7$ 350   **8.** $98 \cdot 478 + 2 \cdot 478$ 47,800

9. $18 \cdot 4 \cdot 2 \cdot \frac{1}{4}$ 36   **10.** $19 \cdot 3 \cdot 2 \cdot \frac{1}{3}$ 38   **11.** $(120)(300)$ 36,000   **12.** $150 \cdot 200$ 30,000

13. $12 \cdot 375(-6 + 6)$ 0   **14.** $(342 \cdot 7)\frac{1}{7}$ 342   **15.** $(354^2 + 17)[2(-5) + 10]$ 0 **16.** $217 + 6 + 8 + 4 + 2$ 237

Combine like terms where possible. See Example 2.

17. $-5w + 6w$ w

18. $-4a + 10a$ $6a$

19. $2x - (-3x)$ $5x$

20. $2b - (-5b)$ $7b$

21. $5mw^2 - 12mw^2$ $-7mw^2$

22. $4ab^2 - 19ab^2$ $-15ab^2$

23. $-3a - (-2a)$ $-a$

24. $10 - 6m$ $10 - 6m$

25. $9 - 4w$ $9 - 4w$

26. $-10m - (-6m)$ $-4m$

27. $3x^2 + 5x^2$ $8x^2$

28. $3b + 4b^2$ $3b + 4b^2$

29. $-4x + 2x^2$ $-4x + 2x^2$

30. $6w - w$ $5w$

31. $4x - x$ $3x$

32. $a - 6a$ $-5a$

33. $-a - a$ $-2a$

34. $a - a$ 0

Simplify the following products or quotients. See Examples 3 and 4.

35. $3(4h)$ $12h$

36. $-2(5h)$ $-10h$

37. $(-3d)(-4d)$ $12d^2$

38. $(-5t)(-2t)$ $10t^2$

39. $-3a(5b)$ $-15ab$

40. $-7w(3r)$ $-21rw$

41. $-3a(2 + b)$ $-6a - 3ab$

42. $-2x(3 + y)$ $-6x - 2xy$

43. $(-y)(-y)$ y^2

44. $y(-y)$ $-y^2$

45. $6b(-3)$ $-18b$

46. $-3m(-1)$ $3m$

47. $-k(1 - k)$ $-k + k^2$

48. $-f(f - 1)$ $-f^2 + f$

49. $(3m)(3m)$ $9m^2$

50. $(-2x)(-2x)$ $4x^2$

51. $\dfrac{3y}{3}$ y

52. $\dfrac{-15y}{5}$ $-3y$

53. $\dfrac{-12ab}{2}$ $-6ab$

54. $\dfrac{-9t}{9}$ $-t$

55. $2\left(\dfrac{y}{2}\right)$ y

56. $8y\left(\dfrac{y}{4}\right)$ $2y^2$

57. $10\left(\dfrac{2a}{5}\right)$ $4a$

58. $6\left(\dfrac{m}{3}\right)$ $2m$

59. $\dfrac{6a - 3}{3}$ $2a - 1$

60. $\dfrac{-8x + 6}{2}$ $-4x + 3$

61. $\dfrac{-9x + 6}{-3}$ $3x - 2$

62. $\dfrac{10 - 5x}{-5}$ $-2 + x$

Simplify each expression. See Example 5.

63. $x - (3x - 1)$ $-2x + 1$

64. $4x - (2x - 5)$ $2x + 5$

65. $5 - (y - 3)$ $8 - y$

66. $8 - (m - 6)$ $-m + 14$

67. $2m + 3 - (m + 9)$ $m - 6$

68. $7 - 8t - (2t + 6)$ $-10t + 1$

69. $-3 - (-w + 2)$ $w - 5$

70. $-5x - (-2x + 9)$ $-3x - 9$

Simplify the following expressions by combining like terms. See Example 6.

71. $3x + 5x + 6 + 9$ $8x + 15$

72. $2x + 6x + 7 + 15$ $8x + 22$

73. $2x + 3 + 7x - 4$ $5x - 1$

74. $-3x + 12 + 5x - 9$ $2x + 3$

75. $3a - 7 - (5a - 6)$ $-2a - 1$

76. $4m - 5 - (m - 2)$ $3m - 3$

77. $2(a - 4) - 3(-2 - a)$ $5a - 2$

78. $2(w + 6) - 3(-w - 5)$ $5w + 27$

79. $-5m + 6(m - 3) + 2m$ $3m - 18$

80. $-3a + 2(a - 5) + 7a$ $6a - 10$

81. $5 - 3(x + 2) - 6$ $-3x - 7$

82. $7 + 2(k - 3) - k + 6$ $k + 7$

83. $x - .05(x + 10)$ $.95x - .5$

84. $x - .02(x + 300)$ $.98x - 6$

85. $4.5 - 3.2(x - 5.3) - 8.75$ $-3.2x + 12.71$

86. $.03(4.5x - 3.9) + .06(9.8x - 45)$ $.723x - 2.817$

Simplify each expression.

87. $3x - (4 - x)$ $4x - 4$

88. $2 + 8x$ $2 + 8x$

89. $5a(-4a)$ $-20a^2$

90. $5a - 4a$ a

91. $.2(x + 3) - .05(x + 20)$ $.15x - .4$

92. $.08x + .12(x + 100)$ $.2x + 12$

93. $y - 5 - (-y - 9)$ $2y + 4$

94. $a - (b - c - a)$ $2a - b + c$

95. $7 - (8 - 2y - m)$ $2y + m - 1$

96. $x - 8 - (-3 - x)$ $2x - 5$

97. $2k + 1 - 3(5k - 6) - k + 4$ $-14k + 23$

98. $2w - 3 + 3(w - 4) - 5(w - 6)$ 15

Wrap-up

CHAPTER 1

SUMMARY

	Fractions	**Examples**
Reducing fractions	$\dfrac{a \cdot b}{a \cdot c} = \dfrac{b}{c}$	$\dfrac{4}{6} = \dfrac{2 \cdot 2}{2 \cdot 3} = \dfrac{2}{3}$
Building up fractions	$\dfrac{b}{c} = \dfrac{b \cdot a}{c \cdot a}$	$\dfrac{3}{8} = \dfrac{3 \cdot 5}{8 \cdot 5} = \dfrac{15}{40}$
Multiplying fractions	$\dfrac{a}{b} \cdot \dfrac{c}{d} = \dfrac{ac}{bd}$	$\dfrac{2}{3} \cdot \dfrac{4}{5} = \dfrac{8}{15}$
Dividing fractions	$\dfrac{a}{b} \div \dfrac{c}{d} = \dfrac{a}{b} \cdot \dfrac{d}{c}$	$\dfrac{2}{3} \div \dfrac{4}{5} = \dfrac{2}{3} \cdot \dfrac{5}{4} = \dfrac{10}{12} = \dfrac{5}{6}$
Adding or subtracting fractions	$\dfrac{a}{b} + \dfrac{c}{b} = \dfrac{a+c}{b}, \quad \dfrac{a}{b} - \dfrac{c}{b} = \dfrac{a-c}{b}$	$\dfrac{1}{5} + \dfrac{2}{5} = \dfrac{3}{5}, \quad \dfrac{3}{5} - \dfrac{2}{5} = \dfrac{1}{5}$
Least common denominator	The smallest number that is a multiple of all denominators	$\dfrac{1}{2} + \dfrac{2}{5} = \dfrac{5}{10} + \dfrac{4}{10} = \dfrac{9}{10}$

	Sets of Numbers	**Examples**
Counting or natural numbers	$\{1, 2, 3, \ldots\}$	
Whole numbers	$\{0, 1, 2, 3, \ldots\}$	
Integers	$\{\ldots, -3, -2, -1, 0, 1, 2, 3, \ldots\}$	
Rational numbers	$\{a/b \mid a \text{ and } b \text{ are integers with } b \neq 0\}$	$3/2, 5, -6, 0$
Irrational numbers	$\{x \mid x \text{ is a real number that is not rational}\}$	$\sqrt{2}, \sqrt{3}, \pi$
Real numbers	$\{x \mid x \text{ is the coordinate of a point on the number line}\}$	

	Operations with Real Numbers	**Examples**
Absolute value	$|a| = \begin{cases} a & \text{if } a \text{ is positive or zero} \\ -a & \text{if } a \text{ is negative} \end{cases}$	$|3| = 3, \ |0| = 0,$ $|-3| = 3$

Sum of two negative numbers	To find the sum of two negative numbers, add their absolute values and then put a negative sign on the result.	$(-3) + (-4) = -7$
Additive inverse property	For any number a, $a + (-a) = 0$, and $(-a) + a = 0$. (The sum of numbers that are opposites of each other is 0.)	$-5 + 5 = 0$ $6 + (-6) = 0$
Sum of two numbers with opposite signs (and different absolute values)	If a and b have opposite signs, the sum of a and b is found by subtracting the absolute values of the numbers. The answer is positive if the number with the larger absolute value is positive. The answer is negative if the number with the larger absolute value is negative.	$-4 + 7 = 3$ $-7 + 4 = -3$
Subtraction of signed numbers	$a - b = a + (-b)$ To do any subtraction, change it to addition of the opposite.	$3 - 5 = 3 + (-5) = -2$ $4 - (-3) = 4 + 3 = 7$
Product or quotient of signed numbers	Same sign $\longleftrightarrow$ Positive result Opposite sign $\longleftrightarrow$ Negative result	$(-3)(-2) = 6$ $(4)(-2) = -8$ $(-8) \div 2 = -4$
Definition of exponents	For any counting number n, $a^n = a \cdot a \cdot a \cdot \ldots \cdot a$ (n factors of a).	$2^3 = 2 \cdot 2 \cdot 2 = 8$
Order of operations	No parentheses or absolute value present: 1. Exponentiation 2. Multiplication and division 3. Addition and subtraction With parentheses or absolute value: First evaluate within each set of parentheses or absolute value, using the order of operations.	$5 + 2^3 = 13$ $2 + 3 \cdot 5 = 17$ $4 + 5 \cdot 3^2 = 49$ $(2 + 3)(5 - 7) = -10$ $2 + 3|2 - 5| = 11$

	Properties	Examples
Commutative properties	$a + b = b + a$ $a \cdot b = b \cdot a$	$5 + 7 = 7 + 5$ $6 \cdot 3 = 3 \cdot 6$
Associative properties	$a + (b + c) = (a + b) + c$ $a \cdot (b \cdot c) = (a \cdot b) \cdot c$	$1 + (2 + 3) = (1 + 2) + 3$ $2 \cdot (3 \cdot 4) = (2 \cdot 3) \cdot 4$
Distributive properties	$a(b + c) = ab + ac$ $a(b - c) = ab - ac$	$2(3 + x) = 6 + 3x$ $-2(x - 5) = -2x + 10$
Identity properties	$a + 0 = a$ and $0 + a = a$ Zero is the additive identity. $1 \cdot a = a$ and $a \cdot 1 = a$ One is the multiplicative identity.	$5 + 0 = 0 + 5 = 5$ $7 \cdot 1 = 1 \cdot 7 = 7$

Inverse properties	For any real number a there is a number $-a$ (additive inverse of a) such that $a + (-a) = 0$ and $(-a) + a = 0$.	$3 + (-3) = 0$ $-3 + 3 = 0$
	For any real number a $(a \neq 0)$ there is a number $1/a$ (multiplicative inverse of a) such that $a \cdot (1/a) = 1$ and $(1/a) \cdot a = 1$.	$3 \cdot (1/3) = 1$ $(-6) \cdot (-1/6) = 1$
Multiplication property of 0	$a \cdot 0 = 0$ and $0 \cdot a = 0$	$5 \cdot 0 = 0$ $0 \cdot (-7) = 0$

| | **Definition** | **Examples** |
| Like terms | Contain the same variables with the same exponents. | $-3x + 7x = 4x$ $5ab - 7ab = -2ab$ |

REVIEW EXERCISES

1.1 *Perform the indicated operations.*

1. $\dfrac{1}{3} + \dfrac{3}{8}$ $\dfrac{17}{24}$

2. $\dfrac{2}{3} - \dfrac{1}{4}$ $\dfrac{5}{12}$

3. $\dfrac{3}{5} \cdot 10$ 6

4. $\dfrac{3}{5} \div 10$ $\dfrac{3}{50}$

5. $\dfrac{2}{5} \cdot \dfrac{15}{14}$ $\dfrac{3}{7}$

6. $7 \div \dfrac{1}{2}$ 14

7. $4 + \dfrac{2}{3}$ $\dfrac{14}{3}$

8. $\dfrac{7}{12} - \dfrac{1}{4}$ $\dfrac{1}{3}$

9. $\dfrac{1}{2} + \dfrac{1}{3} + \dfrac{1}{4}$ $\dfrac{13}{12}$

10. $\dfrac{3}{4} \div 9$ $\dfrac{1}{12}$

1.2 *Which of the numbers* $-\sqrt{5}$, -2, 0, 1, 2, 3.14, π, 10 *are*

11. whole numbers? 0, 1, 2, 10

12. natural numbers? 1, 2, 10

13. integers? −2, 0, 1, 2, 10

14. rational numbers? −2, 0, 1, 2, 3.14, 10

15. irrational numbers? $-\sqrt{5}$, π

16. real numbers? All of them

True or false?

17. Every whole number is a rational number. T

18. Zero is not a rational number. F

19. The counting numbers between −4 and 4 are −3, −2, −1, 0, 1, 2, and 3. F

20. There are infinitely many integers. T

21. The set of counting numbers smaller than the national debt is infinite. F

22. The decimal number .25 is a rational number. T

23. Every integer greater than −1 is a whole number. T

24. The number zero is the only number that is neither rational nor irrational. F

1.3 *Evaluate.*

25. $-5 + 7$ 2

26. $-9 + (-4)$ −13

27. $35 - 48$ −13

28. $-3 - 9$ −12

29. $-12 + 5$ −7

30. $-12 - 5$ −17

31. $-12 - (-5)$ −7

32. $-9 - (-9)$ 0

33. $-.05 + 12$ 11.95

34. $-.03 + (-2)$ −2.03

35. $-.1 - (-.05)$ −.05

36. $-.3 + .3$ 0

37. $\dfrac{1}{3} - \dfrac{1}{2}$ $-\dfrac{1}{6}$

38. $-\dfrac{2}{3} + \dfrac{1}{4}$ $-\dfrac{5}{12}$

39. $-\dfrac{1}{3} + \left(-\dfrac{2}{5}\right)$ $-\dfrac{11}{15}$

40. $\dfrac{1}{3} - \left(-\dfrac{1}{4}\right)$ $\dfrac{7}{12}$

1.4 *Evaluate.*

41. $(-3)(5)$ -15 **42.** $(-9)(-4)$ 36   **43.** $(-8) \div (-2)$ 4   **44.** $50 \div (-5)$ -10

45. $\dfrac{-20}{-4}$ 5   **46.** $\dfrac{30}{-5}$ -6 **47.** $\left(-\dfrac{1}{2}\right)\left(-\dfrac{1}{3}\right)$ $\dfrac{1}{6}$ **48.** $8 \div \left(-\dfrac{1}{3}\right)$ -24

49. $-.09 \div .3$ $-.3$ **50.** $4.2 \div (-.3)$ -14 **51.** $(.3)(-.8)$ $-.24$ **52.** $0 \div (-.0538)$ 0

53. $(-5)(-.2)$ 1   **54.** $(1/2)(-12)$ -6

1.5 *Evaluate.*

55. $3 + 7(9)$ 66   **56.** $(3 + 7)9$ 90   **57.** $(3 + 4)^2$ 49   **58.** $3 + 4^2$ 19

59. $3 + 2 \cdot |5 - 6 \cdot 4|$ 41   **60.** $3 - (8 - 9)$ 4   **61.** $(3 - 7) - (4 - 9)$ 1   **62.** $3 - 7 - 4 - 9$ -17

63. $-2 - 4(2 - 3 \cdot 5)$ 50   **64.** $3^2 - 7 + 5^2$ 27   **65.** $3^2 - (7 + 5)^2$ -135 **66.** $|4 - 6 \cdot 3| - |7 - 9|$ 12

67. $\dfrac{9}{3} + 3$ 6   **68.** $\dfrac{9 + 3}{3}$ 4   **69.** $\dfrac{-3 - 5}{2 - (-2)}$ -2 **70.** $\dfrac{1 - 9}{4 - 6}$ 4

1.6 *Let $a = -1$, $b = -2$, and $c = 3$. Find the value of each algebraic expression.*

71. $b^2 - 4ac$ 16   **72.** $a^2 - 4b$ 9   **73.** $(c - b)(c + b)$ 5

74. $(a + b)(a - b)$ -3 **75.** $a^2 + 2ab + b^2$ 9   **76.** $a^2 - 2ab + b^2$ 1

77. $a^3 - b^3$ 7   **78.** $a^3 + b^3$ -9 **79.** $\dfrac{b + c}{a + b}$ $-\dfrac{1}{3}$

80. $\dfrac{b - c}{2b - a}$ $\dfrac{5}{3}$ **81.** $|a - b|$ 1   **82.** $|b - a|$ 1

83. $(a + b)c$ -9 **84.** $ac + bc$ -9

Find the solution to each equation.

85. $3x - 2 = 10$ 4   **86.** $5(x + 3) = 20$ 1   **87.** $\dfrac{3x}{2} = 9$ 6   **88.** $\dfrac{x}{3} - 4 = 6$ 30

89. $\dfrac{x + 3}{2} = 9$ 15   **90.** $\dfrac{12}{2x + 1} = 4$ 1   **91.** $-x - 3 = 1$ -4 **92.** $-x + 1 = 6$ -5

1.7 *Name the property that justifies each statement.*

93. $a(x + y) = ax + ay$ Distributive   **94.** $3(4y) = (3 \cdot 4)y$ Associative

95. $(.001)(1000) = 1$ Inverse   **96.** $xy = yx$ Commutative

97. $0 + y = y$ Identity   **98.** $325 \cdot 1 = 325$ Identity

99. $3 + (2 + x) = (3 + 2) + x$ Associative   **100.** $2x - 6 = 2(x - 3)$ Distributive

101. $5 \cdot 200 = 200 \cdot 5$ Commutative   **102.** $3 + (x + 2) = (x + 2) + 3$ Commutative

103. $-50 + 50 = 0$ Inverse   **104.** $43 \cdot 59 \cdot 82 \cdot 0 = 0$ Multiplication property of 0

105. $12 \cdot 1 = 12$ Identity   **106.** $3x + 1 = 1 + 3x$ Commutative

1.8 *Simplify by combining like terms.*

107. $3a + 7 - (4a - 5)$ $-a + 12$ **108.** $2m + 6 - (m - 2)$ $m + 8$

109. $2a(3a - 5) + 4a$ $6a^2 - 6a$ **110.** $3a(a - 5) + 5a(a + 2)$ $8a^2 - 5a$

111. $3(t - 2) - 5(3t - 9)$ $-12t + 39$

112. $2(m + 3) - 3(3 - m)$ $5m - 3$

113. $.1(a + .3) - (a + .6)$ $-.9a - .57$

114. $.1(x + .3) - (x - .9)$ $-.9x + .93$

115. $.05(x - 20) - .1(x + 30)$ $-.05x - 4$

116. $.02(x - 100) + .2(x - 50)$ $.22x - 12$

117. $5 - 3x(-5x - 2)$ $15x^2 + 6x + 5$

118. $7 - 2x(3x - 7) - x^2$ $-7x^2 + 14x + 7$

119. $-(a - 2) - 2 - a$ $-2a$

120. $-(w - y) - 3(y - w)$ $-2y + 2w$

121. $x(x + 1) + 3(x - 1)$ $x^2 + 4x - 3$

122. $y(y - 2) + 3(y + 1)$ $y^2 + y + 3$

Miscellaneous

Evaluate each expression.

123. $752(-13) + 752(13)$ 0

124. $75 - (-13)$ 88

125. $|15 - 23|$ 8

126. $4^2 - 6^2$ -20

127. $-6^2 + 3(5)$ -21

128. $(.03)(-200)$ -6

129. $\dfrac{2}{5} + \dfrac{1}{10}$ $\dfrac{1}{2}$

130. $\dfrac{2 + 1}{5 + 10}$ $\dfrac{1}{5}$

131. $(.05) \div (-.1)$ $-.5$

132. $(4 - 9)^2 + (2 \cdot 3 - 1)^2$ 50

133. $2\left(-\dfrac{1}{2}\right)^2 + \left(-\dfrac{1}{2}\right) - 1$ -1

134. $\left(-\dfrac{6}{7}\right)\left(\dfrac{21}{26}\right)$ $-\dfrac{9}{13}$

Simplify each expression if possible.

135. $\dfrac{2x + 4}{2}$ $x + 2$

136. $4(2x)$ $8x$

137. $4 + 2x$ $4 + 2x$

138. $4(2 + x)$ $8 + 4x$

139. $4 \cdot \dfrac{x}{2}$ $2x$

140. $4 - (x - 2)$ $-x + 6$

141. $-4(x - 2)$ $-4x + 8$

142. $(4x)(2x)$ $8x^2$

143. $4x + 2x$ $6x$

144. $2 + (x + 4)$ $x + 6$

145. $4 \cdot \dfrac{x}{4}$ x

146. $4 \cdot \dfrac{3x}{2}$ $6x$

147. $2 \cdot x \cdot 4$ $8x$

148. $4 - 2(2 - x)$ $2x$

CHAPTER 1 TEST

Which of the numbers $-3, -\sqrt{3}, -1/4, 0, \sqrt{5}, \pi, 8$ *are*

1. whole numbers? $0, 8$

2. integers? $-3, 0, 8$

3. rational numbers? $-3, -\dfrac{1}{4}, 0, 8$

4. irrational numbers? $-\sqrt{3}, \sqrt{5}, \pi$

Evaluate the following expressions.

5. $6 + 3(-9)$ -21

6. $(-2)^2 - 4(-2)(-1)$ -4

7. $\dfrac{-3 - 9}{3 - 5}$ 6

8. $-5 + 6 - 12 + 4$ -7

9. $.05 - 1$ $-.95$

10. $(5 - 9)(5 + 9)$ -56

11. $(878 + 89) + 11$ 978

12. $6 + |3 - 5(2)|$ 13

13. $8 - 3|7 - 10|$ -1

14. $(839 + 974)[3(-4) + 12]$ 0

15. $974(7) + 974(3)$ 9740

16. $-\dfrac{2}{3} + \dfrac{3}{8}$ $-\dfrac{7}{24}$

17. $(-.05)(400)$ -20

18. $\left(-\dfrac{3}{4}\right)\left(\dfrac{2}{9}\right)$ $-\dfrac{1}{6}$

19. $13 \div \left(-\dfrac{1}{3}\right)$ -39

Identify the property that justifies each equality.

20. $2(x + 7) = 2x + 14$ Distributive

21. $48 \cdot 1000 = 1000 \cdot 48$ Commutative

22. $2 + (6 + x) = (2 + 6) + x$ Associative

23. $-348 + 348 = 0$ Inverse

24. $1 \cdot (-6) = -6$ Identity

25. $0 \cdot 388 = 0$ Multiplication property of 0

Use the distributive property to factor each of the following.

26. $3x - 30$ 3(x − 10)

27. $7w - 7$ 7(w − 1)

Simplify the following expressions.

28. $6 + 4x + 2x$ 6x + 6

29. $6 + 4(x - 2)$ 4x − 2

30. $5x - (3 - 2x)$ 7x − 3

31. $x + 10 - .1(x + 25)$ .9x + 7.5

32. $2a(4a - 5) - 3a(-2a - 5)$ 14a² + 5a

33. $\dfrac{6x + 12}{6}$ x + 2

34. $8 \cdot \dfrac{t}{2}$ 4t

Evaluate each expression if $a = -2$, $b = 3$, and $c = 4$.

35. $b^2 - 4ac$ 41

36. $\dfrac{a - b}{b - c}$ 5

37. $(a - c)(a + c)$ −12

Solve each equation.

38. $3x - 4 = 2$ 2

39. $\dfrac{x + 3}{8} = 2$ 13

40. $-x + 5 = 8$ −3

CHAPTER 2
Linear Equations and Inequalities

In 1972 a federal judge ordered the integration of the schools in Jefferson County. Wilson High School had 400 students, of whom 20% were black. The school board merged Wilson with Jefferson High to create one school that had a 44% black student population. If Jefferson High had a 60% black student population before the merger, then how many students did Jefferson High have before the merger?

The number of students at Jefferson High is unknown to us. However, the methods of algebra can be used to write an equation that involves the unknown number and then solve the equation to find that number. One of our main goals in algebra is to learn to solve problems. In this chapter we will use algebra to solve many applied problems. This problem is an example of a mixture problem. When we see it again as Exercise 37 of Section 2.5, we will have the necessary skills to solve it.

2.1 Linear Equations in One Variable

IN THIS SECTION:
- **Basic Ideas and Definitions**
- **Using the Addition-Subtraction Property of Equality**
- **Multiplication-Division Property of Equality**

One of the most important techniques that we study in algebra is solving equations. In Section 1.6 we learned to solve some simple equations. In this section we will use a new method to solve more complicated equations.

Basic Ideas and Definitions

In Chapter 1 we solved equations by decoding. Recall how to solve the equation $2x + 4 = 10$:

$$2x + 4 = 10$$

$\quad 2x = 6$ **Decode: the sum of $2x$ and 4 is 10. Since $6 + 4 = 10$, $2x = 6$.**

$\quad\quad x = 3$ **Decode: the product of 2 and x is 6. Since $2 \cdot 3 = 6$, $x = 3$.**

Note that 3 satisfies the equations $x = 3$, $2x = 6$, and $2x + 4 = 10$. Equations that have the same solution are called **equivalent equations.** All three of the above equations are equivalent. When solving an equation, we try to write a sequence of simpler and simpler equivalent equations until we *get an equation with x isolated on one side of the equation.* The solution to an equation such as $x = 3$ is 3, since $3 = 3$ is correct.

While decoding works well on simple equations, we need to use the properties of equality to solve more complicated equations (such as the equation of Example 3).

Using the Addition-Subtraction Property of Equality

An equation says that the two expressions are equal. If we add the same number to each of the two equal expressions, we must get equal results. Likewise, if we subtract the same number from each of the two equal expressions, we get equal results. This idea is called **the addition-subtraction property of equality,** and we can use it to isolate the variable in an equation.

PROPERTY
Addition-Subtraction
Property of Equality

Adding the same number to each side of an equation or subtracting the same number from each side of an equation gives an equation equivalent to the original equation.

EXAMPLE 1   Solve $x - 3 = -7$.

Solution   Note that the left-hand side of this equation is a difference. We can remove the 3 from the left-hand side of the equation by adding 3 to each side of the equation:

$$x - 3 = -7$$
$$x - 3 + 3 = -7 + 3 \qquad \text{Add 3 to each side.}$$
$$x = -4 \qquad \text{Simplify each side.}$$

The solution to the last equation is -4. We check -4 in the original equation. Because $-4 - 3 = -7$ is correct, we can be sure that -4 is the solution to the equation.   ◄

EXAMPLE 2   Solve $5 - x = -9$.

Solution   The 5 can be removed from the left-hand side of the equation by subtracting 5 from each side:

$$5 - x = -9$$
$$5 - x - 5 = -9 - 5 \qquad \text{Subtract 5 from each side.}$$
$$-x = -14 \qquad \text{Simplify each side.}$$
$$x = 14 \qquad \text{Since the opposite of } x \text{ is } -14, x \text{ must be 14.}$$

Check: $5 - 14 = -9$. The solution to the equation is 14.   ◄

Remember that the idea is to isolate x on one side of the equation. In the next example, x appears on both sides of the original equation. *We use the addition-subtraction property of equality to get like terms from opposite sides onto the same side so that they may be combined.*

EXAMPLE 3   Solve $2x - 3 = x + 5$.

Solution   Note that the left-hand side of the equation is a difference. To eliminate the 3 from the left-hand side, we can add 3 to each side:

$$2x - 3 = x + 5$$
$$2x - 3 + 3 = x + 5 + 3 \qquad \text{Add 3 to each side.}$$
$$2x = x + 8 \qquad \text{Simplify each side.}$$
$$2x - x = x + 8 - x \qquad \text{Subtract } x \text{ from each side.}$$
$$x = 8 \qquad \text{Simplify each side.}$$

If we let $x = 8$ in the original equation, we get

$$2(8) - 3 = 8 + 5$$

or

$$13 = 13.$$

Since each side of the original equation has the same value when $x = 8$, the correct solution to the equation is 8.   ◄

Multiplication-Division Property of Equality

Equations involving products or quotients can be solved by using the **multiplication-division property of equality**.

PROPERTY
Multiplication-Division
Property of Equality

> Multiplying or dividing both sides of an equation by the same nonzero number gives an equation equivalent to the original equation.

EXAMPLE 4   Solve $\dfrac{x}{2} = -6$.

Point out that an equation is like a balance scale. To keep the balance we must perform the same operation on each side.

Solution   Note that the left-hand side is the quotient of x and 2. We can eliminate the 2 from the left-hand side by multiplying each side of the equation by 2:

$$\frac{x}{2} = -6$$

$$2 \cdot \frac{x}{2} = 2(-6) \qquad \text{Multiply each side by 2.}$$

$$x = -12 \qquad \text{Simplify: } 2 \cdot \frac{x}{2} = x$$

Check -12 in the original equation to be sure that -12 is the solution.   ◄

EXAMPLE 5   Solve the equation $-5x = 30$.

Solution

$$-5x = 30$$

$$\frac{-5x}{-5} = \frac{30}{-5} \qquad \text{Divide each side by } -5.$$

$$x = -6 \qquad \text{Simplify: } \frac{-5x}{-5} = x$$

Check -6 in the original equation to see that -6 is the solution.   ◄

EXAMPLE 6 Solve the equation $\frac{2}{3}x = 40$.

Solution

$$\frac{2}{3}x = 40$$

$$\frac{3}{2} \cdot \frac{2}{3}x = \frac{3}{2} \cdot 40 \qquad \text{Multiply each side by 3/2, the reciprocal of 2/3.}$$

$$1 \cdot x = 60 \qquad \frac{3}{2} \cdot 40 = \frac{120}{2} = 60$$

$$x = 60$$

The solution is 60. Check it.   ◀

In the next example we use both properties of equality.

EXAMPLE 7 Solve $3x - 5 = 0$.

Solution

$$3x - 5 = 0$$

$$3x - 5 + 5 = 0 + 5 \qquad \text{Add 5 to each side.}$$

$$3x = 5 \qquad \text{Combine like terms.}$$

$$\frac{3x}{3} = \frac{5}{3} \qquad \text{Divide each side by 3.}$$

$$x = \frac{5}{3} \qquad \text{Simplify.}$$

Check: $3 \cdot \frac{5}{3} - 5 = 0$. The solution is 5/3.   ◀

Note that in the next example we *combine like terms as much as possible before we use any properties of equality,* and we use the division property of equality last.

EXAMPLE 8 Solve the equation $2(x - 3) + 5x = 15$.

Emphasize that each side should be simplified as much as possible before using the properties of equality.

Solution

$$2(x - 3) + 5x = 15$$

$$2x - 6 + 5x = 15 \qquad \text{The distributive property.}$$

$$7x - 6 = 15 \qquad \text{Combine like terms.}$$

$$7x - 6 + 6 = 15 + 6 \qquad \text{Add 6 to each side.}$$

$$7x = 21 \qquad \text{Combine like terms.}$$

$$\frac{7x}{7} = \frac{21}{7} \qquad \text{Divide each side by 7.}$$

$$x = 3 \qquad \text{Simplify: } \frac{7x}{7} = x$$

To check our work, we replace x by 3 in the original equation and simplify:

$$2(3 - 3) + 5(3) = 15$$
$$2 \cdot 0 + 15 = 15$$
$$15 = 15$$

Since both sides have a value of 15, we can be certain that 3 is the correct solution.

◄

All of the above equations are examples of linear equations in one variable.

DEFINITION
Linear Equation

> A linear equation in one variable x is any equation that can be written in the form $ax + b = 0$, where a and b are real numbers.

It is certainly not necessary to write a linear equation in the form of the definition in order to solve it. We should always try to solve equations with a minimum number of steps. It is not necessary to solve an equation exactly as was done in the above examples. We can perform the steps in a different order and still get the solution. However, we must have equivalent equations at every step. When solving equations, you should keep in mind the following strategy for isolating the variable.

STRATEGY
Solving Equations

> 1. Remove parentheses and combine like terms to simplify each side as much as possible.
> 2. Use the addition-subtraction property of equality to get like terms from opposite sides onto the same side so that they may be combined.
> 3. The multiplication-division property of equality is generally used last.

Warm-ups

Encourage students to explain why each statement is either true or false. Have students correct each false statement. Answer all 10 questions before looking at the correct answers in the back of the text.

True or false?

1. The solution to $2x - 5 = 5$ is 5. T

2. The equation $\dfrac{x}{2} = 4$ is equivalent to the equation $x - 8 = 0$. T

3. To solve $\dfrac{3}{4}x = 12$, we should multiply each side by 3/4. F

4. If $x + 5 = 0$, then $x = 5$. F

5. Multiplying each side of an equation by the same real number will result in an equation that is equivalent to the original equation. F

6. To solve $3x - 7 = 9$, we should first divide each side by 3 and then add 7 to each side. F

7. To solve $\dfrac{x}{3} = 30$, we should divide each side by 3. F

8. The equation $2x + 3 = 0$ is equivalent to $2x = 3$. F

9. The equation $x - (x - 3) = 5x$ is equivalent to $3 = 5x$. T

10. The solution to $4 - x = -2x$ is -4. T

2.1 EXERCISES

Use the properties of equality to solve each equation. Show your work and check your answer. See Examples 1, 2, and 3.

1. $x - 6 = -5$ 1

2. $-7 + x = -2$ 5

3. $x + 13 = -4$ −17

4. $x + 8 = -12$ −20

5. $-x + 6 = 5$ 1

6. $-x - 2 = 9$ −11

7. $9 - x = -3$ 12

8. $-4 - x = 6$ −10

9. $-6 = -3 + x$ −3

10. $-8 = x - 9$ 1

11. $2x + 5 = x - 7$ −12

12. $3x - 6 = 2x - 7$ −1

13. $-3x + 1 = 5 - 4x$ 4

14. $7 + x = 2x$ 7

15. $-6 + x = 2x$ −6

16. $-9 + 2x = 3x$ −9

17. $-8 + 5x = 6x$ −8

18. $5 - 2x = -x$ 5

19. $-12 - 5x = -4x$ −12

20. $-2x - 4 = -3x + 8$ 12

21. $x + .3 = 2$ 1.7

22. $2x - .05 = x + 1$ 1.05

23. $2x - .6 = x - .5$ .1

24. $2.3x + 6 = 1.3x - 1$ −7

25. $.2x - 4 = .6 - .8x$ 4.6

26. $.3x = 1 - .7x$ 1

27. $.46x - 4.58 = -.54x - 3.55$ 1.03

28. $-2.34x + 3.87 = 4.56 - 3.34x$ .69

Solve each linear equation. Show your work and check your answer. See Examples 4 and 5.

29. $-3x = 15$ −5

30. $-5x = -20$ 4

31. $\dfrac{x}{2} = -4$ −8

32. $\dfrac{x}{3} = -6$ −18

33. $20 = 4x$ 5

34. $18 = -3x$ −6

35. $-x = \dfrac{2}{3}$ $-\dfrac{2}{3}$

36. $-x = -\dfrac{3}{5}$ $\dfrac{3}{5}$

37. $.03 = \dfrac{x}{60}$ 1.8

38. $.05 = \dfrac{x}{80}$ 4

39. $.5x = 10$ 20

40. $.4x = -8$ −20

41. $2x = \dfrac{1}{2}$ $\dfrac{1}{4}$

42. $3x = \dfrac{1}{4}$ $\dfrac{1}{12}$

43. $2x = \dfrac{4}{5}$ $\dfrac{2}{5}$

44. $3x = \dfrac{6}{7}$ $\dfrac{2}{7}$

45. $\dfrac{x}{2} = \dfrac{1}{3}$ $\dfrac{2}{3}$

46. $\dfrac{x}{2} = \dfrac{1}{5}$ $\dfrac{2}{5}$

47. $\dfrac{1}{3}x = \dfrac{1}{6}$ $\dfrac{1}{2}$

48. $\dfrac{1}{3}x = \dfrac{1}{4}$ $\dfrac{3}{4}$

49. $.35x = 1.68$ 4.8

50. $.13x = 7.176$ 55.2

51. $-.76x = 60.8$ −80

52. $-.88x = 22$ −25

Solve each linear equation. Show your work and check your answer. See Example 6.

53. $-\dfrac{1}{4}x = 3$ −12

54. $-\dfrac{1}{5}x = -5$ 25

55. $\dfrac{5x}{7} = 10$ 14

56. $\dfrac{7x}{12} = -24$ $-\dfrac{288}{7}$

57. $\frac{2}{3}x = 8$ 12

58. $\frac{3}{4}x = 9$ 12

59. $\frac{3}{5}x = 6$ 10

60. $\frac{4}{5}x = 12$ 15

61. $-\frac{3}{4}x = 18$ −24

62. $-\frac{4}{3}x = 20$ −15

63. $-10 = -\frac{5}{6}x$ 12

64. $-15 = -\frac{3}{5}x$ 25

Solve each of these linear equations. Show your work and check your answer. See Examples 7 and 8.

65. $2x - 3 = 0$ $\frac{3}{2}$

66. $5x - 7 = 0$ $\frac{7}{5}$

67. $-2x + 5 = 7$ −1

68. $-3x + 4 = 13$ −3

69. $-3(x - 6) = 2 - x$ 8

70. $-2(x - 5) = 3 - x$ 7

71. $-2x - 5 = 7$ −6

72. $-3x - 7 = -1$ −2

73. $2(x + 1) - x = 36$ 34

74. $3(x + 1) - x = 23$ 10

75. $12 = 4x + 3$ $\frac{9}{4}$

76. $14 = 5x - 21$ 7

77. $-3x - 1 = 5 - 2x$ −6

78. $-3x - 2 = -5 - 4x$ −3

79. $.03x + 21 = 27$ 200

80. $.05x - 9 = 6$ 300

81. $\frac{x}{5} - 4 = -6$ −10

82. $\frac{x}{2} + 13 = -22$ −70

83. $\frac{2}{3}x - 5 = 7$ 18

84. $\frac{3}{4}x - 9 = -6$ 4

85. $4 - \frac{2x}{5} = 12$ −20

86. $9 - \frac{2x}{7} = 19$ −35

87. $-x - \frac{1}{2} = \frac{1}{2}$ −1

88. $-x - \frac{2}{3} = \frac{1}{3}$ −1

89. $3.5x - 23.7 = -38.75$ −4.3

90. $3(x - .87) - 2x = 4.98$ 7.59

2.2 More Linear Equations

IN THIS SECTION:

- Identities
- Conditional Equations
- Inconsistent Equations
- Equations Involving Fractions
- Equations Involving Decimals
- Mental Algebra

In this section we will solve more equations of the type that we solved in Section 2.1. However, some equations in this section have infinitely many solutions, and some have no solution.

Identities

Have students write some identities of their own, where both sides are not exactly identical; e.g., $x + 3 = 3 + x$.

It is easy to find equations that are satisfied by any real number that we choose as a replacement for the variable. For example, the equations

$$x \div 2 = \frac{1}{2}x, \qquad x + x = 2x, \qquad \text{and} \qquad x + 1 = x + 1$$

Review why division by 0 is undefined. How would you explain to your child why division by 0 is undefined?

are satisfied by all real numbers. The equation

$$\frac{5}{x} = \frac{5}{x}$$

is satisfied by any real number except 0, because division by 0 is undefined.

DEFINITION
Identity

An equation that is satisfied by every number for which both sides of the equation are defined is called an **identity.**

Emphasize that an identity does not have to be true for every real number.

We cannot recognize that the equation in the next example is an identity until we have simplified each side.

EXAMPLE 1   Solve $7 - 5(x - 6) + 4 = 3 - 2(x - 5) - 3x + 28$.

Solution   We first use the distributive property to remove the parentheses:

$$7 - 5x + 30 + 4 = 3 - 2x + 10 - 3x + 28$$
$$41 - 5x = 41 - 5x \qquad \text{Combine like terms.}$$

This last equation is true for any value of x, because the two sides are identical. All real numbers satisfy this equation. This last equation is an identity, and so is the original equation.   ◄

Conditional Equations

The statement $2x + 4 = 10$ is true only on condition that we choose $x = 3$. For this reason it is called a conditional equation.

DEFINITION
Conditional Equation

A **conditional equation** is an equation that is satisfied by at least one real number but is not an identity.

Every equation that we solved in Section 2.1 is a conditional equation.

Inconsistent Equations

It is easy to find equations that are false no matter what number we use to replace the variable. Consider the equation $x = x + 1$. If we replace x by 3, we get

$$3 = 3 + 1.$$

This is false. If we replace x by 4, we get

$$4 = 4 + 1.$$

This is also false. Clearly, there is no number that will satisfy $x = x + 1$. Other examples of equations with no solutions include

$$x = x - 2, \qquad x - x = 5, \qquad \text{and} \qquad 0 \cdot x + 6 = 7.$$

**DEFINITION
Inconsistent Equation**

An equation that has no solution is called an **inconsistent equation.**

EXAMPLE 2   Solve $2 - 3(x - 4) = 4(x - 7) - 7x$.

Solution   Use the distributive property to remove the parentheses:

$$2 - 3x + 12 = 4x - 28 - 7x$$
$$14 - 3x = -28 - 3x \qquad \text{Combine like terms on each side.}$$
$$14 - 3x + 3x = -28 - 3x + 3x \qquad \text{Add } 3x \text{ to each side.}$$
$$14 = -28 \qquad \text{Simplify.}$$

This last equation is not true for any choice of x. So there is no solution to the original equation, and the equation is inconsistent.   ◄

Recognizing Identities and Inconsistent Equations

1. An equation that is equivalent to an equation in which both sides are identical is an identity.
2. An equation that is equivalent to an equation that is always false is inconsistent.

Equations Involving Fractions

Some equations involving fractions were solved in Section 2.1. Here, we will solve equations with fractions by eliminating all fractions in the first step.

EXAMPLE 3   Solve $\dfrac{x}{2} - 1 = \dfrac{x}{3} + 1$.

Solution   The least common denominator (LCD) for the denominators 2 and 3 is 6. Since both 2 and 3 divide into 6 evenly, multiplying each side by 6 will eliminate the fractions:

$$6\left(\frac{x}{2} - 1\right) = 6\left(\frac{x}{3} + 1\right) \qquad \text{Multiply each side by 6.}$$

$$6 \cdot \frac{x}{2} - 6 \cdot 1 = 6 \cdot \frac{x}{3} + 6 \cdot 1 \qquad \text{Distributive property}$$

$$3x - 6 = 2x + 6 \qquad \text{Simplify: } 6 \cdot \frac{x}{2} = 3x$$

$$3x = 2x + 12 \qquad \text{Add 6 to each side.}$$

$$x = 12 \qquad \text{Subtract } 2x \text{ from each side.}$$

Check: $\dfrac{12}{2} - 1 = \dfrac{12}{3} + 1$. The solution to the equation is 12.   ◄

Equations involving fractions are usually easier to solve if we first multiply each side by the LCD of the fractions.

Equations Involving Decimals

When an equation involves decimal numbers, we can work with the decimal numbers or we can eliminate all of the decimal numbers by multiplying both sides by 10, or 100, or 1000, etc. Multiplying a decimal number by 10 moves the decimal point one place to the right. Multiplying by 100 moves the decimal point two places to the right, and so on.

EXAMPLE 4 Solve $.3x + 8.04 = 12.6$.

Solution The largest number of decimal places appearing in the decimal numbers of the equation is two (in the number 8.04). Therefore we multiply each side of the equation by 100, because multiplying by 100 moves decimal points two places to the right:

$$100(.3x + 8.04) = 100(12.6)$$

$$100(.3x) + 100(8.04) = 100(12.6) \qquad \text{Distributive property}$$

$$30x + 804 = 1260$$

$$30x + 804 - 804 = 1260 - 804$$

$$30x = 456$$

$$\frac{30x}{30} = \frac{456}{30}$$

$$x = 15.2$$

The solution is 15.2. Check in the original equation.   ◄

EXAMPLE 5   Solve $.5x + .04(x + 100) = 9.4$.

Solution   First use the distributive property to remove the parentheses:

$$.5x + .04x + 4 = 9.4 \qquad \text{Distributive property}$$
$$100(.5x + .04x + 4) = 100(9.4) \qquad \text{Multiply each side by 100.}$$
$$50x + 4x + 400 = 940 \qquad \text{Simplify.}$$
$$54x + 400 = 940 \qquad \text{Combine like terms.}$$
$$54x = 540 \qquad \text{Subtract 400 from each side.}$$
$$x = 10 \qquad \text{Divide each side by 54.}$$

Check 10 in the original equation:

$$.5(10) + .04(10 + 100) = 9.4$$
$$5 + .04(110) = 9.4$$
$$5 + 4.4 = 9.4$$

Since both sides have the same value, 10 is the correct solution.   ◄

Have students do this problem without multiplying by 100. The only way to see which way is easier is to try both ways.

Mental Algebra

I emphasize that we want to solve simple equations mentally, because later on we must solve simple equations as part of more complicated problems such as finding intercepts.

It is very important to develop the skill of solving equations in a systematic way, writing down every step as we have been doing. It is also important to be able to solve simple equations quickly. In Chapter 1 we learned to solve simple equations mentally by decoding. The method of solving equations that we have been using in this chapter can also be done mentally if the solution involves only one or two steps. For example, to solve

$$x + 5 = 0,$$

we can subtract 5 from each side mentally and write

$$x = -5.$$

To solve the equation

$$3x - 7 = 0,$$

we add 7 to each side mentally and then divide each side by 3 to get

$$x = \frac{7}{3}.$$

EXAMPLE 6   Solve each equation mentally.

a) $x - 3 = 0$ **b)** $2x + 5 = 0$ **c)** $3x = \dfrac{1}{2}$

Solution

a) Add 3 to each side to get $x = 3$. The solution is 3.

b) Subtract 5 from each side and then divide by 2 to get $x = -5/2$. The solution is $-5/2$.

c) Multiply each side by 1/3. Since

$$\frac{1}{3} \cdot \frac{1}{2} = \frac{1}{6}$$

we get $x = 1/6$. The solution is 1/6.   ◀

Warm-ups

True or false?

1. The equation $x - x = 99$ has no solution. T

2. The equation $2x + 3x = 5x$ is an identity. T

3. All real numbers satisfy the equation $1 \div x = \dfrac{1}{x}$. F

4. Every equation involving fractions should be multiplied by 6 on each side because the equation of Example 3 was multiplied by 6 on each side. F

5. The equation $5x + 3 = 0$ is an inconsistent equation. F

6. The equation $2x = x$ is a conditional equation. T

7. The equation $x - .1x = .9x$ is an identity. T

8. The equation $.2x + .03x = 8$ is equivalent to $20x + 3x = 8$. F

9. The equation $\dfrac{x}{x} = 1$ is an identity. T

10. The solution to $3x - 8 = 0$ is 8/3. T

2.2 EXERCISES

Solve each equation. Identify each as a conditional equation, an inconsistent equation, or an identity. See Examples 1 *and* 2.

1. $3(x + 1) = 3(x + 3)$ No solution, inconsistent

2. $3x + 4x = 12x$ 0, conditional

3. $x + x = 2x$ All real numbers, identity

4. $2x - x = x$ All real numbers, identity

5. $x + x = 2$ 1, conditional

6. $9x - 8x = 7$ 7, conditional

7. $x - 1 = x + 1$ No solution, inconsistent

8. $2 - 3(5 - x) = 3x$ No solution, inconsistent

9. $3 - 3(5 - x) = 0$ 4, conditional

10. $(3 - 3)(5 - x) = 0$ All real numbers, identity

11. $3x - 5 = 0$ $\dfrac{5}{3}$, conditional

12. $\dfrac{0}{x} = 0$ All nonzero real numbers, identity

13. $\dfrac{2x}{2} = x$ All real numbers, identity

14. $5x \div 5 = x$ All real numbers, identity

15. $x \cdot x = x^2$ All real numbers, identity

16. $\dfrac{2x}{2x} = 1$ All nonzero real numbers, identity

Solve each equation by first eliminating the fractions. See Example 3.

17. $\dfrac{1}{2}x + 3 = x - \dfrac{1}{2}$ 7

18. $13 - \dfrac{1}{2}x = x - \dfrac{1}{2}$ 9

19. $\dfrac{x}{2} + \dfrac{x}{3} = 20$ 24

20. $\dfrac{x}{2} - \dfrac{x}{3} = 5$ 30

21. $\dfrac{x}{2} + \dfrac{x}{4} = 12$ 16

22. $\dfrac{x}{4} - \dfrac{x}{2} = -5$ 20

23. $\dfrac{3x}{2} - \dfrac{2x}{3} = -10$ −12

24. $\dfrac{3x}{4} + \dfrac{x}{2} = -5$ −4

25. $\dfrac{x}{3} - 5 = \dfrac{x}{4}$ 60

26. $\dfrac{1}{2}x - 6 = \dfrac{1}{5}x$ 20

27. $\dfrac{x}{6} + 1 = \dfrac{x}{4} - 1$ 24

28. $\dfrac{x}{15} + 5 = \dfrac{x}{6} - 10$ 150

Solve each equation by first eliminating the decimal numbers. See Examples 4 and 5.

29. $.1x - .3 = .2x - 8.3$ 80

30. $.5x + 3.4 = .2x + 12.4$ 30

31. $.05x + .4x = 27$ 60

32. $.08x + 28.3 = .5x - 9.5$ 90

33. $.3x + 1.2 = .5x$ 6

34. $.4x - 1.6 = .6x$ −8

35. $.02x - 1.56 = .8x$ −2

36. $.6x + 10.4 = .08x$ −20

37. $.05x + .03(x + 50) = 17.5$ 200

38. $.07x + .08(x - 100) = 44.5$ 350

39. $.06(x + 200) + .07x = 90$ 600

40. $.12(x + 100) + .1x = 210$ 900

41. $.03(x + 200) + .05x = 86$ 1000

42. $.02(x - 100) + .06x = 62$ 800

43. $.1x + .05(x - 300) = 105$ 800

44. $.2x - .05(x - 100) = 35$ 200

45. $x - .2x = 72$ 90

46. $x - .1x = 63$ 70

Solve each equation mentally. Write down only the solution. See Example 6. If you cannot do some of these mentally now, do them on paper and try them mentally later.

47. $x - 9 = 0$ 9

48. $x - 7 = 0$ 7

49. $x + 5 = 0$ −5

50. $x + 2 = 0$ −2

51. $3x = 9$ 3

52. $2x = -8$ −4

53. $\dfrac{x}{2} = -6$ −12

54. $\dfrac{x}{3} = -5$ −15

55. $x - 3 = -7$ −4

56. $x + 2 = -5$ −7

57. $x - 1 = -1$ 0

58. $x - 3 = -3$ 0

59. $5 - x = 0$ 5

60. $17 - x = 0$ 17

61. $12 + x = 0$ −12

62. $-3 + x = 0$ 3

63. $2x - 4 = 0$ 2

64. $2x + 4 = 0$ −2

65. $3x - 4 = 0$ $\dfrac{4}{3}$

66. $2x - 3 = 0$ $\dfrac{3}{2}$

67. $-2x + 1 = 0$ $\dfrac{1}{2}$

68. $2x - 1 = 0$ $\dfrac{1}{2}$

69. $2x = \dfrac{1}{2}$ $\dfrac{1}{4}$

70. $2x = \dfrac{1}{3}$ $\dfrac{1}{6}$

71. $3x = \dfrac{1}{2}$ $\dfrac{1}{6}$

72. $4x = \dfrac{1}{3}$ $\dfrac{1}{12}$

73. $\dfrac{x}{3} = 2$ 6

74. $\dfrac{x}{5} = -4$ −20

75. $\dfrac{x}{5} + 1 = 6$ 25

76. $\dfrac{x}{2} + 2 = 5$ 6

77. $\dfrac{x}{2} - 3 = -4$ −2

78. $\dfrac{x}{3} - 4 = -7$ −9

Solve each equation.

79. $3 + 2(x - 3) = 5(x - 4) + 2$ 5

80. $-3 - 4(x - 5) = -2(x + 3) + 11$ 6

81. $3x - 5 = 2x - 9$ −4

82. $5x - 9 = 0$ $\dfrac{9}{5}$

83. $5x = 5x + 1$ No solution

84. $x + 2(x + 4) = 3(x + 3) - 1$ All real numbers

85. $-3 - 5(3 - x) = -4(x - 2) + 10$ 4

86. $x + 3(x - 4) = 7x - 12$ 0

87. $.05x + 30 = .4x - 5$ 100

88. $x - .08x = 460$ 500

89. $-\dfrac{2}{3}x + 1 = 2$ $-\dfrac{3}{2}$

90. $-\dfrac{3}{4}x = \dfrac{1}{2}$ $-\dfrac{2}{3}$

91. $\dfrac{x}{2} + \dfrac{x}{6} = 20$ 30

92. $\dfrac{3x}{5} - 1 = \dfrac{x}{2} + 1$ 20

93. $5x = \dfrac{1}{5}$ $\dfrac{1}{25}$

94. $.08x + .5(x + 100) = 73.2$ 40

95. $436x - 789 = -571$ .5

96. $.08x + 4533 = 10x + 69$ 450

97. $\dfrac{x}{344} + 235 = 292$ 19,608

98. $34(x - 98) = \dfrac{x}{2} + 475$ 113.642

2.3 Formulas

IN THIS SECTION:

- **Solving for a Variable**
- **Finding the Value of a Variable**

In this section we learn to rewrite formulas using the same properties of equality that we used to solve equations. Formulas are often called **literal equations.** We will also learn how to find the value of one of the variables in a formula when we know the value of all of the others.

Solving for a Variable

A **formula** is an equation involving two or more variables. The formula

$$D = R \cdot T$$

expresses the relationship between distance, rate, and time of a moving object. The formula

$$C = \frac{5}{9}(F - 32)$$

expresses the relationship between the Fahrenheit and Celsius measurements of temperature. It is often necessary to rewrite a formula for one variable in terms of the other variables. We refer to this as **solving for a certain variable.**

EXAMPLE 1 Solve the formula $D = R \cdot T$ for T.

Solution

$$D = R \cdot T$$

$$\frac{D}{R} = \frac{R \cdot T}{R} \qquad \text{Divide each side by } R.$$

$$\frac{D}{R} = T \qquad \text{Simplify.}$$

$$T = \frac{D}{R} \qquad \text{It is customary to write the single variable on the left.} \qquad \blacktriangleleft$$

EXAMPLE 2 Solve the formula $C = \dfrac{5}{9}(F - 32)$ for F.

Solution We could apply the distributive property to the right side of the equation, but it is simpler to proceed as follows:

$$C = \frac{5}{9}(F - 32)$$

$$\frac{9}{5}C = \frac{9}{5} \cdot \frac{5}{9}(F - 32) \qquad \text{Multiply each side by } \frac{9}{5}, \text{ the reciprocal of } \frac{5}{9}.$$

$$\frac{9}{5}C = F - 32 \qquad \text{Simplify.}$$

$$\frac{9}{5}C + 32 = F - 32 + 32 \qquad \text{Add 32 to each side.}$$

$$\frac{9}{5}C + 32 = F \qquad \text{Simplify.}$$

This is usually written as $F = \dfrac{9}{5}C + 32$. $\qquad \blacktriangleleft$

When solving for a variable that appears more than once in the equation, we must combine the terms to obtain a single occurrence of the variable. *When a formula has been solved for a certain variable, that variable will not occur on both sides of the equation.*

EXAMPLE 3 Solve for x: $5x - b = 3x + d$.

Solution First get all terms involving x onto one side and all other terms on the other side:

$$5x - b = 3x + d$$

$$5x - 3x - b = d \qquad \text{Subtract } 3x \text{ from each side.}$$

$$5x - 3x = b + d \qquad \text{Add } b \text{ to each side.}$$

$$2x = b + d \qquad \text{Combine like terms.}$$

$$x = \frac{b + d}{2} \qquad \text{Divide each side by 2.} \qquad \blacktriangleleft$$

In Chapter 7 it will be necessary to solve an equation involving x and y for y.

EXAMPLE 4   Solve $x + 2y = 6$ for y.

Solution

Note that we are writing a linear equation in slope-intercept form in this example. That is why we prefer the answer given rather than $y = \dfrac{6-x}{2}$.

$$x + 2y = 6$$

$$2y = 6 - x \qquad \text{Subtract } x \text{ from each side.}$$

$$\left(\frac{1}{2}\right) \cdot 2y = \left(\frac{1}{2}\right)(6 - x) \qquad \text{Multiply each side by 1/2.}$$

$$y = \frac{1}{2} \cdot 6 - \frac{1}{2} \cdot x \qquad \text{Distributive property on the right-hand side.}$$

$$y = 3 - \frac{1}{2}x \qquad \text{Simplify.}$$

$$y = -\frac{1}{2}x + 3 \qquad \text{It is customary to write the term involving } x \text{ first.} \qquad \blacktriangleleft$$

Notice that in Example 4 we multiplied each side of the equation by $1/2$, and so we multiplied each term on the right-hand side by $1/2$. Instead of multiplying by $1/2$, we could divide each side of the equation by 2. We would then divide each term on the right-hand side by 2. This idea is illustrated in the next example.

EXAMPLE 5   Solve $2x - 3y = 9$ for y.

Solution

$$2x - 3y = 9$$

$$-3y = -2x + 9 \qquad \text{Subtract } 2x \text{ from each side.}$$

$$\frac{-3y}{-3} = \frac{-2x}{-3} + \frac{9}{-3} \qquad \text{Divide each side by } -3.$$

$$y = \frac{2}{3}x - 3 \qquad \text{Simplify.} \qquad \blacktriangleleft$$

Finding the Value of a Variable

In many situations we know the values of all variables in a formula except one. We use the formula to determine the unknown value.

EXAMPLE 6   If $2x - 3y = 9$, find y when $x = 6$.

Solution   *(Method 1)* First solve the equation for y. Since we have already solved this equation for y in Example 5, we will not repeat that process in this example. We have

$$y = \frac{2}{3}x - 3.$$

Now replace x by 6 in this equation:

$$y = \frac{2}{3}(6) - 3$$

$$= 4 - 3 = 1 \qquad \frac{2}{3}(6) = \frac{12}{3} = 4$$

Thus when $x = 6$, we have $y = 1$.

Math at Work

Food chemists and manufacturers want to know: How hot is hot?

The spiciness of hot peppers could not be measured systematically until W. L. Scoville invented a method in 1912. Using the Scoville method, a panel of tasters sipped measured concentrations of hot pepper diluted with sugar water and determined how long it took for the pepper to stop burning. The longer the time, the higher the pepper's rating on what is now called the Scoville scale.

As spicy foods cooked with hot peppers have become more common in the United States, scientists and food manufacturers have searched for a more accurate way to measure spiciness. Using a method of chemical analysis called chromatography, food chemists can now measure the amount of the powerful chemical called capsaicin that gives peppers their heat.

Today, most food manufacturers use chromotography rather than the Scoville method to measure the hotness of pepper products. Advocates claim that chromatography is more efficient and creates more uniform results than the Scoville method. However, critics point out that chromatography ignores the other chemicals in the pepper that contribute to the overall burning sensation.

A food chemist has determined through experimentation that the relationship between Scoville units (SU) and chromatography measurements (in parts per million) is given by the formula

$$\text{ppm} = \frac{\text{SU}}{15}.$$

How many parts per million of capsaicin are contained in the following foods?

Chili con carne	20 Scoville units	1.3 ppm
Chile pepper, dry ground	975 Scoville units	6.5 ppm
Jalapeno pepper, fresh and green	1700 Scoville units	111.3 ppm
Pure capsaicin	15,000,000 Scoville units	1,000,000 ppm

(Method 2) First replace x by 6 in the original equation, then solve for y:

$$2x - 3y = 9$$
$$2(6) - 3y = 9 \qquad \text{Replace } x \text{ by 6.}$$
$$12 - 3y = 9 \qquad \text{Simplify.}$$
$$-3y = -3 \qquad \text{Subtract 12 from each side.}$$
$$y = 1 \qquad \text{Divide each side by } -3.$$

Thus when $x = 6$, we have $y = 1$. ◄

If we had to find the value of y for many different values of x, it would be best to solve the equation for y, then insert the various values of x. Method 1 of Example 6 would be the better method. If we must find only one value of y, it does not matter which method we use. When doing the exercises corresponding to this example, you should try both methods.

EXAMPLE 7 If the simple interest is $120, the principal is $400, and the time is 2 years, find the rate.

Point out that simple interest is the basis for all interest computation. However, depositors do not like to wait two years to get interest as in this example. More often the time is 1/12 year or 1/365 year. Simple interest is computed every month or every day, giving us compound interest.

Solution Simple interest is calculated by using the formula $I = Prt$ (inside the front cover), where I is the interest, P is the principal, r is the rate, and t is the time in years. First, we solve the formula for r; then we insert the values of P, I, and t:

$$Prt = I$$
$$\frac{Prt}{Pt} = \frac{I}{Pt} \qquad \text{Divide each side by } Pt.$$
$$r = \frac{I}{Pt} \qquad \text{Simplify.}$$
$$r = \frac{120}{400(2)} \qquad \text{Substitute the values of } I, P, \text{ and } t.$$
$$r = .15 \qquad \text{Simplify.}$$
$$r = 15\% \qquad \text{Move the decimal point two places to the right to convert a decimal to a percent.} \quad ◄$$

In solving a geometric problem, it is always helpful to draw a diagram, as we do in the next example.

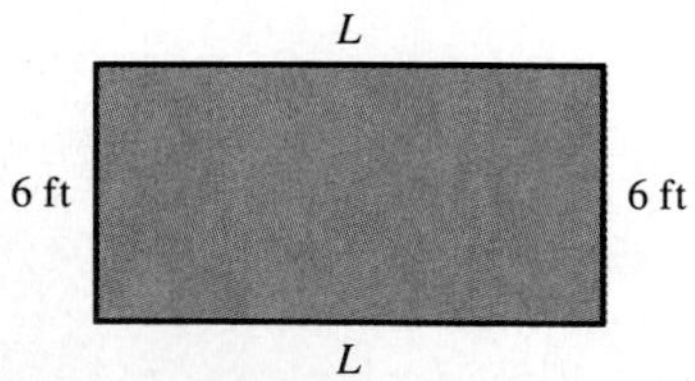

EXAMPLE 8 The perimeter of a rectangle is 36 feet. If the width is 6 feet, then what is the length?

Solution First, put the given information on a diagram as shown in Fig. 2.1. Substitute the given values into the formula found on the front endpaper for the

L

6 ft   6 ft

L

Figure 2.1

perimeter of a rectangle and then solve for L. (We could solve for L first and then insert the given values.)

$$P = 2L + 2W$$
$$36 = 2L + 2(6) \quad \text{Substitute 36 for } P \text{ and 6 for } W.$$
$$36 = 2L + 12 \quad \text{Simplify.}$$
$$24 = 2L \quad \text{Subtract 12 from each side.}$$
$$12 = L \quad \text{Divide each side by 2.}$$

Emphasize the importance of checking answers.

If the length is 12 feet and the width is 6 feet, then the perimeter is $2(12) + 2(6) = 36$ feet. So we can be certain that 12 feet is the correct length. ◄

EXAMPLE 9 What was the original price of a stereo that sold for $560 after a 20% discount?

Solution We can use the formula $S = L - rL$, where S is the selling price, r is the rate of discount, and L is the list price:

$$560 = L - .2L$$
$$10(560) = 10(L - .2L) \quad \text{Multiply each side by 10.}$$
$$5600 = 10L - 2L$$
$$5600 = 8L$$
$$\frac{5600}{8} = \frac{8L}{8}$$
$$700 = L$$

The original price was $700. Check it. ◄

Warm-ups

True or false?
1. If we solve $D = R \cdot T$ for T, we get $T \cdot R = D$. F
2. If we solve $a - b = 3a - m$ for a, we get $a = 3a - m + b$. F
3. Solving $A = L \cdot W$ for L, we get $L = W/A$. F
4. Solving $D = R \cdot T$ for R, we get $R = d/t$. F
5. The perimeter of a rectangle is found by multiplying its length and width. F
6. The volume of a rectangular box is found by multiplying its length, width, and height. T
7. The length of an NFL football field, excluding the end zones, is 99 yards. F

8. Solving $y - x = 5$ for y gives us $y = x + 5$.   T

9. If $x = -1$ and $y = -3x + 6$, then $y = 3$.   F

10. The perimeter of a rectangle is a measurement of the total distance around the outside edge.   T

2.3 EXERCISES

Solve each formula for the specified variable. See Examples 1 and 2.

1. $D = R \cdot T$ for R $R = D/T$

2. $A = L \cdot W$ for W $W = A/L$

3. $I = Prt$ for P $P = \dfrac{I}{rt}$

4. $I = Prt$ for t $t = \dfrac{I}{Pr}$

5. $F = \dfrac{9}{5}C + 32$ for C $C = \dfrac{5}{9}(F - 32)$

6. $C = 2\pi r$ for r $r = \dfrac{C}{2\pi}$

7. $A = \dfrac{1}{2}bh$ for h $h = 2A/b$

8. $A = \dfrac{1}{2}bh$ for b $b = 2A/h$

9. $P = 2L + 2W$ for L $L = \dfrac{P - 2W}{2}$

10. $P = 2L + 2W$ for W $W = \dfrac{P - 2L}{2}$

11. $C = \pi D$ for π $\pi = C/D$

12. $F = ma$ for a $a = F/m$

13. $A = \dfrac{1}{2}(a + b)$ for a $a = 2A - b$

14. $A = \dfrac{1}{2}(a + b)$ for b $b = 2A - a$

15. $S = P + Prt$ for r $r = \dfrac{S - P}{Pt}$

16. $S = P + Prt$ for t $t = \dfrac{S - P}{Pr}$

17. $A = \dfrac{1}{2}h(a + b)$ for a $a = \dfrac{2A - hb}{h}$

18. $A = \dfrac{1}{2}h(a + b)$ for b $b = \dfrac{2A - ah}{h}$

Solve each equation for x. See Example 3.

19. $5x + a = 3x + b$ $x = \dfrac{b - a}{2}$

20. $2c - x = 4x + c - 5b$ $x = \dfrac{c + 5b}{5}$

21. $4(a + x) - 3(x - a) = 0$ $x = -7a$

22. $-2(x - b) - (5a - x) = a + b$ $x = b - 6a$

23. $3x - 2(a - 3) = 4x - 6 - a$ $x = 12 - a$

24. $2(x - 3w) + 3(x + w) = 0$ $x = 3w/5$

25. $3x + 2ab = 4x - 5ab$ $x = 7ab$

26. $x - a = -x + a + 4b$ $x = a + 2b$

Solve for y. See Examples 4 and 5.

27. $x + y = -9$ $y = -x - 9$

28. $3x + y = -5$ $y = -3x - 5$

29. $x + y - 6 = 0$ $y = -x + 6$

30. $4x + y - 2 = 0$ $y = -4x + 2$

31. $2x - y = 2$ $y = 2x - 2$

32. $x - y = -3$ $y = x + 3$

33. $3x - y + 4 = 0$ $y = 3x + 4$

34. $-2x - y + 5 = 0$ $y = -2x + 5$

35. $y - x = 7$ $y = x + 7$

36. $y - x + 3 = 0$ $y = x - 3$

37. $x + 2y = 4$ $y = -\dfrac{1}{2}x + 2$

38. $3x + 2y = 6$ $y = -\dfrac{3}{2}x + 3$

39. $2x - 2y = 1$ $y = x - \dfrac{1}{2}$

40. $3x - 2y = -6$ $y = \dfrac{3}{2}x + 3$

41. $y + 2 = 3(x - 4)$ $y = 3x - 14$

42. $y - 1 = \dfrac{1}{2}(x - 2)$ $y = \dfrac{1}{2}x$

43. $\dfrac{1}{2}x - \dfrac{1}{3}y = -2$ $y = \dfrac{3}{2}x + 6$

44. $\dfrac{x}{2} + \dfrac{y}{4} = \dfrac{1}{2}$ $y = -2x + 2$

45. $y - 3 = -3(x - 1)$ $y = -3x + 6$

46. $y + 9 = -1(x - 1)$ $y = -x - 8$

47. $y - 5 = -6(x + 4)$ $y = -6x - 19$

48. $y - 4 = -\dfrac{2}{3}(x - 9)$ $y = -\dfrac{2}{3}x + 10$

49. $y - 7 = -\dfrac{1}{2}(x - 4)$ $y = -\dfrac{1}{2}x + 9$

50. $\dfrac{2}{3}x - \dfrac{1}{2}y = \dfrac{1}{6}$ $y = \dfrac{4}{3}x - \dfrac{1}{3}$

For each equation that follows, find y given that x = 2. See Example 6.

51. $y = 3x - 4$ 2

52. $y = -2x + 5$ 1

53. $3x - 2y = -8$ 7

54. $4x + 6y = 8$ 0

55. $3x - 5y = 6$ 0

56. $3y - 4x = 4$ 4

57. $y - 3 = \dfrac{1}{2}(x - 6)$ 1

58. $y - 6 = -\dfrac{3}{4}(x - 2)$ 6

59. $y - 4.3 = .45(x - 8.6)$ 1.33

60. $y + 33.7 = .78(x - 45.6)$ −67.708

Solve each of the following problems. A list of common formulas and their meanings is given inside the front cover. They may be helpful for doing the exercises. See Examples 7–9.

61. The area of a rectangle is 28 square yards. The width is 4 yards. Find the length.   7 yards

62. The area of a rectangle is 60 square feet. The length is 4 feet. Find the width.   15 feet

63. If it takes 600 feet of fencing to fence a rectangular feed lot that has a width of 75 feet, then what is the length of the lot?   225 feet

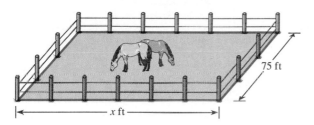

75 ft

x ft

Figure for Exercise 63

64. If it takes 500 feet of fencing to enclose a rectangular lot that is 104 feet wide, then how deep is the lot?   146 feet

65. The perimeter of a football field in the NFL, excluding the end zones, is 920 feet. How wide is the field?   160 feet

66. If a picture frame is 16 inches by 20 inches, then what is its perimeter?   72 inches

67. A rectangular box measures 2 feet wide, 3 feet long, and 4 feet deep. What is its volume?   24 cubic feet

68. The volume of a rectangular refrigerator is 20 cubic feet.

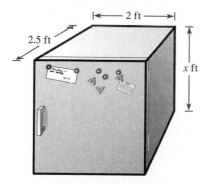

2 ft

2.5 ft

x ft

Figure for Exercise 68

If the top measures 2 feet by 2.5 feet, then what is the depth?   4 feet

69. If the circumference of a pizza is 8π inches, then what is the radius?   4 inches

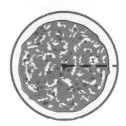

x

Figure for Exercise 69

70. If the circumference of a circle is 4π meters, then what is the diameter?   4 meters

71. If the simple interest on $5000 for 3 years is $600, then what is the rate?   4%

72. Wayne paid $420 in simple interest on a loan of $1000 for 7 years. What was the rate?   6%

73. Kathy paid $500 in simple interest on a loan of $2500. If the rate was 5%, then what was the time?   4 years

74. Robert paid $240 in simple interest on a loan of $1000. If the rate was 8%, then what was the time?   3 years

75. If a banner in the shape of a triangle has an area of 16 square feet with a base of 4 feet, then what is the height of the banner?   8 feet

4 ft

1991 Division Champs

x

Figure for Exercise 75

76. If a right triangle has an area of 14 square meters and one leg is 4 meters in length, then what is the length of the other leg?   7 meters

77. The area of a trapezoid is 200 square inches. If the height is 20 inches and the lower base is 8 inches, then what is the length of the upper base?   12 inches

78. The end of a flower box forms the shape of a trapezoid. The area of the trapezoid is 300 square centimeters. The bases are 16 centimeters and 24 centimeters. Find the height of the flower box.   15 centimeters

79. Find the rate of discount if the discount is $40 and the original price is $200.   20%

80. Find the rate of discount if the discount is $20 and the original price is $250.   8%

81. Find the original price if there is a 15% discount and the sale price is $255. $300

82. Find the list price if there is a 12% discount and the sale price is $4400. $5000

Figure for Exercise 78

2.4 English to Algebra

IN THIS SECTION:

- **Consecutive Integers**
- **Pairs of Numbers**
- **Using Formulas**
- **Encoding Phrases**
- **Writing Equations**

Many real-life problems can be described by equations like those we have been working with in this chapter. To solve these problems, we must be able to translate verbal phrases and sentences into algebraic expressions and equations. We call this encoding. We studied encoding in Section 1.6, but in this section we study it in more detail.

Consecutive Integers

Consider the three consecutive integers 3, 4, and 5. Note that each integer is 1 larger than the previous integer. We represent three *unknown* consecutive integers as follows.

Three Consecutive Integers

Let

x represent the first integer,

$x + 1$ represent the second integer,

$x + 2$ represent the third integer.

Consider the three consecutive odd integers 5, 7, and 9. Note that each odd integer is 2 larger than the previous odd integer. We represent three *unknown* consecutive odd integers as follows.

Three Consecutive Odd Integers

Let

x = the first odd integer,

$x + 2$ = the second odd integer,

$x + 4$ = the third odd integer.

Ask students to tell what kind of integers x, $x + 1$, and $x + 3$ represent

Note that consecutive even integers as well as consecutive odd integers differ by 2. To represent three unknown consecutive even integers we would also use x, $x + 2$, and $x + 4$.

Pairs of Numbers

If one unknown number is 5 more than another, we can use

$$x \quad \text{and} \quad x + 5$$

to represent them. Note that x and $x + 5$ can also be used to represent two unknown numbers that differ by 5, because if two numbers differ by 5, one of the numbers must be 5 more than the other.

Point out to students that x and $x - 10$ do not have a sum of 10.

How would we represent two numbers that have a sum of 10? If one of the numbers is 2, the other is certainly $10 - 2 = 8$. Thus if x is one of the numbers, then $10 - x$ is the other. The expressions

$$x \quad \text{and} \quad 10 - x$$

have a sum of 10 for any value of x. Check this by adding: $(10 - x) + x = 10$.

EXAMPLE 1 Encode the following:

a) Two numbers that differ by 12
b) Two numbers with a sum of -1

Solution

Ask students to name two numbers that do actually differ by 12.

Ask students to name two numbers that have a sum of -1.

a) The expressions x and $x + 12$ differ by 12. Note that we could also use x and $x - 12$ for two numbers that differ by 12.

b) The expressions x and $-1 - x$ have a sum of -1. We can check this by addition: $(-1 - x) + x = -1$. ◀

Using Formulas

To make the encoding more meaningful, we may use standard formulas such as those found inside the front cover.

EXAMPLE 2 Find an algebraic expression for

a) the distance, if the rate is 30 miles per hour and the time is T hours,
b) the discount, if the rate is 40% and the original price is p.

Solution

a) We could use x to represent the distance and be done with this problem. However, it is better to use the formula $D = R \cdot T$ and write

$$30T = \text{the distance.}$$

b) Since discount is the rate times the original price, we can write

$$.40p = \text{the discount.} \quad ◀$$

Encoding Phrases

Many verbal phrases occur repeatedly in applications. The following box contains a list of some frequently occurring verbal phrases and their equivalent algebraic expressions.

Encoding Verbal Phrases

Verbal Phrase	*Algebraic Expression*
Addition	
The sum of a number and 8	$x + 8$
Five is added to a number	$x + 5$
Two more than a number	$x + 2$
A number increased by 3	$x + 3$

Subtraction	
Four is subtracted from a number	$x - 4$
Three less than a number	$x - 3$
The difference between 7 and a number	$7 - x$
Some number decreased by 2	$x - 2$
Multiplication	
The product of 5 and a number	$5x$
Twice a number	$2x$
One half of a number	$\frac{1}{2}x$
Five percent of a number	$.05x$
Division	
The ratio of a number and 6	$x/6$
The quotient of 5 and a number	$5/x$
Three divided by some number	$3/x$

Writing Equations

To solve a problem using algebra, we describe the problem with an equation.

EXAMPLE 3 Identify the variables and write an equation describing the situation.

a) Find two numbers that have a sum of 12 and a product of 32.

b) A coat is on sale for 25% off the list price. If the sale price is $87, then what is the list price?

c) The value of x dimes and $x - 3$ quarters is $2.05.

Solution

Don't worry about solving these equations. We will give complete solutions to problems in the next section.

a) Let

$$x = \text{one of the numbers}$$

and

$$12 - x = \text{the other number.}$$

Since their product is 32, we have

$$x(12 - x) = 32.$$

b) Let

$$x = \text{the list price}$$

and

$$.25x = \text{the amount of discount.}$$

Since the list price minus the discount gives the selling price, we have

$$x - .25x = 87.$$

c) The value of x dimes at 10 cents each is $10x$ cents. The value of $x - 3$ quarters at 25 cents each is $25(x - 3)$ cents. We can write an equation expressing the fact that the total value of the coins is 205 cents:

$$10x + 25(x - 3) = 205 \qquad \blacktriangleleft$$

Warm-ups

True or false?

1. For any value of x, the numbers x and $x + 6$ differ by 6.   T

2. For any value of a, a and $10 - a$ have a sum of 10.   T

3. If Jack travels x miles per hour for 12 hours, then the distance that he travels is $12x$ miles.   T

4. If Jill travels x miles per hour for 300 miles, then she has traveled for $300x$ hours.   F

5. If the realtor gets 6% of the selling price, and the house sells for x dollars, then the owner gets $x - .06x$ dollars.   T

6. If the owner got \$50,000, and the realtor got 10% of the selling price, then the house must have sold for \$55,000.   F

7. Three consecutive odd integers can be represented by x, $x + 1$, and $x + 3$.   F

8. The value in cents of n nickels and d dimes is $.05n + .10d$.   F

9. If the sales tax rate is 5%, and x represents the amount of goods purchased, then the total bill is $1.05x$.   T

10. If the length of a rectangle is 4 feet more than the width, then the perimeter is equal to $x + (x + 4)$.   F

2.4 EXERCISES

Find algebraic expressions for each of the following. See Example 1.

1. The sum of a number and 3 $x + 3$

2. Two more than a number $x + 2$

3. Three less than a number $x - 3$

4. Four subtracted from a number $x - 4$

5. The product of a number and 5 $5x$

6. Five divided by some number $5/x$

7. Ten percent of a number $.1x$

8. Eight percent of a number $.08x$

9. Two consecutive even integers $x, x + 2$

10. Two consecutive odd integers $x, x + 2$

11. Two numbers with a sum of 6 $x, 6 - x$

12. Two numbers with a sum of -4 $x, -4 - x$

13. Two numbers with a difference of 15 $x, x + 15$

14. Two numbers that differ by 9 $x, x + 9$

15. Two consecutive integers $x, x + 1$

16. Four consecutive odd integers $x, x + 2, x + 4, x + 6$

21. $\dfrac{x - 100}{12}$

Find algebraic expressions for the following. See Example 2.

17. The distance, if the rate is x and the time is 3 $3x$

18. The distance, if the rate is $x + 10$ and the time is 5 $5(x + 10)$

19. The time, if the distance is x and the rate is 20 $x/20$

20. The time, if the distance is 300 and the rate is $x + 30$ $\dfrac{300}{x + 30}$

21. The rate, if the distance is $x - 100$ and the time is 12.

22. The rate, if the distance is 200 and the time is $x + 3$ $\dfrac{200}{x + 3}$

23. The area, if the length of the rectangle is x and the width is 5 $5x$

24. The area, if the length of the rectangle is b and the width is $b - 6$ $b(b - 6)$

25. The perimeter, if the length of the rectangle is $w - 3$ and the width is w $2w + 2(w - 3)$

26. The perimeter, if the length of the rectangle is r and the width is $r + 1$ $2r + 2(r + 1)$

27. The width, if the length of the rectangle is x and the perimeter is 300 $150 - x$

28. The length, if the width of the rectangle is w and the area is 200 $200/w$

29. The length, if the width of the rectangle is x and the length is 1 foot longer than twice the width $2x + 1$

30. The length, if the width of the rectangle is w and the length is 3 feet shorter than twice the width $2w - 3$

31. The area of a rectangle, if the width is x meters and the length is 5 meters longer than the width $x(x + 5)$

32. The perimeter of a rectangle, if the length is x yards and the width is 10 yards shorter $2x + 2(x - 10)$

33. The simple interest, if the principle is $x + 1000$, the rate is 18%, and the time is 1 year $.18(x + 1000)$

34. The simple interest, if the principle is $3x$, the rate is 6%, and the time is 1 year $.06(3x)$

35. The price per pound of peaches if x pounds are sold for $16.50 $16.50/x$

Photo for Exercise 35

36. The rate per hour of a mechanic who gets $480 for working x hours $480/x$

Encode each situation described below into an equation. Do not solve the equation. See Example 3.

37. The sum of a number and 5 is 13. $x + 5 = 13$

38. Twelve subtracted from a number is -6. $x - 12 = -6$

39. Two numbers differ by 5 and have a product of 8.

40. Two numbers differ by 6 and have a product of -9.

41. The sum of three consecutive integers is 42.

42. The sum of three consecutive odd integers is 27.

43. The product of two consecutive integers is 182.

44. The product of two consecutive even integers is 168. $x(x + 2) = 168$

45. Twelve percent of Harriet's income is $3000. $.12x = 3000$

46. If nine percent of the members buy tickets, then we should sell 252 tickets. $.09x = 252$

47. Thirteen is 5% of what number? $.05x = 13$

48. What percent of 500 is 100? $500x = 100$

39. $x(x + 5) = 8$ **40.** $x(x + 6) = -9$ **41.** $x + (x + 1) + (x + 2) = 42$ **42.** $x + (x + 2) + (x + 4) = 27$ **43.** $x(x + 1) = 182$

49. What percent of 40 is 120? $40x = 120$

50. Three hundred is 8% of what number? $.08x = 300$

51. The length of a rectangle is 5 feet longer than the width, and the area is 126 square feet. $x(x + 5) = 126$

52. The length of a rectangle is 1 yard shorter than twice the width, and the perimeter is 298 yards. $2x + 2(2x - 1) = 298$

53. The value of x nickels and $x + 2$ dimes is \$3.80.

54. The value of d dimes and $d - 3$ quarters is \$6.75.

54. $.10d + .25(d - 3) = 6.75$ **53.** $.05x + .10(x + 2) = 3.80$

55. The value of n nickels and $n - 1$ dimes is 95 cents.

56. The value of q quarters, $q + 1$ dimes, and $2q$ nickels is 90 cents. $25q + 10(q + 1) + 5(2q) = 90$

57. Herman's house sold for x dollars. The real estate agent received 7% of the selling price, and Herman received \$84,532. $x - .07x = 84,532$

58. Gwen sold her car on consignment for x dollars. The saleswoman's commission was 10% of the selling price, and Gwen received \$6570. $x - .10x = 6570$

55. $5n + 10(n - 1) = 95$

2.5 Verbal Problems

IN THIS SECTION:

- **Number Problems**
- **General Strategy for All Word Problems**
- **Geometric Problems**
- **Percent Problems**
- **Investment Problems**
- **Mixture Problems**
- **Uniform Motion Problems**
- **Commission Problems**

For some classes it may be advisable to spend more than one class period on this section.

In this section we will use the ideas of Section 2.4 to solve verbal problems. Many of the problems are easy and can be solved without the aid of algebra. However, remember that we are not just trying to find the answer, we are trying to learn the algebraic method. So even if the answer is obvious to you, set the problem up and solve it by using algebra as shown in the examples.

Number Problems

EXAMPLE 1   The sum of three consecutive integers is 48. Find the integers.

Solution   We first represent the unknown quantities with variables.

$$x = \text{first integer}$$ **Consecutive integers differ by one.**
$$x + 1 = \text{second integer}$$
$$x + 2 = \text{third integer}$$

We now write an equation that describes the problem and solve it. The equation expresses the fact that the sum of the integers is 48.

$$x + (x + 1) + (x + 2) = 48$$

$$3x + 3 = 48 \qquad \textbf{Combine like terms.}$$

$$3x = 45 \qquad \textbf{Subtract 3 from each side.}$$

$$x = 15 \qquad \textbf{Divide each side by 3.}$$

$$x + 1 = 16 \qquad \textbf{If } x \textbf{ is 15, then } x + 1 \textbf{ is 16 and } x + 2 \textbf{ is 17.}$$

$$x + 2 = 17$$

The three consecutive integers that add up to 48 are 15, 16, and 17. To check this, add the integers:

$$15 + 16 + 17 = 48 \qquad\qquad \blacktriangleleft$$

General Strategy for All Word Problems

The steps to follow in providing a complete solution to a word problem can be stated as follows.

STRATEGY
Solving Word Problems

1. Read the problem.
2. If possible, draw a diagram to illustrate the problem.
3. Choose a variable and write down what it represents.
4. Represent any other unknowns in terms of that variable.
5. Write an equation that fits the situation.
6. Solve the equation.
7. Be sure that your solution answers the question posed in the original problem.
8. Check your answer by using it to solve the original problem (not the equation).

Geometric Problems

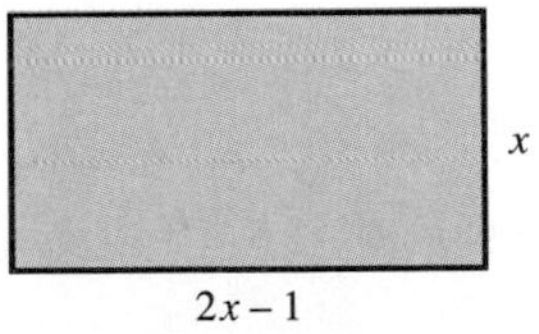

x

$2x - 1$

Figure 2.2

EXAMPLE 2 The length of a rectangular piece of property is one foot less than twice the width. If the perimeter is 748 feet, find the length and width.

Solution Let $x =$ the width. Since the length is one foot less than twice the width, $2x - 1 =$ the length. Draw a diagram as shown in Fig. 2.2.

We know that $2L + 2W = P$ is the formula for perimeter of a rectangle. Replacing $2x - 1$ for L and x for W in this formula yields the following equation:

$$
\begin{array}{c c c}
L & W & P
\end{array}
$$

$$2(2x - 1) + 2(x) = 748$$

$4x - 2 + 2x = 748$	**Remove the parentheses.**
$6x - 2 = 748$	**Combine like terms.**
$6x = 750$	**Add 2 to each side.**
$x = 125$	**Divide each side by 6.**
$2x - 1 = 249$	**If $x = 125$, then $2(125) - 1 = 249$.**

The width is 125 feet, and the length is 249 feet. Check by computing $2L + 2W$: $2(249) + 2(125) = 748$.   ◀

Percent Problems

EXAMPLE 3   Major Motors is discounting all of its deluxe models by 12%. When Ralph bought the deluxe model, he got a discount of $4500. What was the original price of Ralph's car?

Solution   Let x represent the original price. The discount is found by taking 12% of the original price. This gives the following equation:

$$12\% \text{ of original price} = \text{discount}$$

$$.12x = 4500$$

$$x = \frac{4500}{.12} \qquad \text{Divide each side by .12.}$$

$$x = 37{,}500$$

The original price of Ralph's car was $37,500. To check this, we find 12% of $37,500.

$$.12(\$37{,}500) = \$4500 \qquad\qquad ◀$$

EXAMPLE 4   When Susan bought her deluxe model at Major Motors, she also got a discount of 12% and paid $17,600 for her car. What was the original price of Susan's car?

Solution   Let x represent the original price for Susan's car. The amount of discount is 12% of x, or $.12x$. We can write an equation expressing the fact that the original price minus the discount is the price Susan paid:

$$x - .12x = 17{,}600$$

$$.88x = 17{,}600 \qquad \text{Combine like terms: } 1.00x - .12x = .88x$$

$$x = \frac{17{,}600}{.88} \qquad \text{Divide each side by .88.}$$

$$x = \$20{,}000$$

The original price of Susan's car was $20,000. A 12% discount is $2,400, and $20,000 − $2,400 = $17,600. ◄

Investment Problems

EXAMPLE 5 Ruth Ann invested some money in a certificate of deposit with an annual yield of 9%. She invested twice as much in a mutual fund with an annual yield of 10%. Her interest from the two investments at the end of the year was $232. How much was invested at each rate?

Solution When there are many unknown quantities, it is often helpful to identify them in a table. Recall the formula $I = Prt$.

Interest rate	9%	10%
Amount invested	x	$2x$
Interest for one year	$.09x$	$.10(2x)$

Since the total interest from the investments was $232, we can express this in the following equation:

$$.09x + .10(2x) = 232$$
$$.09x + .20x = 232$$
$$.29x = 232$$
$$x = \frac{232}{.29}$$
$$x = \$800$$
$$2x = \$1600$$

Ruth Ann invested $800 at 9% and $1600 at 10%. To check this, we find that $.09(\$800) = \$72$ and $.10(\$1600) = \$160$. Now

$$\$72 + \$160 = \$232.$$ ◄

Mixture Problems

EXAMPLE 6 How many gallons of milk containing 4% butterfat must be mixed with 80 gallons of 1% milk to obtain 2% milk?

Solution We again make a table of the unknown quantities.

Percentage of fat	4%	1%	2%
Amount of milk	x	80	$x + 80$
Amount of fat	$.04x$	$.01(80)$	$.02(x + 80)$

Note that in mixture problems a high concentration combines with a low concentration to give a moderate concentration.

In all mixture problems we write an equation that accounts for one of the quantities being combined. The equation that we write here expresses the fact that the total fat from the first two milks is the same as the fat in the mixture:

$$.04x + .01(80) = .02(x + 80)$$
$$.04x + .8 = .02x + 1.6$$
$$100(.04x + .8) = 100(.02x + 1.6) \quad \text{Multiply each side by 100.}$$
$$4x + 80 = 2x + 160 \quad \text{Distributive property}$$
$$2x + 80 = 160 \quad \text{Subtract } 2x \text{ from each side.}$$
$$2x = 80 \quad \text{Subtract 80 from each side.}$$
$$x = 40 \quad \text{Divide each side by 2.}$$

Use 40 gallons of 4% milk. To check this, calculate the total fat:

$$.04(40) + .01(80) = 1.6 + .8 = 2.4 \text{ gallons}$$
$$.02(120) = 2.4 \text{ gallons} \qquad \blacktriangleleft$$

Uniform Motion Problems

EXAMPLE 7   Bridgette drove her car for 2 hours on an icy road. When the road cleared up, she increased her speed by 35 miles per hour and drove 3 more hours, completing her 255-mile trip. How fast did she travel on the icy road?

Solution   It is helpful to make a table to classify the information given. Remember that $D = RT$.

	Rate	Time	Distance
Icy road	x	2	$2x$
Clear road	$x + 35$	3	$3(x + 35)$

The equation expresses the fact that her total distance traveled was 255 miles.

$$\text{Distance icy road} + \text{distance clear road} = \text{total distance}$$
$$2x + 3(x + 35) = 255$$
$$2x + 3x + 105 = 255$$
$$5x + 105 = 255$$
$$5x = 150$$
$$x = 30$$

She traveled 30 mph on the icy road. *Check:* Two hours at 30 mph is 60 miles, and 3 hours at 65 mph is 195 miles. Since $60 + 195 = 255$, this is the correct answer.

$\blacktriangleleft$

Commission Problems

EXAMPLE 8 Sarah is selling her house through a real estate agent whose commission is 7% of the selling price. What should the selling price be so that Sarah can get the $83,700 she needs to pay off the mortgage?

Solution Let x be the selling price. The commission is 7% of x, not 7% of $83,700. Sarah receives the selling price less the sales commission.

$$\text{Selling price} - \text{commission} = \text{Sarah's share}$$
$$x - .07x = \$83,700$$
$$.93x = 83,700 \qquad 1.00 - .07 = .93$$
$$x = \frac{83,700}{.93} = \$90,000$$

The house should sell for $90,000. *Check:* The sales commission is 7% of $90,000, or $6,300, and $90,000 - \$6,300 = \$83,700$. ◄

Warm-ups

True or false?

1. The first step in solving a word problem is to write the equation. F
2. To represent two consecutive odd integers, we use x and $x + 1$. F
3. When solving word problems, always write down what the variable stands for. T
4. If $5x$ is two miles more than $3(x + 20)$, then $5x + 2 = 3(x + 20)$. F
5. If x is the selling price and the commission is 8% of the selling price, then the commission is .08x. T
6. If you need $40,000 for your house, and the agent gets 10% of the selling price, then the agent gets $4,000, and the house sells for $44,000. F
7. When we mix a 10% acid solution with a 14% acid solution, we can get as high as 24% acid. F
8. We can represent two numbers with a sum of 6 by x and $6 - x$. T
9. Word problems are a mental challenge and must be solved without the benefit of a diagram. F
10. Two numbers that differ by 7 can be represented by x and $x + 7$. T

2.5 EXERCISES

Show a complete solution to each problem. See Example 1.

1. The sum of three consecutive integers is 54. Find the integers.   17, 18, 19

2. Find three consecutive integers whose sum is 141.

3. Find three consecutive even integers whose sum is 114.

4. Find three consecutive even integers whose sum is 78.

5. Two consecutive odd integers have a sum of 152. What are the integers?   75, 77

6. Four consecutive odd integers have a sum of 120. What are the integers?   27, 29, 31, 33

   2. 46, 47, 48   **3.** 36, 38, 40   **4.** 24, 26, 28

Show a complete solution to each problem. See Example 2.

7. If a rectangle is twice as long as it is wide, and the perimeter is 78 inches, then what are the length and width?

8. If the perimeter of a rectangle is 112 meters, and the length is one meter shorter than twice the width, then what are the length and width?   *W* = 19 inches, *L* = 37 inches

9. Julia framed an oil painting that her uncle gave her. The painting was 4 inches longer than it was wide, and it took 176 inches of frame molding. What were the dimensions of the picture?   42 inches by 46 inches

10. While traveling around the perimeter of the Bermuda triangle, Geraldo observed that the second side was just 5 nautical miles short of being twice as long as the first side, and the third side was exactly 66 nautical miles longer than the first side. If Geraldo traveled a total of 413 nautical miles, then how long is each side?   88, 154, and 171 nautical miles

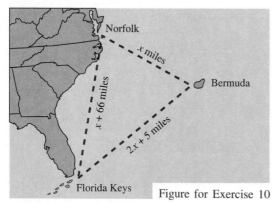

Figure for Exercise 10

   7. 26 inches, 13 inches

11. An isosceles triangle has a base that is 5 inches shorter than either of the equal sides. If the perimeter of the triangle is 34 inches, then what is the length of the equal sides?

12. It has been determined that people wait more patiently in a room that is 8 feet longer than it is wide. When Vincent wallpapered Dr. Wright's waiting room, he used 88 feet of border paper. What are the dimensions of Dr. Wright's waiting room if it was built according to the recommended guidelines?   18 feet by 26 feet

   11. 13 inches

Show a complete solution to each problem. See Examples 3 and 4.

13. At a 25% off sale, Jose saved $80 on a color television. What was the original price of the television? $320

14. If 60% of the registered voters of Lancaster County voted in the November election, and 33,420 votes were cast, then how many registered voters are there in Lancaster County?   55,700

Photo for Exercise 14

15. After getting a 20% discount, Robert paid $320 for a chain saw. What was the original price of the chain saw? $400

16. After getting a 15% discount on the price of a new convertible, Helen paid $17,000. What was the original price of the convertible? $20,000

17. At an 8% sales tax rate, the sales tax on Peter's new car was $1,200. What was the price of the car? $15,000

18. Last year, Faye paid 24% of her income to taxes. If she paid $9,600 in taxes, then what was her income? $40,000

19. A retail store buys shirts for $8 and sells them for $14. What percent increase is this? 75%

20. If 28 new AIDS cases were reported in Landon County this year and 35 new cases were reported last year, then what percent decrease in new cases is this? 20%

Show a complete solution to each problem. See Example 5.

21. Wiley invested some money at 8% simple interest and some money at 12% simple interest. In the second investment he put $3000 more than he put in the first. If the income from the two investments totaled $760 for one year, then how much did he invest at each rate?

22. Becky lent her brother some money at 8% simple interest, and she lent her sister twice as much at twice the interest rate. If she received a total of 20 cents interest, then how much did she lend to each of them?

23. David split his $25,000 inheritance between two investments, one paying 9% simple interest and the other paying 12% simple interest. If his total income for one year on the two investments was $2550, then how much did he invest at each rate? $15,000 at 9%, $10,000 at 12%

24. Of the $50,000 that Natasha pocketed on her last real estate deal, $20,000 went to charity. She invested part of the remainder in her brother's laundromat at 30% simple interest and lent the rest to a friend in the restaurant business at 20% simple interest. In one year she made $6500 on her investments. How much did she invest at each rate?

21. $2000 at 8%, $5000 at 12% 22. $.50 at 8%, $1.00 at 16%

24. $5000 at 30%, $25,000 at 20%

Show a complete solution to each problem. See Example 6.

25. How many gallons of milk containing 1% butterfat must be mixed with 30 gallons of milk containing 3% butterfat to obtain a mixture containing 2% butterfat? 30 gallons

26. How many gallons of a 5% acid solution should be mixed with 30 gallons of a 10% acid solution to obtain a mixture that is 8% acid? 20 gallons

27. Gus has on hand a 5% alcohol solution and a 20% alcohol solution. He needs 30 liters of a 10% alcohol solution.

How many liters of each solution should he mix together to obtain the 30 liters? 20 liters of 5%, 10 liters of 20%

28. Angela needs 20 quarts of 50% antifreeze solution in her radiator. She plans to obtain this by mixing some pure antifreeze with an appropriate amount of a 40% antifreeze solution. How many quarts of each should she use? $\frac{10}{3}$ quarts of pure antifreeze, $\frac{50}{3}$ quarts of 40% solution

Show a complete solution to each problem. See Example 7.

29. Bret drove for 4 hours on the freeway, then decreased his speed by 20 miles per hour and drove for 5 more hours on a country road. If his total trip was 485 miles, then what was his speed on the freeway? 65 mph

30. On Saturday morning, Lynn walked for 2 hours and then ran for 30 minutes. If she ran twice as fast as she walked and covered 12 miles altogether, then how fast did she walk? 4 mph

31. Kathryn drove her rig 5 hours before dawn and 6 hours after dawn. If her average speed was 5 miles per hour more in the dark and she covered 630 miles altogether, then what was her speed after dawn? 55 mph

32. On Monday, Roger drove to work in 45 minutes. On Tuesday he averaged 12 miles per hour more, and it took him 9 minutes less to get to work. How far does he travel to work? 36 miles

Show a complete solution to each problem. See Example 8.

33. Kirk wants to get $72,000 for his house. The real estate agent gets a commission equal to 10% of the selling price for selling the house. What should the selling price be?

34. Merilee sells tomatoes at a roadside stand. Her total receipts including the 7% sales tax were $462.24. What amount of sales tax did she collect? $30.24

35. Gwen bought a new car. The selling price plus the 8% state sales tax was $10,638. What was the selling price?

36. Gene is selling his horse at an auction. The auctioneer's commission is 10% of the selling price. If Gene still owes $810 on the horse, then what must the horse sell for in order for Gene to pay off his loan? $900

33. $80,000   35. $9850

Miscellaneous

37. Wilson High School has 400 students, of whom 20% are black. The school board plans to merge Wilson High with Jefferson High. This one school will then have a 44% black student population. If Jefferson currently has a 60% black student population, then how many students are at Jefferson? 600

38. The school board plans to merge two junior high schools into one school of 800 students in which 40% of the students will be black. One of the schools currently has 58% black students, while the other school has only 10% black students. How many students are in each of the two schools? 500 in the 58% school, 300 in the 10% school

39. When Memorial Hospital is filled to capacity, it has 18 more people in semiprivate rooms (two patients to a room) than in private rooms. The room rates are $200 per day for a private room and $150 per day for a semiprivate room. If the total receipts for rooms are $17,400 per day when all are full, then how many rooms of each type does the hospital have? 42 private, 30 semiprivate

40. Cashews sell for $4.80 per pound, and pistachios sell for $6.40 per pound. How many pounds of pistachios should be mixed with 20 pounds of cashews to get a mixture that sells for $5.40 per pound? 12 pounds

41. Candice has ten coins consisting of nickels and dimes. The total value of the coins is $.80. How many of each type of coin does she have? 4 nickels, 6 dimes

42. Jeremy has 36 coins consisting of dimes and quarters. If the total value of his coins is $4.50, then how many of each type of coin does he have? 30 dimes, 6 quarters

---

2.6 Inequalities

IN THIS SECTION:

- **Basic Ideas**
- **Graphing Inequalities**
- **Writing Inequalities**

In Chapter 1 we defined inequality in terms of the number line. One number is greater than another if it lies to the right of the other on the number line. In this section we will study inequality in greater depth.

Basic Ideas

Recall that we defined an equation to be a statement that two algebraic expressions are equal. An **inequality** is a statement that two algebraic expressions are not equal. The symbols used to express inequality and their meanings are given in the following box.

DEFINITION
Inequality Symbols

Symbol	Meaning
$<$	is less than
$\leq$	is less than or equal to
$>$	is greater than
$\geq$	is greater than or equal to

Make sure that students understand that $3 < x$ means x is greater than 3.

The statement $a < b$ means that a is to the left of b on the number line. The statement $a > b$ means that a is to the right of b on the number line. Of course, *$a < b$ has the same meaning as $b > a$.* The statement $a \leq b$ is true if a is less than b or if a is equal to b. The statement $a \leq b$ has the same meaning as the statement $b \geq a$.

EXAMPLE 1 Determine whether each of the following statements is true or false.

a) $3 < 4$ **b)** $-3 < -4$ **c)** $-2 \leq 0$
d) $0 \geq 0$ **e)** $2(-3) + 8 > 9$ **f)** $(-2)(-5) \leq 10$

Solution

a) True
b) False, because -3 is to the right of -4 on the number line
c) True
d) True, because 0 is equal to 0
e) False, since $(2)(-3) + 8 = 2$ and 2 is not greater than 9
f) True, since $(-2)(-5) = 10$ ◀

Figure 2.3

Graphing Inequalities

A number satisfies the inequality

$$x > 2$$

if it is to the right of 2 on the number line. When we show on a number line the numbers that satisfy an inequality, we say that we are **graphing the inequality.** Figure 2.3 shows the graph of $x > 2$.

We use an open circle on the number line at 2 to indicate that 2 is not a solution. A solid circle is used to indicate that a number is a solution. Figure 2.4 shows the graph of the inequality $x \leq -1$.

The inequality

$$3 < x < 6$$

Figure 2.4

is called a **compound inequality.** It is used to indicate that x is *in between* 3 and 6. The first inequality, $3 < x$, means that x is greater than 3; the second inequality, $x < 6$, means that x is also less than 6. The meaning of the inequality $3 < x < 6$ is

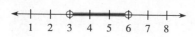

Figure 2.5

clear when we read the variable first:

$$\text{``}x \text{ is greater than 3 } and \text{ } x \text{ is less than 6.''}$$

The graph of $3 < x < 6$ is shown in Fig. 2.5.

EXAMPLE 2 Sketch the graph of each inequality on the number line:

a) $-2 < x$ **b)** $x \le 5$ **c)** $-2 \le x < 1$

Solution

a) A number satisfies the inequality $-2 < x$ if the number is greater than -2 or, in other words, is to the right of -2 on the number line. The graph of $-2 < x$ is shown in Fig. 2.6.

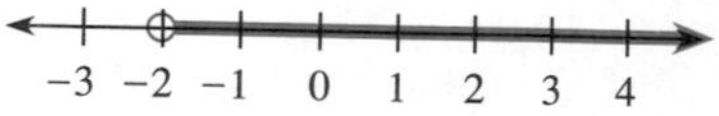

Figure 2.6

b) A number satisfies $x \le 5$ if the number is 5 or is to the left of 5 on the number line. The graph of $x \le 5$ is shown in Fig. 2.7.

c) Numbers that satisfy the inequality $-2 \le x < 1$ are between -2 and 1, including -2 and not including 1. The graph of $-2 \le x < 1$ is shown in Fig. 2.8.

◀

It is easy to determine the solution to inequalities such as

$$x > -2, \quad x \ge 4, \quad 5 > x, \quad \text{and} \quad 3 < x \le 9.$$

Figure 2.7

However, more complicated inequalities frequently occur in applications. We must learn to solve inequalities such as

$$2x - 3 \le -5, \quad x - 5 > 2x + 1, \quad \text{and} \quad 6 < 3x - 5 < 14.$$

In Section 2.7, we will learn to solve inequalities like these systematically. At this point we will concern ourselves with determining whether or not a number is a solution to an inequality of this type.

Figure 2.8

EXAMPLE 3 Determine whether or not the given number satisfies the inequality following it:

a) $0, 2x - 3 \le -5$
b) $-4, x - 5 > 2x + 1$
c) $13/3, 6 < 3x - 5 < 14$

Solution

a) Replace x by 0 in the inequality and simplify:

$$2x - 3 \le -5$$
$$2(0) - 3 \le -5$$
$$-3 \le -5 \quad \textbf{Incorrect}$$

Since this last inequality is incorrect, 0 is not a solution to the inequality.

b) Replace x by -4 and simplify:

$$x - 5 > 2x + 1$$
$$-4 - 5 > 2(-4) + 1$$
$$-9 > -7 \qquad \textbf{Incorrect}$$

Since this last inequality is incorrect, -4 is not a solution to the inequality.

c) Replace x by $13/3$ and simplify:

$$6 < 3x - 5 < 14$$
$$6 < 3\left(\frac{13}{3}\right) - 5 < 14$$
$$6 < 13 - 5 < 14$$
$$6 < 8 < 14 \qquad \textbf{Correct}$$

Since 8 is greater than 6 and less than 14, this inequality is correct. Thus 8 satisfies the inequality. ◀

Writing Inequalities

Inequalities occur in applications just as equations do. To use inequalities, we must be able to encode a verbal expression as an algebraic inequality. If you must be at least 18 years old to vote, then you can vote if you are 18 or older. The phrase

"at least"   means   "greater than or equal to."

If an elevator has a capacity of at most 20 people, then it can hold 20 people or fewer. The phrase

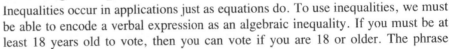

"at most"   means   "less than or equal to."

EXAMPLE 4 Encode each situation as an inequality.

a) Lois plans to spend at most $500 on a washing machine, including the 9% sales tax.

b) The length of a certain rectangle must be 4 meters more than the width, and the perimeter must be at least 120 meters.

c) Fred made a 76 on the midterm exam. To get a B, the average of his midterm and his final exam must be between 80 and 90.

Solution

a) If x is the price of the washing machine, then $.09x$ is the amount of sales tax. Since the total must be less than or equal to $500, the inequality is

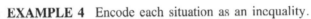

$$x + .09x \le 500.$$

b) If x represents the width of the rectangle, then $x + 4$ represents the length. Since

the perimeter $(2L + 2W)$ must be greater than or equal to 120, the inequality is

$$2(x) + 2(x + 4) \geq 120.$$

c) If we let x represent Fred's final exam score, then his average is $(x + 76)/2$. To indicate that the average is between 80 and 90, we use the compound inequality

$$80 < \frac{x + 76}{2} < 90. \qquad \blacktriangleleft$$

Warm-ups

True or false?

1. $-2 \leq -2$ T
2. The inequalities $7 < x$ and $x > 7$ have the same graph. T
3. The graph of $x < -3$ includes the point at -3. F
4. The number 5 satisfies the inequality $x > 2$. T
5. The number -3 is a solution to $-2 < x$. F
6. $-5 < 4 < 6$ T
7. $-3 < 0 < -1$ F
8. The number 4 satisfies the inequality $2x - 1 < 4$. F
9. The number 0 is a solution to the inequality $2x - 3 \leq 5x - 3$. T
10. The inequalities $2x - 1 < x$ and $x < 2x - 1$ have the same solutions. F

2.6 EXERCISES

Determine whether each of the following statements is true or false. See Example 1.

1. $-3 < 5$ T
2. $-6 < 0$ T
3. $4 \leq 4$ T
4. $-3 \geq -3$ T
5. $-6 > -5$ F
6. $-2 < -9$ F
7. $-4 \leq -3$ T
8. $-5 \geq -10$ T
9. $(-3)(4) - 1 < 0 - 3$ T
10. $2(4) - 6 \leq -3(5) + 1$ F
11. $-4(5) - 6 \geq 5(-6)$ T
12. $4(8) - 30 > 7(5) - 2(17)$ T
13. $7(4) - 12 \leq 3(9) - 2$ T
14. $-3(4) + 12 \leq 2(3) - 6$ T

Sketch the graph of each inequality on the number line. See Example 2.   Graphs in Answers sections.

15. $x < 3$
16. $x < -7$
17. $x > -2$
18. $x > 4$
19. $-1 > x$
20. $0 > x$
21. $-2 \leq x$
22. $-5 \geq x$
23. $x \leq -1$
24. $x \leq 4$
25. $x \geq 5$
26. $x \geq -3$

27. $x < 400$ **28.** $x > 240$ **29.** $x \le 5.3$ **30.** $x \le -3.4$
31. $-3 < x < 1$ **32.** $0 < x < 5$ **33.** $3 \le x \le 7$ **34.** $-3 \le x \le -1$
35. $-5 \le x < 0$ **36.** $-2 < x \le 2$ **37.** $4 < x \le 10$ **38.** $0 \le x < 6$

Determine whether or not the given number satisfies the inequality following it. See Example 3.

39. $-9, -x > 3$ Yes **40.** $5, -3 < -x$ No **41.** $-2, 5 \le x$ No
42. $-6, -9 \ge x$ No **43.** $4, 4 \ge x$ Yes **44.** $-1, 0 < x$ No
45. $-6, 2x - 3 > -11$ No **46.** $4, 3x - 5 < 7$ No **47.** $3, -3x + 4 > -7$ Yes
48. $-4, -3x + 1 > 2x - 5$ Yes **49.** $0, 3x - 7 \le 5x - 7$ Yes **50.** $0, -10x + 9 \le 3(x + 3)$ Yes
51. $2.5, 2x + 6 \ge 4x - 9$ Yes **52.** $1.5, 2x - 3 \le 4(x - 1)$ Yes **53.** $-7, -5 < x < 9$ No
54. $-9, -6 \le x \le 40$ No **55.** $-2, -3 \le 2x + 5 \le 9$ Yes **56.** $-5, -3 < -3x - 7 \le 8$ Yes
57. $-3.4, -4.25x - 13.29 < .89$ No **58.** $4.8, 3.25x - 14.78 \le 1.3$ Yes

Encode each situation as an inequality. See Example 4.

59. At an 8% sales tax rate, Susan paid more than $1500 sales tax when she purchased her new car. $.08x > 1500$

60. At Burger Brothers, the price of a hamburger is twice the price of an order of French fries, and the price of a Coke is $.25 more than the price of the fries. Burger Brothers advertises that you can get a complete meal (burger, fries, and Coke) for under two dollars. $x + 2x + (x + .25) < 2.00$

61. Travis made 44 and 72 on the first two tests in algebra and has one test remaining. The average on the three tests must be at least 60 in order for Travis to pass the course. $\frac{44 + 72 + x}{3} \ge 60$

62. The Concerned Mothers Group claims that 60% of the videos shown on a certain music television station have eight or more acts of violence. This subjects the viewer to more than 2568 acts of violence per day. $8(.60x) > 2568$

63. On Howard's recent trip from Bangor to San Diego, he drove for 8 hours each day and traveled between 396 and 453 miles each day. $396 < 8x < 453$

64. Bart and Betty are looking at color televisions that range in price from $399.99 to $579.99. Bart can afford more than Betty and has agreed to spend $100 more than Betty when they purchase this gift for their mother.
$399.99 < x + x + 100 < 579.99$

By using trial and error, find one number that satisfies each of the following inequalities. Do not solve the inequalities.

65. $-x > 0$ Any number less than 0 **66.** $-x \le 3$ Any number greater than or equal to -3
67. $5 < x$ Any number greater than 5 **68.** $6 > x$ Any number less than 6
69. $4 \ge x + 1$ Any number less than or equal to 3 **70.** $5 < 1 - x$ Any number less than -4
71. $2x - 9 < -4$ Any number less than $\frac{5}{2}$ **72.** $-3x + 3 < 5$ Any number greater than $-\frac{2}{3}$
73. $5 - 4x > -5$ Any number less than $\frac{5}{2}$ **74.** $6 - 5x > 7$ Any number less than $-\frac{1}{5}$
75. $4x + 1 \le x - 8$ Any number less than -3 **76.** $-2x + 6 \le 5 - 4x$ Any number less than or equal to $-\frac{1}{2}$
77. $4 - 2x \ge 3 - x$ Any number less than 1 **78.** $4x - 9 \ge x + 19$ Any number greater than or equal to $\frac{28}{3}$
79. $-3 < 3x - 1 < 2$ Any number between $-\frac{2}{3}$ and 1 **80.** $-5 < -2x + 1 < 6$ Any number between $-\frac{5}{2}$ and 3
81. $3 < 2x < 4$ Any number between $\frac{3}{2}$ and 2 **82.** $5 < 2x < 6$ Any number between $\frac{5}{2}$ and 3

2.7 Solving Inequalities

IN THIS SECTION:

- Rules for Inequalities
- Solving Inequalities
- Applications of Inequalities

In Section 2.1 we learned to solve equations by writing a sequence of equivalent equations that ends in a very simple equation whose solution is obvious. In this section we will learn that the procedure for solving inequalities is the same. However, the rules for performing operations on each side of an inequality are slightly different from the rules for equations.

Rules for Inequalities

It is a good idea to have students perform these operations for themselves so that they can see what kind of inequalities result.

Consider the inequalities $x > 8$ and $x + 2 > 10$. Certainly, any number larger than 8 satisfies $x + 2 > 10$, and any number that solves $x + 2 > 10$ also solves $x > 8$. These two inequalities are equivalent. **Equivalent inequalities** are inequalities that have exactly the same solutions.

We can get equivalent inequalities by performing operations on each side of an inequality just as we do for solving equations. If we start with the inequality $6 < 10$ and add 2 to each side, we get the true statement $8 < 12$. Examine the results of performing the same operation on each side of $6 < 10$.

Perform these operations on each side:

	Add 2	Subtract 2	Multiply by 2	Divide by 2
Start with $6 < 10$	$8 < 12$	$4 < 8$	$12 < 20$	$3 < 5$

All of the resulting inequalities are correct. Now if we repeat these operations using -2, we get the results shown below.

Perform these operations on each side:

	Add -2	Subtract -2	Multiply by -2	Divide by -2
Start with $6 < 10$	$4 < 8$	$8 < 12$	$-12 > -20$	$-3 > -5$

Notice that the direction of the inequality symbol is the same for all of the results except the last two. When we multiplied each side by -2 and when we divided each side by -2, we had to reverse the inequality symbol to get a correct result. We can now state the rules to follow when solving inequalities.

PROPERTY
Addition-Subtraction
Property of Inequality

> If we add *any* number to or subtract *any* number from each side of an inequality, we get an equivalent inequality.

PROPERTY
Multiplication-Division
Property of Inequality

> If we multiply or divide each side of an inequality by a *positive* number, we get an equivalent inequality. If we multiply or divide each side of an inequality by a *negative* number and *reverse the inequality symbol*, we get an equivalent inequality.

Note that we need to be most careful of the last rule. *When we multiply or divide an inequality by a negative number, we must reverse the inequality symbol*.

Solving Inequalities

We use the properties of inequality just as we use the properties of equality to solve equations. We simplify the inequality until we get the variable isolated on one side.

EXAMPLE 1   Solve and graph the inequality $4x - 5 > 19$.

Solution

$$4x - 5 > 19$$
$$4x - 5 + 5 > 19 + 5 \qquad \text{Add 5 to each side.}$$
$$4x > 24 \qquad \text{Simplify.}$$
$$x > 6 \qquad \text{Divide each side by 4.}$$

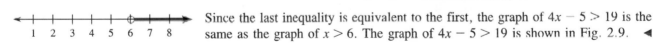

Figure 2.9

Since the last inequality is equivalent to the first, the graph of $4x - 5 > 19$ is the same as the graph of $x > 6$. The graph of $4x - 5 > 19$ is shown in Fig. 2.9. ◄

EXAMPLE 2   Solve and graph the inequality $5 - 3x \leq 11$.

Solution

$$5 - 3x \leq 11$$
$$5 - 3x - 5 \leq 11 - 5 \qquad \text{Subtract 5 from each side.}$$
$$-3x \leq 6 \qquad \text{Simplify.}$$
$$x \geq -2 \qquad \text{Divide each side by } -3 \text{ and reverse the inequality.}$$

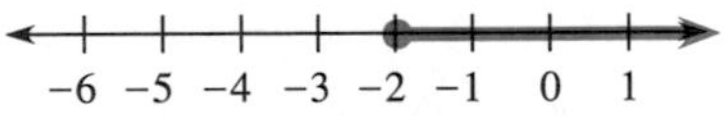

Figure 2.10

The graph of $5 - 3x \leq 11$ is the same as the graph of $x \geq -2$ (see Fig. 2.10). ◄

We can use the rules for solving inequalities on the compound inequalities that we studied in Section 2.6.

EXAMPLE 3   Solve and graph the inequality $-9 \leq 2x - 3 < 5$.

Solution

$$-9 \leq 2x - 3 < 5$$

$$-9 + 3 \leq 2x - 3 + 3 < 5 + 3 \qquad \text{Add 3 to each part.}$$

$$-6 \leq 2x < 8 \qquad \text{Simplify.}$$

$$\frac{-6}{2} \leq \frac{2x}{2} < \frac{8}{2} \qquad \text{Divide each part by 2.}$$

$$-3 \leq x < 4 \qquad \text{Simplify.}$$

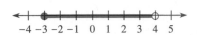

Figure 2.11

The graph of the inequality $-9 \leq 2x - 3 < 5$ is the same as the graph of the inequality $-3 \leq x < 4$, shown in Fig. 2.11.   ◄

EXAMPLE 4   Solve and graph the inequality $-3 \leq 5 - x \leq 5$.

Solution

It is best to write compound inequalities like this one with the smallest number on the left, because that is how the graph looks.

$$-3 \leq 5 - x \leq 5$$

$$-3 - 5 \leq 5 - x - 5 \leq 5 - 5 \qquad \text{Subtract 5 from each part.}$$

$$-8 \leq -x \leq 0 \qquad \text{Simplify.}$$

$$(-1)(-8) \geq (-1)(-x) \geq (-1)0 \qquad \text{Multiply each part by } -1, \text{ reversing the inequality symbols.}$$

$$8 \geq x \geq 0$$

It is customary to write $8 \geq x \geq 0$ with the smallest number on the left:

$$0 \leq x \leq 8$$

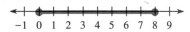

Figure 2.12

The graph of $-3 \leq 5 - x \leq 5$ is shown in Fig. 2.12.   ◄

Applications of Inequalities

The following example shows how inequalities can be used in applications.

EXAMPLE 5   Mei Lin made a 76 on the midterm exam in history. To get a B, the average of her midterm and her final exam must be between 80 and 90. For what range of scores on the final exam will she get a B?

Solution   Let x represent the final exam score. Her average is then $(x + 76)/2$. The inequality expresses the fact that the average must be between 80 and 90:

$$80 < \frac{x + 76}{2} < 90$$

$$2(80) < 2\left(\frac{x + 76}{2}\right) < 2(90) \qquad \text{Multiply each part by 2.}$$

$$160 < x + 76 < 180 \qquad \text{Simplify.}$$

$$160 - 76 < x + 76 - 76 < 180 - 76 \qquad \text{Subtract 76 from each side.}$$

$$84 < x < 104 \qquad \text{Simplify.}$$

The last inequality indicates that Mei Lin's final exam score must be between 84 and 104. ◄

Warm-ups

True or false?

1. The inequality $2x > 18$ is equivalent to $x > 9$. T
2. The inequality $x - 5 > 0$ is equivalent to $x < 5$. F
3. We can divide each side of an inequality by any real number, as long as we use the correct inequality symbol. F
4. The inequality $-2x \leq 6$ is equivalent to $-x \leq 3$. T
5. The statement ''x is at most 7'' is written as $x < 7$. F
6. The statement ''The total of x and $.05x$ is at least 76'' is written as $x + .05x \geq 76$. T
7. The statement ''x is not more than 85'' is written as $x < 85$. F
8. The inequality $-3 > x > -9$ is equivalent to $-9 < x < -3$. T
9. If x is the sale price of Glen's truck, the sales tax rate is 8%, and the title fee is $50, then the total he pays is $1.08x + 50$ dollars. T
10. If the selling price of the house, x, less the sales commission of 6% must be at least $60,000, then this is described by the inequality $x - .06x \leq 60,000$. F

2.7 EXERCISES

Solve and graph each of the following inequalities. See Examples 1 *and* 2.   Graphs in Answers sections.

1. $x + 3 > 0$ $x > -3$
2. $x - 9 \leq -8$ $x \leq 1$
3. $3 < x + 1$ $x > 2$
4. $8 > 2x$ $x < 4$
5. $3x - 2 < 7$ $x < 3$
6. $2x - 5 > -9$ $x > -2$
7. $3 - 7x \leq 17$ $x \geq -2$
8. $5 - 3x \geq 20$ $x \leq -5$

9. $-\dfrac{5}{6}x \geq -20$ $x \leq 24$ **10.** $-\dfrac{2}{3}x \geq -4$ $x \leq 6$ **11.** $1 - \dfrac{1}{4}x \geq -2$ $x \leq 12$ **12.** $2 - \dfrac{1}{3}x > 0$ $x < 6$

13. $2x + 5 < x - 6$ $x < -11$ **14.** $3x - 4 < 2x + 9$ $x < 13$ **15.** $5 - x > -8$ $x < 13$ **16.** $3 \geq -x$ $x \geq -3$

17. $6 \leq -3x$ $x \leq -2$ **18.** $7 - x > 11$ $x < -4$ **19.** $x - 4 < 2(x + 3)$ $x > -10$

20. $2x + 3 < 3(x - 5)$ $x > 18$ ▤ **21.** $.52x - 35 < .45x + 8$ $x < 614.3$ ▤ **22.** $8455(x - 3.4) > 4320$ $x > 3.91$

Solve and graph each compound inequality. See Examples 3 and 4.   Graphs in Answers sections.

23. $5 < x - 3 < 7$ $8 < x < 10$

24. $2 < x - 5 < 6$ $7 < x < 11$

25. $3 < 2x + 1 < 10$ $1 < x < \dfrac{9}{2}$

26. $-3 < 3x + 4 < 7$ $-\dfrac{7}{3} < x < 1$

27. $5 < 2x - 3 < 17$ $4 < x < 10$

28. $-4 < 3x - 1 < 11$ $-1 < x < 4$

29. $-2 < 7 - 3x \leq 22$ $-5 \leq x < 3$

30. $-1 \leq 1 - 2x < 3$ $-1 < x \leq 1$

31. $-7 < \dfrac{3m + 1}{2} \leq 8$ $-5 < m \leq 5$

32. $0 \leq \dfrac{3 - 2x}{2} < 9$ $-\dfrac{15}{2} < x \leq \dfrac{3}{2}$

33. $-4 \leq 5 - x \leq 7$ $-2 \leq x \leq 9$

34. $2 \leq 3 - x \leq 8$ $-5 \leq x \leq 1$

▤ **35.** $.02 < .54 - .0048x < .05$ $102.1 < x < 108.3$

▤ **36.** $.44 < \dfrac{34.55 - 22.3x}{124.5} < .76$ $-2.69 < x < -.91$

Solve each of the following problems by using an inequality. See Example 5.

37. The length of a rectangular boat storage shed must be 4 meters more than the width, and the perimeter must be at least 120 meters. What is the range of values for the width? $w \geq 28$

38. Harold Ivan is shopping for a new car. In addition to the price of the car, there is a 5% sales tax and a $144 title and license fee. If Harold Ivan decides that he will spend less than $9970 total, then what is the price range for the car? $x < \$9358$

39. Sherie is going to buy a microwave in a city with an 8% sales tax. She has at most $594 to spend. In what price range should she look? $x \leq \$550$

40. Albert is shopping for a van in a city with a 9% sales tax. There is also a $100 title and license fee to pay. He wants to get a good one and plans to spend at least $8820 but not over $12,090. What is the price range for the truck? $\$8000 < x < \$11,000$

41. Professor Williams gives only a midterm exam and a final exam. The semester average is computed by taking 1/3 of the midterm exam score plus 2/3 of the final exam score. To get a C, Stacy must have a semester average between 70 and 79 inclusive. If Stacy scored only 48 on the midterm, then for what range of scores on the final exam will Stacy get a C? $81 \leq x \leq 94.5$

42. Professor Williamson counts the midterm as 2/3 of the grade and the final as 1/3 of the grade. Wendy scored only 48 on the midterm. What range of scores on the final exam would put Wendy's average between 70 and 79 inclusive? $114 \leq x \leq 141$

43. Ronald wants to sell his car through a broker who charges a commission of 10% of the selling price. Ronald still owes $11,025 on the car. Ronald must get enough to at least pay off the loan. What is the range of the selling price? $x \geq \$12,250$

44. At Burger Brothers, the price of a hamburger is twice the price of an order of French fries, and the price of a Coke is $.25 more than the price of the fries. Burger Brothers advertises that you can get a complete meal (burger, fries, and Coke) for under two dollars. What is the price range of an order of fries? $x < \$.44$

45. Tilak made 44 and 72 on the first two tests in algebra and has one test remaining. The average on the three tests must be at least 60 in order for Tilak to pass the course. For what range of scores on the last test will Tilak pass the course? $x \geq 64$

46. On Halley's recent trip from Bangor to San Diego, she drove for 8 hours each day and traveled between 396 and 453 miles each day. In what range was her average speed for each day of the trip? $49.5 < x < 56.625$

For each graph, write an inequality that describes the graph.

47.
$$-3\,-2\,-1\ \ 0\ \ 1\ \ 2\ \ 3\ \ 4\ \ 5\ \ 6\ \ 7$$ $x > 3$

48.
$$-4\,-3\,-2\,-1\ \ 0\ \ 1\ \ 2\ \ 3\ \ 4\ \ 5\ \ 6$$ $x \le 4$

49.
$$-6\,-5\,-4\,-3\,-2\,-1\ \ 0\ \ 1\ \ 2\ \ 3\ \ 4$$ $x \le 2$

50.
$$-5\,-4\,-3\,-2\,-1\ \ 0\ \ 1\ \ 2\ \ 3\ \ 4\ \ 5$$ $0 < x \le 3$

51.
$$-5\,-4\,-3\,-2\,-1\ \ 0\ \ 1\ \ 2\ \ 3\ \ 4\ \ 5$$ $0 < x < 2$

52.
$$-5\,-4\,-3\,-2\,-1\ \ 0\ \ 1\ \ 2\ \ 3\ \ 4\ \ 5$$ $-1 \le x < 3$

53.
$$-6\ \ -4\ \ -2\ \ \ 0\ \ \ 2\ \ \ 4\ \ \ 6\ \ \ 8$$ $-5 < x \le 7$

54.
$$-5\,-4\,-3\,-2\,-1\ \ 0\ \ 1\ \ 2\ \ 3\ \ 4\ \ 5$$ $x < 4$

55.
$$-5\ \ -4\ \ -3\,-2\,-1\ \ 0\ \ 1\ \ 2\ \ 3\ \ 4\ \ 5$$ $x > -4$

56.
$$-5\,-4\,-3\,-2\,-1\ \ 0\ \ 1\ \ 2\ \ 3\ \ 4\ \ 5$$ $0 < x \le 2$

Wrap-up

CHAPTER 2

SUMMARY

	Equations	Examples
Linear equation	An equation that can be written in the form $ax + b = 0$	$2(x - 3) + 6 = 5(x + 4)$ $-3x - 20 = 0$
Identity	An equation that is satisfied by every number for which both sides of the equation are defined	$x + x = 2x$
Conditional equation	An equation that has at least one solution but is not an identity	$5x - 10 = 0$
Inconsistent equation	An equation that has no solution	$x = x + 1$
Equivalent equations	Equations that have exactly the same solutions	$2x + 1 = 5$ $2x = 4$
Properties of equality	If the same number is added to or subtracted from each side of an equation, the resulting equation is equivalent to the original equation.	$x + 5 = 7$ $x = 2$

If each side of an equation is multiplied
or divided by the same nonzero number,
the resulting equation is equivalent
the original equation.

$$3x = 6$$
$$x = 2$$

Solving equations

1. Remove parentheses and combine like terms to simplify each side as much as possible.
2. Use the addition-subtraction property of equality to get like terms from opposite sides onto the same side so that they may be combined.
3. The multiplication-division property of equality is generally used last.

Verbal Problems

Steps in solving word problems

1. Read the problem.

2. If possible, draw a diagram to illustrate the problem.

3. Choose a variable and write down what it represents.

4. Represent any other unknowns in terms of that variable.

5. Write an equation that fits the situation.

6. Solve the equation.

7. Be sure that your solution answers the question posed in the original problem.

8. Check your answer by using it to solve the original problem (not the equation).

Inequalities

Examples

Properties of inequality

Addition, subtraction, multiplication, and division
may be performed on each side of an inequality, just
as we do in solving equations, with one exception.
When multiplying or dividing by a negative number,
the inequality symbol is reversed.

$$-3x > 6$$
$$x < -2$$

REVIEW EXERCISES

2.1 *Solve each equation.*

1. $2x - 5 = 9$ 7

2. $5x - 8 = 38$ $\dfrac{46}{5}$

3. $3x - 7 = 0$ $\dfrac{7}{3}$

4. $3x + 5 = 0$ $-\dfrac{5}{3}$

5. $3 - 4x = 11$ -2

6. $4 - 3x = -8$ 4

7. $2(x - 7) = -14$ 0

8. $2(x - 7) = 0$ 7

9. $3x - 5 = 6x + 7$ -4

10. $2(x - 4) + 4 = 5(9 - x)$ 7

11. $-\dfrac{2}{3}x = 20$ -30

12. $\dfrac{3}{4}x = -6$ -8

13. $.24x + 1 = 97$ 400

14. $1.05x = 420$ 400

15. $x - .1x = 90$ 100

16. $x + .05x = 2.1$ 2

17. $\dfrac{1}{3}x = \dfrac{1}{7}$ $\dfrac{3}{7}$

18. $\dfrac{x}{2} = -\dfrac{2}{5}$ $-\dfrac{4}{5}$

2.2 *Solve each equation.*

19. $2(x-7)-5=5-(3-2x)$

20. $2(x-7)+5=-(9-2x)$

21. $2(x-x)=0$ All real numbers

22. $2x-x=0$ 0

23. $.06x+14=.3x-5.2$ 80

24. $.05(x+20)=.1x-.5$ 30

25. $.05(x+100)+.06x=115$ 1000

26. $.06x+.08(x+1)=.41$ $\dfrac{33}{14}$

27. $\dfrac{1}{2}x-5=\dfrac{1}{3}x-1$ 24

28. $\dfrac{1}{2}x-\dfrac{1}{2}=\dfrac{1}{4}x$ 2

29. $\dfrac{3x}{3x}=1$ All real numbers except 0

30. $\dfrac{3x}{3}=1$ 1

19. No solution   **20.** All real numbers

Solve each equation mentally. Write down only the answer.

31. $3x=24$ 8

32. $4x=12$ 3

33. $x-7=2$ 9

34. $x-6=-6$ 0

35. $\dfrac{1}{2}x=-3$ −6

36. $\dfrac{1}{3}x=\dfrac{1}{4}$ $\dfrac{3}{4}$

37. $-\dfrac{1}{2}x=10$ −20

38. $-\dfrac{1}{3}x=20$ −60

39. $2x-3=0$ $\dfrac{3}{2}$

40. $5x-1=0$ $\dfrac{1}{5}$

41. $-2x=8$ −4

42. $-3x=30$ −10

43. $2x-5=1$ 3

44. $3x-4=2$ 2

45. $-2x=1$ $-\dfrac{1}{2}$

46. $-3x=2$ $-\dfrac{2}{3}$

47. $-\dfrac{3}{2}x=1$ $-\dfrac{2}{3}$

48. $-\dfrac{2}{5}x=1$ $-\dfrac{5}{2}$

49. $\dfrac{31}{67}x=0$ 0

50. $-\dfrac{3}{5}x=0$ 0

51. $4x-2x=2x$

52. $.01x=.01$ 1

53. $.01x=.02$ 2

54. $3x+4x=7x$

55. $x+7=x$ No solution

56. $x+x=7$ $\dfrac{7}{2}$

51. All real numbers   **54.** All real numbers

2.3 *Solve each of the following for x.*

57. $ax+b=0$ $x=-\dfrac{b}{a}$

58. $mx+e=t$ $x=\dfrac{t-e}{m}$

59. $ax-2=b$ $x=\dfrac{b+2}{a}$

60. $b=5-x$ $x=5-b$

61. $LWx=V$ $x=\dfrac{V}{LW}$

62. $3xy=6$ $x=\dfrac{2}{y}$

63. $2x-b=5x$ $x=-\dfrac{b}{3}$

64. $t-5x=4x$ $x=\dfrac{t}{9}$

Solve each equation for y.

65. $5x+2y=6$ $y=-\dfrac{5}{2}x+3$

66. $5x-3y+9=0$ $y=\dfrac{5}{3}x+3$

67. $y-1=-\dfrac{1}{2}(x-6)$ $y=-\dfrac{1}{2}x+4$

68. $y+6=\dfrac{1}{2}(x+8)$ $y=\dfrac{1}{2}x-2$

69. $\dfrac{1}{2}x+\dfrac{1}{4}y=4$ $y=-2x+16$

70. $-\dfrac{x}{3}+\dfrac{y}{2}=1$ $y=\dfrac{2}{3}x+2$

Find the value of y in each of the following formulas if $x=-3$.

71. $y=3x-4$ −13

72. $2x-3y=-7$ $\dfrac{1}{3}$

73. $5xy=6$ $-\dfrac{2}{5}$

74. $3xy-2x=-12$ 2

75. $y-3=-2(x-4)$ 17

76. $y+1=2(x-5)$ −17

2.4 *Encode each of the following phrases or sentences as an algebraic expression or an equation.*

77. The sum of a number and 9 $x+9$

78. The product of a number and 7 $7x$

79. Two numbers that differ by 8 $x, x+8$

80. Two numbers with a sum of 12 $x, 12-x$

81. Sixty-five percent of a number $.65x$

82. One half of a number $\dfrac{1}{2}x$

83. One side of a rectangle is 5 feet longer than the other, and the area is 98 square feet. $x(x+5)=98$

84. One side of a rectangle is one foot longer than twice the other side, and the perimeter is 56 feet. $2x + 2(2x + 1) = 56$

85. By driving 10 miles per hour slower than Jim, Barbara travels the same distance in 3 hours as Jim does in 2 hours. $2(x + 10) = 3x$

86. The sum of three consecutive even integers is 88.
$$x + x + 2 + x + 4 = 88$$

2.5 *Solve each word problem.*

87. If the sum of three consecutive odd integers is 237, then what are the integers?   77, 79, 81

88. Lawanda and Betty both drive the same distance to the shore. By driving 15 miles per hour faster than Betty, Lawanda can get there in 3 hours, while Betty takes 4 hours. How fast does each of them drive?

89. The length of a rectangular lot is 50 feet more than the width. If the perimeter is 500 feet, then what are the length and width?   150 feet, 100 feet

90. Wanda makes $6000 more per year than her husband. Wanda saves 10% of her income for retirement, and her husband saves 6%. If together they save $5400 per year, then how much does each of them make per year?

91. The owners of ABC Video discovered that they had no movies in common with XYZ Video and bought XYZ's entire stock. Although XYZ had 200 titles, they had no children's movies, while 60% of ABC's titles were children's movies. If 40% of the movies in the combined stock are children's movies, then how many movies did ABC have before the merger?   400

92. Gary has figured that he needs to take home $30,400 a year to live comfortably. If the government takes 24% of Gary's income, then what must his income be in order for him to live comfortably? $40,000

88. Lawanda: 60 mph, Betty: 45 mph

90. Wanda: $36,000, her husband: $30,000

2.6 *Determine whether or not the given number is a solution to the inequality following it.*

93. $3, -2x + 5 \leq x - 6$ No

94. $-2, 5 - x > 4x + 3$ Yes

95. $-1, -2 \leq 6 + 4x < 0$ No

96. $0, 4x + 9 \geq 5(x - 3)$ Yes

For each graph, write an inequality that has the solution shown by the graph.

97. $x > 1$

98. $x < 2$

99. $x \geq 2$

100. $3 < x < 5$

101. $-3 \leq x < 3$

102. $x \leq 1$

103. $x < -1$

104. $-2 \leq x < 2$

2.7 *Solve and graph each inequality.*   Graphs in Answers sections.

105. $3 - 2x < 11$ $x > -4$

106. $5 - 3x > 35$ $x < -10$

107. $x + 2 > 1$ $x > -1$

108. $x - 3 > 7$ $x > 10$

109. $-\dfrac{3}{4}x \geq 3$ $x \leq -4$

110. $-\dfrac{2}{3}x \leq 10$ $x \geq -15$

111. $-3 < 2x - 1 < 9$ $-1 < x < 5$

112. $0 \leq 1 - 2x < 5$ $-2 < x \leq \dfrac{1}{2}$

113. $-1 \leq \dfrac{2x - 3}{3} \leq 1$ $0 \leq x \leq 3$

114. $-3 < \dfrac{4 - x}{2} < 2$ $0 < x < 10$

Use an inequality to solve each problem.

115. One side of a triangle is 1 foot longer than the shortest side, and the third side is twice as long as the shortest side. If the perimeter is less than 25 feet, then what is the range of the length of the shortest side? $x < 6$

116. Alana makes $5.80 per hour working in the library. To keep her job, she must make at least $116 per week; but to keep her scholarship, she must not earn more than $145 per week. What is the range of the number of hours per week that she may work? $20 \leq x \leq 25$

CHAPTER 2 TEST

Solve each equation.

1. $-10x - 6 + 4x = -4x + 8$ -7

2. $5(2x - 3) = x + 3$ 2

3. $-\dfrac{2}{3}x + 1 = 7$ -9

4. $x + .06x = 742$ 700

Solve for the indicated variable.

5. $2x - 3y = 9$ for y $y = \dfrac{2}{3}x - 3$

6. $m = aP - w$ for a $a = \dfrac{m + w}{P}$

Write an inequality that has the solution shown by the graph.

7. ![number line from -5 to 5] $-3 < x \leq 2$
$\quad -5\ -4\ -3\ -2\ -1\ \ 0\ \ 1\ \ 2\ \ 3\ \ 4\ \ 5$

8. ![number line from -2 to 8] $x > 1$
$\quad -2\ -1\ \ 0\ \ 1\ \ 2\ \ 3\ \ 4\ \ 5\ \ 6\ \ 7\ \ 8$

Solve and graph each inequality.   Graphs in Answers sections.

9. $4 - 3(w - 5) < -2w$ $w > 19$

10. $1 < \dfrac{1 - 2x}{3} < 5$ $-7 < x < -1$

11. $1 < 3x - 2 < 7$ $1 < x < 3$

12. $-\dfrac{2}{3}y < 4$ $y > -6$

Solve each equation.

13. $2(x + 6) = 2x - 5$ No solution

14. $x + 7x = 8x$ All real numbers

15. $x - .03x = .97$ 1

16. $6x - 7 = 0$ $\dfrac{7}{6}$

Write a complete solution to each of the following problems.

17. The perimeter of a rectangle is 72 meters. If the width is 8 meters less than the length, then what is the width of the rectangle?   14 meters

18. If the area of a triangle is 54 square inches and the base is 12 inches, then what is the height?   9 inches

19. Brandon bought a diamond for his fiancee at 40% discount. If he paid $720, then what was the original price? $1200

20. How many liters of a 20% alcohol solution should Maria mix with 50 liters of a 60% alcohol solution to obtain a 30% solution?   150 liters

Photo for Exercise 19

Tying It All Together

CHAPTERS 1–2

Simplify each expression.

1. $3x + 5x$ $8x$
2. $3x \cdot 5x$ $15x^2$
3. $\dfrac{4x + 2}{2}$ $2x + 1$
4. $5 - 4(3 - x)$ $4x - 7$

5. $3x + 8 - 5(x - 1)$ $-2x + 13$
6. $(-6)^2 - 4(-3)2$ 60
7. $3^2 \cdot 2^3$ 72

8. $4(-7) - (-6)(3)$ -10
9. $-2x \cdot x \cdot x$ $-2x^3$
10. $(-1)(-1)(-1)(-1)(-1)$ -1

Solve each equation.

11. $3x + 5x = 8$ 1
12. $3x + 5x = 8x$ All real numbers
13. $3x + 5x = 7x$ 0

14. $3x + 5 = 8$ 1
15. $3x + 1 = 7$ 2
16. $5 - 4(3 - x) = 1$ 2

17. $3x + 8 = 5(x - 1)$ $\dfrac{13}{2}$
18. $x - .05x = 190$ 200

CHAPTER 3

Exponents and Polynomials

Rose's garden is currently a square with sides of length x feet. Next spring, she plans to make it a rectangular shape by lengthening one side by 5 feet and shortening the other side by 5 feet. Write a polynomial in x that gives the new area. Will the new garden be bigger or smaller than the old garden? How much bigger or smaller?

Since the area of the garden is the product of the length and width, we must multiply $x + 5$ and $x - 5$ to get an expression for the area. The product of these two expressions is a polynomial. Even though we don't know the value of x, we will be able to determine from the polynomial how much the new garden will differ in area from the old garden. We will solve this problem as Exercise 59 of Section 3.4.

3.1   Addition and Subtraction of Polynomials

IN THIS SECTION:

- ● **Polynomials**
- ● **Addition of Polynomials**
- ● **Subtraction of Polynomials**

We first used polynomials in Chapter 1, but we did not identify them as polynomials. Polynomials also occurred in the equations and inequalities of Chapter 2. In this section we will define polynomials and begin a systematic study of polynomials.

Polynomials

In Chapter 3 we learn the basic operations with polynomials. We introduce the product and quotient rules for exponents because they are necessary for multiplying and dividing polynomials. The introduction to rules of exponents in this chapter gives the student some experience before the student sees the more extensive treatment of exponents in Chapter 6.

In Chapter 1 we defined a **term** as an expression containing a number or the product of a number and one or more variables. Some examples of terms are

$$4x^3, \quad -x^2y^2, \quad 6ab, \quad \text{and} \quad -2.$$

A **polynomial** is defined as a single term or a finite sum of terms. Thus

$$4x^3 + (-15x^2) + 7x + (-2)$$

is a polynomial. Since it is simpler to write addition of a negative as subtraction, this polynomial is written as

$$4x^3 - 15x^2 + 7x - 2.$$

The number preceding the variable in each term is called the **coefficient** of that variable or the coefficient of that term. In the above polynomial, 4 is the coefficient of x^3, -15 is the coefficient of x^2, and 7 is the coefficient of x. In algebra a number is frequently called a **constant.** The term -2 is called the **constant term.**

EXAMPLE 1   Determine the coefficient of x^3 in each polynomial:

a) $x^3 + 5x^2 - 6$ **b)** $4x^6 - x^3 + x$ **c)** $x^4 - x$

Solution

a) Since $x^3 = 1 \cdot x^3$, the coefficient of x^3 is 1.
b) The coefficient of x^3 is -1.
c) Since x^3 is missing in $x^4 - x$, the coefficient of x^3 is 0.   ◀

For simplicity we generally write polynomials with the exponents decreasing from left to right and the constant term last. Thus we write

$$x^3 - 4x^2 + 5x + 1 \qquad \text{rather than} \qquad -4x^2 + 1 + 5x + x^3.$$

When a polynomial is written this way, the coefficient of the first term is called the **leading coefficient.**

Certain polynomials are given special names. A **monomial** is a polynomial that has one term, a **binomial** is a polynomial that has two terms, and a **trinomial** is a polynomial that has three terms. The **degree** of a polynomial in one variable is the highest power of the variable in the polynomial. If a polynomial consists of only a nonzero constant term, we say that it has **degree 0.** The number 0 is considered a polynomial without degree, because $0 = 0 \cdot x^n$ for any value of n.

EXAMPLE 2 For each polynomial, state the degree and whether it is a monomial, binomial, or trinomial:

a) $5x^2 - 7x^3 + 2$ **b)** $x^{43} - x^2$ **c)** $5x$

Solution

a) This is a third-degree trinomial.
b) This is a binomial whose degree is 43.
c) This is a monomial with degree 1.   ◀

Addition of Polynomials

If we assign a value to the variable in a polynomial, then the value of the polynomial is a real number. Thus the operations that we perform with real numbers can be performed with polynomials. Actually, we were adding polynomials when we learned to add like terms in Chapter 1.

RULE
Addition of Polynomials

> To add two polynomials, add the like terms.

To add the polynomials $x^3 + 2x^2 - 7x$ and $-5x^3 + 4x + 3$, we can use the commutative and associative properties to get the like terms next to each other, and then add them as follows:

$$(x^3 + 2x^2 - 7x) + (-5x^3 + 4x + 3) = x^3 - 5x^3 + 2x^2 - 7x + 4x + 3$$
$$= -4x^3 + 2x^2 - 3x + 3$$

We may also write the addition vertically. When we do this, we must line up the like terms as shown.

Add:

$$
\begin{array}{r}
x^3 + 2x^2 - 7x \\
-5x^3 \qquad\; + 4x + 3 \\
\hline
-4x^3 + 2x^2 - 3x + 3
\end{array}
$$

EXAMPLE 3 Perform the indicated additions of the polynomials:

a) $(x^2 - 6x + 5) + (-3x^2 + 5x - 9)$
b) $(-5a^3 + 3a - 7) + (4a^2 - 3a + 7)$

Point out that each of these
equations is an identity.

Solution

a) $(x^2 - 6x + 5) + (-3x^2 + 5x - 9) = x^2 - 3x^2 - 6x + 5x + 5 - 9$

$$= -2x^2 - x - 4$$

b) $(-5a^3 + 3a - 7) + (4a^2 - 3a + 7) = -5a^3 + 4a^2 + 3a - 3a - 7 + 7$

$$= -5a^3 + 4a^2 \qquad \blacktriangleleft$$

Subtraction of Polynomials

When we subtract polynomials, we subtract the like terms. However, it is usually easier to do this by addition. Since $a - b = a + (-b)$, we can add the first polynomial and the opposite of the second polynomial. Remember that a negative sign in front of a polynomial changes the sign of each term of the polynomial. For example,

$$-(x^2 - 2x + 8) = -x^2 + 2x - 8.$$

Thus to subtract

$$x^2 - 2x + 8 \qquad \text{from} \qquad 4x^2 + 6x - 9,$$

we write

$$(4x^2 + 6x - 9) - (x^2 - 2x + 8) = (4x^2 + 6x - 9) + (-x^2 + 2x - 8)$$

Change signs.

$$= 3x^2 + 8x - 17. \qquad \textbf{Add.}$$

If we write this subtraction vertically, we still change the signs of the terms on the bottom and add. Subtract:

$$
\begin{array}{r}
4x^2 + 6x - 9 \\
\underline{x^2 - 2x + 8}
\end{array}
\quad \longleftarrow \text{ Change signs and add. } \longrightarrow \quad
\begin{array}{r}
4x^2 + 6x - 9 \\
\underline{-x^2 + 2x - 8} \\
3x^2 + 8x - 17
\end{array}
$$

EXAMPLE 4 Perform the indicated operation:

a) $(x^2 - 5x - 3) - (4x^2 + 8x - 9)$

b) $(4y^3 - 3y + 2) - (5y^2 - 7y - 6)$

Point out that each of these
equations is an identity.

Solution

a) $(x^2 - 5x - 3) - (4x^2 + 8x - 9) = x^2 - 5x - 3 - 4x^2 - 8x + 9$

Change signs.

$$= -3x^2 - 13x + 6 \qquad \textbf{Add.}$$

b) $(4y^3 - 3y + 2) - (5y^2 - 7y - 6) = 4y^3 - 3y + 2 - 5y^2 + 7y + 6$

Change signs.

$$= 4y^3 - 5y^2 + 4y + 8 \qquad \textbf{Add.} \qquad \blacktriangleleft$$

It is certainly not necessary to write out all of the steps shown in these examples. For addition the like terms should be observed and combined mentally. For subtraction, both changing the signs and adding can be done mentally.

EXAMPLE 5 Write the addition or subtraction vertically and perform the operation:

a) $(2x^2 - 3x - 7) + (x^3 - 5x^2 + 6)$

b) $(x^3 - 5x^2 - 6) - (x^4 - 5x^2 - 9)$

Solution

a) Add:

$$
\begin{array}{r}
2x^2 - 3x - 7 \\
x^3 - 5x^2 \phantom{{} - 3x} + 6 \\
\hline
x^3 - 3x^2 - 3x - 1
\end{array}
$$

Note that the like terms are lined up.

Math at Work

What should be your ideal weight if you want to be as healthy as possible? Finding out might not be as easy as you think.

Many of us step onto the bathroom scale to determine whether we are overweight. However, medical evidence shows that weight is a poor measure of general health because it ignores factors such as age, body composition, heredity, diet, and exercise. For example, healthy, active muscles weigh more than unconditioned ones; therefore an athlete may weigh more than an inactive person of the same size and be equally healthy or more so.

In this era of fitness awareness, many of us know that being greatly overweight is unhealthy. However, it is also unhealthy to be too thin. What weight is healthy for you? Medical researchers have shown that the ideal weight for every height may have a range of nearly 15 pounds and that a person's ideal weight may change as the person ages.

Researchers use a variety of methods for determining a person's ideal weight. One method, hydrostatic immersion testing, involves immersing the subject in a tank of water to determine the volume of the body. Pound for pound, muscle has less volume than fat. A less accurate but more practical method, called the "pinch test," involves measurements of skin folds at the arms, waist, and back.

Below are two formulas for determining ideal weight. The first takes into account the part age plays in weight. The second distinguishes between sex but not age. Use both formulas to see how your ideal weight varies with the method chosen.

1. Ideal weight $= (\text{age} + 100)\left(\dfrac{\text{height in inches}}{66}\right)^2$

2. Ideal weight for men $= 4\,(\text{height in inches}) - 128$
   Ideal weight for women $= 3.5\,(\text{height in inches}) - 108$

b) Subtract:

$$
\begin{array}{l}
x^3 - 5x^2 - 6 \\
\underline{x^4 - 5x^2 - 9} \qquad \textbf{Change signs.}
\end{array}
$$

Add:

$$
\begin{array}{l}
 \; x^3 - 5x^2 - 6 \\
\underline{-x^4 + 5x^2 + 9} \\
-x^4 + x^3 + 3
\end{array}
\qquad \blacktriangleleft
$$

Note that the results of adding or subtracting two polynomials are often expressed as equations. For example,

$$(3x - 2) + (5x - 6) = 8x - 8,$$

and

$$(3x - 2) - (5x - 6) = -2x + 4.$$

Since we have correctly performed these operations, these equations are satisfied by all real numbers. They are identities. If we make a mistake, then the resulting equation is not an identity. For example,

$$(3x - 2) - (5x - 6) = 2x + 4$$

is a conditional equation.

Warm-ups

True or false?

1. In the polynomial $2x^2 - 4x + 7$, the coefficient of x is 4. F
2. The degree of the polynomial $x^2 + 5x - 9x^3 + 6$ is 2. F
3. $(3x^2 - 8x + 6) + (x^2 + 4x - 9) = 4x^2 - 4x - 3$ for any value of x. T
4. $(x^2 - 4x) - (x^2 - 3x) = -7x$ for any value of x. F
5. In the polynomial $x^2 - x$, the coefficient of x is -1. T
6. The degree of the polynomial $x^2 - x$ is 2. T
7. A binomial always has a degree of 2. F
8. The polynomial $3x^2 - 5x + 9$ is a trinomial. T
9. Every trinomial has degree 2. F
10. $x^2 - 7x^2 = -6x^2$ for any value of x. T

3.1 EXERCISES

Determine the coefficient of x^2 in each polynomial. See Example 1.

1. $x^5 - x^3 + 3$ 0

2. $10 - x^2$ -1

3. $x^3 + 5x^5 - 5x^2$ -5

4. $\dfrac{x^3}{3} + \dfrac{7x^2}{2} - 4$ $\dfrac{7}{2}$

5. $\dfrac{x^3}{2} - \dfrac{x^2}{4} + 2x + 1$ $-\dfrac{1}{4}$

6. $3x^4 - 7x + 4$ 0

7. $x^2 - 9$ 1

8. $1 - 16x^2$ -16

Identify each polynomial as a monomial, binomial, or trinomial. Give the degree of each. See Example 2.

9. $4x + 7$ Binomial, 1

10. -1 Monomial, 0

11. $x^{10} - 3x^2 + 2$ Trinomial, 10

12. $y^6 - 6y^3 + 9$ Trinomial, 6

13. m^3 Monomial, 3

14. $a + 6$ Binomial, 1

15. $x^6 + 1$ Binomial, 6

16. $3a^8$ Monomial, 8

17. 5   Monomial, 0

18. $a^3 - a^2 + 5$ Trinomial, 3

19. $-x^2 + 4x - 9$ Trinomial, 2

20. $b^2 - 4$ Binomial, 2

Perform the indicated operations. See Example 3.

21. $(x - 3) + (3x - 5)$ $4x - 8$

22. $(4x - 1) + (x^3 + 5x - 6)$ $x^3 + 9x - 7$

23. $(3x + 2) + (x^2 - 4)$ $x^2 + 3x - 2$

24. $(3x - 7) + (x^2 - 4x + 6)$ $x^2 - x - 1$

25. $(x - 3) + (x + 3)$ $2x$

26. $(x - 2) + (x + 3)$ $2x + 1$

27. $(5x^2 - 2) + (-3x^2 - 1)$ $2x^2 - 3$

28. $(w + 4) + (w + 6)$ $2w + 10$

29. $(a^2 - 3a + 1) + (2a^2 - 4a - 5)$ $3a^2 - 7a - 4$

30. $(w^2 - 2w + 1) + (2w - 5 + w^2)$ $2w^2 - 4$

31. $(w^2 - 9w - 3) + (w - 4w^2 + 8)$ $-3w^2 - 8w + 5$

32. $(a^3 - 5a) + (6 - a - 3a^2)$ $a^3 - 3a^2 - 6a + 6$

33. $(5.76x^2 - 3.14x - 7.09) + (3.9x^2 + 1.21x + 5.6)$ $9.66x^2 - 1.93x - 1.49$

34. $(8.5x^2 + 3.27x - 9.33) + (x^2 - 4.39x - 2.32)$ $9.5x^2 - 1.12x - 11.65$

Perform the indicated operations. See Example 4.

35. $(2x^2 - 3x) - (3x^2 - 5x)$ $-x^2 + 2x$

36. $(x^2 - 4x) - (5x^2 - 3x)$ $-4x^2 - x$

37. $(x^2 - 3x + 4) - (x^2 - 5x - 9)$ $2x + 13$

38. $(x^2 - 6x + 7) - (5x^2 - 3x - 2)$ $-4x^2 - 3x + 9$

39. $(x - 2) - (5x - 8)$ $-4x + 6$

40. $(x - 7) - (3x - 1)$ $-2x - 6$

41. $(x^5 - x^3) - (x^4 - x^2)$ $x^5 - x^4 - x^3 + x^2$

42. $(x^6 - x^3) - (x^2 - x)$ $x^6 - x^3 - x^2 + x$

43. $(9 - 3x + x^2) - (2 - 5x - x^2)$ $2x^2 + 2x + 7$

44. $(4 - 5x + x^3) - (2 - 3x + x^2)$ $x^3 - x^2 - 2x + 2$

45. $(x - 2) - (x - 3)$ 1

46. $(x + 5) - (x + 9)$ -4

47. $(3.55x - 879) - (26.4x - 455.8)$ $-22.85x - 423.2$

48. $(345.56x - 347.4) - (56.6x + 433)$ $288.96x - 780.4$

Add or subtract the polynomials as indicated. See Example 5.

49. Add: $4a + 2$
  $3a - 4$
  $\underline{a + 6}$

50. Add: $3w - 5$
  $2w - 8$
  $\underline{w + 3}$

51. Subtract: $-2x + 4$
  $3x + 11$
  $\underline{5x + 7}$

52. Subtract: $2x - 6$
  $4x + 3$
  $\underline{2x + 9}$

53. Add: $7x^2 - 4x - 14$
  $x^3 + 4x^2 - 6x - 5$
  $\underline{-x^3 + 3x^2 + 2x - 9}$

54. Add: $-2x^2 - 11x + 14$
  $x^2 - 4x + 9$
  $\underline{-3x^2 - 7x + 5}$

55. Subtract: $x + 7$
  $3x + 1$
  $\underline{2x - 6}$

56. Subtract: $2x + 6$
  $5x + 2$
  $\underline{3x - 4}$

57. Add: $3x + 1$
$2x - 3$
$x + 4$

58. Add: $2x^2 + 3x - 11$
$-x^2 + 4x - 6$
$3x^2 - x - 5$

59. Subtract: $a^3 - 9a^2 + 2a + 7$
$3a^3 - 5a^2 + 7$
$2a^3 + 4a^2 - 2a$

60. Subtract: $-3x^2 + 11x - 7$
$-2x^2 + 7x - 9$
$x^2 - 4x - 2$

61. Subtract: $-3x + 9$
$x^2 - 3x + 6$
$x^2 - 3$

62. Subtract: $-2x^4 - x^2 + 2$
$x^4 - 3x^2 + 2$
$3x^4 - 2x^2$

63. Add: $2a$
$a - b$
$a + b$

64. Add: $s + 5$
$-s + 6$
$2s - 1$

65. Subtract: $-2b$
$a - b$
$a + b$

66. Subtract: 4
$t - 3$
$t - 7$

67. Add: $2p + 2q$
$p + q$
$p + q$

68. Add: $2a - 2b$
$a - b$
$a - b$

Find the polynomial that answers each of the following.

69. If Jessica traveled $2x + 50$ miles in the morning and $3x - 10$ miles in the afternoon, then what was the total distance she traveled?   5x + 40 miles

70. If the width of a tennis court is $2x - 5$ feet and the length is $3x + 9$ feet, then what is the perimeter?   10x + 8 feet

Figure for Exercise 70

71. If the shortest side of a triangle is x meters and the other two sides are $3x - 1$ and $2x + 4$ meters, then what is the perimeter?   6x + 3 meters

72. A red ball and a green ball are simultaneously tossed into the air. The red ball is given an initial velocity of 48 feet per second, and its altitude at any time t is $-16t^2 + 48t$ feet. The green ball is given an initial velocity of 30 feet per second, and its altitude at any time t is $-16t^2 + 30t$ feet. How much higher is the red ball at any time t?   18t feet

73. Ace Manufacturing has determined that the cost of labor for producing x constant velocity joints is $300x^2 + 400x - 550$ dollars, while the cost of materials is $100x^2 - 50x + 800$ dollars. What is the total cost of materials and labor for producing x constant velocity joints?   400x² + 350x + 250

74. Walter Waterman, of Walter's Water Pumps, has found that when he produces x water pumps per month, his revenue is $-x^2 + 400x + 300$ dollars. His cost for producing x water pumps per month is $x^2 + 300x - 100$ dollars. What is his monthly profit for x water pumps?   −2x² + 100x + 400

75. Donald received $.08(x + 554)$ dollars interest on one investment and $.09(x + 335)$ interest on another investment. What was the total interest he received?   .17x + 74.47

76. Deborah figured that the amount of acid in one bottle of solution was $.12x$ milliliters and the amount of acid in another bottle of solution was $.22(75 - x)$ milliliters. What was the total amount of acid?   ₋x + 16.5 milliliters

---

3.2   Multiplication of Polynomials

IN THIS SECTION:

- **Multiplying Monomials**
- **Multiplying Polynomials**

We learned to multiply some polynomials in Chapter 1. In this section we will learn how to multiply any two polynomials.

Multiplying Monomials

When we add, subtract, or multiply polynomials, the results are polynomials. When we add, subtract, or multiply integers, the results are integers.

To multiply two monomials such as x^3 and x^5, we must remember that

$$x^3 = x \cdot x \cdot x \qquad \text{and} \qquad x^5 = x \cdot x \cdot x \cdot x \cdot x.$$

Thus

$$x^3 \cdot x^5 = \underbrace{\overbrace{(x \cdot x \cdot x)}^{3 \text{ factors}} \cdot \overbrace{(x \cdot x \cdot x \cdot x \cdot x)}^{5 \text{ factors}}}_{8 \text{ factors of } x} = x^8.$$

The exponent of the product of x^3 and x^5 is the sum of the exponents 3 and 5. As another example, consider the product

$$x^7 \cdot x^9.$$

In this product there are seven factors of x followed by nine more factors of x, for a total of 16 factors of x. Thus

$$x^7 \cdot x^9 = x^{16}.$$

These examples illustrate the **product rule** for multiplying exponential expressions.

Product Rule

> If x is any real number, and m and n are any positive integers, then
>
> $$x^m \cdot x^n = x^{m+n}.$$

EXAMPLE 1 Find the indicated products:

a) $3x^2 \cdot 5x^3$ b) $(-2ab)(-3ab)$
c) $(-4x^2y^2)(3xy^5)$ d) $(3a)^2$

Solution
a) $3x^2 \cdot 5x^3 = 3 \cdot 5 \cdot x^2 \cdot x^3$
   $\qquad\qquad = 15x^5$ Apply the product rule.
b) $(-2ab)(-3ab) = (-2)(-3) \cdot a \cdot a \cdot b \cdot b$
   $\qquad\qquad\qquad = 6a^2b^2$ Apply the product rule.
c) $(-4x^2y^2)(3xy^5) = (-4)(3)x^2 \cdot x \cdot y^2 \cdot y^5$
   $\qquad\qquad\qquad\quad = -12x^3y^7$ Apply the product rule.
d) $(3a)^2 = 3a \cdot 3a = 9a^2$ ◄

Multiplying Polynomials

To multiply a monomial and a polynomial, we can apply the distributive property.

EXAMPLE 2 Find each product:

a) $3x^2(x^3 - 4x)$ **b)** $(y^2 - 3y + 4)(-2y)$ **c)** $-a(b - c)$

Solution

a) $3x^2(x^3 - 4x) = 3x^2(x^3) - 3x^2(4x)$ **Distributive property**
$\qquad\qquad\qquad = 3x^5 - 12x^3$ **Product rule**

b) $(y^2 - 3y + 4)(-2y) = y^2(-2y) - 3y(-2y) + 4(-2y)$ **Distributive property**
$\qquad\qquad\qquad\qquad = -2y^3 - (-6y^2) + (-8y)$
$\qquad\qquad\qquad\qquad = -2y^3 + 6y^2 - 8y$

c) $-a(b - c) = (-a)b - (-a)c$ **Distributive property**
$\qquad\qquad = -ab + ac$
$\qquad\qquad = ac - ab$

Note that either of the last two binomials is the correct answer. The last one is just a little simpler to write.   ◀

In the next example we apply the distributive property to find the product of two binomials and the product of a binomial and a trinomial.

EXAMPLE 3 Use the distributive property to find each product:

a) $(x + 2)(x + 5)$ **b)** $(x + 3)(x^2 + 2x - 7)$

Solution

a) First we multiply $x + 2$ by each term of $x + 5$:

$\underline{(x + 2)}(x + 5) = \underline{(x + 2)}x + \underline{(x + 2)}5$ **Apply the distributive property.**
$\qquad\qquad\quad = x^2 + 2x + 5x + 10$ **Apply the distributive property again.**
$\qquad\qquad\quad = x^2 + 7x + 10$ **Combine like terms.**

b) First multiply $x + 3$ and each term of the trinomial:

$\underline{(x + 3)}(x^2 + 2x - 7) = \underline{(x + 3)}x^2 + \underline{(x + 3)}2x + \underline{(x + 3)}(-7)$ **Apply the distributive property.**

$\qquad\qquad\qquad\quad = x^3 + 3x^2 + 2x^2 + 6x - 7x - 21$ **Apply the distributive property again.**

$\qquad\qquad\qquad\quad = x^3 + 5x^2 - x - 21$ **Combine like terms.**   ◀

Note that in Example 3, every term of the first polynomial was multiplied by every term of the second polynomial. This can also be accomplished by arranging the multiplication like multiplication of whole numbers.

EXAMPLE 4 Multiply the polynomials:

a) $(x - 2)(3x + 7)$ **b)** $(x + y)(a + 3)$

It is important for students to multiply polynomials in this manner, before they learn shortcuts.

Solution

a)
$$3x + 7$$
$$x - 2$$
$$-6x - 14 \longleftarrow -2 \text{ times } 3x + 7$$
$$3x^2 + 7x \longleftarrow x \text{ times } 3x + 7$$
$$3x^2 + x - 14$$

b)
$$x + y$$
$$a + 3$$
$$3x + 3y$$
$$ax + ay$$
$$ax + ay + 3x + 3y \quad \blacktriangleleft$$

These examples illustrate the following rule.

RULE
Multiplication of Polynomials

To multiply polynomials, multiply each term of the first polynomial by every term of the second polynomial, then combine like terms.

Note the result of multiplying the difference $a - b$ by -1:

$$-1(a - b) = -a + b = b - a$$

Since multiplying by -1 is the same as taking the opposite, we can write this equation as

$$-(a - b) = b - a.$$

This equation says that $a - b$ and $b - a$ are opposites or additive inverses of each other. Note that the opposite of $a + b$ is $-a - b$, *not* $a - b$.

EXAMPLE 5 Find the opposite of each binomial:

a) $x - 2$
b) $9 - y^2$
c) $a + 4$
d) $-x - 3$

Solution

a) $2 - x$
b) $y^2 - 9$
c) $-a - 4$
d) $x + 3$ $\blacktriangleleft$

Warm-ups

True or false?
1. $3x^3 \cdot 5x^4 = 15x^{12}$ for any value of x.   F
2. $3x^2 \cdot 2x^7 = 5x^9$ for any value of x.   F
3. $(3y^3)^2 = 9y^6$ for any value of y.   T
4. $-3x(5x - 7x^2) = -15x^3 + 21x^2$ for any value of x.   F
5. $2x(x^2 - 3x + 4) = 2x^3 - 6x^2 + 8x$ for any number x.   T

> **6.** $-2(3 - x) = 2x - 6$ for any value of x. T
> **7.** $(a + b)(c + d) = ac + ad + bc + bd$ for any values of a, b, c, and d. T
> **8.** $-(x - 7) = 7 - x$ for any value of x. T
> **9.** $83 - 37 = -(37 - 83)$ T
> **10.** The opposite of $x + 3$ is $x - 3$ for any number x. F

3.2 EXERCISES

Find the products. See Example 1.

1. $3x^2 \cdot 9x^3$ $27x^5$ **2.** $-6x^2 \cdot 5x^2$ $-30x^4$ **3.** $-2x^2 \cdot 8x^5$ $-16x^7$ **4.** $5x^7 \cdot 3x^5$ $15x^{12}$
5. $(-9x^{10})(-3x^7)$ $27x^{17}$ **6.** $(-2x^2)(-8x^9)$ $16x^{11}$ **7.** $2a^3 \cdot 7a^8$ $14a^{11}$ **8.** $3y^{12} \cdot 5y^{15}$ $15y^{27}$
9. $3wt \cdot 8w^7t^6$ $24t^7w^8$ **10.** $-6st \cdot 9st$ $-54s^2t^2$ **11.** $-12sq \cdot 3s$ $-36qs^2$ **12.** $h^8k^3 \cdot 5h$ $5h^9k^3$
13. $(5y)^2$ $25y^2$ **14.** $(6x)^2$ $36x^2$ **15.** $(2x^3)^2$ $4x^6$ **16.** $(3y^2)^2$ $9y^4$

Find the products. See Example 2.

17. $7x^2(x^3 - 5x^2 - 1)$ $7x^5 - 35x^4 - 7x^2$ **18.** $-3y(6y - 4)$ $-18y^2 + 12y$ **19.** $ab(a^2 - b^2)$ $a^3b - ab^3$
20. $(3ab^3 - a^2b^2 - 2a^3b)(5a^3)$ **21.** $(y^2 - 5y + 6)(-3)$ $-3y^2 + 15y - 18$ **22.** $-9y(y^2 - 1)$ $-9y^3 + 9y$
23. $-x(y^2 - x^2)$ $-xy^2 + x^3$ **24.** $c(c - d + 1)$ $c^2 - cd + c$
20. $15a^4b^3 - 5a^5b^2 - 10a^6b$

Use the distributive property to find each product. See Example 3.

25. $(x + 1)(x + 2)$ $x^2 + 3x + 2$ **26.** $(x + 6)(x + 3)$ $x^2 + 9x + 18$
27. $(x - 3)(x + 5)$ $x^2 + 2x - 15$ **28.** $(y - 2)(y + 4)$ $y^2 + 2y - 8$
29. $(x + 1)(x^2 + 2x + 2)$ $x^3 + 3x^2 + 4x + 2$ **30.** $(x - 1)(x^2 + x + 1)$ $x^3 - 1$
31. $(y + 2)(2y^2 - y + 3)$ $2y^3 + 3y^2 + y + 6$ **32.** $(y + 3)(y^2 + 3y + 1)$ $y^3 + 6y^2 + 10y + 3$
33. $3y(y^2 + 3y + 1)$ $3y^3 + 9y^2 + 3y$ **34.** $2x(x^2 - x + 3)$ $2x^3 - 2x^2 + 6x$
35. $(t - 4)(t - 9)$ $t^2 - 13t + 36$ **36.** $(w - 3)(w - 5)$ $w^2 - 8w + 15$

Multiply the following polynomials. See Example 4.

37. $2a - 3$
$\underline{a + 5}$ $2a^2 + 7a - 15$

38. $2w - 6$
$\underline{w + 5}$ $2w^2 + 4w - 30$

39. $7x + 30$
$\underline{2x + 5}$ $14x^2 + 95x + 150$

40. $5x + 7$
$\underline{3x + 6}$ $15x^2 + 51x + 42$

41. $x^3 + 3x^2 - 5x - 2$
$\underline{\quad\quad 3x}$

42. $x^2 - 3x + 7$
$\underline{\quad -5x + 2}$

43. $5x + 2$
$\underline{4x - 3}$ $20x^2 - 7x - 6$

44. $4x + 3$
$\underline{2x - 6}$ $8x^2 - 18x - 18$

45. $2x - 3$
$\underline{x - 4}$ $2x^2 - 11x + 12$

46. $-x^2 + 3x - 5$
$\underline{\quad -2x - 7}$ $2x^3 + x^2 - 11x + 35$

47. $2a^3 - 3a^2 + 4$
$\underline{\quad\quad -2a}$ $-4a^4 + 6a^3 - 8a$

48. $-3x^2 + 5x - 2$
$\underline{\quad -5x - 6}$

41. $3x^4 + 9x^3 - 15x^2 - 6x$ **42.** $-5x^3 + 17x^2 - 41x + 14$ **48.** $15x^3 - 7x^2 - 20x + 12$

49. $x^2 - 2x + 5$
$\underline{x^2 - 3}$
$x^4 - 2x^3 + 2x^2 + 6x - 15$

50. $x^4 - 5x^2 + 6$
$\underline{2x^2}$
$2x^6 - 10x^4 + 12x^2$

51. $x - y$
$\underline{x + y}$ $x^2 - y^2$

52. $-w + 4$
$\underline{2w - 3}$ $-2w^2 + 11w - 12$

Match each polynomial with its opposite from the following list **(a)** *through* **(j)**. *See Example 5.*

a) $-3t^2 + t + 6$ **b)** $-3t - u$ **c)** $3t^2 + t - 6$ **d)** $-3tu$ **e)** $3t^2 - t + 6$

f) $u - 3t$ **g)** $-3t^2 - t - 6$ **h)** $u + 3t$ **i)** $3t - u$ **j)** $-3t^2 - t + 6$

53. $3t - u$ f   **54.** $-3t - u$ h   **55.** $3t + u$ b   **56.** $u - 3t$ i

57. $-3t^2 - t + 6$ c   **58.** $3t^2 - t - 6$ a   **59.** $3t^2 + t - 6$ j   **60.** $-3t^2 + t - 6$ e

Perform the indicated operations.

61. $-3x(2x - 9)$ $-6x^2 + 27x$ **62.** $(2 - 3x)(2x - 9)$ $-6x^2 + 31x - 18$ **63.** $(2 - 3x) + (2x - 9)$ $-x - 7$

64. $(2 - 3x) - (2x - 9)$ $-5x + 11$ **65.** $-1(2 - 3x)$ $3x - 2$ **66.** $2 - 3x(2x - 9)$ $-6x^2 + 27x + 2$

67. $2x^2(3x^5 - 4x^2)$ $6x^7 - 8x^4$ **68.** $(x + 6)(x + 6)$ $x^2 + 12x + 36$ **69.** $(x - 6)(x - 6)$ $x^2 - 12x + 36$

70. $(2x - 9)(2x - 9)$ $4x^2 - 36x + 81$ **71.** $(2x - 9)(2x + 9)$ $4x^2 - 81$ **72.** $(x - 6)(x + 6)$ $x^2 - 36$

73. $3ab^3 \cdot (-2a^2b^7)$ $-6a^3b^{10}$ **74.** $(6x^6)^2$ $36x^{12}$ **75.** $4xst \cdot 8xs$ $32s^2tx^2$

76. $3ab^3 - 2ab^3$ ab^3 **77.** $(-3a^3b)^2$ $9a^6b^2$ **78.** $(m - 1)(m^2 + m + 1)$ $m^3 - 1$

Find the polynomial that answers each of the following.

79. If the length of a professor's office is x feet and the width is $x + 4$ feet, then what is the area? $x^2 + 4x$ square feet

80. If the length of a rectangular swimming pool is $2x - 3$ meters and the width is $x + 5$ meters, then what is the area? $2x^2 + 7x - 15$

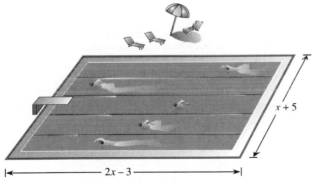

$x + 5$

$2x - 3$

Figure for Exercise 80

81. If a roof truss is in the shape of a triangle with a base of x feet and a height of $2x + 1$ feet, then what is the area of the triangle? $x^2 + \dfrac{1}{2}x$

82. If the length, width, and height of a box are x, $2x$, and $3x - 5$ inches, respectively, then what is its volume?

83. If two numbers differ by 5, then what is their product? $x^2 + 5x$

82. $6x^3 - 10x^2$ cubic inches

84. If two numbers have a sum of 9, then what is their product? $9x - x^2$

85. If the length of a rectangle is $2.3x + 1.2$ meters and its width is $3.5x + 5.1$ meters, then what is its area?

86. If a quilt patch cut in the shape of a triangle has a base of $5x$ inches and a height of $1.732x$ inches, then what is its area? $4.33x^2$ square inches

85. $8.05x^2 + 15.93x + 6.12$ square meters

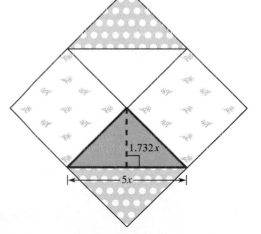

$1.732x$

$5x$

Figure for Exercise 86

3.3 Multiplication of Binomials

IN THIS SECTION:

- The FOIL Method
- Multiplying Binomials Mentally

In Section 3.2 we learned to multiply polynomials. In this section we will learn a rule to make multiplication of binomials simpler.

The FOIL Method

We can use the distributive property to find the product of two binomials. For example,

$$(x + 2)(x + 3) = (x + 2)x + (x + 2)3 \qquad \text{The distributive property}$$
$$= x^2 + 2x + 3x + 6 \qquad \text{The distributive property again}$$
$$= x^2 + 5x + 6. \qquad \text{Combine like terms.}$$

There are four terms in the product. The term x^2 is the product of the *first* term of each binomial, x and x. The term $3x$ is the product of the two *outer* terms, 3 and x. The term $2x$ is the product of the two *inner* terms, 2 and x. The term 6 is the product of the *last* term of each binomial, 2 and 3. It may be helpful to connect the terms multiplied by lines as follows:

$$(x + 2)(x + 3)$$

F = First terms
O = Outer terms
I = Inner terms
L = Last terms

This is simply a way of organizing the multiplication of binomials. It is called the **FOIL method.** The name should make it easier to remember.

Now we apply FOIL to the product $(x + 5)(x - 3)$:

$$(x + 5)(x - 3) = x^2 - 3x + 5x - 15 = x^2 + 2x - 15$$

Remember that the equation $(x + 5)(x - 3) = x^2 + 2x - 15$ is true for all values of x.

EXAMPLE 1 Use FOIL to find the products:

a) $(x + 2)(x - 4)$ b) $(2x + 5)(3x - 4)$
c) $(a - b)(2a - b)$ d) $(x + 3)(y + 5)$

Solution

Remind students that all of these equations are identities.

                    F   O   I   L

a) $(x + 2)(x - 4) = x^2 - 4x + 2x - 8$
                $= x^2 - 2x - 8$ **Add the like terms.**

b) $(2x + 5)(3x - 4) = 6x^2 - 8x + 15x - 20$
                   $= 6x^2 + 7x - 20$

c) $(a - b)(2a - b) = 2a^2 - ab - 2ab + b^2$
                 $= 2a^2 - 3ab + b^2$

d) $(x + 3)(y + 5) = xy + 5x + 3y + 15$ **There are no like terms to combine.** ◄

FOIL can be used to multiply any two binomials. The binomials in the next example have higher powers than those of Example 1.

EXAMPLE 2 Use FOIL to find each product:

a) $(x^3 - 3)(x^3 + 6)$ b) $(2a^2 + 1)(a^2 + 5)$

Solution

a) $(x^3 - 3)(x^3 + 6) = x^6 + 6x^3 - 3x^3 - 18$
                 $= x^6 + 3x^3 - 18$

b) $(2a^2 + 1)(a^2 + 5) = 2a^4 + 10a^2 + a^2 + 5$
                   $= 2a^4 + 11a^2 + 5$ ◄

Multiplying Binomials Mentally

The outer and inner products in the FOIL method are often like terms, and we can learn to add them mentally. We can then find the product of two binomials without writing anything except the answer.

EXAMPLE 3 Use FOIL to find each product mentally:

a) $(x + 3)(x + 4)$ b) $(2x - 1)(x + 5)$ c) $(a - 6)(a + 6)$

Solution

a) $(x + 3)(x + 4) = x^2 + 7x + 12$ **Combine the like terms mentally: $3x + 4x = 7x$.**

b) $(2x - 1)(x + 5) = 2x^2 + 9x - 5$ **Combine the like terms mentally: $10x - x = 9x$.**

c) $(a - 6)(a + 6) = a^2 - 36$ *Note: $6a - 6a = 0$.* ◄

Warm-ups

True or false? Answer true only if the equation is true for all values of the variable or variables.

1. $(x + 3)(x + 2) = x^2 + 6$ F
2. $(x + 2)(y + 1) = xy + x + 2y + 2$ T
3. $(3a - 5)(2a + 1) = 6a^2 + 3a - 10a - 5$ T
4. $(y + 3)(y - 2) = y^2 + y - 6$ T
5. $(x^2 + 2)(x^2 + 3) = x^4 + 5x^2 + 6$ T
6. $(3a^2 - 2)(3a^2 + 2) = 9a^2 - 4$ F
7. $(t + 3)(t + 5) = t^2 + 8t + 15$ T
8. $(y - 9)(y - 2) = y^2 - 11y - 18$ F
9. $(x + 4)(x - 7) = x^2 + 4x - 28$ F
10. It is not necessary to learn FOIL as long as you can get the answer.   F

3.3 EXERCISES

Use FOIL *to find each product. See Example* 1.

1. $(x + 2)(x + 4)$ $x^2 + 6x + 8$
2. $(x + 3)(x + 5)$ $x^2 + 8x + 15$
3. $(a - 3)(a + 2)$ $a^2 - a - 6$
4. $(b - 1)(b + 2)$ $b^2 + b - 2$
5. $(2x - 1)(x + 2)$ $2x^2 + 3x - 2$
6. $(2y - 5)(y + 2)$ $2y^2 - y - 10$
7. $(2a - 3)(a + 1)$ $2a^2 - a - 3$
8. $(3x - 5)(x + 4)$ $3x^2 + 7x - 20$
9. $(w + 5)(w - 1)$ $w^2 + 4w - 5$
10. $(w + 3)(w - 2)$ $w^2 + w - 6$
11. $(2m - 3)(5m + 3)$ $10m^2 - 9m - 9$
12. $(2x - 5)(x + 1)$ $2x^2 - 3x - 5$
13. $(a + 2)(a - 7)$ $a^2 - 5a - 14$
14. $(x - 3)(x + 4)$ $x^2 + x - 12$
15. $(y - a)(y + 5)$ $y^2 - ay + 5y - 5a$
16. $(a + t)(3 - y)$ $3a + 3t - ay - ty$
17. $(5 - w)(w + m)$ $5w - w^2 + 5m - mw$
18. $(a - h)(b + t)$ $ab - hb + at - ht$

Use FOIL *to find each product. See Example* 2.

19. $(x^2 - 5)(x^2 - 2)$ $x^4 - 7x^2 + 10$
20. $(y^2 + 1)(y^2 - 2)$ $y^4 - y^2 - 2$
21. $(3b^3 + 2)(b^3 + 4)$ $3b^6 + 14b^3 + 8$
22. $(h^3 - 5)(h^3 + 5)$ $h^6 - 25$
23. $(5n^4 - 1)(n^4 + 3)$ $5n^8 + 14n^4 - 3$
24. $(y^6 + 1)(y^6 - 4)$ $y^{12} - 3y^6 - 4$
25. $(y^2 - 3)(y + 2)$ $y^3 + 2y^2 - 3y - 6$
26. $(x + 1)(x^2 - 1)$ $x^3 + x^2 - x - 1$

Find each product mentally by using FOIL. *Write down only the answer. See Example* 3.

27. $(b + 4)(b + 5)$ $b^2 + 9b + 20$
28. $(y + 8)(y + 4)$ $y^2 + 12y + 32$
29. $(x - 3)(x + 9)$ $x^2 + 6x - 27$
30. $(m + 7)(m - 8)$ $m^2 - m - 56$
31. $(a + 5)(a + 5)$ $a^2 + 10a + 25$
32. $(t - 4)(t - 4)$ $t^2 - 8t + 16$
33. $(2x - 1)(2x - 1)$ $4x^2 - 4x + 1$
34. $(3y + 4)(3y + 4)$ $9y^2 + 24y + 16$
35. $(z - 10)(z + 10)$ $z^2 - 100$
36. $(3h - 5)(3h + 5)$ $9h^2 - 25$
37. $(a + b)(a + b)$ $a^2 + 2ab + b^2$
38. $(x - y)(x - y)$ $x^2 - 2xy + y^2$
39. $(a - 1)(a - 2)$ $a^2 - 3a + 2$
40. $(b - 8)(b - 1)$ $b^2 - 9b + 8$
41. $(2x - 1)(x + 3)$ $2x^2 + 5x - 3$
42. $(3y + 5)(y - 3)$ $3y^2 - 4y - 15$
43. $(5t - 2)(t - 1)$ $5t^2 - 7t + 2$
44. $(2t - 3)(2t - 1)$ $4t^2 - 8t + 3$
45. $(h - 7)(h - 9)$ $h^2 - 16h + 63$
46. $(h - 7)(h - 7)$ $h^2 - 14h + 49$
47. $(h + 7)(h + 7)$ $h^2 + 14h + 49$
48. $(h - 7)(h + 7)$ $h^2 - 49$
49. $(2h - 1)(2h - 1)$ $4h^2 - 4h + 1$
50. $(3h + 1)(3h + 1)$ $9h^2 + 6h + 1$

Find a polynomial that answers each of the following.

51. What is the area of a rectangular rug whose sides are $x + 3$ feet and $2x - 1$ feet? $2x^2 + 5x - 3$ square feet

52. What is the area of a lot that is in the shape of a parallelogram whose base is $3x + 2$ meters and whose height is $2x + 3$ meters? $6x^2 + 13x + 6$ square meters

53. If the sail of a tall ship is a triangle with a base of $4.57x + 3$ meters and a height of $2.3x - 1.33$ meters, then what is the area of the triangle? $5.2555x^2 + .41095x - 1.995$ square meters

54. If a square has a side of length $1.732x + 1.414$ meters, then what is its area? $2.9998x^2 + 4.898x + 1.999$ square meters

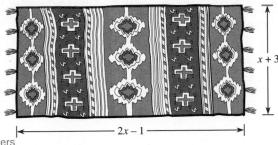

$x + 3$

$2x - 1$

Figure for Exercise 51

3.4 Special Products

IN THIS SECTION:

- **The Square of a Binomial**
- **Product of a Sum and a Difference**

In Section 3.3 we learned the FOIL method to make multiplying binomials simpler. In this section we will learn three rules that are even simpler than the FOIL method for certain special products.

The Square of a Binomial

To compute $(a + b)^2$, the square of a binomial, we can write it as $(a + b)(a + b)$ and use FOIL:

$$(a + b)(a + b) = a^2 + ab + ab + b^2 = a^2 + 2ab + b^2$$

The square of a binomial occurs so frequently that it is better to learn a new rule to find it. There are two rules that make squaring a binomial easier. To square a sum, we use the following rule.

Note that the square of $a + b$ is the area of the square shown here:

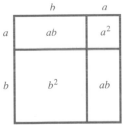

From this diagram it is clear that $(a + b)^2 = a^2 + b^2$ is not true for all values of a and b.

RULE
The Square of a Sum

$$(a + b)^2 = a^2 + 2ab + b^2$$

In words: To square $a + b$, we square the first term (a^2), add twice the product of the two terms ($2ab$), and then add the square of the last term (b^2).

Ask students to figure out how to mentally calculate the square of a number that ends in 5:
$(15)^2 = 225,$
$(25)^2 = 625,$
$(35)^2 = 1225,$
$(45)^2 = 2025,$
$(55)^2 = 3025,$
etc.
The reason this trick works is found in the square of a binomial:
$(30 + 5)^2 = 30^2 + 2 \cdot 30 \cdot 5 + 25$
$= 30(40) + 25$
$= 1200 + 25$
$= 1225$
$(40 + 5)^2 = 40 \cdot 50 + 25$
$= 2025$

EXAMPLE 1 Square each sum, using the new rule:

a) $(x + 3)^2$

b) $(2a + 5)^2$

Solution

a) $(x + 3)^2 = \underset{\substack{\uparrow \\ \text{Square} \\ \text{of} \\ \text{first}}}{x^2} + \underset{\substack{\uparrow \\ \text{Twice} \\ \text{the} \\ \text{product}}}{2(x)(3)} + \underset{\substack{\uparrow \\ \text{Square} \\ \text{of} \\ \text{last}}}{3^2} = x^2 + 6x + 9$

b) $(2a + 5)^2 = (2a)^2 + 2(2a)(5) + 5^2$
$= 4a^2 + 20a + 25$ ◄

Note that the equation $(x + 3)^2 = x^2 + 6x + 9$ is an identity. This equation is true for any value of x. The equation $(x + 3)^2 = x^2 + 9$ is *not* true for all real numbers. For example, if $x = 1$, we get the false equation

$$(1 + 3)^2 = 1^2 + 9.$$

When we use FOIL to compute $(a - b)^2$, we see that

$$(a - b)(a - b) = a^2 - ab - ab + b^2 = a^2 - 2ab + b^2.$$

We can use instead the following rule for squaring a difference.

RULE
The Square of a Difference

$$(a - b)^2 = a^2 - 2ab + b^2$$

In words: Square the first term, subtract twice the product of the two terms, and add the square of the last term.

EXAMPLE 2 Square each difference, using the new rule:

a) $(x - 4)^2$

b) $(4b - 5y)^2$

Solution

a) $(x - 4)^2 = x^2 - 2(x)(4) + 4^2$
$= x^2 - 8x + 16$

b) $(4b - 5y)^2 = (4b)^2 - 2(4b)(5y) + (5y)^2$
$= 16b^2 - 40by + 25y^2$ ◄

Product of a Sum and a Difference

If we multiply the sum $a + b$ and the difference $a - b$ by using FOIL, we get

$$(a + b)(a - b) = a^2 - ab + ab - b^2$$
$$= a^2 - b^2.$$

Ask students to mentally find products such as
$49 \cdot 51 = 2499.$

The inner and outer products add up to 0, cancelling each other out. So *the product of a sum and a difference of the same two terms is equal to the difference of two squares*.

RULE
The Product of a Sum and a Difference

$$(a + b)(a - b) = a^2 - b^2$$

Ask students to figure out how to mentally find products such as
$24 \cdot 26 = 624$,
$34 \cdot 36 = 1224$,
$44 \cdot 46 = 2024$,
etc.

EXAMPLE 3 Find the products by using the rule for the product of a sum and a difference:

a) $(x + 2)(x - 2)$ b) $(b + 7)(b - 7)$ c) $(3x - 5)(3x + 5)$

Solution
a) $(x + 2)(x - 2) = x^2 - 4$
b) $(b + 7)(b - 7) = b^2 - 49$
c) $(3x - 5)(3x + 5) = 9x^2 - 25$ ◄

Warm-ups

True or false?

1. $(2 + 3)^2 = 2^2 + 3^2$ F
2. $(x + 3)^2 = x^2 + 6x + 9$ for any value of x. T
3. $(3 + 5)^2 = 9 + 25 + 30$ T
4. $(2x + 7)^2 = 4x^2 + 28x + 49$ for any value of x. T
5. $(y + 8)^2 = y^2 + 64$ for any value of y. F
6. The product of a sum and a difference of the same two terms is equal to the difference of two squares. T
7. $(40 - 1)(40 + 1) = 1599$ T
8. $49 \cdot 51 = 2499$ T
9. $(x - 3)^2 = x^2 - 3x + 9$ for any value of x. F
10. The square of a sum is equal to a sum of two squares. F

3.4 EXERCISES

Square each binomial mentally. Write down only the answer. See Example 1.

1. $(x + 1)^2$ $x^2 + 2x + 1$ 2. $(y + 2)^2$ $y^2 + 4y + 4$ 3. $(y + 4)^2$ $y^2 + 8y + 16$
4. $(z + 3)^2$ $z^2 + 6z + 9$ 5. $(x + 8)^2$ $x^2 + 16x + 64$ 6. $(m + 7)^2$ $m^2 + 14m + 49$

7. $(s + t)^2$ $s^2 + 2st + t^2$ **8.** $(x + 5)^2$ $x^2 + 10x + 25$ **9.** $(2x + .5)^2$ $4x^2 + 2x + .25$

10. $(t + 10)^2$ $t^2 + 20t + 100$ **11.** $(2t + 1)^2$ $4t^2 + 4t + 1$ **12.** $(3z + 5)^2$ $9z^2 + 30z + 25$

Square each binomial mentally. See Example 2.

13. $(a - 3)^2$ $a^2 - 6a + 9$ **14.** $(w - 4)^2$ $w^2 - 8w + 16$ **15.** $(2t - 1)^2$ $4t^2 - 4t + 1$

16. $(3t - 4)^2$ $9t^2 - 24t + 16$ **17.** $(t - 2)^2$ $t^2 - 4t + 4$ **18.** $(a - 6)^2$ $a^2 - 12a + 36$

19. $(s - t)^2$ $s^2 - 2st + t^2$ **20.** $(3r - 2)^2$ $9r^2 - 12r + 4$ **21.** $(a - .4)^2$ $a^2 - .8a + .16$

22. $(w - 7)^2$ $w^2 - 14w + 49$ **23.** $(3z - 5)^2$ $9z^2 - 30z + 25$ **24.** $(2z - 3)^2$ $4z^2 - 12z + 9$

Find the products mentally. See Example 3.

25. $(a - 5)(a + 5)$ $a^2 - 25$ **26.** $(x - 6)(x + 6)$ $x^2 - 36$ **27.** $(y - 1)(y + 1)$ $y^2 - 1$

28. $(p + 2)(p - 2)$ $p^2 - 4$ **29.** $(r + s)(r - s)$ $r^2 - s^2$ **30.** $(b - y)(b + y)$ $b^2 - y^2$

31. $(3x - 8)(3x + 8)$ $9x^2 - 64$ **32.** $(5x^2 - 2)(5x^2 + 2)$ $25x^4 - 4$ **33.** $(6x + 1)(6x - 1)$ $36x^2 - 1$

34. $(3y^2 + 1)(3y^2 - 1)$ $9y^4 - 1$ **35.** $(x^2 - 1)(x^2 + 1)$ $x^4 - 1$ **36.** $(y^3 - 1)(y^3 + 1)$ $y^6 - 1$

Find each of the following products.

37. $(x + 8)(x + 7)$ $x^2 + 15x + 56$ **38.** $(x - 9)(x + 5)$ $x^2 - 4x - 45$ **39.** $(t - 5)(t + 5)$ $t^2 - 25$

40. $(n + 1)^2$ $n^2 + 2n + 1$ **41.** $(y - 11)^2$ $y^2 - 22y + 121$ **42.** $(z - 1)(z + 1)$ $z^2 - 1$

43. $(a - 20)(a + 20)$ $a^2 - 400$ **44.** $(1 - x)(1 + x)$ $1 - x^2$ **45.** $(4x - 1)(4x + 1)$ $16x^2 - 1$

46. $(9y - 1)(9y + 1)$ $81y^2 - 1$ **47.** $(9y - 1)^2$ $81y^2 - 18y + 1$ **48.** $(4x - 1)^2$ $16x^2 - 8x + 1$

49. $(2t - 5)(3t + 4)$ $6t^2 - 7t - 20$ **50.** $(2t + 5)(3t - 4)$ $6t^2 + 7t - 20$ **51.** $(2t - 5)^2$ $4t^2 - 20t + 25$

52. $(2t + 5)^2$ $4t^2 + 20t + 25$ **53.** $(2t + 5)(2t - 5)$ $4t^2 - 25$ **54.** $(3t - 4)(3t + 4)$ $9t^2 - 16$

55. $(1.5x + 3.8)^2$ $2.25x^2 + 11.4x + 14.44$ **56.** $(3.45a - 2.3)^2$ $11.9025a^2 - 15.87a + 5.29$

57. $(3.5t - 2.5)(3.5t + 2.5)$ $12.25t^2 - 6.25$ **58.** $(4.5h + 5.7)(4.5h - 5.7)$ $20.25h^2 - 32.49$

Find a polynomial that answers each of the following.

59. Rose's garden is currently a square with sides of length x feet. Next spring, she plans to make it a rectangular shape by lengthening one side 5 feet and shortening the other side by 5 feet. What will the new area be? By how much will the area of the new garden differ from that of the old garden?

60. Sam lives on a lot that he thought was a square, 157 feet by 157 feet. When he had it surveyed, he discovered that one side was actually 2 feet longer than he thought and the other side was actually 2 feet shorter than he thought. How much less area does he have than he thought he had? 4 square feet

59. $x^2 - 25$ square feet, 25 square feet smaller

61. Find the area of a circle whose radius is $b + 1$ meters. Use the value 3.14 for π. $3.14b^2 + 6.28b + 3.14$ square meters

62. A toy store sells two sizes of dartboards, both in the shape of circles. The larger of the two has a radius that is 3 centimeters greater than the radius of the other. If the radius of the smaller is t centimeters, then how much more area does the larger dartboard have?

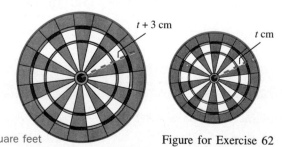

$t + 3$ cm t cm

Figure for Exercise 62

62. $6\pi t + 9\pi$ square centimeters

3.5   Division of Polynomials

IN THIS SECTION:

- Dividing Monomials
- Dividing a Polynomial by a Monomial
- Dividing a Polynomial by a Binomial

We learned to multiply polynomials in Section 3.2. In this section we will learn to divide polynomials.

Dividing Monomials

In Chapter 1 we used the definition of division to develop rules for dividing signed numbers. Since the definition of division is used for any division, we will restate it here.

DEFINITION
Division of Real Numbers

> If a, b, and c are any numbers with $b \neq 0$, then
>
> $$a \div b = c \qquad \text{provided that} \qquad c \cdot b = a.$$

We divided some monomials in Chapter 1 when we did problems such as $10x \div 2 = 5x$. To check that $5x$ is the quotient, we use the definition of division: $5x \cdot 2 = 10x$. We can use the same idea to divide more complicated monomials. For example,

$$x^{10} \div x^3 = x^7 \qquad \text{because} \qquad x^7 \cdot x^3 = x^{10},$$

and

$$x^4 \div x^4 = 1 \qquad \text{because} \qquad 1 \cdot x^4 = x^4.$$

Note that the exponent 7 is obtained by subtracting 10 and 3:

$$x^{10} \div x^3 = x^{10-3} = x^7$$

If we subtract the exponents in $x^4 \div x^4$, we get an expression that we have not yet defined:

$$x^4 \div x^4 = x^{4-4} = x^0$$

Since $x^4 \div x^4 = 1$ if $x \neq 0$, it makes sense to define x^0 to be 1 if $x \neq 0$. We do not define the expression 0^0.

DEFINITION
Zero Exponent

For any nonzero number a,

$$a^0 = 1.$$

With this definition of zero exponent, we can give the quotient rule for exponents.

RULE
Quotient Rule

If m and n are positive integers with $m \geq n$ and $a \neq 0$, then

$$\frac{a^m}{a^n} = a^{m-n}.$$

EXAMPLE 1   Find the quotients of the monomials:

a) $\dfrac{y^9}{y^5}$ b) $\dfrac{12b^4}{3b}$

c) $-6x^3 \div 2x^2$ d) $\dfrac{x^8y^2}{x^2y^2}$

Solution

a) $\dfrac{y^9}{y^5} = y^{9-5} = y^4$ *Check:* $y^4 \cdot y^5 = y^9$.

b) $\dfrac{12b^4}{3b} = \dfrac{12}{3} \cdot \dfrac{b^4}{b} = 4 \cdot b^3 = 4b^3$ *Check:* $4b^3 \cdot 3b = 12b^4$.

c) $-6x^3 \div 2x^2 = -3x$ *Check:* $-3x \cdot 2x^2 = -6x^3$.

d) $\dfrac{x^8y^2}{x^2y^2} = \dfrac{x^8}{x^2} \cdot \dfrac{y^2}{y^2} = x^6 \cdot y^0 = x^6$ *Check:* $x^6 \cdot x^2y^2 = x^8y^2$.

We have shown more steps in this example than is necessary. Division problems like these and the check should be done mentally.   ◀

Dividing a Polynomial by a Monomial

We divided a polynomial by a monomial in Chapter 1 when we did problems such as

$$\frac{2x + 6}{2} = x + 3.$$

The distributive property is used in checking: $2(x + 3) = 2x + 6$. Since the distributive property distributes the 2 over each term of $x + 3$, 2 is divided into each term of $2x + 6$ to get $x + 3$.

To divide $9x^3 + 3x^2 - 6x$ by the monomial $3x$, we divide each term of the trinomial by $3x$:

$$\frac{9x^3 + 3x^2 - 6x}{3x} = \frac{9x^3}{3x} + \frac{3x^2}{3x} - \frac{6x}{3x} = 3x^2 + x - 2$$

We call $3x$ the **divisor,** $9x^3 + 3x^2 - 6x$ the **dividend,** $3x^2 + x - 2$ the **quotient,** and 0 the **remainder.** The remainder is not always 0. For example, if we divide 25 by 7, we get

$$\text{Divisor} \longrightarrow 7\overline{)25} \begin{matrix} 3 & \longleftarrow & \text{Quotient} \\ & \longleftarrow & \text{Dividend} \\ \underline{21} & & \\ 4 & \longleftarrow & \text{Remainder} \end{matrix}$$

Note that $3 \cdot 7 + 4 = 25$. It is always true that

$$(\text{quotient})(\text{divisor}) + (\text{remainder}) = \text{dividend}.$$

EXAMPLE 2   Find the quotient: $(-8x^6 + 12x^4 - 4x^2) \div 4x^2$

Solution

$$\frac{-8x^6 + 12x^4 - 4x^2}{4x^2} = \frac{-8x^6}{4x^2} + \frac{12x^4}{4x^2} - \frac{4x^2}{4x^2} = -2x^4 + 3x^2 - 1$$

The quotient is $-2x^4 + 3x^2 - 1$. *Check:*

$$4x^2(-2x^4 + 3x^2 - 1) = -8x^6 + 12x^4 - 4x^2. \qquad \blacktriangleleft$$

Dividing a Polynomial by a Binomial

We know that

$$(x + 2)(x - 5) = x^2 - 3x - 10.$$

So if we divide $x^2 - 3x - 10$ by $x + 2$, we should get $x - 5$. However, dividing by a binomial is not as simple a process as dividing by a monomial. We can perform this division like long division of whole numbers. We get the first term of the quotient by dividing the first term of $x + 2$ into the first term of $x^2 - 3x - 10$. Divide x^2 by x to get x.

$$\begin{array}{r} x \\ x + 2\overline{)x^2 - 3x - 10} \\ \underline{x^2 + 2x} \\ -5x \end{array}$$

Multiply: $\qquad x^2 \div x = x$

$\qquad x \cdot (x + 2) = x^2 + 2x$

Subtract: $\qquad -3x - 2x = -5x$

Now bring down -10 and continue the division. We get the second term of the quotient (below) by dividing the first term of $x + 2$ into the first term of $-5x - 10$. Divide $-5x$ by x to get -5.

$$
\begin{array}{r}
x - 5 \\
x + 2)\overline{x^2 - 3x - 10} \\
\underline{x^2 - 2x} \\
-5x - 10 \\
\underline{-5x - 10} \\
0
\end{array}
$$

Multiply:

Subtract:

$5x \div x = 5$

$-5 \cdot (x + 2) = -5x - 10$

$-10 - (-10) = 0$

It is usually best to write the terms of the divisor and the dividend in descending order of the exponents. In the next example the x term is missing, so we use $0 \cdot x$ in its place.

EXAMPLE 3 Divide $2x^3 - 4 - 7x^2$ by $2x - 3$, and identify the quotient and the remainder.

Solution Rearrange the dividend as $2x^3 - 7x^2 - 4$. Since the x term in the dividend is missing, we write $0 \cdot x$ for it.

$$
\begin{array}{r}
x^2 - 2x - 3 \\
2x - 3)\overline{2x^3 - 7x^2 + 0 \cdot x - 4} \\
\underline{2x^3 - 3x^2} \\
-4x^2 + 0 \cdot x \\
\underline{-4x^2 + 6x} \\
-6x - 4 \\
\underline{-6x + 9} \\
-13
\end{array}
$$

$-7x^2 - (-3x^2) = -4x^2$

$0 \cdot x - (6x) = -6x$

$-4 - (9) = -13$

The quotient is $x^2 - 2x - 3$, and the remainder is -13. The degree of the remainder is 0, and the degree of the divisor is 1. When dividing polynomials, we do not stop until the remainder is 0 or the degree of the remainder is smaller than the degree of the divisor. To check the answer, we must verify that

$$(2x - 3)(x^2 - 2x - 3) - 13 = 2x^3 - 7x^2 - 4.$$ ◀

If we divide 19 by 5, we get a quotient of 3 and a remainder of 4:

$$
\begin{array}{r}
3 \\
5)\overline{19} \\
\underline{15} \\
4
\end{array}
$$

We can express this result as

$$\frac{19}{5} = 3 + \frac{4}{5}.$$

It is always true that

$$\frac{dividend}{divisor} = quotient + \frac{remainder}{divisor}.$$

EXAMPLE 4   Express $\dfrac{-3x}{x-2}$ in the form: quotient + $\dfrac{remainder}{divisor}$.

Solution   We can use long division to get the required form:

$$
\begin{array}{r}
-3 \\
x-2\overline{)-3x+0} \\
\underline{-3x+6} \\
-6
\end{array}
$$

Since the quotient is -3 and the remainder is -6, we can write

$$\frac{-3x}{x-2} = -3 + \frac{-6}{x-2}.$$

To check, we must verify that $-3(x-2) - 6 = -3x$.   ◀

Note that when we divide two integers we may get a remainder, and when we divide two polynomials we may get a remainder.

Warm-ups

True or false?

1. $y^{10} \div y^2 = y^5$ for any nonzero value of y.   F

2. $\dfrac{7x+2}{7} = x + 2$ for any value of x.   F

3. $\dfrac{7x^2}{7} = x^2$ for any value of x.   T

4. If $3x^2 + 6$ is divided by 3, the quotient is $x^2 + 6$.   F

5. If $4y^2 - 6y$ is divided by $2y$, the quotient is $2y - 3$.   T

6. The quotient times the remainder plus the dividend equals the divisor.   F

7. $(x+2)(x+1) + 3 = x^2 + 3x + 5$ for any value of x.   T

8. The quotient of $(x^2 + 3x + 5) \div (x + 2)$ is $x + 1$.   T

9. If $x^2 + 3x + 5$ is divided by $x + 2$, the remainder is 3.   T

10. If the remainder is zero, then the dividend is equal to the divisor times the quotient.   T

3.5 EXERCISES

Find the quotients of the monomials. Do these mentally. Write only the answer. See Example 1.

1. $12x^2 \div 3$ $4x^2$

2. $-6y \div (-3)$ $2y$

3. $x^8 \div x^2$ x^6

4. $y^9 \div y^3$ y^6

5. $\dfrac{3a^{12}}{a^7}$ $3a^5$

6. $\dfrac{30b^{10}}{b^6}$ $30b^4$

7. $\dfrac{-6x^3y^2}{2x^2y^2}$ $-3x$

8. $\dfrac{-4h^2k^4}{-2hk^3}$ $2hk$

9. $-6y^2 \div 6y$ $-y$

10. $-3a^2b \div 3ab$ $-a$

11. $-x^2y^2 \div xy^2$ $-x$

12. $-y^2 \div y^2$ -1

Find the quotients. See Example 2.

13. $\dfrac{3x - 6}{3}$ $x - 2$

14. $\dfrac{5y - 10}{-5}$ $-y + 2$

15. $\dfrac{x^5 + 3x^4 - x^3}{x^2}$ $x^3 + 3x^2 - x$

16. $\dfrac{6y^6 - 9y^4 + 12y^2}{3y^2}$ $2y^4 - 3y^2 + 4$

17. $\dfrac{-8x^2y^2 + 4x^2y - 2xy^2}{-2xy}$ $4xy - 2x + y$

18. $\dfrac{-9ab^2 - 6a^3b^3}{-3ab^2}$ $3 + 2a^2b$

19. $(x^2 - 3x^3) \div x^2$ $1 - 3x$

20. $(4h^5k - 6h^2k^2) \div (-2h^2k)$ $-2h^3 + 3k$

Find the quotient and remainder for each division. Check by using the fact: dividend = (divisor)(quotient) + remainder. See Example 3.

21. $(x^2 + 5x + 13) \div (x + 3)$ $x + 2, 7$

22. $(x^2 + 3x + 6) \div (x + 3)$ $x, 6$

23. $(a^3 + 4a - 3) \div (a - 2)$ $a^2 + 2a + 8, 13$

24. $(w^3 + 2w^2 - 3) \div (w - 2)$ $w^2 + 4w + 8, 13$

25. $(x^2 - 3x) \div (x + 1)$ $x - 4, 4$

26. $(2x) \div (x + 5)$ $2, -10$

27. $(5x) \div (x - 1)$ $5, 5$

28. $(3x^2) \div (x + 1)$ $3x - 3, 3$

29. $(8x^3 + 27) \div (2x + 3)$ $4x^2 - 6x + 9, 0$

30. $(8y^3 - 1) \div (2y - 1)$ $4y^2 + 2y + 1, 0$

31. $(6x^2 - 13x + 7) \div (3x - 2)$ $2x - 3, 1$

32. $(4b^2 + 25b - 3) \div (4b + 1)$ $b + 6, -9$

33. $(x^3 - x^2 + x - 2) \div (x - 1)$ $x^2 + 1, -1$

34. $(a^2 - 4) \div (a - 2)$ $a + 2, 0$

Write each expression in the form: quotient $+ \dfrac{\text{remainder}}{\text{divisor}}$. See Example 4.

35. $\dfrac{3x}{x - 5}$ $3 + \dfrac{15}{x - 5}$

36. $\dfrac{2x}{x - 1}$ $2 + \dfrac{2}{x - 1}$

37. $\dfrac{x^2}{x + 1}$ $x - 1 + \dfrac{1}{x + 1}$

38. $\dfrac{x^2 + 4}{x + 2}$ $x - 2 + \dfrac{8}{x + 2}$

39. $\dfrac{x^2 + 1}{x - 1}$ $x + 1 + \dfrac{2}{x - 1}$

40. $\dfrac{-x}{x + 3}$ $-1 + \dfrac{3}{x + 3}$

41. $\dfrac{-3x}{x + 1}$ $-3 + \dfrac{3}{x + 1}$

42. $\dfrac{x^2}{x + 1}$ $x - 1 + \dfrac{1}{x + 1}$

43. $\dfrac{x^3}{x - 2}$ $x^2 + 2x + 4 + \dfrac{8}{x - 2}$

44. $\dfrac{x^3 - 1}{x + 1}$ $x^2 - x + 1 + \dfrac{-2}{x + 1}$

45. $\dfrac{x^3 + 3}{x}$ $x^2 + \dfrac{3}{x}$

46. $\dfrac{2x^2 + 4}{2x}$ $x + \dfrac{2}{x}$

47. $\dfrac{x - 1}{x}$ $1 - \dfrac{1}{x}$

48. $\dfrac{3x + 1}{x}$ $3 + \dfrac{1}{x}$

49. $\dfrac{2y + 1}{y}$ $2 + \dfrac{1}{y}$

50. $\dfrac{a - 5}{a}$ $1 - \dfrac{5}{a}$

Find the quotients.

51. $-6a^3b \div 2a^2b$ $-3a$

52. $-x^7 \div (-x^2)$ x^5

53. $(3a - 12) \div (-3)$ $-a + 4$

54. $(-6z + 3z^2) \div (-3z)$ $2 - z$

55. $(3x^2 - 9x) \div (3x)$ $x - 3$

56. $(5x^3 + 15x^2 - 25x) \div (5x)$ $x^2 + 3x - 5$

57. $(h^3 - 27) \div (h - 3)$ $h^2 + 3h + 9$

58. $(w^3 + 1) \div (w + 1)$ $w^2 - w + 1$

59. $(12x^4 - 4x^3 + 6x^2) \div (-2x^2)$ $-6x^2 + 2x - 3$

60. $(-9x^3 + 3x^2 - 15x) \div (-3x)$ $3x^2 - x + 5$

61. $(x^6 - 8) \div (x^2 - 2)$ $x^4 + 2x^2 + 4$

62. $(x^9 + a^3) \div (x^3 + a)$ $x^6 - ax^3 + a^2$

Solve each problem.

63. The area of a rectangle is $x^2 + x - 30$ square meters. If the length is $x + 6$ meters, find a binomial that represents the width. $x - 5$ meters

64. The perimeter of a rectangle is $6x + 6$ yards. If the width is x yards, find a binomial that represents the length. $2x + 3$ yards

Wrap-up

CHAPTER 3

SUMMARY

	Concepts	Examples
Term of a polynomial	A number or the product of a number and one or more variables	$5x^3$, $-4x$, 7
Polynomial	A single term or a finite sum of terms	$2x^5 - 4x^2 + 5$
Add or subtract polynomials	Add or subtract the like terms.	$(x + 1) + (x - 4) = 2x - 3$ $(x^2 - 3x) - (4x^2 - x) = -3x^2 - 2x$
Product rule	$x^m \cdot x^n = x^{m+n}$	$a \cdot a^3 = a^4$ $3x^6 \cdot 4x^9 = 12x^{15}$
Multiply polynomials	Multiply each term of one polynomial by every term of the other polynomial, then combine like terms.	$\begin{array}{r} x^2 + 2x + 5 \\ x - 1 \\ \hline -x^2 - 2x - 5 \\ x^3 + 2x^2 + 5x \\ \hline x^3 + x^2 + 3x - 5 \end{array}$
Multiply two binomials	FOIL: A method for multiplying two binomials quickly Square of a sum: $(a + b)^2 = a^2 + 2ab + b^2$	$(x - 2)(x + 3) = x^2 + x - 6$ $(x + 3)^2 = x^2 + 6x + 9$

Square of a difference:
$$(a - b)^2 = a^2 - 2ab + b^2$$ $(m - 5)^2 = m^2 - 10m + 25$

Product of a sum and a difference:
$$(a - b)(a + b) = a^2 - b^2$$ $(x + 2)(x - 2) = x^2 - 4$

Definition of zero exponent   If $a \neq 0$, then $a^0 = 1$.

$2^0 = 1$

$(-34)^0 = 1$

Quotient rule   If $a \neq 0$ and $m \geq n$, then $\dfrac{a^m}{a^n} = a^{m-n}$.

$x^8 \div x^2 = x^6$

$\dfrac{x^3}{x^3} = x^0 = 1$

$\dfrac{3x^5 + 9x}{3x} = x^4 + 3$

Divide polynomials   When dividing a polynomial by a monomial, each term of the polynomial is divided by the monomial.

If the divisor is a binomial, use long division.

(divisor)(quotient) + (remainder) = dividend

$$
\begin{array}{r}
x - 7 \quad \longleftarrow \text{Quotient} \\
\text{Divisor} \longrightarrow x + 2\overline{)x^2 - 5x - 4} \quad \longleftarrow \text{Dividend} \\
\underline{x^2 + 2x} \\
-7x - 4 \\
\underline{-7x - 14} \\
10 \quad \longleftarrow \text{Remainder}
\end{array}
$$

REVIEW EXERCISES

3.1 *Perform the indicated operations.*

1. $(2w - 6) + (3w + 4)$ $5w - 2$

2. $(1 - 3x) + (4x - 6)$ $x - 5$

3. $(x^2 - 2x - 5) - (x^2 + 4x - 9)$ $-6x + 4$

4. $(3 - 5x - x^2) - (x^2 - 7x + 8)$ $-2x^2 + 2x - 5$

5. $(5 - 3x + x^2) + (x^2 - 4x - 9)$ $2x^2 - 7x - 4$

6. $(-2x^2 + 3x - 4) + (x^2 - 7x + 2)$ $-x^2 - 4x - 2$

7. $(4 - 3x - x^2) - (x^2 - 6x + 5)$ $-2x^2 + 3x - 1$

8. $(x^3 - x) - (x^2 + 5)$ $x^3 - x^2 - x - 5$

3.2 *Perform the indicated operations.*

9. $5x^2 \cdot (-10x^9)$ $-50x^{11}$

10. $(-11a^7)^2$ $121a^{14}$

11. $(12b^3)^2$ $144b^6$

12. $3h^3t \cdot 2h^2t^5$ $6h^5t^6$

13. $(x^2 - 2x + 4)(3x - 2)$ $3x^3 - 8x^2 + 16x - 8$

14. $(x - 5)(x^2 - 2x + 10)$ $x^3 - 7x^2 + 20x - 50$

15. $(x + 2)(x^2 - 2x + 4)$ $x^3 + 8$

16. $(5x + 3)(x^2 - 5x + 4)$ $5x^3 - 22x^2 + 5x + 12$

17. $3m^2(5m^3 - m + 2)$ $15m^5 - 3m^3 + 6m^2$

18. $(a - 2)(a^2 + 2a + 4)$ $a^3 - 8$

19. $x - 5(x - 3)$ $-4x + 15$

20. $x - 4(x - 3)$ $-3x + 12$

21. $5x + 3(x^2 - 5x + 4)$ $3x^2 - 10x + 12$

22. $5 + 4x^2(x - 5)$ $4x^3 - 20x^2 + 5$

3.3 *Perform the indicated operations.*

23. $(q - 6)(q + 8)$ $q^2 + 2q - 48$

24. $(w + 5)(w + 12)$ $w^2 + 17w + 60$

25. $(2t - 3)(t - 9)$ $2t^2 - 21t + 27$

26. $(5r + 1)(5r + 2)$ $25r^2 + 15r + 2$

27. $(4y - 3)(5y + 2)$ $20y^2 - 7y - 6$

28. $(11y + 1)(y + 2)$ $11y^2 + 23y + 2$

29. $(3x^2 + 5)(2x^2 + 1)$ $6x^4 + 13x^2 + 5$

30. $(x^3 - 7)(2x^3 + 7)$ $2x^6 - 7x^3 - 49$

3.4 *Perform the following operations mentally. Write down the answers only.*

31. $(x + 3)(x + 7)$ $x^2 + 10x + 21$ **32.** $(k + 5)(k + 4)$ $k^2 + 9k + 20$ **33.** $(z - 7)(z + 7)$ $z^2 - 49$

34. $(a - 4)(a + 4)$ $a^2 - 16$ **35.** $(y + 7)^2$ $y^2 + 14y + 49$ **36.** $(a + 5)^2$ $a^2 + 10a + 25$

37. $(t - 3)(t - 4)$ $t^2 - 7t + 12$ **38.** $(w - 3)^2$ $w^2 - 6w + 9$ **39.** $(a - 6)^2$ $a^2 - 12a + 36$

40. $(x^2 - 3)(x^2 + 3)$ $x^4 - 9$ **41.** $(2w + 3)(w - 6)$ $2w^2 - 9w - 18$ **42.** $(3x + 5)(2x - 6)$ $6x^2 - 8x - 30$

43. $(3a + 1)^2$ $9a^2 + 6a + 1$ **44.** $(1 - 3c)^2$ $1 - 6c + 9c^2$ **45.** $(4 - y)^2$ $y^2 - 8y + 16$

46. $(t + 7)(t + 6)$ $t^2 + 13t + 42$ **47.** $(3x - 7)(x + 3)$ $3x^2 + 2x - 21$ **48.** $(t - 4)^2$ $t^2 - 8t + 16$

3.5 *Find the quotient and remainder.*

49. $10x^2 \div 2x$ $5x, 0$ **50.** $-6x^4y^2 \div (-2x^2y^2)$ $3x^2, 0$ **51.** $\dfrac{6a^5b^7}{-3a^3b^4}$ $-2a^2b^3, 0$

52. $\dfrac{-9h^5t^9}{3h^2t^6}$ $-3h^3t^3, 0$ **53.** $\dfrac{3x - 9}{-3}$ $-x + 3, 0$ **54.** $\dfrac{9x^3 - 6x^2 + 3x}{-3x}$ $-3x^2 + 2x - 1, 0$

55. $\dfrac{-3x + 5}{-1}$ $3x - 5, 0$ **56.** $\dfrac{7 - y}{-1}$ $y - 7, 0$

57. $(x^3 + x^2 - 11x + 10) \div (x - 1)$ $x^2 + 2x - 9, 1$ **58.** $(x^9 - 8) \div (x^3 - 2)$ $x^6 + 2x^3 + 4, 0$

59. $(m^4 - 16) \div (m - 2)$ $m^3 + 2m^2 + 4m + 8, 0$ **60.** $(a^2 - b^2) \div (a - b)$ $a + b, 0$

61. $(3m^3 - 9m^2 + 18m) \div (3m)$ $m^2 - 3m + 6, 0$ **62.** $(a - 1) \div (1 - a)$ $-1, 0$

63. $(t - 3) \div (3 - t)$ $-1, 0$ **64.** $(x^4 - 1) \div (x - 1)$ $x^3 + x^2 + x + 1, 0$

65. $(8x^3 - 4x^2 - 18x) \div (2x)$ $4x^2 - 2x - 9, 0$ **66.** $(2x^3 + 6x^2 - 8x) \div (2x)$ $x^2 + 3x - 4, 0$

Express each of the following in the form: $\text{quotient} + \dfrac{\text{remainder}}{\text{divisor}}$.

67. $\dfrac{x^2 - 3}{x + 1}$ $x - 1 - \dfrac{2}{x + 1}$ **68.** $\dfrac{x^2 + 3x + 1}{x - 3}$ $x + 6 + \dfrac{19}{x - 3}$ **69.** $\dfrac{2x}{x - 3}$ $2 + \dfrac{6}{x - 3}$ **70.** $\dfrac{3x}{x - 4}$ $3 + \dfrac{12}{x - 4}$

71. $\dfrac{2x}{1 - x}$ $-2 + \dfrac{2}{1 - x}$ **72.** $\dfrac{a}{a + 1}$ $1 - \dfrac{1}{a + 1}$ **73.** $\dfrac{x^2}{x + 1}$ $x - 1 + \dfrac{1}{x + 1}$ **74.** $\dfrac{3x}{5 - x}$ $-3 + \dfrac{15}{5 - x}$

75. $\dfrac{-3x}{x + 4}$ $-3 + \dfrac{12}{x + 4}$ **76.** $\dfrac{-2x}{x + 3}$ $-2 + \dfrac{6}{x + 3}$ **77.** $\dfrac{2x - 3}{x}$ $2 - \dfrac{3}{x}$ **78.** $\dfrac{1 - x}{x}$ $-1 + \dfrac{1}{x}$

CHAPTER 3 TEST

Perform the indicated operations.   **1.** $7x^3 + 4x^2 + 2x - 11$ **2.** $-x^2 - 9x + 2$

1. $(7x^3 - x^2 - 6) + (5x^2 + 2x - 5)$ **2.** $(x^2 - 3x - 5) - (2x^2 + 6x - 7)$ **3.** $-5x^3 \cdot 7x^5$ $-35x^8$

4. $3x^3 \cdot 3x^3$ $9x^6$ **5.** $3x^3 + 3x^3$ $6x^3$ **6.** $3x^3 \div 3x^3$ 1

7. $-4a^6b^5 \div (-2a^5b)$ $2ab^4$ **8.** $\dfrac{-6a^7b^6}{-2a^3b^4}$ $3a^4b^2$ **9.** $\dfrac{6y^3 - 9y^2}{-3y}$ $-2y^2 + 3y$

10. $(x^2 - 5x + 3)(x - 2)$ **11.** $(x^3 - 2x^2 - 4x + 3) \div (x - 3)$ **12.** $3x^2(5x^3 - 7x^2 + 4x - 1)$

13. $(x - 2) \div (2 - x)$ -1 **10.** $x^3 - 7x^2 + 13x - 6$ **11.** $x^2 + x - 1$ **12.** $15x^5 - 21x^4 + 12x^3 - 3x^2$

Find the products.

14. $(x + 5)(x - 2)$ $x^2 + 3x - 10$

15. $(a - 7)^2$ $a^2 - 14a + 49$

16. $(b - 3)(b + 3)$ $b^2 - 9$

17. $(4x + 3)^2$ $16x^2 + 24x + 9$

18. $(4x^2 - 3)(x^2 + 2)$ $4x^4 + 5x^2 - 6$

19. $(3t^2 - 7)(3t^2 + 7)$ $9t^4 - 49$

Express in the form: $quotient + \dfrac{remainder}{divisor}$.

20. $\dfrac{2x}{x - 3}$ $2 + \dfrac{6}{x - 3}$

21. $\dfrac{x^2 - 3x + 5}{x + 2}$ $x - 5 + \dfrac{15}{x + 2}$

Tying It All Together

CHAPTERS 1–3

Simplify each expression.

1. $-16 \div (-2)$ 8

2. $(-2)^3 - 1$ -9

3. $(-5)^2 - 3(-5) + 1$ 41

4. $2^{10} \cdot 2^{15}$ 2^{25}

5. $2^{15} \div 2^{10}$ 2^5

6. $2^{10} - 2^5$ 992

7. $3^2 \cdot 4^2$ 144

8. $(172 - 85) \div (85 - 172)$ -1

9. $(5 + 3)^2$ 64

10. $5^2 + 3^2$ 34

11. $(30 - 1)(30 + 1)$ 899

12. $(30 + 1)^2$ 961

Perform the indicated operations.

13. $(x + 3)(x + 5)$ $x^2 + 8x + 15$

14. $(x^2 + 8x + 15) \div (x + 5)$ $x + 3$

15. $x + 3(x + 5)$ $4x + 15$

16. $(x^2 + 8x + 15)(x + 5)$ $x^3 + 13x^2 + 55x + 75$

17. $-5t^3v \cdot 3t^2v^6$ $-15t^5v^7$

18. $(-10t^3v^2) \div (-2t^2v)$ $5tv$

19. $(-6y^3 + 8y^2) \div (-2y^2)$ $3y - 4$

20. $(y^2 - 3y - 9) - (-3y^2 + 2y - 6)$ $4y^2 - 5y - 3$

Solve each equation.

21. $2x + 1 = 0$ $-\dfrac{1}{2}$

22. $x - 7 = 0$ 7

23. $2x - 3 = 0$ $\dfrac{3}{2}$

24. $3x - 7 = 5$ 4

25. $8 - 3x = x + 20$ -3

26. $4 - 3(x + 2) = 0$ $-\dfrac{2}{3}$

A glider-pilot is participating in a contest in which he must attempt to drop a sandbag onto a target from a moving glider. How does he know when to release the sandbag? He certainly cannot wait until he is directly above the target. The sandbag has the same forward motion as the glider, and if released directly above the target, it will surely miss. There are many factors to be taken into account in solving this problem. The two most important are the speed of the glider and the altitude of the glider.

If the sandbag takes 4 seconds to fall to the ground, then the sandbag must be released 4 seconds before the glider is above the target. The sandbag continues the same forward motion as the glider, and it will hit the target when the glider is directly over the target. The time it takes an object to fall to the ground from a given altitude can be found by solving an equation involving a polynomial. In Exercise 41 of Section 4.6, we will determine how long it takes for a sandbag to fall from an altitude of 784 feet.

4.1 Factoring Out Common Factors

IN THIS SECTION:

- Prime Factorization of Integers
- Greatest Common Factor
- Finding the Greatest Common Factor for Monomials
- Factoring Out the Greatest Common Factor
- Factoring Out a Factor with a Negative Sign

This chapter depends heavily on Chapter 3. Students who did well in Chapter 3 will find Chapter 4 relatively easy. The exercises in Sections 4.2 through 4.5 provide reviews of all factoring taught prior to those sections.

In Chapter 3 we learned how to multiply a monomial and a polynomial. In this section we will learn how to reverse that multiplication. We will find the greatest common factor for the terms of a polynomial and then factor the polynomial.

Prime Factorization of Integers

To **factor** an expression means to write the expression as a product. If we start with 12 and write $12 = 4 \cdot 3$, we have factored 12. Both 4 and 3 are **factors** or **divisors** of 12. The number 3 is a prime number, but 4 is not a prime.

DEFINITION
Prime Number

> A positive integer larger than 1 that has no integral factors other than itself and 1 is called a **prime number.**

The numbers 2, 3, 5, 7, 11, 13, 17, 19, and 23 are the first nine prime numbers. There are other factorizations of 12:

$$12 = 2 \cdot 6 \qquad 12 = 1 \cdot 12 \qquad 12 = 2 \cdot 2 \cdot 3 = 2^2 \cdot 3$$

The factorization that is most useful to us is $12 = 2^2 \cdot 3$. This factorization is called the **prime factorization,** because it is a product of prime numbers.

EXAMPLE 1 Find the prime factorization for 36.

Solution

$$36 = 2 \cdot 18 \qquad \text{Replace 36 by } 2 \cdot 18.$$
$$= 2 \cdot 2 \cdot 9 \qquad \text{Replace 18 by } 2 \cdot 9.$$
$$= 2 \cdot 2 \cdot 3 \cdot 3 \qquad \text{Replace 9 by } 3 \cdot 3.$$
$$= 2^2 \cdot 3^2 \qquad \text{Use exponential notation.} \quad \blacktriangleleft$$

For larger numbers it is helpful to use the method shown in the next example.

EXAMPLE 2 Find the prime factorization for 420.

Solution Start by dividing 420 by the smallest prime number that will divide into it evenly (without remainder). The smallest prime divisor of 420 is 2.

$$\begin{array}{r} 210 \\ 2\overline{)420} \end{array}$$

Now find the smallest prime that will divide into the quotient, 210. The smallest prime divisor of 210 is 2. Continue this procedure, as follows, until the quotient is a prime number:

$$\begin{array}{r} 7 \\ 5\overline{)35} \\ 3\overline{)105} \\ 2\overline{)210} \\ 2\overline{)420} \end{array}$$

$105 \div 3 = 35$

$210 \div 2 = 105$

Start here $\longrightarrow$

The prime factorization of 420 is $420 = 2 \cdot 2 \cdot 3 \cdot 5 \cdot 7 = 2^2 \cdot 3 \cdot 5 \cdot 7$. Note that it is really not necessary to divide by the smallest prime divisor at each step. We obtain the same factorization if we divide by any prime divisor at each step. ◄

Greatest Common Factor

Students often find the GCF for small integers without understanding the method. That is why we use larger integers in Example 3.

The largest integer that is a factor of two or more integers is called the **greatest common factor (GCF)** of the integers. If the integers are not too large, the GCF can be found by inspection. For example, the GCF of 8 and 12 is 4. The number 4 is the largest number that is a factor of both 8 and 12. The GCF of 6 and 11 is 1. The next example shows a systematic procedure for finding the GCF for larger integers.

EXAMPLE 3 Find the GCF for each group of numbers:

a) 150, 225

b) 216, 360, 504

Solution

a) First find the prime factorization for each number:

$$150 = 2 \cdot 3 \cdot 5^2 \qquad 225 = 3^2 \cdot 5^2$$

Both 3 and 5 are common factors of 150 and 225. To form the GCF, write the product of the common factors in which the exponent on each common factor equals the smallest power that appears on that factor in any of the prime factorizations. The smallest power of 5 is 2, and the smallest power of 3 is 1. Thus the GCF of 150 and 225 is $3 \cdot 5^2 = 75$.

b) First find the prime factorization for each number:

$$216 = 2^3 \cdot 3^3 \qquad 360 = 2^3 \cdot 3^2 \cdot 5 \qquad 504 = 2^3 \cdot 3^2 \cdot 7$$

The only common factors are 2 and 3. The smallest power of 2 in the factorizations is 3, and the smallest power of 3 is 2. Thus the GCF is $2^3 \cdot 3^2 = 72$.   ◄

Finding the Greatest Common Factor for Monomials

To find the GCF for a group of monomials, we use the same strategy as that used for integers.

Math at Work

Cryptograms have been used widely throughout history by diplomats, politicians, military agents, criminals, lovers, and business people for sending secret messages. One method, called a polyalphabetic cipher, uses an alphabetic grid system such as that shown. The standard alphabet (called plaintext) runs across the top, while 26 scrambled alphabets (called the cipher text) are positioned below the plaintext.

The users of the code agree on a keyword to use in encoding and decoding messages. For example, if the keyword is FACTOR, the encipherer uses the ciphertext beginning with F to encipher the first letter of the message, the ciphertext beginning with A to encipher the second letter, and so on, repeating the keyword as necessary. Since the keyword contains six letters, six separate ciphers are used to encode the message. Consider the following example:

Key	FACT	OR	FAC	TOR	FACT	ORFA	CTOR	FAC	TORF
Plaintext	This	is	the	day	that	will	make	the	news
Ciphertext	EUQH	CD	EUT	AOG	EUCZ	UMLO	WTIJ	EUT	VHVQ

Note that in the plaintext the combination of letters TH occurs frequently. In the ciphertext we get EU for those letters every time FA appears above TH. To break a code, the cryptologist searches the cryptogram for combinations of letters that are repeated. By counting the number of letters occurring between each combination of repeated letters the cryptologist can tell the length of the keyword. For example, in the coded message above, EU is found three times. The second E is the sixth letter after the first E, and the third E is the twelfth letter after the second E. The greatest common factor of 6 and 12 is 6, and that is the length of the keyword.

Suppose a spy intercepts this message:

N R V B R W O D E S K <u>H</u> <u>U</u> T Y <u>H</u> <u>U</u> Q C S F X S J J T <u>H</u> <u>U</u> W E

STRATEGY
Finding the GCF for a Group of Monomials

1. Find the GCF for the coefficients of the monomials.
2. Form the product of the GCF of the coefficients and each variable that is common to all of the monomials, where the exponent on each variable equals the lowest power of that variable in any of the monomials.

EXAMPLE 4 Find the greatest common factor for each group of monomials:

a) $15x^2$, $9x^3$

b) $12x^2y^2$, $30x^2yz$, $42x^3y$

4 letters.

Notice the HU repeats three times. How many letters are in the keyword? Once we determine how long the keyword is, we still need to guess what word is used. The keyword for this message is given in the answer section. Use it to decode the message. Do not forget that the enemy is within.

MATH

Keys	a	b	c	d	e	f	g	h	i	j	k	l	m	n	o	p	q	r	s	t	u	v	w	x	y	z
1 N	N	A	Q	E	B	S	W	D	U	Y	F	V	O	G	X	R	H	Z	K	J	C	L	I	M	T	P
2 A	A	Q	E	B	S	W	D	U	Y	F	V	O	G	X	R	H	Z	K	J	C	L	I	M	T	P	N
3 Q	Q	E	B	S	W	D	U	Y	F	V	O	G	X	R	H	Z	K	J	C	L	I	M	T	P	N	A
4 E	E	B	S	W	D	U	Y	F	V	O	G	X	R	H	Z	K	J	C	L	I	M	T	P	N	A	O
5 D	D	S	W	D	U	Y	F	V	O	G	X	R	H	Z	K	J	C	L	I	M	I	P	N	A	Q	E
6 S	S	W	D	U	Y	F	V	O	G	X	R	H	Z	K	J	C	L	I	M	T	P	N	A	Q	E	B
7 W	W	D	U	Y	F	V	O	G	X	R	H	Z	K	J	C	L	I	M	T	P	N	A	Q	E	B	S
8 D	D	U	Y	F	V	O	G	X	R	H	Z	K	J	C	L	I	M	T	P	N	A	Q	E	B	S	W
9 U	U	Y	F	V	O	G	X	R	H	Z	K	J	C	L	I	M	T	P	N	A	Q	E	B	S	W	D
10 Y	Y	F	V	O	G	X	R	H	Z	K	J	C	L	I	M	T	P	N	A	Q	E	B	S	W	D	U
11 F	F	V	O	G	X	R	H	Z	K	J	C	L	I	M	T	P	N	A	Q	E	B	S	W	D	U	Y
12 V	V	O	G	X	R	H	Z	K	J	C	L	I	M	T	P	N	A	Q	E	B	S	W	D	U	Y	F
13 O	O	G	X	R	H	Z	K	J	C	L	I	M	T	P	N	A	Q	E	B	S	W	D	U	Y	F	V
14 G	G	X	R	H	Z	K	J	C	L	I	M	T	P	N	A	Q	E	B	S	W	D	U	Y	F	V	O
15 X	X	R	H	Z	K	J	C	L	I	M	T	P	N	A	Q	E	B	S	W	D	U	Y	F	V	O	G
16 R	R	H	Z	K	J	C	L	I	M	T	P	N	A	Q	E	B	S	W	D	U	Y	F	V	O	G	X
17 H	H	Z	K	J	C	L	I	M	T	P	N	A	Q	E	B	S	W	D	U	Y	F	V	O	G	X	R
18 Z	Z	K	J	C	L	I	M	T	P	N	A	Q	E	B	S	W	D	U	Y	F	V	O	G	X	R	H
19 K	K	J	C	L	I	M	T	P	N	A	Q	E	B	S	W	D	U	Y	F	V	O	G	X	R	H	Z
20 J	J	C	L	I	M	T	P	N	A	Q	E	B	S	W	D	U	Y	F	V	O	G	X	R	H	Z	K
21 C	C	L	I	M	T	P	N	A	Q	E	B	S	W	D	U	Y	F	V	O	G	X	R	H	Z	K	J
22 L	L	I	M	T	P	N	A	Q	E	B	S	W	D	U	Y	F	V	O	G	X	R	H	Z	K	J	C
23 I	I	M	T	P	N	A	Q	E	B	S	W	D	U	Y	F	V	O	G	X	R	H	Z	K	J	C	L
24 M	M	T	P	N	A	Q	E	B	S	W	D	U	Y	F	V	O	G	X	R	H	Z	K	J	C	L	I
25 T	T	P	N	A	Q	E	B	S	W	D	U	Y	F	V	O	G	X	R	H	Z	K	J	C	L	I	M
26 P	P	N	A	Q	E	B	S	W	D	U	Y	F	V	O	G	X	R	H	Z	K	J	C	L	I	M	T

Solution

a) The GCF for 15 and 9 is 3, and the smallest power of x is 2. So the GCF for the monomials is $3x^2$.
b) The GCF for 12, 30, and 42 is 6. The common variables are x and y. So the GCF for the monomials is $6x^2y$.   ◄

Factoring Out the Greatest Common Factor

In Chapter 3 we used the distributive property to multiply monomials and polynomials. For example,

$$6(5x - 3) = 30x - 18.$$

If we start with $30x - 18$ and write

$$30x - 18 = 6(5x - 3),$$

we have factored $30x - 18$. Since 6 is the GCF of 30 and 18, we have **factored out** the GCF.

If we start with $2x^4 - 14x$ and reverse the multiplication of the distributive property, we get

$$2x^4 - 14x = 2x(x^3 - 7).$$

We have factored out the monomial $2x$. The monomial $2x$ is the GCF of $2x^4$ and $-14x$.

EXAMPLE 5   Factor the following polynomials by factoring out the greatest common factor:

a) $6x^4 - 12x^3 + 15x^2$ b) $xy^3 + x^3y$ c) $(a + b)w + (a + b)6$

Solution

a) The GCF of 6, 12, and 15 is 3. We can factor x^2 out of each term since the lowest power of x in the three terms is 2. So factor $3x^2$ out of each term as follows:

$$6x^4 - 12x^3 + 15x^2 = 3x^2(2x^2 - 4x + 5)$$

Check this by multiplying $3x^2$ and $2x^2 - 4x + 5$.

b) The GCF of the numerical coefficients is 1. Both x and y are common to each term. Using the lowest powers of x and y, we get

$$xy^3 + x^3y = xy(y^2 + x^2)$$

Check by multiplying.

c) Even though this expression looks different from the others, we can factor it in the same way. The binomial $a + b$ is a common factor, and we can factor it out just as we factor out a monomial:

$$(a + b)w + (a + b)6 = (a + b)(w + 6)$$ ◄

Factoring Out a Factor with a Negative Sign

Generally, we factor out a common factor with no sign in front of it, but we could factor out a common factor with a negative sign in front of it. It will be necessary to do this when we learn factoring by grouping in Section 4.2. Remember that we can check all factoring by multiplying the factors to see whether we get the original polynomial.

EXAMPLE 6   Factor out the greatest common factor, first using a factor with no sign and then a factor with a negative sign:

a) $3x - 3y$ **b)** $a - b$ **c)** $-x^3 + 2x^2 - 8x$

Solution

a) $3x - 3y = 3(x - y)$ Check by multiplying.

$\qquad\qquad = -3(-x + y)$ Check by multiplying.

b) $a - b = 1(a - b)$ The greatest common factor of a and b is 1.

$\qquad\quad = -1(-a + b)$ We can also write $a - b = -1(b - a)$.

c) $-x^3 + 2x^2 - 8x = x(-x^2 + 2x - 8)$ Check by multiplying.

$\qquad\qquad\qquad\quad = -x(x^2 - 2x + 8)$ Check by multiplying. ◄

Warm-ups

True or false?

1. There are only nine prime numbers. F
2. The prime factorization of 32 is $2^3 \cdot 3$. F
3. The integer 51 is a prime number. F
4. The GCF of the integers 12 and 16 is 4. T
5. The GCF of the integers 10 and 21 is 1. T
6. The GCF for the polynomial $x^5y^3 - x^4y^7$ is x^4y^3. T
7. For the polynomial $2x^2y - 6xy^2$ we can factor out either $2xy$ or $-2xy$. T
8. The greatest common factor for the polynomial $8a^3b - 12a^2b$ is $4ab$. F
9. $x - 7 = 7 - x$ for any value of x. F
10. $-3x^2 + 6x = -3x(x - 2)$ for any value of x. T

4.1 EXERCISES

Find the prime factorization of each integer. See Examples 1 and 2.

1. 18 $2 \cdot 3^2$ **2.** 20 $2^2 \cdot 5$ **3.** 52 $2^2 \cdot 13$ **4.** 76 $2^2 \cdot 19$ **5.** 98 $2 \cdot 7^2$

6. 100 $2^2 \cdot 5^2$ **7.** 460 $2^2 \cdot 5 \cdot 23$ **8.** 345 $3 \cdot 5 \cdot 23$ **9.** 924 $2^2 \cdot 3 \cdot 7 \cdot 11$ **10.** 585 $3^2 \cdot 5 \cdot 13$

Find the greatest common factor (GCF) for each group of integers. See Example 3.

11. 8, 20   4
12. 18, 42   6
13. 36, 60   12
14. 42, 70   14
15. 40, 48, 88   8

16. 15, 35, 45   5
17. 76, 84, 100   4
18. 66, 72, 120   6
19. 39, 68, 77   1
20. 200, 500   100

Find the greatest common factor (GCF) for each group of monomials. See Example 4.

21. $6x$, $8x^3$ $2x$
22. $12x^2$, $4x^3$ $4x^2$
23. $3x^2y$, $2xy^2$ xy
24. $7x^3a^2$, $5xa^3$ a^2x

25. $24a^2bc$, $60ab^2$ $12ab$
26. $30x^2$, $75y$ 15
27. $12x^3$, $4x^2$, $6x$ $2x$
28. $3y^5$, $9y^4$, $15y^3$ $3y^3$

29. $18a^3b$, $30a^2b^2$, $54ab^3$ $6ab$ **30.** $16x^2z$, $40xz^2$, $72z^3$ $8z$

Factor out the GCF in each expression. See Example 5.

31. $x^3 - 6x$ $x(x^2 - 6)$
32. $10x^2 - 30y^2$ $10(x^2 - 3y^2)$

33. $5ax + 5ay$ $5a(x + y)$
34. $6wz + 15wa$ $3w(2z + 5a)$

35. $2x^3 - 6x^2 + 8x$ $2x(x^2 - 3x + 4)$
36. $6x^3 + 18x^2 - 24x$ $6x(x^2 + 3x - 4)$

37. $12x^4 + 30x^3 - 24x^2$ $6x^2(2x^2 + 5x - 4)$
38. $15x^2y^2 - 9xy^2 + 6x^2y$ $3xy(5xy - 3y + 2x)$

39. $h^5 - h^3$ $h^3(h^2 - 1)$
40. $k^7m^4 + k^3m^6$ $k^3m^4(k^4 + m^2)$

41. $-6h^5t^2 + 3h^3t^6$ $3h^3t^2(-2h^2 + t^4)$
42. $-9y^6 + 3y^5$ $3y^5(-3y + 1)$

43. $(x - 3)a + (x - 3)b$ $(x - 3)(a + b)$
44. $(y + 4)3 + (y + 4)b$ $(y + 4)(3 + b)$

45. $(y + 1)^2a + (y + 1)^2b$ $(y + 1)^2(a + b)$
46. $(w + 2)^2 \cdot w + (w + 2)^2 \cdot 8$ $(w + 2)^2(w + 8)$

Factor each binomial, first using no sign on the greatest common factor and then using a negative sign on the greatest common factor. See Example 6.

47. $3x - 3y$ $3(x - y)$, $-3(-x + y)$
48. $-5x + 10$ $5(-x + 2)$, $-5(x - 2)$

49. $-4x + 8x^2$ $4x(-1 + 2x)$, $-4x(1 - 2x)$
50. $2x + 6x^2$ $2x(1 + 3x)$, $-2x(-1 - 3x)$

51. $-a^3 + 5a^2$ $a^2(-a + 5)$, $-a^2(a - 5)$
52. $-3b^4 + 6b^3$ $3b^3(-b + 2)$, $-3b^3(b - 2)$

53. $x^2 + x$ $x(x + 1)$, $-x(-x - 1)$
54. $-2b^2 - 8b$ $2b(-b - 4)$, $-2b(b + 4)$

55. $3x - 5$ $1(3x - 5)$, $-1(-3x + 5)$
56. $a - 6$ $1(a - 6)$, $-1(-a + 6)$

57. $b + 4$ $1(b + 4)$, $-1(-b - 4)$
58. $2t + 5$ $1(2t + 5)$, $-1(-2t - 5)$

59. $4 - a$ $1(4 - a)$, $-1(-4 + a)$
60. $7 - b$ $1(7 - b)$, $-1(-7 + b)$

Solve each problem by factoring.

61. Helen traveled $20x + 40$ miles at 20 miles per hour. Find a binomial that represents the time that she traveled. $x + 2$ hours

62. A rectangular painting has an area of $x^2 + 3x$ square meters and a width of x meters. Find a binomial that represents the length. $x + 3$ meters

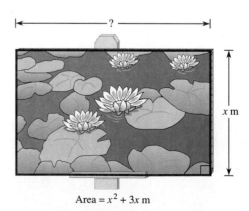

Area $= x^2 + 3x$ m

Figure for Exercise 62

4.2 Factoring the Special Products

IN THIS SECTION:

- **Factoring the Difference of Two Squares**
- **Factoring a Perfect Square Trinomial**
- **Factoring Completely**
- **Factoring by Grouping**

In Section 3.4 we learned how to square a sum, square a difference, and find the product of a sum and a difference. In this section we will learn how to reverse those operations.

Factoring the Difference of Two Squares

We introduce square roots here so that we can discuss the factoring of the difference of two squares. We will discuss square roots extensively in Chapter 6.

Consider the binomial $x^2 - 9$. Since $9 = 3^2$, this binomial is a difference of two squares:

$$x^2 - 9 = x^2 - 3^2$$

We say that x is the **square root** of x^2 and 3 is the square root of 9. To factor $x^2 - 9$, we recall that in Section 3.3 we multiplied two binomials and got $x^2 - 9$ as the product. The two binomials were the sum and the difference of the square roots of x^2 and 9:

$$(x + 3)(x - 3) = x^2 - 9$$

So we factor $x^2 - 9$ by writing

$$x^2 - 9 = (x + 3)(x - 3).$$

EXAMPLE 1   Factor each of the following polynomials:

a) $y^2 - 81$ **b)** $9m^2 - 16$ **c)** $4x^2 - 9y^2$

Solution   Each of these binomials is the difference of two squares. Each binomial factors into a product of a sum and a difference.

a) The square root of y^2 is y, and the square root of 81 is 9.

$$y^2 - 81 = y^2 - 9^2 = (y + 9)(y - 9)$$

Check by multiplying.

b) The square root of $9m^2$ is $3m$, and the square root of 16 is 4.

$$9m^2 - 16 = (3m + 4)(3m - 4)$$

Check by multiplying.

c) $4x^2 - 9y^2 = (2x + 3y)(2x - 3y)$ Check.   ◄

Any difference of two squares can be factored into the product of a sum and a difference. This is expressed in the following rule.

**RULE
Difference of Two Squares**

For any real numbers a and b,

$$a^2 - b^2 = (a + b)(a - b).$$

Factoring a Perfect Square Trinomial

To factor

$$x^2 + 6x + 9,$$

we recall the rule

$$(a + b)^2 = a^2 + 2ab + b^2.$$

Notice that $x^2 + 6x + 9$ can be written as $x^2 + 2 \cdot x \cdot 3 + 3^2$. We see that this fits the rule for $(a + b)^2$ if $a = x$ and $b = 3$:

$$x^2 + 6x + 9 = (x + 3)^2$$

The trinomial $x^2 + 6x + 9$ is called a **perfect square trinomial,** because it is the square of a binomial. Perfect square trinomials are easy to identify by using the following strategy.

**STRATEGY
Identifying a Perfect Square
Trinomial**

A trinomial is a perfect square trinomial if:

1. the first and last terms are perfect squares and
2. the middle term is $+2$ or -2 times the product of the square roots of the first and last terms.

EXAMPLE 2 Identify each polynomial as a difference of two squares, a perfect square trinomial, or neither of these:

a) $x^2 - 10x + 25$ **b)** $4x^2 - 25$ **c)** $4a^2 + 24a + 25$

Solution

a) The first term, x^2, and the last term, 25, are perfect squares. The middle term, $-10x$, is -2 times the product of x and 5 (the square roots of the first and last terms). This trinomial is a perfect square trinomial.

b) The terms $4x^2$ and 25 are both perfect squares. This binomial is the difference of two squares.

c) The terms $4a^2$ and 25 are perfect squares. However, 2 times the product of $2a$ and 5 is $20a$. Since the middle term is $24a$, this trinomial is not a perfect square trinomial. ◄

Note that the middle term in a perfect square trinomial may have a positive or a negative coefficient, while the first and last terms must be positive.

EXAMPLE 3 Factor the following polynomials:

a) $x^2 - 4x + 4$ **b)** $a^2 + 16a + 64$
c) $b^2 - 49$ **d)** $4x^2 - 12x + 9$

Solution

a) $x^2 - 4x + 4 = x^2 - 2 \cdot 2x + 2^2 = (x - 2)^2$ **This is a perfect square trinomial.**

−2 times ⌐ Square root of first term / Square root of last term

Check by finding $(x - 2)^2$.

b) $a^2 + 16a + 64 = (a + 8)^2$ **Check by finding $(a + 8)^2$.**
c) $b^2 - 49 = (b + 7)(b - 7)$ **Check by multiplying.**
d) $4x^2 - 12x + 9 = (2x)^2 - 2(3)(2x) + 3^2 = (2x - 3)^2$ **Check.**

−2 times ⌐ Square root of first term / Square root of last term ◄

Factoring Completely

A polynomial is factored when it is written as a product of simpler polynomials. Polynomials that cannot be factored are called **prime polynomials.** Since binomials such as $x + 2$, $b - 7$, and $2x + 1$ cannot be factored, they are prime polynomials. Any monomial is also considered a prime polynomial. A polynomial is **factored completely** when it is written as a product of prime polynomials.

Some polynomials have a factor common to all terms. To factor such polynomials completely, it is simpler to factor out the greatest common factor (GCF) and then factor the remaining polynomial. The following example will illustrate factoring completely.

EXAMPLE 4 Factor each polynomial completely:

a) $2x^3 - 50x$ **b)** $8x^2y - 32xy + 32y$

Solution

a) The greatest common factor of $2x^3$ and $50x$ is $2x$:

$$2x^3 - 50x = 2x(x^2 - 25)$$ **Check this step by multiplying.**
$$= 2x(x + 5)(x - 5)$$ **Factor the difference of two squares.**

b) $8x^2y - 32xy + 32y = 8y(x^2 - 4x + 4)$ Check this step by multiplying.
$\qquad\qquad\qquad\qquad = 8y(x - 2)^2$ Factor the perfect square trinomial.

Note that even though 8 is not a prime number, $8y$ is a prime polynomial because it is a monomial.   ◄

Remember that factoring reverses multiplication and *every step of factoring can be checked by multiplication.*

Factoring by Grouping

Since we use the FOIL method so often, students may have forgotten how to multiply two binomials using the distributive property. Have them do a few problems of this type before teaching them factoring by grouping.

The product of two binomials may be a polynomial with four terms. For example,

$$(x + a)(x + 3) = (x + a)x + (x + a)3$$
$$= x^2 + ax + 3x + 3a.$$

We can factor a polynomial of this type by simply reversing the steps we used to find the product. For example, to factor $xy + 2y + 3x + 6$, we notice that the first two terms have a common factor of y and the last two have a common factor of 3:

$$xy + 2y + 3x + 6 = (x + 2)y + (x + 2)3 \qquad \text{Factor first two terms and last two terms.}$$
$$= (x + 2)(y + 3) \qquad \text{Factor out the common factor } x + 2.$$

Since we factored common factors from groups of two terms, this procedure is called **factoring by grouping.**

EXAMPLE 5   Use grouping to factor each polynomial completely:

a) $ax - 3x - ay + 3y$ $\qquad\qquad$ **b)** $2x^3 - 3x^2 - 2x + 3$

Solution

a) The first two terms can be factored as $x(a - 3)$. To get another factor of $a - 3$, we must factor out $-y$ from the last two terms:

$$ax - 3x - ay + 3y = x(a - 3) - y(a - 3) \qquad \text{Factor the first two and last two terms.}$$
$$= (x - y)(a - 3) \qquad \text{Factor out } a - 3.$$

b) If we factor x^2 out of the first two terms, we get $x^2(2x - 3)$. To get another factor of $2x - 3$, we must factor out -1 from the last two terms:

$$2x^3 - 3x^2 - 2x + 3 = x^2(2x - 3) - 1(2x - 3)$$
$$= (x^2 - 1)(2x - 3) \qquad \text{Factor out } 2x - 3.$$
$$= (x - 1)(x + 1)(2x - 3) \qquad \text{Factor completely.} \quad ◄$$

Warm-ups

True or false?

1. The polynomial $x^2 + 16$ is called a difference of two squares.   F
2. The polynomial $x^2 - 8x + 16$ is called a perfect square trinomial.   T
3. The polynomial $9x^2 + 21x + 49$ is a perfect square trinomial.   F
4. $4x^2 + 4 = (2x + 2)^2$ for any value of x.   F
5. The difference of two squares factors into a product of a sum and a difference.   T
6. Any trinomial in which the first and last terms are perfect squares is a perfect square trinomial.   F
7. The polynomial $x^2 + 9$ can be factored as $(x + 3)(x + 3)$.   F
8. $4x^2 - 4 = 4(x^2 - 1)$ for any value of x.   T
9. $y^2 - 2y + 1 = (y - 1)^2$ for any value of y.   T
10. $2x^2 - 18 = 2(x - 3)(x + 3)$ for any value of x.   T

4.2 EXERCISES

Factor each polynomial. See Example 1.

12. $(12w - 11a)(12w + 11a)$

1. $a^2 - 4$ $(a - 2)(a + 2)$
2. $h^2 - 9$ $(h - 3)(h + 3)$
3. $x^2 - 49$ $(x - 7)(x + 7)$
4. $y^2 - 36$ $(y - 6)(y + 6)$
5. $4y^2 - 9x^2$ $(2y + 3x)(2y - 3x)$
6. $16x^2 - y^2$ $(4x - y)(4x + y)$
7. $25a^2 - b^2$ $(5a + b)(5a - b)$
8. $9a^2 - 64b^2$ $(3a - 8b)(3a + 8b)$
9. $m^2 - 1$ $(m - 1)(m + 1)$
10. $4n^2 - 1$ $(2n - 1)(2n + 1)$
11. $9w^2 - 25c^2$ $(3w - 5c)(3w + 5c)$
12. $144w^2 - 121a^2$

Identify each polynomial as a difference of two squares, a perfect square trinomial, or neither of these. See Example 2.

13. $x^2 - 20x + 100$ Perfect square trinomial.
14. $x^2 - 10x - 25$ Neither
15. $y^2 - 40$ Neither
16. $a^2 - 49$ Difference of two squares
17. $4y^2 + 12y + 9$ Perfect square trinomial
18. $9a^2 - 30a - 25$ Neither
19. $x^2 - 8x + 64$ Neither
20. $x^2 + 4x + 4$ Perfect square trinomial
21. $y^2 + 25$ Neither
22. $x^2 - 50$ Neither
23. $9x^2 + 6x + 1$ Perfect square trinomial
24. $4x^2 - 4xy + y^2$ Perfect square trinomial

Factor each of the following perfect square trinomials. See Example 3.

25. $x^2 + 12x + 36$ $(x + 6)^2$
26. $y^2 + 14y + 49$ $(y + 7)^2$
27. $a^2 - 4a + 4$ $(a - 2)^2$
28. $b^2 - 6b + 9$ $(b - 3)^2$
29. $4w^2 + 4w + 1$ $(2w + 1)^2$
30. $9m^2 + 6m + 1$ $(3m + 1)^2$

31. $16x^2 - 8x + 1$ $(4x - 1)^2$

32. $25y^2 - 10y + 1$ $(5y - 1)^2$

33. $4t^2 + 20t + 25$ $(2t + 5)^2$

34. $9y^2 - 12y + 4$ $(3y - 2)^2$

35. $n^2 + 2nt + t^2$ $(n + t)^2$

36. $x^2 - 2xy + y^2$ $(x - y)^2$

37. $9w^2 + 42w + 49$ $(3w + 7)^2$

38. $144x^2 + 24x + 1$ $(12x + 1)^2$

Factor each polynomial completely. See Example 4.

39. $5x^2 - 125$ $5(x - 5)(x + 5)$

40. $3y^2 - 27$ $3(y - 3)(y + 3)$

41. $-2x^2 + 18$ $-2(x - 3)(x + 3)$

42. $-5y^2 + 20$ $-5(y - 2)(y + 2)$

43. $3x^2 + 6x + 3$ $3(x + 1)^2$

44. $-5y^2 + 50y - 125$ $-5(y - 5)^2$

45. $x^3 - 2x^2y + xy^2$ $x(x - y)^2$

46. $x^3y + 2x^2y^2 + xy^3$ $xy(x + y)^2$

47. $32x^2y - 2y^3$ $2y(4x - y)(4x + y)$

48. $2ax^2 - 98a$ $2a(x - 7)(x + 7)$

49. $3ab^2 - 18ab + 27a$ $3a(b - 3)^2$

50. $-2a^2b + 8ab - 8b$ $-2b(a - 2)^2$

51. $12w^2 - 3$ $3(2w - 1)(2w + 1)$

52. $10a^3 - 20a^2b + 10ab^2$ $10a(a - b)^2$

53. $a^3 - ab^2$ $a(a - b)(a + b)$

54. $x^2y - y$ $y(x - 1)(x + 1)$

55. $-3x^2 + 3y^2$ $-3(x - y)(x + y)$

56. $-8a^2 + 8b^2$ $-8(a - b)(a + b)$

Use grouping to factor each polynomial completely. See Example 5.

57. $bx + by + cx + cy$ $(b + c)(x + y)$

58. $3x + 3z + ax + az$ $(3 + a)(x + z)$

59. $x^3 + x^2 - 4x - 4$ $(x - 2)(x + 2)(x + 1)$

60. $x^3 + x^2 - x - 1$ $(x - 1)(x + 1)^2$

61. $3a - 3b - xa + xb$ $(3 - x)(a - b)$

62. $xa - ay + 3x - 3y$ $(a + 3)(x - y)$

63. $a^3 + 3a^2 + a + 3$ $(a^2 + 1)(a + 3)$

64. $ax - bx - 4a + 4b$ $(x - 4)(a - b)$

65. $y^3 - 5y^2 + 8y - 40$ $(y^2 + 8)(y - 5)$

66. $x^3 + ax + 3a + 3x^2$ $(x + 3)(x^2 + a)$

67. $abc + c - 3 - 3ab$ $(c - 3)(ab + 1)$

68. $xa + ba + tx + tb$ $(a + t)(x + b)$

69. $x^2a - a + bx^2 - b$ $(a + b)(x - 1)(x + 1)$

70. $a^2m - b^2m + a^2n - b^2n$ $(m + n)(a - b)(a + b)$

71. $y^2 + y + by + b$ $(y + b)(y + 1)$

72. $ac + mc + aw^2 + mw^2$ $(c + w^2)(a + m)$

4.3 Factoring $ax^2 + bx + c$ with $a = 1$

IN THIS SECTION:

- **Factoring Trinomials with a Leading Coefficient of 1**
- **Factoring Completely**

In this section we will factor the type of trinomials that result from multiplying two binomials. We will factor only trinomials in which the coefficient of x^2, the leading coefficient, is 1. Factoring trinomials with leading coefficients not equal to 1 will be explained in Section 4.4.

Factoring Trinomials with a Leading Coefficient of 1

Let's look closely at an example of finding the product of two binomials using the distributive property:

$$(x + 2)(x + 3) = (x + 2)x + (x + 2)3 \qquad \text{The distributive property}$$
$$= x^2 + 2x + 3x + 6 \qquad \text{The distributive property again}$$
$$= x^2 + 5x + 6 \qquad \text{Combining like terms}$$

To factor $x^2 + 5x + 6$, we need to reverse these steps. First observe that the coefficient 5 is the sum of two numbers that have a product of 6. The only numbers that have a product of 6 and a sum of 5 are 2 and 3. So write $5x$ as $2x + 3x$:

$$x^2 + \underline{5x} + 6 = x^2 + \underline{2x + 3x} + 6$$

Now apply factoring by grouping to the polynomial on the right:

Factor out x.   Factor out 3.

$$x^2 + 5x + 6 = \overbrace{x^2 + 2x} + \overbrace{3x + 6}$$
$$= (x + 2)x + (x + 2)3$$
$$= (x + 2)(x + 3) \qquad \text{Factor out } x + 2.$$

To factor $x^2 + 8x + 12$, we must find two numbers that have a product of 12 and a sum of 8. The pairs of numbers with a product of 12 are 1 and 12, 2 and 6, and 3 and 4. Only 2 and 6 have a sum of 8. So write $8x$ as $2x + 6x$ and factor by grouping:

$$x^2 + 8x + 12 = x^2 + 2x + 6x + 12$$
$$= (x + 2)x + (x + 2)6$$
$$= (x + 2)(x + 6)$$

Check by using FOIL.

We can actually skip most of the steps shown in these examples. For example, to factor $x^2 + x - 6$, we must find a pair of numbers with a product of -6 and a sum of 1. Since the numbers are 3 and -2, we can write

$$x^2 + x - 6 = (x + 3)(x - 2).$$

Check by using FOIL.

EXAMPLE 1 Factor each trinomial:

a) $x^2 + 5x + 4$
b) $x^2 + 2x - 8$
c) $a^2 - 7ab + 10b^2$

Solution

a) To get a sum of 5 and a product of 4, use 1 and 4:

$$x^2 + 5x + 4 = (x + 1)(x + 4)$$

Check by using FOIL on the right side.

b) To get a sum of 2 and a product of -8, use -2 and 4:

$$x^2 + 2x - 8 = (x + 4)(x - 2)$$

Check by multiplying.

c) To get a sum of -7 and a product of 10, use -2 and -5:

$$a^2 - 7ab + 10b^2 = (a - 5b)(a - 2b)$$

Check by multiplying.   ◀

EXAMPLE 2   Factor $x^2 + 7x - 6$.

Solution   The 6 can be factored only as $1 \cdot 6$ or $2 \cdot 3$. Since the last term is negative, we must have a positive factor and a negative factor. Check the sums:

$$-1 + 6 = 5 \qquad -2 + 3 = 1$$
$$1 + (-6) = -5 \qquad 2 + (-3) = -1$$

Since none of these possible factors of -6 have a sum of 7, we can be certain that $x^2 + 7x - 6$ cannot be factored. It is a prime polynomial.   ◀

Factoring Completely

In Section 4.2 we learned that any monomial is a prime polynomial. We also learned that binomials such as $3x - 5$ (with no common factor) are prime polynomials. In Example 2 of this section we saw a trinomial that is a prime polynomial. There are many prime trinomials. When factoring a polynomial completely, we may have a factor that is a prime trinomial.

EXAMPLE 3   Factor each polynomial completely:

a) $x^3 - 6x^2 - 16x$ **b)** $4x^3 + 4x^2 + 4x$

Solution

a) $x^3 - 6x^2 - 16x = x(x^2 - 6x - 16)$ **First factor out the GCF.**
$$= x(x - 8)(x + 2) \qquad \text{Then factor } x^2 - 6x - 16.$$

b) $4x^3 + 4x^2 + 4x = 4x(x^2 + x + 1)$ **First factor out the GCF.**

If we try to factor $x^2 + x + 1$, we will find that $x^2 + x + 1$ is a prime polynomial. So the factorization is complete.   ◀

Warm-ups

True or false? Answer true if the correct factorization is given and false if the factorization is incorrect.

1. $x^2 - 6x + 9 = (x - 3)^2$ T
2. $x^2 + 6x + 9 = (x + 3)^2$ T
3. $x^2 + 10x + 9 = (x - 9)(x - 1)$ F
4. $x^2 - 8x - 9 = (x - 8)(x - 9)$ F
5. $x^2 + 8x - 9 = (x + 9)(x - 1)$ T
6. $x^2 + 8x + 9 = (x + 3)^2$ F
7. $x^2 - 10xy + 9y^2 = (x - y)(x - 9y)$ T
8. $x^2 + x + 1 = (x + 1)(x + 1)$ F
9. $x^2 + xy + 20y^2 = (x + 5y)(x - 4y)$ F
10. $x^2 + 1 = (x + 1)(x + 1)$ F

4.3 EXERCISES

Factor the trinomials. See Examples 1 and 2. Watch for prime polynomials.

1. $x^2 + 4x + 3$ $(x + 3)(x + 1)$
2. $y^2 + 6y + 5$ $(y + 5)(y + 1)$
3. $x^2 + 9x + 18$ $(x + 3)(x + 6)$
4. $y^2 + 6y + 8$ $(y + 2)(y + 4)$
5. $y^2 + 7y + 10$ $(y + 2)(y + 5)$
6. $x^2 + 8x + 15$ $(x + 3)(x + 5)$
7. $b^2 - 5b - 6$ $(b - 6)(b + 1)$
8. $a^2 + 5a - 6$ $(a + 6)(a - 1)$
9. $a^2 - 2a - 12$ Prime
10. $x^2 + 3x + 3$ Prime
11. $a^2 - 4a + 12$ Prime
12. $y^2 + 3y - 10$ $(y + 5)(y - 2)$
13. $y^2 - 6y - 8$ Prime
14. $w^2 + 9w - 10$ $(w + 10)(w - 1)$
15. $w^2 - 2w - 8$ $(w - 4)(w + 2)$
16. $a^2 - 6a + 8$ $(a - 4)(a - 2)$
17. $t^2 - 3t + 10$ Prime
18. $x^2 - 5x - 3$ Prime
19. $m^2 + 10m + 16$ $(m + 2)(m + 8)$
20. $m^2 - 10m + 16$ $(m - 2)(m - 8)$
21. $m^2 - 6m - 16$ $(m - 8)(m + 2)$
22. $m^2 + 6m - 16$ $(m + 8)(m - 2)$
23. $m^2 - 17m + 16$ $(m - 16)(m - 1)$
24. $m^2 + 17m + 16$ $(m + 16)(m + 1)$
25. $m^2 - 15m - 16$ $(m - 16)(m + 1)$
26. $m^2 + 15m - 16$ $(m + 16)(m - 1)$
27. $m^2 - 15m + 16$ Prime
28. $m^2 + 15m + 16$ Prime
29. $t^2 + 5t - 24$ $(t + 8)(t - 3)$
30. $t^2 - 10t - 24$ $(t - 12)(t + 2)$
31. $t^2 - 2t - 24$ $(t - 6)(t + 4)$
32. $t^2 + 14t + 24$ $(t + 12)(t + 2)$
33. $t^2 - 10t - 200$ $(t - 20)(t + 10)$
34. $t^2 + 30t + 200$ $(t + 10)(t + 20)$
35. $x^2 - 5x - 150$ $(x - 15)(x + 10)$
36. $x^2 - 25x + 150$ $(x - 10)(x - 15)$
37. $10y + 24 + y^2$ $(y + 6)(y + 4)$
38. $18z + 45 + z^2$ $(z + 15)(z + 3)$
39. $x^2 - 4xy - 12y^2$ $(x - 6y)(x + 2y)$
40. $y^2 + yt - 12t^2$ $(y + 4t)(y - 3t)$
41. $x^2 - 13xy + 12y^2$ $(x - 12y)(x - y)$
42. $h^2 - 9hs + 9s^2$ Prime
43. $x^2 - 5xs - 24s^2$ $(x - 8s)(x + 3s)$
44. $x^2 + 4xz - 32z^2$ $(x + 8z)(x - 4z)$

Factor each polynomial completely. Use the methods discussed in Sections 4.1–4.3. See Example 3.

45. $2w^2 - 162$ $2(w - 9)(w + 9)$
46. $w^2 - 18w + 81$ $(w - 9)^2$
47. $w^2 + 30w + 81$ $(w + 3)(w + 27)$
48. $6w^2 - 12w - 18$ $6(w - 3)(w + 1)$
49. $20w^2 + 100w + 40$ $20(w^2 + 5w + 2)$
50. $w^2 - 8w$ $w(w - 8)$

51. $w^3 - 3w^2 - 18w$ $w(w - 6)(w + 3)$

52. $3w^2 + 27w + 54$ $3(w + 3)(w + 6)$

53. $w^2 + 18w + 36$ Prime

54. $6w^4 - 54w^2$ $6w^2(w - 3)(w + 3)$

55. $a^3b + a^2b^2$ $a^2b(a + b)$

56. $x^4 - x^3$ $x^3(x - 1)$

57. $4 - w^2$ $(2 - w)(2 + w)$

58. $x^2w^2 + 9x^2$ $x^2(w^2 + 9)$

59. $8vw^2 + 32vw + 32v$ $8v(w + 2)^2$

60. $3h^2t + 6ht + 3t$ $3t(h + 1)^2$

61. $6x^3y + 30x^2y^2 + 36xy^3$
$6xy(x + 3y)(x + 2y)$

62. $3x^3y^2 - 3x^2y^2 + 3xy^2$
$3xy^2(x^2 - x + 1)$

Use factoring to solve each problem.

63. A rectangle has an area of $x^2 + 6x + 8$ square feet and a width of $x + 2$ feet. Find a binomial that represents the length. $x + 4$ feet

64. A triangle has an area of $x^2 + 5x + 6$ square meters and a height of $x + 3$ meters. Find a binomial that represents the length of the base. $2x + 4$ meters

4.4   Factoring $ax^2 + bx + c$ with $a \neq 1$

IN THIS SECTION:

● Factoring Trinomials with a Leading Coefficient That Is Not 1

● Factoring Completely

In Section 4.3 we factored trinomials with a leading coefficient of 1. In this section we will use a slightly different technique to factor trinomials with leading coefficients that are not equal to 1.

Factoring Trinomials with a Leading Coefficient That Is Not 1

If the leading coefficient of a trinomial is not 1, we can again use grouping to factor the trinomial. However, the procedure is slightly different.

Consider the trinomial $2x^2 + 7x + 6$. First find the product of the leading coefficient and the constant term. In this case it is $2 \cdot 6 = 12$. Now find two numbers with a product of 12 and a sum of 7. The pairs of numbers with a product of 12 are 1 and 12, 2 and 6, 3 and 4. Only 3 and 4 have a product of 12 and a sum of 7. Now replace $7x$ by $3x + 4x$ and factor by grouping:

$$2x^2 + 7x + 6 = 2x^2 + 3x + 4x + 6$$
$$= (2x + 3)x + (2x + 3)2$$
$$= (2x + 3)(x + 2)$$

The strategy for factoring a trinomial is summarized in the following box. The steps listed here work whether or not the leading coefficient is 1. This method is called the *ac* **method.**

STRATEGY
Factoring $ax^2 + bx + c$ by the ac Method

To factor the trinomial $ax^2 + bx + c$:

1. Find two numbers that have a product equal to ac and a sum equal to b.
2. Replace bx by two terms using the two new numbers as coefficients.
3. Factor the resulting four-term polynomial by grouping.

EXAMPLE 1   Factor each trinomial:

a) $2x^2 + x - 6$ **b)** $10x^2 + 13x - 3$

Solution

a) Since $2 \cdot (-6) = -12$, we need two numbers with a product of -12 and a sum of 1. The numbers are -3 and 4. Replace x by $-3x + 4x$ and factor by grouping:

$$2x^2 + x - 6 = 2x^2 - 3x + 4x - 6$$
$$= (2x - 3)x + (2x - 3)2$$
$$= (2x - 3)(x + 2)$$

Check by FOIL.

b) Since $10 \cdot (-3) = -30$, we need two numbers with a product of -30 and a sum of 13. The pairs of numbers with a product of 30 are 1 and 30, 2 and 15, 3 and 10, and 5 and 6. To get a product of -30, one of the numbers must be negative and the other positive. To get a sum of positive 13, we need -2 and 15:

$$10x^2 + 13x - 3 = 10x^2 - 2x + 15x - 3$$
$$= (5x - 1)2x + (5x - 1)3$$
$$= (5x - 1)(2x + 3)$$

Check by FOIL.   ◄

Factoring Completely

We can use the latest factoring technique along with the techniques that we learned earlier to factor polynomials completely.

EXAMPLE 2   Factor each polynomial completely:

a) $4x^3 + 14x^2 + 6x$ **b)** $12x^2y + 6xy + 6y$

Solution

a) $4x^3 + 14x^2 + 6x = 2x(2x^2 + 7x + 3)$ **First factor out the GCF $2x$.**
$$= 2x(2x + 1)(x + 3)$$ **Factor the trinomial.**

b) $12x^2y + 6xy + 6y = 6y(2x^2 + x + 1)$ **First factor out the GCF $6y$.**

Since the trinomial $2x^2 + x + 1$ is prime, the factorization is complete.   ◄

Warm-ups

True or false? Answer true if the correct factorization is given and false if the factorization is incorrect.

1. $2x^2 + 3x + 1 = (2x + 1)(x + 1)$ T

2. $2x^2 + 5x + 3 = (2x + 1)(x + 3)$ F

3. $3x^2 + 10x + 3 = (3x + 1)(x + 3)$ T

4. $15x^2 + 31x + 14 = (3x + 7)(5x + 2)$ F

5. $2x^2 - 7x - 9 = (2x - 9)(x + 1)$ T

6. $2x^2 + 3x - 9 = (2x + 3)(x - 3)$ F

7. $2x^2 - 16x - 9 = (2x - 9)(2x + 1)$ F

8. $8x^2 - 22x - 5 = (4x - 1)(2x + 5)$ F

9. $9x^2 + x - 1 = (5x - 1)(4x + 1)$ F

10. $12x^2 - 13x + 3 = (3x - 1)(4x - 3)$ T

4.4 EXERCISES

Factor each trinomial. See Example 1.

1. $2x^2 + 3x + 1$ $(2x + 1)(x + 1)$

2. $2x^2 + 11x + 5$ $(2x + 1)(x + 5)$

3. $2x^2 + 9x + 4$ $(2x + 1)(x + 4)$

4. $2h^2 + 7h + 3$ $(2h + 1)(h + 3)$

5. $3t^2 + 7t + 2$ $(3t + 1)(t + 2)$

6. $3t^2 + 8t + 5$ $(3t + 5)(t + 1)$

7. $2x^2 + 5x - 3$ $(2x - 1)(x + 3)$

8. $3x^2 - x - 2$ $(3x + 2)(x - 1)$

9. $6x^2 + 7x - 3$ $(3x - 1)(2x + 3)$

10. $21x^2 + 2x - 3$ $(3x - 1)(7x + 3)$

11. $2x^2 - 7x + 6$ $(2x - 3)(x - 2)$

12. $3a^2 - 14a + 15$ $(3a - 5)(a - 3)$

13. $5b^2 - 13b + 6$ $(5b - 3)(b - 2)$

14. $7y^2 + 16y - 15$ $(7y - 5)(y + 3)$

15. $4y^2 - 11y - 3$ $(4y + 1)(y - 3)$

16. $35x^2 - 2x - 1$ $(7x + 1)(5x - 1)$

17. $3x^2 + 2x + 1$ Prime

18. $6x^2 - 4x - 5$ Prime

19. $8x^2 - 2x - 1$ $(4x + 1)(2x - 1)$

20. $8x^2 - 10x - 3$ $(4x + 1)(2x - 3)$

21. $8x^2 - 6x + 1$ $(4x - 1)(2x - 1)$

22. $8x^2 - 22x + 5$ $(4x - 1)(2x - 5)$

23. $9t^2 - 9t + 2$ $(3t - 1)(3t - 2)$

24. $9t^2 + 5t - 4$ $(9t - 4)(t + 1)$

25. $15x^2 + 13x + 2$ $(5x + 1)(3x + 2)$

26. $15x^2 - 7x - 2$ $(5x + 1)(3x - 2)$

27. $15x^2 - 13x + 2$ $(5x - 1)(3x - 2)$

28. $15x^2 + x - 2$ $(5x + 2)(3x - 1)$

29. $15x^2 - x - 2$ $(5x - 2)(3x + 1)$

30. $15x^2 + 13x - 2$ $(15x - 2)(x + 1)$

31. $15x^2 - 31x + 2$ $(15x - 1)(x - 2)$

32. $15x^2 + 31x + 2$ $(15x + 1)(x + 2)$

33. $2x^2 + 18x - 90$ $2(x^2 + 9x - 45)$

34. $3x^2 + 11x + 10$ $(3x + 5)(x + 2)$

35. $3x^2 + x - 10$ $(3x - 5)(x + 2)$

36. $3x^2 - 17x + 10$ $(3x - 2)(x - 5)$

Factor each polynomial completely. See Example 2.

37. $4w^2 + 2w - 30$ $2(2w - 5)(w + 3)$

38. $81w^3 - w$ $w(9w - 1)(9w + 1)$

39. $81w^3 - w^2$ $w^2(81w - 1)$

40. $12x^2 + 36x + 27$ $3(2x + 3)^2$

41. $10x^2y^2 + xy^2 - 9y^2$ $y^2(10x - 9)(x + 1)$

42. $6w^2 - 11w - 35$ $(3w + 5)(2w - 7)$

43. $2x^2y^2 + xy^2 + 3y^2$ $y^2(2x^2 + x + 3)$

44. $18x^2 - 6x + 6$ $6(3x^2 - x + 1)$

45. $2x^2 - 28x + 98$ $2(x - 7)^2$

46. $12y^2 + 24y + 12$ $12(y + 1)^2$

47. $3x^2z - 3zx - 18z$ $3z(x - 3)(x + 2)$

48. $a^2b + 2ab - 15b$ $b(a + 5)(a - 3)$

49. $a^2 + 2ab - 15b^2$ $(a + 5b)(a - 3b)$ **50.** $a^2b^2 - 2a^2b - 15a^2$ $a^2(b - 5)(b + 3)$ **51.** $6t^3 + t^2 - 2t$ $t(3t + 2)(2t - 1)$

52. $36t^2 + 6t - 12$ $6(3t + 2)(2t - 1)$ **53.** $12t^4 - 2t^3 - 4t^2$ $2t^2(3t - 2)(2t + 1)$ **54.** $12t^3 + 14t^2 + 4t$ $2t(3t + 2)(2t + 1)$

55. $4x^2y - 8xy^2 + 3y^3$ $y(2x - y)(2x - 3y)$ **56.** $9x^2 + 24xy - 9y^2$ $3(3x - y)(x + 3y)$

4.5 More Factoring

IN THIS SECTION:

- **The Factoring Strategy**
- **Using Division in Factoring**

In previous sections we established the general idea of factoring and some special cases. In this section we will summarize all of the factoring that we have done with a factoring strategy and see how division relates to factoring.

The Factoring Strategy

The following strategy is a summary of the ideas that we use to factor a polynomial completely.

STRATEGY
Factoring Polynomials
Completely

> 1. If there are any common factors, factor them out first.
> 2. When factoring a binomial, check to see whether it is the difference of two squares. Remember that a sum of two squares does not factor.
> 3. When factoring a trinomial, check to see whether it is a perfect square trinomial.
> 4. When factoring a trinomial that is not a perfect square, use the *ac* method.
> 5. If the polynomial has four terms, try factoring by grouping.

We will use the factoring strategy in the next example.

EXAMPLE 1 Factor each polynomial completely:

a) $2a^2b - 24ab + 72b$
b) $3x^3 + 6x^2 - 75x - 150$

Solution

a) $2a^2b - 24ab + 72b = 2b(a^2 - 12a + 36)$ **First factor out the GCF 2b.**

                    $= 2b(a - 6)^2$ **Factor the perfect square trinomial.**

b) $3x^3 + 6x^2 - 75x - 150 = 3[x^3 + 2x^2 - 25x - 50]$ **First factor out the GCF 3.**

                    $= 3[x^2(x + 2) - 25(x + 2)]$ **Factor by grouping.**

                    $= 3(x^2 - 25)(x + 2)$ **Factor out x + 2.**

                    $= 3(x + 5)(x - 5)(x + 2)$ **Factor the difference of two squares.** ◄

Using Division in Factoring

To find the prime factorization for the integer 1001, we try possible factors (prime numbers) until we find one that divides into 1001 without remainder. If we are told that 13 is a factor, then we do not need to waste time trying numbers that are not factors. We divide 1001 by 13 and get 77. Thus

$$1001 = 77 \cdot 13$$
$$= 7 \cdot 11 \cdot 13 \quad \text{**Factor 77 as 7 · 11.**}$$

We can use the same idea with polynomials that are of higher degree than the ones we have been factoring. If we can guess a factor or if we are given a factor, we can use division to find the other factor and then proceed to factor the polynomial completely. Of course, it is harder to guess a factor of a polynomial than it is to guess a factor of an integer. In the next example we will factor a third-degree polynomial completely, given one factor.

EXAMPLE 2   Factor the polynomial $x^3 + 2x^2 - 5x - 6$ completely, given that the binomial $x + 1$ is a factor of the polynomial.

We gave $x + 1$ as a factor in this example, but we could have guessed at a factor and checked by division until we found a factor. Ask the students to give other binomials that would be intelligent guesses.

Solution   Divide the polynomial by the binomial:

$$
\begin{array}{r}
x^2 + x - 6 \\
x + 1 \overline{) x^3 + 2x^2 - 5x - 6} \\
\underline{x^3 + x^2} \\
x^2 - 5x \\
\underline{x^2 + x} \qquad \text{$-5x - x = -6x$} \\
-6x - 6 \\
\underline{-6x - 6} \qquad \text{$-6 - (-6) = 0$} \\
0
\end{array}
$$

Since the remainder is 0, the dividend is the divisor times the quotient:

$$x^3 + 2x^2 - 5x - 6 = (x + 1)(x^2 + x - 6)$$
$$= (x + 1)(x + 3)(x - 2) \quad \text{**Factor the trinomial.**} \quad ◄$$

Warm-ups

True or false?

1. $x^2 - 4 = (x - 2)^2$ for any value of x. F
2. The trinomial $4x^2 + 6x + 9$ is a perfect square trinomial. F
3. The polynomial $4y^2 + 25$ is a prime polynomial. T
4. $3y + ay + 3x + ax = (x + y)(3 + a)$ for any values of the variables. T
5. The polynomial $3x^2 + 51$ cannot be factored. F
6. To factor a polynomial completely, always factor out the GCF first. T
7. $x^2 + 9 = (x + 3)^2$ for any value of x. F
8. The polynomial $x^2 - 3x - 5$ is a prime polynomial. T
9. The polynomial $y^2 - 5y - my + 5m$ can be factored by grouping. T
10. The polynomial $x^2 + ax - 3x + 3a$ can be factored by grouping. F

4.5 EXERCISES

Factor each polynomial completely. Watch for prime polynomials. See Example 1.

1. $2x^2 - 18$ $2(x - 3)(x + 3)$
2. $3x^3 - 12x$ $3x(x - 2)(x + 2)$
3. $4x^2 + 8x - 60$ $4(x + 5)(x - 3)$
4. $3x^2 + 18x + 27$ $3(x + 3)^2$
5. $x^3 + 4x^2 + 4x$ $x(x + 2)^2$
6. $a^3 - 5a^2 + 6a$ $a(a - 3)(a - 2)$
7. $5max^2 + 20ma$ $5am(x^2 + 4)$
8. $3bmw^2 - 12bm$ $3bm(w - 2)(w + 2)$
9. $9x^2 + 6x + 1$ $(3x + 1)^2$
10. $9x^2 + 6x + 3$ $3(3x^2 + 2x + 1)$
11. $6x^2y + xy - 2y$ $y(3x + 2)(2x - 1)$
12. $5x^2y^2 - xy^2 - 6y^2$ $y^2(5x - 6)(x + 1)$
13. $y^2 + 10y - 25$ Prime
14. $8b^2 + 24b + 18$ $2(2b + 3)^2$
15. $16m^2 - 4m - 2$ $2(4m + 1)(2m - 1)$
16. $32a^2 + 4a - 6$ $2(2a + 1)(8a - 3)$
17. $9a^2 + 24a + 16$ $(3a + 4)^2$
18. $3x^2 - 18x - 48$ $3(x - 8)(x + 2)$
19. $24x^2 - 26x + 6$ $2(3x - 1)(4x - 3)$
20. $4x^2 - 6x - 12$ $2(2x^2 - 3x - 6)$
21. $3a^2 - 27a$ $3a(a - 9)$
22. $a^2 - 25a$ $a(a - 25)$
23. $8 - 2x^2$ $2(2 - x)(2 + x)$
24. $x^3 + 6x^2 + 9x$ $x(x + 3)^2$
25. $6x^3 - 5x^2 + 12x$ $x(6x^2 - 5x + 12)$
26. $x^3 + 2x^2 - x - 2$ $(x - 1)(x + 1)(x + 2)$
27. $a^3b - 4ab$ $ab(a - 2)(a + 2)$
28. $2m^2 - 1800$ $2(m - 30)(m + 30)$
29. $x^3 + 2x^2 - 4x - 8$ $(x - 2)(x + 2)^2$
30. $m^2a + 2ma^2 + a^3$ $a(m + a)^2$
31. $3a^2w - 18aw + 27w$ $3w(a - 3)^2$
32. $8a^3 + 4a$ $4a(2a^2 + 1)$
33. $5x^2 - 500$ $5(x - 10)(x + 10)$
34. $25x^2 - 16y^2$ $(5x - 4y)(5x + 4y)$
35. $2m + 2n - wm - wn$ $(2 - w)(m + n)$
36. $aw - bw + 5a - 5b$ $(w + 5)(a - b)$
37. $4w^2 + 4w - 4$ $4(w^2 + w - 1)$
38. $4w^2 + 8w - 5$ $(2w + 5)(2w - 1)$
39. $a^4 + 7a^3 - 30a^2$ $a^2(a + 10)(a - 3)$
40. $2y^5 + 3y^4 - 20y^3$ $y^3(2y - 5)(y + 4)$
41. $4aw^3 - 12aw^2 + 9aw$ $aw(2w - 3)^2$
42. $9bn^3 + 15bn^2 - 14bn$ $bn(3n - 2)(3n + 7)$
43. $t^2 + 6t + 9$ $(t + 3)^2$
44. $t^3 + 12t^2 + 36t$ $t(t + 6)^2$

Factor each polynomial completely, given that the binomial following it is a factor of the polynomial. See Example 2.

45. $x^3 + 3x^2 - 10x - 24, x + 4$ $(x + 4)(x - 3)(x + 2)$
46. $x^3 - 7x + 6, x - 1$ $(x - 1)(x - 2)(x + 3)$
47. $x^3 + 4x^2 + x - 6, x - 1$ $(x - 1)(x + 3)(x + 2)$
48. $x^3 - 5x^2 - 2x + 24, x + 2$ $(x + 2)(x - 3)(x - 4)$

49. $x^3 - 8,\ x - 2$ $(x - 2)(x^2 + 2x + 4)$

51. $x^3 + 4x^2 - 3x + 10,\ x + 5$ $(x + 5)(x^2 - x + 2)$

53. $x^3 + 2x^2 + 2x + 1,\ x + 1$ $(x + 1)(x^2 + x + 1)$

50. $x^3 + 27,\ x + 3$ $(x + 3)(x^2 - 3x + 9)$

52. $2x^3 - 5x^2 - x - 6,\ x - 3$ $(x - 3)(2x^2 + x + 2)$

54. $x^3 + 2x^2 - 5x - 6,\ x + 3$ $(x + 3)(x - 2)(x + 1)$

4.6 Using Factoring to Solve Equations

IN THIS SECTION:

- **The Zero Factor Property**
- **Applications**

The techniques of factoring can be used to solve equations involving polynomials. These equations cannot be solved by the other methods that we have learned. After we learn to solve equations by factoring, we will use this technique to solve some new types of problems.

The Zero Factor Property

Note that the zero factor property gives us an equivalent equation, but it is not a "do something to each side" property.

Consider the equation

$$x^2 + x - 6 = 0.$$

We can factor the left-hand side to get $(x + 3)(x - 2) = 0$. This equation states that the product of $x + 3$ and $x - 2$ is 0. The product of two numbers is 0 only when one or the other of the numbers is 0. Therefore the equation

$$(x + 3)(x - 2) = 0$$

is equivalent to the equation

$$x + 3 = 0 \qquad \text{or} \qquad x - 2 = 0.$$

A sentence made up of two or more equations connected with the word "or" is called a **compound equation.** From the compound equation we see that either -3 or 2 will satisfy the original equation. Check these solutions in the original equation as follows:

$$(-3)^2 + (-3) - 6 = 9 - 3 - 6 = 0$$
$$(2)^2 + 2 - 6 = 4 + 2 - 6 = 0$$

We call the idea used here the **zero factor property** and state it as follows.

PROPERTY
Zero Factor Property

The equation $a \cdot b = 0$ is equivalent to the compound equation

$$a = 0 \quad \text{or} \quad b = 0.$$

EXAMPLE 1 Solve the equation $6x^2 + 3x = 0$.

Solution

$$3x(2x + 1) = 0 \qquad \textbf{First factor the left-hand side.}$$

$$3x = 0 \quad \text{or} \quad 2x + 1 = 0 \qquad \textbf{Apply the zero factor property.}$$

$$x = 0 \quad \text{or} \quad x = -\frac{1}{2} \qquad \textbf{Solve each equation.}$$

There are two solutions to the original equation, 0 and $-1/2$. Check these solutions in the original equation as follows.

$$6(0)^2 + 3(0) = 0$$

$$6\left(-\frac{1}{2}\right)^2 + 3\left(-\frac{1}{2}\right) = 6\left(\frac{1}{4}\right) - \frac{3}{2} = \frac{3}{2} - \frac{3}{2} = 0 \qquad \blacktriangleleft$$

The steps used to solve an equation by factoring are listed in the following box.

STRATEGY
Solving an Equation by
Factoring

1. Write the equation with 0 on the right-hand side.
2. Factor the left-hand side.
3. Use the zero factor property to get two simpler linear equations.
4. Solve the two linear equations.
5. Check the answers in the original equation.

If there are more than two factors, we can write an equivalent equation by setting each factor equal to zero.

EXAMPLE 2 Solve $2x^3 - x^2 - 8x + 4 = 0$.

Solution

$$x^2(2x - 1) - 4(2x - 1) = 0 \qquad \textbf{Factor by grouping.}$$

$$(x^2 - 4)(2x - 1) = 0 \qquad \textbf{Factor out } 2x - 1.$$

$$(x - 2)(x + 2)(2x - 1) = 0 \qquad \textbf{Factor the difference of two squares.}$$

$$x - 2 = 0 \quad \text{or} \quad x + 2 = 0 \quad \text{or} \quad 2x - 1 = 0 \qquad \textbf{Apply the zero factor property.}$$

$$x = 2 \quad \text{or} \quad x = -2 \quad \text{or} \quad x = \frac{1}{2} \qquad \textbf{Solve each equation.}$$

Blackboard proof:

Given

Draw two squares with sides of length $a + b$, but partition them as shown:

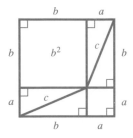

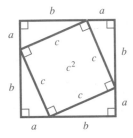

These two big squares have the same area to begin with. After erasing the four triangles from each picture, we are left with equal areas:

$a^2 + b^2 = c^2$

THEOREM
The Pythagorean Theorem

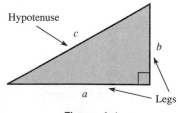

Figure 4.1

The solutions to this equation are -2, $1/2$, and 2. The reader should check that all three numbers satisfy the original equation. ◀

Applications

There are many problems that can be solved by equations like those we have just discussed.

EXAMPLE 3 Merida's garden has a rectangular shape with a length that is 1 foot longer than twice the width. If the area of the garden is 55 square feet, then what are the dimensions of the garden?

Solution If x represents the width of the garden, then $2x + 1$ is the length. Since the area of a rectangle is the length times the width, we can write the equation

$$x(2x + 1) = 55.$$

We must have zero on the right-hand side of the equation to use the zero factor property. The equation can be rewritten as

$$2x^2 + x - 55 = 0$$
$$(2x + 11)(x - 5) = 0 \qquad \text{Factor.}$$

$$2x + 11 = 0 \qquad \text{or} \qquad x - 5 = 0 \qquad \text{Apply the zero factor property.}$$

$$x = -\frac{11}{2} \qquad \text{or} \qquad x = 5 \qquad \text{Solve each equation.}$$

The width is certainly not $-11/2$. We use $x = 5$ to get a width of 5 feet and a length of $2(5) + 1 = 11$ feet. We check by multiplying 11 feet and 5 feet to get the area of 55 square feet. ◀

The next application involves a theorem from geometry called the Pythagorean Theorem. This theorem says that in any right triangle, the sum of the squares of the lengths of the legs is equal to the square of the length of the hypotenuse.

The triangle pictured in Fig. 4.1 is a right triangle if and only if
$$a^2 + b^2 = c^2.$$

EXAMPLE 4 The length of a rectangular carpet is 1 meter longer than the width, and the diagonal measures 5 meters. What are the length and width?

Solution If x represents the width of the rectangle, then $x + 1$ represents the length. Since the two sides are the legs of a right triangle, we can use the Pythago-

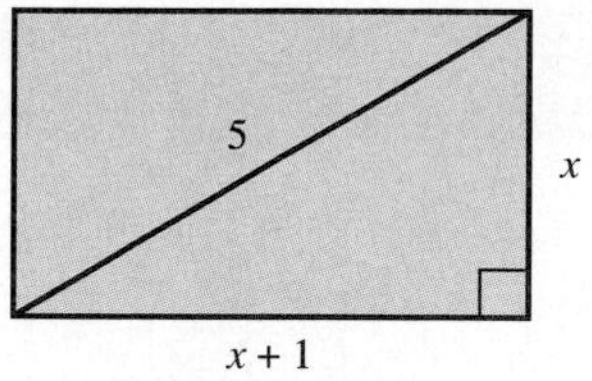

Figure 4.2

rean Theorem to get a relationship between the length, width, and diagonal. See Fig. 4.2.

$$x^2 + (x + 1)^2 = 5^2 \qquad \text{The Pythagorean Theorem}$$
$$x^2 + x^2 + 2x + 1 = 25 \qquad \text{Simplify.}$$
$$2x^2 + 2x - 24 = 0 \qquad \text{Combine like terms.}$$
$$x^2 + x - 12 = 0 \qquad \text{Divide each side by 2.}$$
$$(x - 3)(x + 4) = 0 \qquad \text{Factor.}$$
$$x - 3 = 0 \quad \text{or} \quad x + 4 = 0 \qquad \text{Apply the zero factor property.}$$
$$x = 3 \quad \text{or} \quad x = -4$$
$$x + 1 = 4$$

The length of a side of a rectangle cannot be -4. If $x = 3$, then $x + 1 = 4$, and the rectangle is 3 meters by 4 meters. To check this answer, we compute

$$3^2 + 4^2 = 5^2$$

or

$$9 + 16 = 25.$$ ◄

Warm-ups

True or false?

1. The equation $(x + 2)(x - 1) - 1$ is equivalent to $x + 2 - 0$ or $x - 1 = 0$. F
2. Equations solved by factoring always have two different solutions. F
3. The equation $a \cdot d = 0$ is equivalent to $a = 0$ or $d = 0$. T
4. If the length of a room is 5 feet longer than the width x, then the area could be expressed as $x(x + 5)$ square feet. T
5. Both 1 and -4 are solutions to the equation $(x - 1)(x + 4) = 0$. T
6. In any triangle the sum of the squares of the smaller two sides equals the square of the largest side. F
7. If the perimeter of a rectangular room is 50 feet, then the sum of the length and width is 25 feet. T
8. Equations solved by factoring may have more than two solutions. T
9. Both 0 and 2 are solutions to the equation $x(x - 2) = 0$. T
10. The solutions to the equation $3(x - 2)(x + 5) = 0$ are 3, 2, and -5. F

4.6 EXERCISES

Solve each equation. See Examples 1 and 2.

1. $(x + 5)(x + 4) = 0$ $-4, -5$

2. $(a + 6)(a + 5) = 0$ $-6, -5$

3. $(2x + 5)(3x - 4) = 0$ $-\dfrac{5}{2}, \dfrac{4}{3}$

4. $(3k - 8)(4k + 3) = 0$ $-\dfrac{3}{4}, \dfrac{8}{3}$

5. $w^2 - 9w + 14 = 0$ $2, 7$

6. $t^2 + 6t - 27 = 0$ $-9, 3$

7. $m^2 + 7m = 0$ $0, -7$

8. $h^2 + 5h = 0$ $-5, 0$

9. $a^2 + a = 20$ $-5, 4$

10. $p^2 + p = 42$ $-7, 6$

11. $2x^2 + 5x = 3$ $\dfrac{1}{2}, -3$

12. $3x^2 - 10x = -7$ $1, \dfrac{7}{3}$

13. $(x + 2)(x + 6) = 12$ $0, -8$

14. $(x + 2)(x - 6) = 20$ $-4, 8$

15. $3x^2 + 15x + 18 = 0$ $-3, -2$

16. $-2x^2 - 2x + 24 = 0$ $-4, 3$

17. $z^2 + \dfrac{11}{2}z = -6$ $-\dfrac{3}{2}, -4$

18. $m^2 + \dfrac{8}{3}m = 1$ $-3, \dfrac{1}{3}$

19. $x^2 - 16 = 0$ $-4, 4$

20. $x^2 - 36 = 0$ $-6, 6$

21. $x^2 = 9$ $-3, 3$

22. $x^2 = 25$ $-5, 5$

23. $x^3 - 9x = 0$ $0, -3, 3$

24. $25x - x^3 = 0$ $-5, 0, 5$

25. $a^3 = a$ $0, -1, 1$

26. $x^3 = 4x$ $-2, 0, 2$

27. $w^3 + 4w^2 - 4w - 16 = 0$ $-4, -2, 2$

28. $a^3 + 2a^2 - a - 2 = 0$ $-2, -1, 1$

29. $n^3 - 3n^2 - n + 3 = 0$ $-1, 1, 3$

30. $w^3 + w^2 - 25w - 25 = 0$ $-5, -1, 5$

31. $y^3 - 9y^2 + 20y = 0$ $0, 4, 5$

32. $m^3 + 2m^2 - 3m = 0$ $-3, 0, 1$

Solve each problem. See Examples 3 and 4.

33. The perimeter of a rectangle is 34 feet, and the diagonal is 13 feet long. What are the length and width of the rectangle? 12 feet, 5 feet

34. The perimeter of the cover of an address book is 14 inches, and the diagonal measures 5 inches. What are the length and width of the book? $L = 4$ inches, $W = 3$ inches

Figure for Exercise 34

35. The length of Violla's bathroom is 2 feet longer than twice the width. If the diagonal measures 13 feet, then what are the length and width? 12 feet, 5 feet

36. One side of the rectangular base of a stage is 2 meters longer than the other. If the diagonal is 10 meters, then what are the lengths of the sides? 6 meters, 8 meters

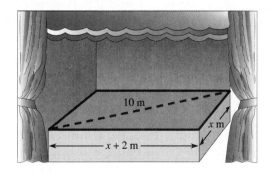

Figure for Exercise 36

37. The sum of the squares of two consecutive integers is 13. Find the integers. 2, 3 or $-3, -2$

38. The sum of the squares of two consecutive even integers is 52. Find the integers. 4, 6 or $-6, -4$

39. The sum of two numbers is 11, and their product is 30. Find the numbers.   5 and 6

40. Molly's age is twice Anita's. If the sum of the squares of their two ages is 80, then what are their ages?   4 and 8

41. If an object is dropped from a height of s_0 feet, then its altitude S after t seconds is given by the formula $S = -16t^2 + s_0$. If a sandbag is dropped by a glider-pilot from a height of 784 feet, then how long does it take for the sandbag to reach the ground? (On the ground, $S = 0$.)   7 seconds

42. If the glider-pilot of Exercise 41 throws his sandbag downward from an altitude of 128 feet with an initial velocity of 32 feet per second, then its altitude after t seconds is given by the formula $S = -16t^2 - 32t + 128$. How long does it take for the sandbag to reach the earth?   2 seconds

43. One end of a glass prism is in the shape of a triangle that has a height that is 1 inch longer than twice the base. If the area of the triangle is 39 square inches, then how long are the base and height?   6 inches, 13 inches

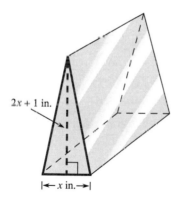

Figure for Exercise 43

44. The radius of a circle is 1 meter longer than the radius of another circle. If their areas differ by 5π square meters, then what is the radius of each?   2 meters and 3 meters

45. Last year Otto's garden was square in shape. This year he plans to make it smaller by shortening one side by 5 feet and the other by 8 feet. If the area of the smaller garden will be 180 square feet, then what was the size of Otto's garden last year?   20 feet by 20 feet

46. Rosita's Christmas present from Carlos is in a box that has a width that is 3 inches shorter than the height. The length of the base is 5 inches longer than the height. If the area of the base is 84 square inches, then what is the height of the package?   9 inches

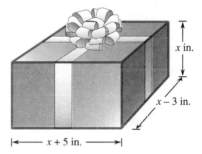

Figure for Exercise 46

47. Imelda and Gordon have designed a new kite. While Imelda is flying the kite, Gordon is standing directly below the kite. The kite is designed so that its altitude is always 20 feet larger than the distance between Imelda and Gordon. What is the altitude of the kite when it is 100 feet from Imelda?   80 feet

48. A car is traveling on a road that is perpendicular to a railroad track. When the car is 30 meters from the crossing, the car's new collision detector warns the driver that there is a train 50 meters from the car and heading toward the same crossing. How far is the train from the crossing?

49. Virginia is buying carpet for two square rooms. One room is 3 yards wider than the other. If she needs 45 square yards of carpet, then what are the dimensions of each room?   3 yards by 3 yards and 6 yards by 6 yards

50. Ahmed has one-half of a treasure map, which indicates that the treasure is buried in the desert $2x + 6$ paces from Castle Rock. Vanessa has the other half, which indicates that to find the treasure one must get to Castle Rock, walk x paces to the north, and then $2x + 4$ paces to the east. If they pool their information to save a lot of digging, then what is the value of x?   10

48. 40 meters

Wrap-up

CHAPTER 4

SUMMARY

	Concepts	Examples
Prime number	A positive integer larger than 1 that has no integral factors other than itself and 1	2, 3, 5, 7, 11
Factor a polynomial	Write a polynomial as a product of simpler polynomials	$2x + 6 = 2(x + 3)$

	Types of Factoring	Examples
Finding the GCF for a group of monomials	1. Find the GCF for the coefficients of the monomials.	The GCF for $2x^3$ and $4x$ is $2x$.
	2. Form the product of the GCF of the coefficients and each variable that is common to all of the monomials, where the exponent on each variable equals the lowest power of that variable in any of the monomials.	$2x^3 - 4x = 2x(x^2 - 2)$
Difference of two squares	$a^2 - b^2 = (a + b)(a - b)$	$m^2 - 9 = (m - 3)(m + 3)$ Sum of 2 squares does not factor.
Identifying a perfect square trinomial	A trinomial is a perfect square trinomial if: 1. the first and last terms are perfect squares, and 2. the middle term is $+2$ or -2 times the product of the square roots of the first and last terms. $$a^2 + 2ab + b^2 = (a + b)^2$$ $$a^2 - 2ab + b^2 = (a - b)^2$$	$x^2 + 6x + 9 = (x + 3)^2$ $x^2 - 6x + 9 = (x - 3)^2$
Grouping	Factor common factors from groups of terms.	$6x + 6w + ax + aw$ $= 6(x + w) + a(x + w)$ $= (6 + a)(x + w)$

Factoring
$ax^2 + bx + c$
by the *ac*
method

1. Find two numbers that have a product equal to *ac* and a sum equal to *b*.
2. Replace *bx* by two terms using the two new numbers as coefficients.
3. Factor the resulting four-term polynomial by grouping.

$6x^2 + 17x + 12$
$= 6x^2 + 9x + 8x + 12$
$= (2x + 3)3x + (2x + 3)4$

$= (2x + 3)(3x + 4)$

Other Important Facts

Examples

Factoring
polynomials
completely

1. If there are any common factors, factor them out first.
2. When factoring a binomial, check to see if it is the difference of two squares. Remember that a sum of two squares does not factor.
3. When factoring a trinomial, check to see whether it is a perfect square trinomial.
4. When factoring a trinomial that is not a perfect square, use the *ac* method.
5. If the polynomial has 4 terms, try factoring by grouping.

Zero factor
property

The equation $a \cdot b = 0$ is equivalent to
$\qquad a = 0 \qquad$ or $\qquad b = 0.$

Solving an
equation
by factoring

1. Write the equation with 0 on the right side.
2. Factor the left side.
3. Use the zero factor property to get two simpler linear equations.
4. Solve the two linear equations.
5. Check the answers in the original equation.

$x^2 + 3x - 18 = 0$
$(x + 6)(x - 3) = 0$
$x + 6 = 0 \quad$ or $\quad x - 3 = 0$

$x = -6 \quad$ or $\qquad x = 3$
$(-6)^2 + 3(-6) - 18 = 0$
$\qquad (3)^2 + 3(3) - 18 = 0$

The
Pythagorean
Theorem

In any right triangle, the sum of the squares of the lengths of the legs is equal to the square of the length of the hypotenuse.

$a^2 + b^2 = c^2$

REVIEW EXERCISES

4.1 *Find the prime factorization for each integer.*

1. 144 $2^4 \cdot 3^2$ **2.** 121 11^2 **3.** 58 $2 \cdot 29$ **4.** 76 $2^2 \cdot 19$ **5.** 150 $2 \cdot 3 \cdot 5^2$ **6.** 200 $2^3 \cdot 5^2$

Find the greatest common factor for each group of integers.

7. 24, 56   8   **8.** 36, 90   18   **9.** 30, 42, 78   6   **10.** 44, 48, 120   4

Complete the factorization by filling in the parentheses.

11. $3x + 6 = 3(\quad)$ $x + 2$ **12.** $7x^2 + x = x(\quad)$ $7x + 1$

13. $2a - 20 = -2(\quad)$ $-a + 10$ **14.** $a^2 - a = -a(\quad)$ $-a + 1$

15. $2a - a^2 = -a(\quad)$ $-2 + a$

16. $3b - 9 = (\quad)(3 - b)$ -3

17. $2m - 10 = (\quad)(m - 5)$ 2

18. $3x - 6 = (\quad)(x - 2)$ 3

19. $m - n = -1(\quad)$ $-m + n$

20. $y - 9 = -1(\quad)$ $-y + 9$

21. $x + b = -1(\quad)$ $-x - b$

22. $x - 6 = (\quad)(6 - x)$ -1

23. $a - 7 = -(\quad)$ $-a + 7$

24. $w + 3 = -(\quad)$ $-w - 3$

25. $6x^2y^2 - 9x^5y = (\quad)(2y - 3x^3)$ $3x^2y$

26. $a^3b^5 + a^3b^2 = a^3b^2(\quad)$ $b^3 + 1$

27. $3x^2y - 12xy - 9y^2 = 3y(\quad)$ $x^2 - 4x - 3y$

28. $-2a^2 - 4ab^2 - 2ab = -2a(\quad)$ $a + 2b^2 + b$

4.2 *Factor each polynomial completely.*

29. $y^2 - 400$ $(y - 20)(y + 20)$

30. $4m^2 - 9$ $(2m - 3)(2m + 3)$

31. $w^2 - 8w + 16$ $(w - 4)^2$

32. $t^2 + 20t + 100$ $(t + 10)^2$

33. $4y^2 + 20y + 25$ $(2y + 5)^2$

34. $2a^2 - 4a - 2$ $2(a^2 - 2a - 1)$

35. $r^2 - 4r + 4$ $(r - 2)^2$

36. $3m^2 - 75$ $3(m - 5)(m + 5)$

37. $8t^3 - 24t^2 + 18t$ $2t(2t - 3)^2$

38. $t^2 - 9w^2$ $(t - 3w)(t + 3w)$

39. $x^2 + 12xy + 36y^2$ $(x + 6y)^2$

40. $9y^2 - 12xy + 4x^2$ $(3y - 2x)^2$

41. $x^2 + 5x - xy - 5y$ $(x - y)(x + 5)$

42. $x^2 + xy + ax + ay$ $(x + a)(x + y)$

4.3 *Factor each polynomial.*

43. $b^2 + 5b - 24$ $(b + 8)(b - 3)$

44. $a^2 - 2a - 35$ $(a - 7)(a + 5)$

45. $r^2 - 4r - 60$ $(r - 10)(r + 6)$

46. $x^2 + 13x + 40$ $(x + 5)(x + 8)$

47. $y^2 - 6y - 55$ $(y - 11)(y + 5)$

48. $a^2 + 6a - 40$ $(a - 4)(a + 10)$

4.4 *Factor each polynomial.*

49. $14t^2 + t - 3$ $(7t - 3)(2t + 1)$

50. $15x^2 - 22x - 5$ $(5x + 1)(3x - 5)$

51. $6x^2 - 19x - 7$ $(3x + 1)(2x - 7)$

52. $2x^2 - x - 10$ $(2x - 5)(x + 2)$

53. $6p^2 + 5p - 4$ $(3p + 4)(2p - 1)$

54. $3p^2 + 2p - 5$ $(3p + 5)(p - 1)$

4.5 *Factor completely.*

55. $5x^3 + 40x$ $5x(x^2 + 8)$

56. $w^2 + 6w + 9$ $(w + 3)^2$

57. $9x^2 + 3x - 2$ $(3x - 1)(3x + 2)$

58. $ax^3 + ax$ $ax(x^2 + 1)$

59. $x^3 + 2x^2 - x - 2$ $(x + 2)(x - 1)(x + 1)$

60. $16x^2 - 2x - 3$ $(8x + 3)(2x - 1)$

61. $x^2y - 16xy^2$ $xy(x - 16y)$

62. $-3x^2 + 27$ $-3(x - 3)(x + 3)$

63. $a^2 + 2a + 1$ $(a + 1)^2$

64. $-2w^2 - 12w - 18$ $-2(w + 3)^2$

65. $x^3 - x^2 + x - 1$ $(x^2 + 1)(x - 1)$

66. $9x^2y^2 - 9y^2$ $9y^2(x + 1)(x - 1)$

67. $a^2 + ab + 2a + 2b$ $(a + 2)(a + b)$

68. $4m^2 + 20m + 25$ $(2m + 5)^2$

69. $-2x^2 + 16x - 24$ $-2(x - 6)(x - 2)$

70. $6x^2 + 21x - 45$ $3(2x - 3)(x + 5)$

Factor each polynomial completely, given that the binomial following it is a factor of the polynomial.

71. $x^3 + x + 10, \ x + 2$ $(x + 2)(x^2 - 2x + 5)$

72. $x^3 - 5x - 12, \ x - 3$ $(x - 3)(x^2 + 3x + 4)$

73. $x^3 + 6x^2 - 7x - 60, \ x + 4$ $(x + 4)(x + 5)(x - 3)$

74. $x^3 - 125, \ x - 5$ $(x - 5)(x^2 + 5x + 25)$

4.6 *Solve each equation.*

75. $x^3 - 5x^2 = 0$ $0, 5$

76. $2m^2 + 10m + 12 = 0$ $-3, -2$

77. $(a - 2)(a - 3) = 6$ $0, 5$

78. $(w - 2)(w + 3) = 50$ $-8, 7$

79. $2m^2 - 9m - 5 = 0$ $-\dfrac{1}{2}, 5$

80. $m^3 + 4m^2 - 9m - 36 = 0$ $-4, -3, 3$

81. $w^3 + 5w^2 - w - 5 = 0$ $-5, -1, 1$

82. $12x^2 + 5x - 3 = 0$ $-\dfrac{3}{4}, \dfrac{1}{3}$

Solve each problem.

83. Two positive numbers differ by 6, and their squares differ by 96. Find the numbers.   5, 11

84. Find three consecutive integers such that the sum of their squares is 77.   4, 5, 6 or $-6, -5, -4$

85. The perimeter of a notebook is 28 inches, and the diagonal measures 10 inches. What are the length and width of the notebook?   6 inches, 8 inches

86. The sum of two numbers is 8.5, and their product is 18. Find the numbers. $4, \dfrac{9}{2}$

CHAPTER 4 TEST

Give the prime factorization for each integer.

1. 66 $2 \cdot 3 \cdot 11$

2. 336 $2^4 \cdot 3 \cdot 7$

Find the greatest common factor (GCF) for each group of integers.

3. 48, 80   16

4. 42, 66, 78   6

Factor each polynomial completely.

5. $5x^2 - 10x$ $5x(x - 2)$

6. $6x^2y^2 + 12xy^2 + 12y^2$ $6y^2(x^2 + 2x + 2)$

7. $3a^3b - 3ab^3$ $3ab(a - b)(a + b)$

8. $a^2 + 2a - 24$ $(a + 6)(a - 4)$

9. $4b^2 - 28b + 49$ $(2b - 7)^2$

10. $3m^3 + 27m$ $3m(m^2 + 9)$

11. $ax - ay + bx - by$ $(a + b)(x - y)$

12. $ax - 2a - 5x + 10$ $(a - 5)(x - 2)$

13. $6b^2 - 7b - 5$ $(3b - 5)(2b + 1)$

14. $m^2 + 4mn + 4n^2$ $(m + 2n)^2$

15. $2a^2 - 13a + 15$ $(2a - 3)(a - 5)$

16. $z^3 + 9z^2 + 18z$ $z(z + 3)(z + 6)$

Factor the polynomial completely, given that $x - 1$ is a factor.

17. $x^3 - 6x^2 + 11x - 6$ $(x - 1)(x - 2)(x - 3)$

Solve each equation.

18. $2x^2 + 5x - 12 = 0$ $\dfrac{3}{2}, -4$

19. $3x^3 - 12x = 0$ $0, -2, 2$

Write a complete solution to the problem.

20. If the length of a rectangle is 3 feet longer than the width and the diagonal is 15 feet, then what are the length and width?   12 foot, 9 foot

Tying It All Together

CHAPTERS 1–4

Simplify each of the following.

1. $\dfrac{91 - 17}{17 - 91}$ -1

2. $\dfrac{4 - 18}{-6 - 1}$ 2

3. $5 - 2(7 - 3)$ -3

4. $3^2 - 4(6)(-2)$ 57

5. $2^5 - 2^4$ 16

6. $.07(37) + .07(63)$ 7

Perform the indicated operations.

7. $x \cdot 2x$ $2x^2$

8. $x + 2x$ $3x$

9. $\dfrac{6 + 2x}{2}$ $3 + x$

10. $\dfrac{6 \cdot 2x}{2}$ $6x$

11. $2 \cdot 3y \cdot 4z$ $24yz$

12. $2(3y + 4z)$ $6y + 8z$

13. $2 - (3 - 4z)$ $4z - 1$

14. $t^8 \div t^2$ t^6

15. $t^8 \cdot t^2$ t^{10}

16. $\dfrac{8t^8}{2t^2}$ $4t^6$

Solve each equation.

17. $.05x + .04(x - 40) = 2$ 40

18. $7x - 15 = 0$ $\dfrac{15}{7}$

19. $2x^2 + 7x - 15 = 0$ $-5, \dfrac{3}{2}$

20. $2x^2 + 7x = 0$ $0, -\dfrac{7}{2}$

21. $2x^2 - 18 = 0$ $-3, 3$

22. $3x^3 + 3x = 0$ 0

Solve and graph each inequality.   Graphs in Answers sections.

23. $2x - 5 > 3x + 4$ $x < -9$

24. $4 - 5x \le -11$ $x \ge 3$

25. $-\dfrac{2}{3}x + 3 < -5$ $x > 12$

Solve each equation mentally.

26. $x - 3 = 0$ 3

27. $x + 5 = 0$ -5

28. $2x - 3 = 0$ $\dfrac{3}{2}$

29. $2x + 1 = 0$ $-\dfrac{1}{2}$

30. $(x - 3)(x + 5) = 0$ $3, -5$

31. $(2x - 3)(2x + 1) = 0$ $\dfrac{3}{2}, -\dfrac{1}{2}$

32. $x(x - 3) = 0$ $0, 3$

33. $8x = 0$ 0

34. $2x - 3 = 9$ 6

35. $2x + 1 = -7$ -4

36. $3x - 3x = 0$ All real numbers

37. $3x - 3x = 1$ No solution

38. $\dfrac{1}{2}x = -5$ -10

39. $-3x = 1$ $-\dfrac{1}{3}$

40. $-6 + 2x = 3x$ -6

CHAPTER 5

Rational Expressions

Tracy, Stacy, and Fred all worked together to assemble a very large puzzle in 40 hours. Tracy observed that she found two pieces for every one found by Fred, while Stacy seemed to be working just as fast as Tracy. How long would it take for Fred to assemble the puzzle by himself?

This problem is called a work problem. In a work problem we assume that the people do not interfere with one another when they work together. In other words, Tracy finds the same number of pieces per hour whether or not the others are working on the puzzle. The same goes for Stacy and Fred. This problem can be solved by using an equation involving rational expressions. In this chapter we will learn the fundamental operations with rational expressions, and we will learn to solve problems using them. When we see this problem as Exercise 55 of the Chapter Review, we will have the required skills to solve it.

5.1 Properties of Rational Expressions

IN THIS SECTION:

- **Rational Expressions**
- **Reducing to Lowest Terms**
- **Reducing with the Quotient Rule**
- **Dividing** $a - b$ **by** $b - a$.
- **Factoring Out a Common Factor with a Negative Sign**

Rational expressions are handled just like the rational numbers of arithmetic. In this section we will learn the basic ideas of rational expressions.

Rational Expressions

Note how the operations that we learned to perform with polynomials parallel the operations performed on whole numbers. The operations performed with rational expressions parallel the operations performed with fractions.

A rational number is the ratio of two integers with the denominator not equal to 0. For example,

$$\frac{3}{4}, \quad \frac{-9}{-6} \quad 7, \quad \text{and} \quad 0$$

are rational numbers. A **rational expression** is the ratio of two polynomials with the denominator not equal to 0. Since an integer is a monomial, a rational number is a rational expression. The following expressions are rational expressions:

$$\frac{x^2 - 1}{x + 8}, \quad \frac{3a^2 + 5a - 3}{a - 9}, \quad \frac{3}{7}, \quad w - 2$$

If the denominator is 1, it may be omitted, as in the expression $w - 2$.

Since the denominator cannot be zero, any number can be used in place of the variable *except* numbers that cause the denominator to be zero.

Here is where solving equations mentally comes in handy. The students should immediately recognize the solution to $x + 8 = 0$ and $2x + 1 = 0$.

EXAMPLE 1   Which numbers cannot be used in place of x in each rational expression?

a) $\dfrac{x^2 - 1}{x + 8}$ **b)** $\dfrac{x + 5}{x^2 - 4}$ **c)** $\dfrac{x + 2}{2x + 1}$

Solution

a) The denominator is 0 if $x + 8 = 0$. So -8 cannot be used in place of x.

b) The denominator is zero if $x^2 - 4 = 0$. Solve this equation:

$$x^2 - 4 = 0$$

$$(x - 2)(x + 2) = 0 \qquad \text{Factor.}$$

$$x - 2 = 0 \quad \text{or} \quad x + 2 = 0 \qquad \text{Apply the zero factor property.}$$

$$x = 2 \quad \text{or} \quad x = -2$$

Thus 2 and -2 cannot be used in place of x.

c) The denominator is zero if $2x + 1 = 0$, or $x = -1/2$. So we cannot use $-1/2$ in place of x. ◀

When dealing with rational expressions, we will generally assume that the variables represent numbers for which the denominator is not zero.

Reducing to Lowest Terms

Rational expressions are a generalization of rational numbers. The operations that we perform on rational numbers can be performed on rational expressions in exactly the same manner.

Each rational number can be written in infinitely many equivalent forms. For example,

$$\frac{3}{5} = \frac{6}{10} = \frac{9}{15} = \frac{12}{20} = \frac{15}{25} = \cdots.$$

Each equivalent form of 3/5 is obtained from 3/5 by multiplying both numerator and denominator by the same nonzero number. For example,

$$\frac{3}{5} = \frac{3 \cdot 2}{5 \cdot 2} = \frac{6}{10} \quad \text{and} \quad \frac{3}{5} = \frac{3 \cdot 3}{5 \cdot 3} = \frac{9}{15},$$

If we start with 6/10 and convert it into 3/5, we say that we are **reducing** 6/10 to **lowest terms.** We write this process as

$$\frac{6}{10} = \frac{\cancel{2} \cdot 3}{\cancel{2} \cdot 5} = \frac{3}{5}.$$

A rational number is expressed in lowest terms when the numerator and denominator have no common factors other than 1. In reducing 6/10 we divide the numerator and denominator by the common factor 2.

Although it is true that

$$\frac{6}{10} = \frac{2 + 4}{2 + 8},$$

we cannot divide out the 2's now. Removing them now would not give us 3/5. The 2's in this expression are not factors. *We can reduce fractions only by dividing the numerator and denominator by a common factor.*

RULE
Reducing Fractions

If $a \neq 0$ and $c \neq 0$, then

$$\frac{ab}{ac} = \frac{b}{c}.$$

To reduce rational expressions to lowest terms, we use exactly the same procedure as we use for fractions:

RULE
Reducing Rational Expressions

1. Factor the numerator and denominator completely, then
2. divide the numerator and denominator by the greatest common factor.

EXAMPLE 2   Reduce to lowest terms:

a) $\dfrac{30}{42}$ b) $\dfrac{x^2 - 9}{6x + 18}$

Solution

a) $\dfrac{30}{42} = \dfrac{2 \cdot 3 \cdot 5}{2 \cdot 3 \cdot 7}$ Factor.

$= \dfrac{5}{7}$ Divide out the GCF: $2 \cdot 3$ or 6.

b) $\dfrac{x^2 - 9}{6x + 18} = \dfrac{(x - 3)(x + 3)}{6(x + 3)}$ Factor.

$= \dfrac{x - 3}{6}$ Divide out the GCF: $x + 3$.   ◄

To say that two rational expressions, such as those of Example 2b, are equivalent means that they have the same numerical value for any replacement of the variables. Of course, the replacement must not give us an undefined expression (0 in the denominator). This means that the equation of Example 2b,

$$\frac{x^2 - 9}{6x + 18} = \frac{x - 3}{6},$$

is an identity.

Reducing with the Quotient Rule

To reduce the rational expression

$$\frac{x^{11}}{x^5},$$

Note that the quotient rule of Chapter 3 applies only when the degree of the numerator is larger than the degree of the denominator.

we could use the quotient rule from Chapter 3 to get x^6. We can also reduce by factoring and dividing out the GCF as follows:

$$\frac{x^{11}}{x^5} = \frac{\cancel{x^5} \cdot x^6}{\cancel{x^5}} = x^6$$

Dividing out the GCF, x^5, leaves six x's in the numerator.

The quotient rule of Chapter 3 does not tell us how to reduce the rational expression

$$\frac{b^8}{b^{12}}.$$

However, we can factor and divide out the GCF:

$$\frac{b^8}{b^{12}} = \frac{\cancel{b^8} \cdot 1}{\cancel{b^8} b^4} = \frac{1}{b^4} \qquad 12 - 8 = 4$$

Dividing out the GCF, b^8, leaves four b's in the denominator.

Since the reducing shown here can be accomplished by subtracting the exponents in either case, we can extend the quotient rule to include both of the above cases.

Quotient Rule

> Suppose $a \neq 0$ and m and n are positive integers.
> If $m \geq n$, then
>
> $$\frac{a^m}{a^n} = a^{m-n}.$$
>
> If $n > m$, then
>
> $$\frac{a^m}{a^n} = \frac{1}{a^{n-m}}.$$

EXAMPLE 3   Reduce to lowest terms:

a) $\dfrac{3a^{15}}{6a^7}$ b) $\dfrac{6x^4 y^2}{4xy^5}$

Solution

a) $\dfrac{3a^{15}}{6a^7} = \dfrac{\cancel{3} a^{15}}{\cancel{3} \cdot 2a^7}$

$\quad = \dfrac{a^8}{2}$ **Apply the quotient rule: $15 - 7 = 8$.**

b) $\dfrac{6x^4y^2}{4xy^5} = \dfrac{\cancel{2} \cdot 3x^4y^2}{\cancel{2} \cdot 2xy^5}$

$\quad\quad = \dfrac{3x^3}{2y^3}$ Apply the quotient rule: $4 - 1 = 3$, $5 - 2 = 3$.   ◄

The essential part of reducing is getting a complete factorization for the numerator and denominator. To get a complete factorization, we must use the techniques for factoring from Chapter 4. If there are large integers in the numerator and denominator, we can use the technique shown in Section 4.1 to get a prime factorization of each integer.

EXAMPLE 4   Reduce to lowest terms: $\dfrac{420}{616}$.

Solution   Use the method of Section 4.1 to get a prime factorization of 420 and 616:

$$
\begin{array}{r}
7 \\
5\overline{)35} \\
3\overline{)105} \\
2\overline{)210} \\
\text{Start here} \longrightarrow \quad 2\overline{)420}
\end{array}
\qquad
\begin{array}{r}
11 \\
7\overline{)77} \\
2\overline{)154} \\
2\overline{)308} \\
\text{Start here} \longrightarrow \quad 2\overline{)616}
\end{array}
$$

The complete factorization for 420 is $2^2 \cdot 3 \cdot 5 \cdot 7$, and the complete factorization for 616 is $2^3 \cdot 7 \cdot 11$. To reduce the fraction, we divide out the common factors:

$$\frac{420}{616} = \frac{2^2 \cdot 3 \cdot 5 \cdot 7}{2^3 \cdot 7 \cdot 11} = \frac{3 \cdot 5}{2 \cdot 11} = \frac{15}{22} \qquad ◄$$

Dividing $a - b$ by $b - a$

In Section 3.2 we learned that

$$a - b = -(b - a) = -1(b - a).$$

Since $a - b = -1(b - a)$, we can write

$$\frac{a - b}{b - a} = -1.$$

We will use this fact in the next example.

EXAMPLE 5   Reduce to lowest terms:

a) $\dfrac{5x - 5y}{4y - 4x}$ **b)** $\dfrac{m^2 - n^2}{n - m}$

Solution

a) $\dfrac{5x - 5y}{4y - 4x} = \dfrac{5(x - y)}{4(y - x)}$ Factor.

$$= \frac{5}{4} \cdot (-1) \qquad \frac{x - y}{y - x} = -1$$

$$= -\frac{5}{4}$$

b) $\dfrac{m^2 - n^2}{n - m} = \dfrac{\overset{-1}{\cancel{(m - n)}}(m + n)}{\cancel{n - m}}$ Factor.

$$= -1(m + n) \qquad \frac{m - n}{n - m} = -1$$

$$= -m - n \qquad\qquad\qquad\qquad\qquad\blacktriangleleft$$

Factoring Out a Common Factor with a Negative Sign

Students will think of this as factoring out a positive or a negative factor. However, if the factor involves a variable, it may be neither positive nor negative. That is why we say negative sign or no sign.

When we factor out any common factor, we can use a factor with no sign or a factor with a negative sign. For example,

$$-3x - 6y = 3(-x - 2y)$$
$$= -3(x + 2y).$$

When reducing an expression that has many negative signs, we usually factor out a common factor with a negative sign.

EXAMPLE 6 Reduce:

$$\frac{-3w - 3w^2}{w^2 - 1}$$

Solution We can factor $3w$ or $-3w$ from the numerator. If we factor out $-3w$, we will get a common factor in the numerator and denominator:

$$\frac{-3w - 3w^2}{w^2 - 1} = \frac{-3w\cancel{(1 + w)}}{(w - 1)\cancel{(w + 1)}} \qquad \text{Factor.}$$

$$= \frac{-3w}{w - 1} \qquad \text{Since } 1 + w = w + 1, \text{ we divide out } w + 1.$$

$$= \frac{3w}{1 - w} \qquad \text{Multiply numerator and denominator by } -1.$$

Note that the last step in this reduction is not absolutely necessary, but we usually perform it to make the answer look a little simpler. $\blacktriangleleft$

The main points to remember for reducing rational expressions are summarized in the following reducing strategy.

STRATEGY
Reducing Rational
Expressions

1. All reducing is done by dividing out common factors.
2. Factor the numerator and denominator completely to see the common factors.
3. Use the quotient rule to reduce a ratio of two monomials involving exponents.
4. We may have to factor out a common factor with a negative sign to get identical factors in the numerator and denominator.
5. Remember that

$$\frac{a - b}{b - a} = -1.$$

Warm-ups

True or false?

1. Any number can be used in place of x in the expression $\frac{x - 2}{5}$.   T

2. We cannot replace x by -1 or 3 in the expression $\frac{x + 1}{x - 3}$.   F

3. The rational expression $\frac{x + 2}{2}$ reduces to x.   F

4. $\frac{2x}{2} = x$ for any real number x.   T

5. $\frac{x^{13}}{x^{20}} = \frac{1}{x^7}$ for any nonzero value of x.   T

6. In reducing $\frac{a^2 + b^2}{a + b}$ to lowest terms we get $a + b$.   F

7. If $a \neq b$, then $\frac{a - b}{b - a} = 1$.   F

8. The expression $\frac{-3x - 6}{x + 2}$ reduces to -3.   T

9. A complete factorization of 3003 is $2 \cdot 3 \cdot 7 \cdot 11 \cdot 13$.   F

10. A complete factorization of 120 is $2^3 \cdot 3 \cdot 5$.   T

5.1 EXERCISES

Which numbers cannot be used in place of the variable in each rational expression? See Example 1.

1. $\dfrac{x}{x+1}$ -1

2. $\dfrac{3x}{x-7}$ 7

3. $\dfrac{2x+3}{x^2-1}$ $-1, 1$

4. $\dfrac{5x-9}{x^2-4}$ $-2, 2$

5. $\dfrac{7a}{3a-5}$ $\dfrac{5}{3}$

6. $\dfrac{84}{3-2a}$ $\dfrac{3}{2}$

7. $\dfrac{2x+1}{x^2-x-6}$ $-2, 3$

8. $\dfrac{3x+5}{x^2+x-20}$ $-5, 4$

9. $\dfrac{a+1}{a}$ 0

10. $\dfrac{3a+1}{2}$ None

Reduce each rational expression to lowest terms. Assume that the variables represent only numbers for which the denominators are nonzero. See Example 2.

11. $\dfrac{6}{27}$ $\dfrac{2}{9}$

12. $\dfrac{14}{21}$ $\dfrac{2}{3}$

13. $\dfrac{42}{90}$ $\dfrac{7}{15}$

14. $\dfrac{42}{54}$ $\dfrac{7}{9}$

15. $\dfrac{36a}{90}$ $\dfrac{2a}{5}$

16. $\dfrac{56y}{40}$ $\dfrac{7y}{5}$

17. $\dfrac{78}{30w}$ $\dfrac{13}{5w}$

18. $\dfrac{68}{44y}$ $\dfrac{17}{11y}$

19. $\dfrac{6x+2}{6}$ $\dfrac{3x+1}{3}$

20. $\dfrac{2w+2}{2}$ $w+1$

21. $\dfrac{2x+4y}{6y+3x}$ $\dfrac{2}{3}$

22. $\dfrac{3m+9w}{3m-6w}$ $\dfrac{m+3w}{m-2w}$

23. $\dfrac{5a-10}{5a}$ $\dfrac{a-2}{a}$

24. $\dfrac{a^2-b^2}{a-b}$ $a+b$

25. $\dfrac{a^2-1}{a^2+2a+1}$ $\dfrac{a-1}{a+1}$

26. $\dfrac{7x^2-7y^2}{7x^2+7y^2}$ $\dfrac{x^2-y^2}{x^2+y^2}$

27. $\dfrac{2x^2+4x+2}{4x^2-4}$ $\dfrac{x+1}{2x-2}$

28. $\dfrac{2x^2+10x+12}{3x^2-27}$ $\dfrac{2x+4}{3x-9}$

29. $\dfrac{3x^2+18x+27}{21x+63}$ $\dfrac{x+3}{7}$

30. $\dfrac{x^3-3x^2-4x}{x^3-16x}$ $\dfrac{x+1}{x+4}$

Reduce to lowest terms. Assume that all denominators are nonzero. See Example 3.

31. $\dfrac{x^{10}}{x^7}$ x^3

32. $\dfrac{y^8}{y^5}$ y^3

33. $\dfrac{z^3}{z^8}$ $\dfrac{1}{z^6}$

34. $\dfrac{w^9}{w^{12}}$ $\dfrac{1}{w^3}$

35. $\dfrac{4x^7}{2x^5}$ $2x^2$

36. $\dfrac{6y^3}{3y^9}$ $\dfrac{2}{y^6}$

37. $\dfrac{-2x^2}{12x^8}$ $\dfrac{-1}{6x^6}$

38. $\dfrac{-8y^{12}}{20y^{19}}$ $-\dfrac{2}{5y^7}$

39. $\dfrac{a^4b^8}{a^7b^3}$ $\dfrac{b^5}{a^3}$

40. $\dfrac{s^3t^6}{s^5t^9}$ $\dfrac{1}{s^2t^3}$

41. $\dfrac{m^9n^{18}}{m^6n^{16}}$ m^3n^2

42. $\dfrac{u^9v^{19}}{u^9v^{14}}$ v^5

43. $\dfrac{6b^{10}c^4}{8b^{10}c^7}$ $\dfrac{3}{4c^3}$

44. $\dfrac{9x^{20}y}{6x^{25}y^3}$ $\dfrac{3}{2x^5y^2}$

45. $\dfrac{30a^3bc}{18a^7b^{17}}$ $\dfrac{5c}{3a^4b^{16}}$

46. $\dfrac{15m^{10}n^3}{24m^{12}np}$ $\dfrac{5n^2}{8m^2p}$

Reduce to lowest terms. See Example 4.

47. $\dfrac{210}{264}$ $\dfrac{35}{44}$

48. $\dfrac{616}{660}$ $\dfrac{14}{15}$

49. $\dfrac{231}{168}$ $\dfrac{11}{8}$

50. $\dfrac{936}{624}$ $\dfrac{3}{2}$

51. $\dfrac{630x^5}{300x^9}$ $\dfrac{21}{10x^4}$

52. $\dfrac{96y^2}{108y^5}$ $\dfrac{8}{9y^3}$

53. $\dfrac{924a^{23}}{448a^{19}}$ $\dfrac{33a^4}{16}$

54. $\dfrac{270b^{75}}{165b^{12}}$ $\dfrac{18b^{63}}{11}$

Reduce to lowest terms. See Example 5.

55. $\dfrac{h^2-t^2}{t-h}$ $-h-t$

56. $\dfrac{r^2-s^2}{s-r}$ $-r-s$

57. $\dfrac{2g-6h}{9h^2-g^2}$ $\dfrac{-2}{3h+g}$

58. $\dfrac{5a-10b}{4b^2-a^2}$ $\dfrac{-5}{2b+a}$

59. $\dfrac{3a - 2b}{2b - 3a}$ -1

60. $\dfrac{5m - 6n}{6n - 5m}$ -1

61. $\dfrac{x^2 - x - 6}{9 - x^2}$ $\dfrac{-x - 2}{x + 3}$

62. $\dfrac{1 - a^2}{a^2 + a - 2}$ $-\dfrac{a + 1}{a + 2}$

Reduce to lowest terms. See Example 6.

63. $\dfrac{-x - 6}{x + 6}$ -1

64. $\dfrac{-5x - 20}{3x + 12}$ $-\dfrac{5}{3}$

65. $\dfrac{-2y - 6y^2}{3 + 9y}$ $\dfrac{-2y}{3}$

66. $\dfrac{y^2 - 16}{-8 - 2y}$ $\dfrac{4 - y}{2}$

67. $\dfrac{-3x - 6}{3x - 6}$ $\dfrac{x + 2}{2 - x}$

68. $\dfrac{8 - 4x}{-8x - 16}$ $\dfrac{x - 2}{2x + 4}$

69. $\dfrac{-12a - 6}{2a^2 + 7a + 3}$ $\dfrac{-6}{a + 3}$

70. $\dfrac{-2b^2 - 6b - 4}{b^2 - 1}$ $\dfrac{-2b - 4}{b - 1}$

Reduce to lowest terms.

71. $\dfrac{2x + 4}{4x}$ $\dfrac{x + 2}{2x}$

72. $\dfrac{2x + 4x^2}{4x}$ $\dfrac{1 + 2x}{2}$

73. $\dfrac{2x - 4}{4 - x^2}$ $\dfrac{-2}{x + 2}$

74. $\dfrac{-2x - 4}{4 - x^2}$ $\dfrac{2}{x - 2}$

75. $\dfrac{2x^{12}}{4x^8}$ $\dfrac{x^4}{2}$

76. $\dfrac{4x^2}{2x^9}$ $\dfrac{2}{x^7}$

77. $\dfrac{x - 4}{4 - x}$ -1

78. $\dfrac{2x - 4}{2x + 4}$ $\dfrac{x - 2}{x + 2}$

79. $\dfrac{x^2 + 4x + 4}{x^2 - 4}$ $\dfrac{x + 2}{x - 2}$

80. $\dfrac{3x - 6}{x^2 - 4x + 4}$ $\dfrac{3}{x - 2}$

81. $\dfrac{-2x - 4}{x^2 + 5x + 6}$ $\dfrac{-2}{x + 3}$

82. $\dfrac{-2x - 8}{x^2 + 2x - 8}$ $\dfrac{-2}{x - 2}$

83. $\dfrac{2x^8 + x^7}{2x^6 + x^5}$ x^2

84. $\dfrac{8x^{12}}{12x^6 - 16x^5}$ $\dfrac{2x^7}{3x - 4}$

85. $\dfrac{x^2 - 6x - 16}{x^2 - 16x + 64}$ $\dfrac{x + 2}{x - 8}$

86. $\dfrac{x^2 + 3x - 18}{x^2 + 12x + 36}$ $\dfrac{x - 3}{x + 6}$

Find a rational expression that answers each of the following.

87. If Sergio drove 300 miles at $x + 10$ miles per hour, then how many hours did he drive?

88. If Carrie walked 40 miles in x hours, then how fast did she walk?

89. If $x + 4$ pounds of peaches cost $4.50, then what is the cost per pound?

90. $\dfrac{x}{9}$ dollars   **91.** $\dfrac{1}{x}$ **92.** $\dfrac{1}{x - 3}$

90. If nine pounds of pears cost x dollars, then what is the price per pound?

91. If Ayesha can clean the entire swimming pool in x hours, then how much of the pool does she clean per hour?

92. If Ramon can mow the entire lawn in $x - 3$ hours, then how much of the lawn does he mow per hour?

87. $\dfrac{300}{x + 10}$ hours **88.** $\dfrac{40}{x}$ mph **89.** $\dfrac{4.50}{x + 4}$

5.2 Multiplication and Division

IN THIS SECTION:

- Multiplying Rational Numbers
- Multiplying Rational Expressions
- Division of Rational Numbers
- Division of Rational Expressions

In Section 5.1 we learned to reduce rational expressions in the same way that we reduce rational numbers. In this section we will multiply and divide rational expressions using the same procedures that we use for rational numbers.

Multiplying Rational Numbers

Two rational numbers are multiplied by multiplying their numerators and multiplying their denominators.

DEFINITION
Multiplication of Rational Numbers

If $b \neq 0$ and $d \neq 0$, then

$$\frac{a}{b} \cdot \frac{c}{d} = \frac{ac}{bd}.$$

Using this definition, we get

$$\frac{6}{7} \cdot \frac{14}{15} = \frac{84}{105}$$

$$= \frac{21 \cdot 4}{21 \cdot 5} \qquad \text{Factor the numerator and denominator.}$$

$$= \frac{4}{5}. \qquad \text{Divide out the GCF 21.}$$

The reducing that we did after multiplying is easier to do before multiplying. First factor all terms, then reduce, then multiply:

$$\frac{6}{7} \cdot \frac{14}{15} = \frac{2 \cdot 3}{7} \cdot \frac{2 \cdot 7}{3 \cdot 5} = \frac{4}{5}$$

Multiplying Rational Expressions

We multiply rational expressions exactly as we multiply rational numbers.

EXAMPLE 1 Multiply the rational expressions:

a) $\dfrac{2x - 2y}{4} \cdot \dfrac{2x}{x^2 - y^2}$

b) $\dfrac{x^2 + 7x + 12}{2x + 6} \cdot \dfrac{x}{x^2 - 16}$

Solution

a) $\dfrac{2x - 2y}{4} \cdot \dfrac{2x}{x^2 - y^2} = \dfrac{2(x - y)}{2 \cdot 2} \cdot \dfrac{2 \cdot x}{(x - y)(x + y)} \qquad$ **Factor.**

$$= \frac{x}{x + y} \qquad \text{Reduce.}$$

b) $\dfrac{x^2 + 7x + 12}{2x + 6} \cdot \dfrac{x}{x^2 - 16} = \dfrac{(x + 3)(x + 4)}{2(x + 3)} \cdot \dfrac{x}{(x - 4)(x + 4)}$

$$= \frac{x}{2x - 8} \qquad 2(x - 4) = 2x - 8 \qquad \blacktriangleleft$$

Division of Rational Numbers

Division of rational numbers can be accomplished by multiplying by the reciprocal of the divisor. For example,

$$5 \div \frac{1}{2} = 5 \cdot \frac{2}{1} = 10.$$

DEFINITION
Division of Rational Numbers

If $b \neq 0$, $c \neq 0$, and $d \neq 0$, then

$$\frac{a}{b} \div \frac{c}{d} = \frac{a}{b} \cdot \frac{d}{c}.$$

Division of Rational Expressions

We divide rational expressions just as we divide rational numbers: Invert the divisor and multiply.

EXAMPLE 2 Divide:

a) $\dfrac{5}{3x} \div \dfrac{5}{6x}$ b) $\dfrac{4 - x^2}{x^2 + x} \div \dfrac{x - 2}{x^2 - 1}$ c) $\dfrac{x^7}{2} \div 2x^2$

Solution

a) $\dfrac{5}{3x} \div \dfrac{5}{6x} = \dfrac{5}{3x} \cdot \dfrac{6x}{5}$ Invert and multiply.

$\qquad = \dfrac{\cancel{5}}{3x} \cdot \dfrac{2 \cdot 3x}{\cancel{5}}$ Factor.

$\qquad = 2$ Reduce.

b) $\dfrac{4 - x^2}{x^2 + x} \div \dfrac{x - 2}{x^2 - 1} = \dfrac{4 - x^2}{x^2 + x} \cdot \dfrac{x^2 - 1}{x - 2}$ Invert and multiply.

$\qquad = \dfrac{\overset{-1}{\cancel{(2 - x)}}(2 + x)}{x\cancel{(x + 1)}} \cdot \dfrac{\cancel{(x + 1)}(x - 1)}{\cancel{x - 2}}$ Factor.

$\qquad = \dfrac{-1(2 + x)(x - 1)}{x} \qquad \dfrac{2 - x}{x - 2} = -1$

$\qquad = \dfrac{-1(x^2 + x - 2)}{x}$ Simplify.

$\qquad = \dfrac{-x^2 - x + 2}{x}$

c) $\dfrac{x^7}{2} \div 2x^2 = \dfrac{x^7}{2} \cdot \dfrac{1}{2x^2}$ Invert and multiply.

$\qquad = \dfrac{x^5}{4}$ ◀

We sometimes write division of rational expressions using the fraction bar. In this case we still invert the divisor and multiply.

EXAMPLE 3   Perform the operations indicated:

a) $\dfrac{\dfrac{a+b}{3}}{\dfrac{1}{6}}$

b) $\dfrac{\dfrac{x^2-1}{2}}{\dfrac{x-1}{3}}$

c) $\dfrac{\dfrac{a^2+5}{3}}{2}$

Solution

a) $\dfrac{\dfrac{a+b}{3}}{\dfrac{1}{6}} = \dfrac{a+b}{3} \cdot \dfrac{6}{1}$ **Invert and multiply.**

$= (a+b)2$ $\dfrac{6}{3} = 2$

$= 2a + 2b$

b) $\dfrac{\dfrac{x^2-1}{2}}{\dfrac{x-1}{3}} = \dfrac{x^2-1}{2} \cdot \dfrac{3}{x-1}$ **Invert and multiply.**

$= \dfrac{(x-1)(x+1)}{2} \cdot \dfrac{3}{x-1}$ **Factor.**

$= \dfrac{3x+3}{2}$ **Reduce.**

c) $\dfrac{\dfrac{a^2+5}{3}}{2} = \dfrac{a^2+5}{3} \cdot \dfrac{1}{2} = \dfrac{a^2+5}{6}$ ◀

Warm-ups

True or false?

1. $\dfrac{2}{3} \cdot \dfrac{5}{3} = \dfrac{10}{3}$ F

2. $\dfrac{x-7}{3} \cdot \dfrac{6}{7-x} = -2$ for any value of x except 7.   T

3. Dividing by 2 is equivalent to multiplying by $1/2$.   T

4. $3 \div x = \dfrac{1}{3} \cdot x$ for any nonzero number x.   F

5. Factoring polynomials is essential to multiplying rational expressions.   T

6. One-half of one-fourth is one-sixth. F

7. One-half divided by three is three-halves. F

8. The quotient of $(839 - 487)$ and $(487 - 839)$ is -1. T

9. $\dfrac{a}{3} \div 3 = \dfrac{a}{9}$ for any value of a. T

10. $\dfrac{a}{b} \cdot \dfrac{b}{a} = 1$ for any nonzero values of a and b. T

5.2 EXERCISES

Perform the indicated operations. See Example 1.

1. $\dfrac{8}{15} \cdot \dfrac{35}{24}$ $\dfrac{7}{9}$

2. $\dfrac{3}{4} \cdot \dfrac{8}{21}$ $\dfrac{2}{7}$

3. $\dfrac{5a}{12b} \cdot \dfrac{3b^2}{55}$ $\dfrac{ab}{44}$

4. $\dfrac{2x}{7a} \cdot \dfrac{21a^2}{6x}$ a

5. $\dfrac{16a + 8}{5a^2 + 5} \cdot \dfrac{2a^2 + a - 1}{4a^2 - 1}$ $\dfrac{8a + 8}{5a^2 + 5}$

6. $\dfrac{b^3 + b}{5} \cdot \dfrac{10}{b^2 + b}$ $\dfrac{2b^2 + 2}{b + 1}$

7. $\dfrac{3x + 3w + bx + bw}{x^2 - w^2} \cdot \dfrac{6 - 2b}{9 - b^2}$ $\dfrac{2}{x - w}$

8. $(x^2 - 9) \cdot \dfrac{3}{x - 3}$ $3x + 9$

9. $\dfrac{a^2 - 2a + 4}{a^2 - 4} \cdot \dfrac{(a + 2)^3}{2a + 4}$ $\dfrac{a^3 + 8}{2a - 4}$

10. $\dfrac{w^2 - 1}{(w - 1)^2} \cdot \dfrac{w - 1}{w^2 + 2w + 1}$ $\dfrac{1}{w + 1}$

Perform the indicated operations. See Example 2.

11. $\dfrac{5}{7} \div \dfrac{10}{7}$ $\dfrac{1}{2}$

12. $\dfrac{1}{4} \div \dfrac{1}{2}$ $\dfrac{1}{2}$

13. $\dfrac{x^2 + 4x + 4}{8} \div \dfrac{(x + 2)^3}{16}$ $\dfrac{2}{x + 2}$

14. $(a - 1) \div \dfrac{a^2 - 1}{a}$ $\dfrac{a}{a + 1}$

15. $\dfrac{y - 6}{2} \div \dfrac{6 - y}{6}$ -3

16. $\dfrac{4 - a}{5} \div \dfrac{a^2 - 16}{3}$ $\dfrac{-3}{5a + 20}$

Perform the indicated operations. See Example 3.

17. $\dfrac{\dfrac{x^2 - 4}{12}}{\dfrac{x - 2}{6}}$ $\dfrac{x + 2}{2}$

18. $\dfrac{\dfrac{6a^2 + 6}{5}}{\dfrac{6a + 6}{5}}$ $\dfrac{a^2 + 1}{a + 1}$

19. $\dfrac{\dfrac{x + 7}{3}}{\dfrac{5}{5}}$ $\dfrac{x + 7}{15}$

20. $\dfrac{\dfrac{1}{a - 3}}{4}$ $\dfrac{1}{4a - 12}$

21. $\dfrac{\dfrac{x^2 - y^2}{x - y}}{9}$ $9x + 9y$

22. $\dfrac{\dfrac{x^2 + 6x + 8}{x + 2}}{\dfrac{x + 1}{}}$ $x^2 + 5x + 4$

Find the following mentally. Write down only the answer.

23. $\dfrac{2x}{3} \div 3$ $\dfrac{2x}{9}$

24. $\dfrac{b}{a} \div a$ $\dfrac{b}{a^2}$

25. $\dfrac{2}{5} \div \dfrac{1}{5}$ 2

26. $\dfrac{1}{2} \div \dfrac{1}{4}$ 2

27. $6 \div \dfrac{1}{2}$ 12

28. $\dfrac{1}{4} \div \dfrac{1}{2}$ $\dfrac{1}{2}$ **29.** $9 \div 3$ 3   **30.** $9 \div \dfrac{1}{3}$ 27   **31.** $\dfrac{1}{3} \div 9$ $\dfrac{1}{27}$ **32.** $\dfrac{1}{3} \div \dfrac{1}{9}$ 3

33. One-half of 1/2 $\dfrac{1}{4}$ **34.** One-half of 1/4 $\dfrac{1}{8}$ **35.** One-third of 4 $\dfrac{4}{3}$ **36.** One-third of 1/2 $\dfrac{1}{6}$

37. One-half of 2x/3 $\dfrac{x}{3}$ **38.** One-half of 3/2 $\dfrac{3}{4}$ **39.** $(x-3) \div (3-x)$ -1 **40.** $(x-3) \div (-1)$ $3-x$

41. $\dfrac{x-1}{3} \cdot \dfrac{9}{1-x}$ -3 **42.** $\dfrac{2x-2y}{3} \cdot \dfrac{1}{x-y}$ $\dfrac{2}{3}$ **43.** $\dfrac{3a+3b}{a} \cdot \dfrac{1}{3}$ $\dfrac{a+b}{a}$ **44.** $\dfrac{a-b}{b-a} \cdot \dfrac{1}{5}$ $-\dfrac{1}{5}$

45. $2xh \div \dfrac{1}{x}$ $2x^2h$ **46.** $6a \div \dfrac{1}{2}$ $12a$ **47.** $\dfrac{x-y}{\frac{1}{2}}$ $2x-2y$ **48.** $\dfrac{x+1}{\frac{1}{3}}$ $3x+3$

49. $\dfrac{\frac{2x}{7}}{x}$ $\dfrac{2}{7}$ **50.** $\dfrac{\frac{b^2-4a}{2}}{a}$ $\dfrac{b^2-4a}{2a}$ **51.** $\dfrac{\frac{x}{5b}}{2}$ $\dfrac{x}{10b}$ **52.** $\dfrac{\frac{2x}{3}}{x}$ $\dfrac{2}{3}$

53. $\dfrac{\frac{b}{a}}{3}$ $\dfrac{b}{3a}$ **54.** $\dfrac{\frac{2g}{3h}}{\frac{1}{h}}$ $\dfrac{2g}{3}$ **55.** $\dfrac{6y}{3} \cdot \dfrac{1}{2x}$ $\dfrac{y}{x}$ **56.** $\dfrac{8x}{9} \cdot \dfrac{18x}{16}$ x^2

57. $\dfrac{x-1}{8} \cdot \dfrac{4}{1-x}$ $-\dfrac{1}{2}$ **58.** $\dfrac{b^3-8}{4} \cdot \dfrac{20}{8-b^3}$ -5 **59.** $\dfrac{x+3}{6} \cdot \dfrac{6}{x-3}$ $\dfrac{x+3}{x-3}$ **60.** $\dfrac{y-2}{10} \cdot \dfrac{5}{y+2}$ $\dfrac{y-2}{2y+4}$

Perform the indicated operations.

61. $\dfrac{3x^2+16x+5}{x} \cdot \dfrac{x^2}{9x^2-1}$ $\dfrac{x^2+5x}{3x-1}$ **62.** $\dfrac{x^2+6x+5}{x} \cdot \dfrac{x^4}{3x+3}$ $\dfrac{x^4+5x^3}{3}$ **63.** $\dfrac{a^3b^4}{-2ab^2} \cdot \dfrac{a^5b^7}{ab}$ $\dfrac{-a^6b^8}{2}$

64. $\dfrac{-2a^2}{3a^2} \cdot \dfrac{20a}{15a^3}$ $\dfrac{-8}{9a^2}$ **65.** $\dfrac{2mn^4}{6mn^2} \div \dfrac{3m^5n^7}{m^2n^4}$ $\dfrac{1}{9m^3n}$ **66.** $\dfrac{rt^2}{rt^2} \div \dfrac{rt^2}{r^3t^2}$ r^2

67. $\dfrac{2x^2+19x-10}{x^2-100} \cdot \dfrac{2x^2-19x-10}{4x^2-1}$ 1   **68.** $\dfrac{x-1}{x^2+1} \cdot \dfrac{5x+5}{x^2-x}$ $\dfrac{5x+5}{x^3+x}$

69. $\dfrac{9+6m+m^2}{9-6m+m^2} \cdot \dfrac{m^2-9}{m^2+mk+3m+3k}$ $\dfrac{(m+3)^2}{(m-3)(m+k)}$ **70.** $\dfrac{a^2+2ab+b^2}{ac+bc-ad-bd} \div \dfrac{3a+3b}{c^2-d^2}$ $\dfrac{c+d}{3}$

5.3 Building Up the Denominator

IN THIS SECTION:

- Changing the Denominator
- Finding the Least Common Denominator
- Converting to the LCD

Every rational expression can be written in infinitely many equivalent forms. Since we can add or subtract only fractions with identical denominators, we must be able to change the denominator of a fraction. We have already learned how to change the denominator of a fraction by reducing. In this section we will learn the opposite of reducing, building up the denominator.

Changing the Denominator

To convert the fraction 2/3 into an equivalent fraction with a denominator of 21, we factor 21 as $21 = 3 \cdot 7$. Since 2/3 already has a 3 in the denominator, multiply the numerator and denominator of 2/3 by the missing factor 7 to get a denominator of 21:

$$\frac{2}{3} = \frac{2 \cdot 7}{3 \cdot 7} = \frac{14}{21}$$

For rational expressions the process is the same. To convert the rational expression

$$\frac{5}{x + 3}$$

into an equivalent rational expression with a denominator of $x^2 - x - 12$, first factor $x^2 - x - 12$:

$$x^2 - x - 12 = (x + 3)(x - 4)$$

From the factorization we can see that the denominator $x + 3$ needs only a factor of $x - 4$ to have the required denominator. So multiply the numerator and denominator by $x - 4$:

$$\frac{5}{x + 3} = \frac{5(x - 4)}{(x + 3)(x - 4)} = \frac{5x - 20}{x^2 - x - 12}$$

EXAMPLE 1 Convert each rational expression into an equivalent rational expression with the indicated denominator:

a) $3 = \dfrac{?}{12}$ **b)** $\dfrac{3}{w} = \dfrac{?}{wx}$

c) $\dfrac{7}{3x - 3y} = \dfrac{?}{6y - 6x}$ **d)** $\dfrac{x - 2}{x + 2} = \dfrac{?}{x^2 + 8x + 12}$

Solution

a) Since $3 = 3/1$, we get a denominator of 12 by multiplying numerator and denominator by 12:

$$3 = \frac{3}{1} = \frac{3 \cdot 12}{1 \cdot 12} = \frac{36}{12}$$

b) Multiply numerator and denominator by x:

$$\frac{3}{w} = \frac{3 \cdot x}{w \cdot x} = \frac{3x}{wx}$$

c) Since $3x - 3y = 3(x - y)$, we factor -6 out of $6y - 6x$. This will give a factor of $x - y$ in each denominator:

$$3x - 3y = 3(x - y)$$
$$6y - 6x = -6(x - y) = -2 \cdot 3(x - y)$$

To get the required denominator, we multiply numerator and denominator by -2 only:

$$\frac{7}{3x - 3y} = \frac{7(-2)}{(3x - 3y)(-2)} = \frac{-14}{6y - 6x}$$

d) Since $x^2 + 8x + 12 = (x + 2)(x + 6)$, we multiply numerator and denominator by $x + 6$, the missing factor:

$$\frac{x - 2}{x + 2} = \frac{(x - 2)(x + 6)}{(x + 2)(x + 6)} = \frac{x^2 + 4x - 12}{x^2 + 8x + 12} \qquad \blacktriangleleft$$

Finding the Least Common Denominator

We can use the idea of building up the denominator to convert two fractions with different denominators into fractions with identical denominators. For example, both

$$\frac{5}{6} \quad \text{and} \quad \frac{1}{4}$$

can be converted into fractions with a denominator of 12, since $12 = 2 \cdot 6$ and $12 = 3 \cdot 4$:

$$\frac{5}{6} = \frac{5 \cdot 2}{6 \cdot 2} = \frac{10}{12} \qquad \frac{1}{4} = \frac{1 \cdot 3}{4 \cdot 3} = \frac{3}{12}$$

The smallest number that is a multiple of all of the denominators is called the **least common denominator (LCD)**. The LCD for the denominators 6 and 4 is 12.

To find the LCD in a systematic way, we look at a complete factorization of each denominator. Consider the denominators 24 and 30:

$$24 = 2 \cdot 2 \cdot 2 \cdot 3 = 2^3 \cdot 3$$
$$30 = 2 \cdot 3 \cdot 5$$

Any multiple of 24 must have three 2's in its factorization, and any multiple of 30 must have one 2 as a factor. So a number with three 2's in its factorization will have enough to be a multiple of both 24 and 30. The LCD must also have one 3 and one 5 in its factorization. *We use each factor the maximum number of times it appears in either factorization.* Thus the LCD is

$$2^3 \cdot 3 \cdot 5 = 2 \cdot 2 \cdot 2 \cdot 3 \cdot 5 = 120.$$

If we were to omit any one of the factors in $2 \cdot 2 \cdot 2 \cdot 3 \cdot 5$, we would not have a multiple of both 24 and 30. That is what makes it the *least* common denominator. To find the LCD for two polynomials, we follow the same strategy.

STRATEGY
Finding the LCD for Two Polynomials

1. Factor each denominator completely. Use exponent notation for repeated factors.
2. Write the product of all of the different factors that appear in the denominators.
3. On each factor, use the highest power that appears on that factor in any of the denominators.

EXAMPLE 2   If the given expressions were used as denominators of rational expressions, then what would be the LCD for each pair of denominators?

a) 20, 50   **b)** $a^2 + 5a + 6,\ a^2 + 4a + 4$ **c)** $x^3yz^2,\ x^5y^2z$

Solution

a) First factor each number completely:

$$20 = 2^2 \cdot 5 \qquad 50 = 2 \cdot 5^2$$

The highest power of 2 is 2, and the highest power of 5 is 2. The LCD of 20 and 50 is $2^2 \cdot 5^2 = 100$.

b) First factor each polynomial:

$$a^2 + 5a + 6 = (a + 2)(a + 3) \qquad a^2 + 4a + 4 = (a + 2)^2$$

The highest power of $(a + 3)$ is 1, and the highest power of $(a + 2)$ is 2. The LCD is $(a + 3)(a + 2)^2$.

c) These expressions are already factored. For the LCD, use the maximum exponent on each variable. Thus the LCD is $x^5y^2z^2$.   ◀

Converting to the LCD

When adding or subtracting two rational expressions, we must convert the expressions into expressions with identical denominators. To keep the computations as simple as possible, we use the least common denominator. The next two examples illustrate this.

EXAMPLE 3   Find the LCD for the rational expressions

$$\frac{4}{9xy} \quad \text{and} \quad \frac{2}{15xz}$$

and convert each expression into an equivalent rational expression with the LCD as denominator.

Solution Factor each denominator completely:

$$9xy = 3^2xy \qquad 15xz = 3 \cdot 5xz$$

The LCD is $3^2 \cdot 5xyz$. Now convert each expression into an expression with this denominator. We must multiply the numerator and denominator of the first rational expression by $5z$ and the second by $3y$:

$$\frac{4}{9xy} = \frac{4}{3 \cdot 3xy} = \frac{4 \cdot 5z}{3 \cdot 3xy \cdot 5z} = \frac{20z}{45xyz} \leftarrow$$

$$\frac{2}{15xz} = \frac{2}{3 \cdot 5xz} = \frac{2 \cdot 3y}{3 \cdot 5xz \cdot 3y} = \frac{6y}{45xyz} \leftarrow$$

Same denominator

◄

EXAMPLE 4 Find the LCD for the rational expressions

$$\frac{5x}{x^2 - 4} \quad \text{and} \quad \frac{3}{x^2 + x - 6}$$

and convert each into an equivalent rational expression with that denominator.

Solution First factor the denominators:

$$x^2 - 4 = (x - 2)(x + 2)$$
$$x^2 + x - 6 = (x - 2)(x + 3)$$

The LCD is $(x - 2)(x + 2)(x + 3)$. Now we multiply the numerator and denominator of the first rational expression by $(x + 3)$ and the second by $(x + 2)$. Since each denominator already has one factor of $(x - 2)$, there is no reason to multiply by $(x - 2)$. We multiply each denominator by the factors in the LCD that are missing from that denominator:

$$\frac{5x}{(x^2 - 4)} = \frac{5x(x + 3)}{(x^2 - 4)(x + 3)} = \frac{5x^2 + 15x}{(x - 2)(x + 2)(x + 3)} \leftarrow$$

$$\frac{3}{x^2 + x - 6} = \frac{3(x + 2)}{(x^2 + x - 6)(x + 2)} = \frac{3x + 6}{(x - 2)(x + 2)(x + 3)} \leftarrow$$

Same denominator

We will generally simplify the numerators but leave the denominators in factored form.

◄

Warm-ups

True or false?

1. To convert $2/3$ into an equivalent fraction with a denominator of 18, we would multiply only the denominator of $2/3$ by 6. F

2. Factoring has nothing to do with determining the least common denominator. F

3. To convert the rational expression $\dfrac{3}{2ab^2}$ into an equivalent rational expression with a denominator of $10a^3b^4$, we would multiply the numerator and denominator by $5a^2b^2$.   T

4. The LCD for the denominators $2^5 \cdot 3$ and $2^4 \cdot 3^2$ is $2^5 \cdot 3^2$.   T

5. The LCD for the fractions $1/6$ and $1/10$ is 60.   F

6. The LCD for the denominators $6a^2b$ and $4ab^3$ is $2ab$.   F

7. The LCD for the denominators $a^2 + 1$ and $a + 1$ is $a^2 + 1$.   F

8. Adding the same number to the numerator and denominator of a rational expression gives us an equivalent rational expression.   F

9. The LCD for the expressions $\dfrac{1}{x-2}$ and $\dfrac{3}{x+2}$ is $x^2 - 4$.   T

10. If we convert the rational expression x into an equivalent rational expression with a denominator of 3, we get $\dfrac{3x}{3}$.   T

5.3 EXERCISES

Convert each rational expression into an equivalent rational expression with the indicated denominator. See Example 1.

1. $\dfrac{1}{3} = \dfrac{?}{30}$ $\dfrac{10}{30}$ 2. $\dfrac{2}{3} = \dfrac{?}{12}$ $\dfrac{8}{12}$ 3. $\dfrac{1}{a} = \dfrac{?}{5a^3}$ $\dfrac{5a^2}{5a^3}$ 4. $\dfrac{4}{xy^2} = \dfrac{?}{x^2y^5}$ $\dfrac{4xy^3}{x^2y^5}$

5. $\dfrac{3}{x+1} = \dfrac{?}{x^2+2x+1}$ $\dfrac{3x+3}{x^2+2x+1}$ 6. $\dfrac{a}{a+3} = \dfrac{?}{a^2-9}$ $\dfrac{a^2-3a}{a^2-9}$

7. $\dfrac{5}{2x+2} = \dfrac{?}{-8x-8}$ $\dfrac{-20}{-8x-8}$ 8. $\dfrac{3}{m-n} = \dfrac{?}{2n-2m}$ $\dfrac{-6}{2n-2m}$

9. $\dfrac{x-6}{x-4} = \dfrac{?}{x^2+x-20}$ $\dfrac{x^2-x-30}{x^2+x-20}$ 10. $\dfrac{2x}{x+3} = \dfrac{?}{x^2-2x-15}$ $\dfrac{2x^2-10x}{x^2-2x-15}$

11. $\dfrac{8}{b-1} = \dfrac{?}{1-b}$ $\dfrac{-8}{1-b}$ 12. $\dfrac{3}{x+2} = \dfrac{?}{x^2-4}$ $\dfrac{3x-6}{x^2-4}$ 13. $5 = \dfrac{?}{3b}$ $\dfrac{15b}{3b}$ 14. $5 = \dfrac{?}{x+4}$ $\dfrac{5x+20}{x+4}$

If the given expressions were used as denominators of rational expressions, then what would be the LCD for each group of denominators? See Example 2.

15. $12, 16$ 48   16. $28, 42$ 84   17. $6a^2, 15a$ $30a^2$

18. $18x^2, 20xy$ $180x^2y$ 19. $24, 40$ 120   20. $12, 18, 20$ 180

21. a^4b, ab^6 a^4b^6 22. m^3nw, mn^5w^8 $m^3n^5w^8$ 23. $x^2 - 16, x^2 + 8x + 16$

24. $x^2 - 9, x^2 + 6x + 9$ $(x-3)(x+3)^2$ 25. $x, x+2, x-2$ $x(x+2)(x-2)$ 26. $y, y-5, y+2$ $y(y-5)(y+2)$

27. $x^2 - 4x, x^2 - 16$ $x(x-4)(x+4)$ 28. $y, y^2 - 3y, 3y$ $3y(y-3)$ 23. $(x-4)(x+4)^2$

Find the LCD for each pair of rational expressions, and convert each rational expression into an equivalent rational expression with the LCD as denominator. See Examples 3 and 4.

29. $\dfrac{3}{84}, \dfrac{5}{63}$ $\dfrac{9}{252}, \dfrac{20}{252}$

30. $\dfrac{4}{75}, \dfrac{6}{105}$ $\dfrac{28}{525}, \dfrac{30}{525}$

31. $\dfrac{1}{3x}, \dfrac{3}{2x}$ $\dfrac{2}{6x}, \dfrac{9}{6x}$

32. $\dfrac{3}{8ab}, \dfrac{5}{6ac}$ $\dfrac{9c}{24abc}, \dfrac{20b}{24abc}$

33. $\dfrac{x}{9yz}, \dfrac{y}{12x}$ $\dfrac{4x^2}{36xyz}, \dfrac{3y^2z}{36xyz}$

34. $\dfrac{a}{12b}, \dfrac{3b}{14a}$ $\dfrac{7a^2}{84ab}, \dfrac{18b^2}{84ab}$

35. $\dfrac{2x}{x-3}, \dfrac{5x}{x+2}$ $\dfrac{2x^2+4x}{(x-3)(x+2)}, \dfrac{5x^2-15x}{(x-3)(x+2)}$

36. $\dfrac{2a}{a-5}, \dfrac{3a}{a+2}$ $\dfrac{2a^2+4a}{(a-5)(a+2)}, \dfrac{3a^2-15a}{(a-5)(a+2)}$

37. $\dfrac{x}{x^2-9}, \dfrac{5x}{x^2-6x+9}$ $\dfrac{x^2-3x}{(x-3)^2(x+3)}, \dfrac{5x^2+15x}{(x-3)^2(x+3)}$

38. $\dfrac{5x}{x^2-1}, \dfrac{4}{x^2-2x+1}$ $\dfrac{5x^2-5x}{(x+1)(x-1)^2}, \dfrac{4x+4}{(x+1)(x-1)^2}$

39. $\dfrac{4}{a-6}, \dfrac{5}{6-a}$ $\dfrac{4}{a-6}, \dfrac{-5}{a-6}$

40. $\dfrac{4}{x-y}, \dfrac{5x}{2y-2x}$ $\dfrac{8}{2x-2y}, \dfrac{-5x}{2x-2y}$

In place of the question mark, put an expression that will make these rational expressions equivalent. Do these mentally.

41. $\dfrac{1}{4} = \dfrac{?}{20}$ 5

42. $4 = \dfrac{?}{5}$ 20

43. $3 = \dfrac{12}{?}$ 4

44. $\dfrac{3}{4} = \dfrac{15}{?}$ 20

45. $\dfrac{6}{x} = \dfrac{?}{x^2}$ $6x$

46. $\dfrac{3}{b} = \dfrac{12}{?}$ $4b$

47. $\dfrac{1}{w-z} = \dfrac{?}{z-w}$ -1

48. $\dfrac{2}{x-7} = \dfrac{?}{7-x}$ -2

49. $\dfrac{x}{x-1} = \dfrac{?}{x^2-1}$ x^2+x

50. $\dfrac{3}{x+2} = \dfrac{?}{x^2-4}$ $3x-6$

51. $\dfrac{?}{x} = \dfrac{5}{5x}$ 1

52. $\dfrac{3}{?} = \dfrac{3y}{2y}$ 2

53. $\dfrac{5x}{?} = \dfrac{10x}{6x}$ $3x$

54. $\dfrac{?}{9} = \dfrac{x}{3}$ $3x$

55. $\dfrac{4}{x-2} = \dfrac{-4}{?}$ $2-x$

56. $\dfrac{-2}{3-4x} = \dfrac{2}{?}$ $4x-3$

57. $\dfrac{2x+6}{4} = \dfrac{?}{2}$ $x+3$

58. $\dfrac{x-3}{2x-6} = \dfrac{1}{?}$ 2

59. $\dfrac{a-1}{a^2-1} = \dfrac{1}{?}$ $a+1$

60. $\dfrac{2a+2}{6} = \dfrac{a+1}{?}$ 3

61. $\dfrac{3x-3}{3a} = \dfrac{?}{a}$ $x-1$

62. $\dfrac{2x-3}{2x^2-3x} = \dfrac{1}{?}$ x

5.4 Addition and Subtraction

IN THIS SECTION:

- ● Addition and Subtraction of Rational Numbers
- ● Addition and Subtraction of Rational Expressions
- ● Shortcuts

In Section 5.3 we learned how to find the LCD and build up the denominators of rational expressions. In this section we will use that knowledge to add and subtract rational expressions with different denominators.

Addition and Subtraction of Rational Numbers

We can add or subtract rational numbers only with identical denominators according to the following definition.

DEFINITION
Addition and Subtraction of Rational Numbers

If $b \neq 0$, then

$$\frac{a}{b} + \frac{c}{b} = \frac{a+c}{b} \quad \text{and} \quad \frac{a}{b} - \frac{c}{b} = \frac{a-c}{b}.$$

EXAMPLE 1 Perform the indicated operations. Reduce answers to lowest terms.

a) $\dfrac{1}{12} + \dfrac{7}{12}$

b) $\dfrac{3}{4} - \dfrac{1}{4}$

Solution

a) $\dfrac{1}{12} + \dfrac{7}{12} = \dfrac{8}{12} = \dfrac{\cancel{4} \cdot 2}{\cancel{4} \cdot 3} = \dfrac{2}{3}$

b) $\dfrac{3}{4} - \dfrac{1}{4} = \dfrac{2}{4} = \dfrac{1}{2}$ ◄

If the rational numbers have different denominators, we must convert them to equivalent rational numbers that have identical denominators and then add or subtract. Of course, it is most efficient to use the least common denominator (LCD), as in the following example.

EXAMPLE 2 Perform the addition $\dfrac{1}{4} + \dfrac{1}{6}$.

Solution The LCD for $4 = 2^2$ and $6 = 2 \cdot 3$ is $2^2 \cdot 3 = 12$. Convert each fraction to an equivalent fraction with a denominator of 12:

$$\frac{1}{4} + \frac{1}{6} = \frac{1 \cdot 3}{4 \cdot 3} + \frac{1 \cdot 2}{6 \cdot 2}$$

$$= \frac{3}{12} + \frac{2}{12} \qquad \text{Get identical denominators.}$$

$$= \frac{5}{12} \qquad \text{Add the fractions.} \qquad ◄$$

Addition and Subtraction of Rational Expressions

Rational expressions are added or subtracted just like rational numbers. We can add or subtract rational expressions only with identical denominators.

EXAMPLE 3 Perform the indicated operations and reduce answers to lowest terms:

a) $\dfrac{2x}{x+2} + \dfrac{4}{x+2}$

b) $\dfrac{x^2+2x}{(x-1)(x+3)} - \dfrac{2x+1}{(x-1)(x+3)}$

Solution

a) $\dfrac{2x}{x+2} + \dfrac{4}{x+2} = \dfrac{2x+4}{x+2}$

$= \dfrac{2\cancel{(x+2)}}{\cancel{x+2}}$ Factor the numerator.

$= 2$ Reduce.

b) $\dfrac{x^2+2x}{(x-1)(x+3)} - \dfrac{2x+1}{(x-1)(x+3)} = \dfrac{x^2+2x-(2x+1)}{(x-1)(x+3)}$

$= \dfrac{x^2-1}{(x-1)(x+3)}$

$= \dfrac{\cancel{(x-1)}(x+1)}{\cancel{(x-1)}(x+3)}$ Factor.

$= \dfrac{x+1}{x+3}$ Reduce.

Note that when we subtract the numerator $2x + 1$, it is treated as if it were in parentheses.   ◀

In the next example the rational expressions have different denominators.

EXAMPLE 4 Perform the indicated operations:

a) $\dfrac{4}{x^3y} + \dfrac{2}{xy^3}$

b) $\dfrac{a+1}{6} - \dfrac{a-2}{8}$

Solution

a) The LCD is x^3y^3.

$\dfrac{4}{x^3y} + \dfrac{2}{xy^3} = \dfrac{4(y^2)}{x^3y(y^2)} + \dfrac{2(x^2)}{xy^3(x^2)}$ Build both denominators into the LCD.

$= \dfrac{4y^2}{x^3y^3} + \dfrac{2x^2}{x^3y^3}$ Simplify numerators and denominators.

$= \dfrac{4y^2+2x^2}{x^3y^3}$ With identical denominators we can add the expressions.

b) The LCD for $6 = 2 \cdot 3$ and $8 = 2^3$ is $2^3 \cdot 3 = 24$.

$$\frac{a+1}{6} - \frac{a-2}{8} = \frac{(a+1)(4)}{6(4)} - \frac{(a-2)(3)}{8(3)} \qquad \text{Build up the denominators to the LCD 24.}$$

$$= \frac{4a+4}{24} - \frac{3a-6}{24} \qquad \text{Simplify.}$$

$$= \frac{4a+4-(3a-6)}{24} \qquad \text{The numerator is always treated as if it were in parentheses.}$$

$$= \frac{4a+4-3a+6}{24} \qquad \text{Remove parentheses.}$$

$$= \frac{a+10}{24} \qquad \text{Combine like terms.} \qquad \blacktriangleleft$$

EXAMPLE 5 Perform the indicated operations:

a) $\dfrac{1}{x^2-9} + \dfrac{2}{x^2+3x}$

b) $\dfrac{4}{a-5} - \dfrac{2}{5-a}$

Solution

a) $\dfrac{1}{x^2-9} + \dfrac{2}{x^2+3x} = \underbrace{\dfrac{1}{(x-3)(x+3)}}_{\text{Needs } x} + \underbrace{\dfrac{2}{x(x+3)}}_{\text{Needs } x-3} \qquad \text{The LCD is } x(x-3)(x+3).$

$$= \frac{1(x)}{(x-3)(x+3)(x)} + \frac{2(x-3)}{x(x+3)(x-3)}$$

$$= \frac{x}{x(x-3)(x+3)} + \frac{2x-6}{x(x-3)(x+3)}$$

$$= \frac{3x-6}{x(x-3)(x+3)}$$

In this type of answer we generally leave the denominator in factored form.

b) Since $-1(5-a) = a-5$, we can get identical denominators by multiplying only the second expression by -1 in the numerator and denominator:

$$\frac{4}{a-5} - \frac{2}{5-a} = \frac{4}{a-5} - \frac{2(-1)}{(5-a)(-1)}$$

$$= \frac{4}{a-5} - \frac{-2}{a-5}$$

$$= \frac{6}{a-5} \qquad 4-(-2)=6 \qquad \blacktriangleleft$$

Shortcuts

Some instructors refer to this shortcut as cross-multiplying. We do not use that term because it can be confused with the procedure for solving proportions, which some instructors also call cross-multiplying.

Consider the following addition:

$$\frac{a}{b} + \frac{c}{d} = \frac{a(d)}{b(d)} + \frac{c(b)}{d(b)} = \frac{ad + bc}{bd}$$

We can use this result to add two rational expressions. We do not find the least common denominator. We just multiply the denominators and use the formula $ad + bc$ for the numerator. For example,

$$\frac{2}{3} + \frac{1}{4} = \frac{2}{3} + \frac{1}{4} = \frac{2 \cdot 4 + 3 \cdot 1}{3 \cdot 4} = \frac{11}{12}.$$

RULE
Adding Simple Rational Expressions

If $b \neq 0$ and $d \neq 0$, then

$$\frac{a}{b} + \frac{c}{d} = \frac{ad + bc}{bd}.$$

It is easy to see that a similar rule works for subtraction.

RULE
Subtracting Simple Rational Expressions

If $b \neq 0$ and $d \neq 0$, then

$$\frac{a}{b} - \frac{c}{d} = \frac{ad - bc}{bd}.$$

These rules can be used to do any addition or subtraction of rational expressions. However, since we are using the product of the two denominators, the denominator of the answer may be too complicated to work with. These rules are best used for doing simple addition and subtraction problems mentally. Since the product of the denominators is not always the LCD, we will often have to reduce the answer.

EXAMPLE 6   Perform the operations mentally:

a) $\dfrac{1}{3} + \dfrac{1}{4}$

b) $\dfrac{2}{a} - \dfrac{1}{3}$

c) $\dfrac{3}{4} - \dfrac{1}{5}$

d) $\dfrac{x}{4} + \dfrac{x}{3}$

e) $\dfrac{1}{a - 1} + \dfrac{1}{a + 1}$

Solution   Use the rules given above to do the operations mentally. Here we will write down a middle step, but that is the step that should be done mentally.

a) $\dfrac{1}{3} + \dfrac{1}{4} = \dfrac{1 \cdot 4 + 3 \cdot 1}{12} = \dfrac{7}{12}$

b) $\dfrac{2}{a} - \dfrac{1}{3} = \dfrac{2 \cdot 3 - a \cdot 1}{3a} = \dfrac{6 - a}{3a}$

c) $\dfrac{3}{4} - \dfrac{1}{5} = \dfrac{3 \cdot 5 - 4 \cdot 1}{20} = \dfrac{11}{20}$

d) $\dfrac{x}{4} + \dfrac{x}{3} = \dfrac{3x + 4x}{12} = \dfrac{7x}{12}$

e) $\dfrac{1}{a - 1} + \dfrac{1}{a + 1} = \dfrac{1(a + 1) + 1(a - 1)}{a^2 - 1} = \dfrac{2a}{a^2 - 1}$ ◀

Warm-ups

True or false?

1. $\dfrac{1}{2} + \dfrac{1}{3} = \dfrac{2}{5}$ F

2. $\dfrac{7}{12} - \dfrac{1}{12} = \dfrac{1}{2}$ T

3. $\dfrac{2}{x} + 1 = \dfrac{3}{x}$ for any nonzero value of x.   F

4. $1 + \dfrac{1}{a} = \dfrac{a + 1}{a}$ for any nonzero value of a.   T

5. $a - \dfrac{1}{4} = \dfrac{3}{4}a$ for any value of a.   F

6. $\dfrac{a}{2} + \dfrac{b}{3} = \dfrac{3a + 2b}{6}$ for any values of a and b.   T

7. The LCD for the rational expressions $\dfrac{1}{x}$ and $\dfrac{3x}{x - 1}$ is $x^2 - 1$.   F

8. $\dfrac{3}{5} + \dfrac{4}{3} = \dfrac{29}{15}$ T

9. $\dfrac{4}{5} - \dfrac{5}{7} = \dfrac{3}{35}$ T

10. $\dfrac{5}{20} + \dfrac{3}{4} = 1$ T

5.4 EXERCISES

Perform the indicated operations. Reduce answers to lowest terms. See Example 1.

1. $\dfrac{1}{10} + \dfrac{1}{10}$ $\dfrac{1}{5}$

2. $\dfrac{1}{8} + \dfrac{3}{8}$ $\dfrac{1}{2}$

3. $\dfrac{7}{8} - \dfrac{1}{8}$ $\dfrac{3}{4}$

4. $\dfrac{4}{9} - \dfrac{1}{9}$ $\dfrac{1}{3}$

5. $\dfrac{1}{6} - \dfrac{5}{6}$ $-\dfrac{2}{3}$

6. $-\dfrac{3}{8} - \dfrac{7}{8}$ $-\dfrac{5}{4}$

7. $-\dfrac{7}{8} + \dfrac{1}{8}$ $-\dfrac{3}{4}$

8. $-\dfrac{9}{20} + \left(-\dfrac{3}{20}\right)$ $-\dfrac{3}{5}$

Perform the indicated operations. Reduce answers to lowest terms. See Example 2.

9. $\dfrac{1}{3} + \dfrac{2}{9}$ $\dfrac{5}{9}$

10. $\dfrac{1}{4} - \dfrac{5}{6}$ $-\dfrac{7}{12}$

11. $\dfrac{7}{16} - \dfrac{5}{18}$ $\dfrac{23}{144}$

12. $\dfrac{7}{6} + \dfrac{4}{15}$ $\dfrac{43}{30}$

13. $\dfrac{1}{8} - \dfrac{9}{10}$ $-\dfrac{31}{40}$

14. $-\dfrac{5}{12} + \dfrac{2}{15}$ $-\dfrac{17}{60}$

15. $-\dfrac{1}{6} - \left(-\dfrac{3}{8}\right)$ $\dfrac{5}{24}$

16. $\dfrac{1}{5} - \left(-\dfrac{1}{7}\right)$ $\dfrac{12}{35}$

Perform the indicated operations. Reduce answers to lowest terms. See Example 3.

17. $\dfrac{3x}{x+5} + \dfrac{15}{x+5}$ 3

18. $\dfrac{3}{2y} + \dfrac{7}{2y}$ $\dfrac{5}{y}$

19. $\dfrac{3}{3} - \dfrac{a}{3}$ $\dfrac{3-a}{3}$

20. $\dfrac{x-1}{x-4} - \dfrac{3x-9}{x-4}$ -2

21. $\dfrac{4x-3}{x(x+1)} - \dfrac{x-6}{x(x+1)}$ $\dfrac{3}{x}$

22. $\dfrac{x^2 - x - 5}{(x+1)(x+2)} + \dfrac{1-2x}{(x+1)(x+2)}$ $\dfrac{x-4}{x+2}$

Perform the indicated operations. Reduce answers to lowest terms. See Example 4.

23. $\dfrac{3}{2a} + \dfrac{1}{5a}$ $\dfrac{17}{10a}$

24. $\dfrac{5}{6y} - \dfrac{3}{8y}$ $\dfrac{11}{24y}$

25. $2a - \dfrac{1}{2}$ $\dfrac{4a-1}{2}$

26. $\dfrac{2}{x+1} - \dfrac{3}{x}$ $\dfrac{-x-3}{x(x+1)}$

27. $\dfrac{1}{3} + m$ $\dfrac{3m+1}{3}$

28. $\dfrac{2}{a-b} + \dfrac{1}{a+b}$ $\dfrac{3a+b}{(a-b)(a+b)}$

29. $\dfrac{1}{a-1} - \dfrac{2}{a}$ $\dfrac{2-a}{a(a-1)}$

30. $\dfrac{2}{w z^2} + \dfrac{3}{w^2 z}$ $\dfrac{2w+3z}{w^2 z^2}$

31. $\dfrac{3}{x+1} + \dfrac{2}{x-1}$ $\dfrac{5x-1}{(x+1)(x-1)}$

32. $\dfrac{3}{4a} - y$ $\dfrac{3-4ay}{4a}$

33. $\dfrac{1}{a^2 b} - \dfrac{5}{ab^2}$ $\dfrac{b-5a}{a^2 b^2}$

34. $3m + \dfrac{1}{2}$ $\dfrac{6m+1}{2}$

35. $\dfrac{b^2}{4a} - c$ $\dfrac{b^2 - 4ac}{4a}$

36. $\dfrac{1}{4} - y$ $\dfrac{1-4y}{4}$

Perform the indicated operations. Reduce answers to lowest terms. See Example 5.

37. $\dfrac{2a}{a^2-9} + \dfrac{a}{a-3}$ $\dfrac{a^2+5a}{(a-3)(a+3)}$

38. $\dfrac{x}{x^2-1} + \dfrac{3}{x-1}$ $\dfrac{4x+3}{(x-1)(x+1)}$

39. $\dfrac{3}{x^2+x-2} + \dfrac{4}{x^2+2x-3}$ $\dfrac{7x+17}{(x+2)(x-1)(x+3)}$

40. $\dfrac{1}{x^2-4} - \dfrac{3}{x^2-3x-10}$ $\dfrac{-2x+1}{(x-5)(x+2)(x-2)}$

41. $\dfrac{2x}{x^2-9} + \dfrac{3x}{x^2+4x+3}$ $\dfrac{5x^2-7x}{(x-3)(x+3)(x+1)}$

42. $\dfrac{x-1}{x^2-x-12} + \dfrac{x+4}{x^2+5x+6}$ $\dfrac{2x^2+x-18}{(x+2)(x+3)(x-4)}$

43. $\dfrac{2}{x} - \dfrac{1}{x-1} + \dfrac{1}{x+2}$ $\dfrac{2x^2-x-4}{x(x-1)(x+2)}$

44. $\dfrac{1}{a} - \dfrac{2}{a+1} + \dfrac{3}{a-1}$ $\dfrac{2a^2+5a-1}{a(a-1)(a+1)}$

45. $\dfrac{3}{x^2+x} - \dfrac{4}{5x+5}$ $\dfrac{15-4x}{5x(x+1)}$

46. $\dfrac{3}{a^2+3a} - \dfrac{2}{5a+15}$ $\dfrac{15-2a}{5a(a+3)}$

47. $\dfrac{4}{a-b} + \dfrac{4}{b-a}$ 0

48. $\dfrac{2}{x-3} + \dfrac{3}{3-x}$ $\dfrac{1}{3-x}$

49. $\dfrac{3}{2a-2} - \dfrac{2}{1-a}$ $\dfrac{7}{2a-2}$

50. $\dfrac{5}{2x-4} - \dfrac{3}{2-x}$ $\dfrac{11}{2x-4}$

Perform the following operations mentally. Write down only the answer. See Example 6.

51. $\dfrac{1}{2} + \dfrac{1}{2}$ 1

52. $\dfrac{3}{4} + \dfrac{1}{4}$ 1

53. $\dfrac{1}{5} + \dfrac{1}{4}$ $\dfrac{9}{20}$

54. $\dfrac{1}{2} + \dfrac{5}{3}$ $\dfrac{13}{6}$

55. $\dfrac{2}{3} + \dfrac{1}{2}$ $\dfrac{7}{6}$

56. $\dfrac{1}{4} - \dfrac{1}{5}$ $\dfrac{1}{20}$

57. $\dfrac{a}{3} - \dfrac{2}{3}$ $\dfrac{a-2}{3}$

58. $\dfrac{x}{9} + \dfrac{1}{9}$ $\dfrac{x+1}{9}$

59. $a + \dfrac{1}{3}$ $\dfrac{3a+1}{3}$

60. $\dfrac{m}{2} + 1$ $\dfrac{m+2}{2}$

61. $\dfrac{1}{a} + 1$ $\dfrac{a+1}{a}$

62. $\dfrac{2}{b} + 1$ $\dfrac{2+b}{b}$

63. $\dfrac{3}{x} - 3$ $\dfrac{3-3x}{x}$

64. $\dfrac{1}{4} + \dfrac{3}{2}$ $\dfrac{7}{4}$

65. $\dfrac{3}{8} - \dfrac{1}{4}$ $\dfrac{1}{8}$

66. $\dfrac{3}{7} - 1$ $-\dfrac{4}{7}$

67. $\dfrac{1}{7} - 2$ $-\dfrac{13}{7}$

68. $\dfrac{3}{5} + \dfrac{2}{3}$ $\dfrac{19}{15}$

69. $\dfrac{1}{2} + \dfrac{6}{7}$ $\dfrac{19}{14}$

70. $\dfrac{2}{3} + 2$ $\dfrac{8}{3}$

71. $\dfrac{a-1}{a} + 1$ $\dfrac{2a-1}{a}$

72. $\dfrac{x}{r} - \dfrac{y}{s}$ $\dfrac{xs-ry}{rs}$

73. $\dfrac{1}{a-1} + \dfrac{1}{a}$ $\dfrac{2a-1}{a(a-1)}$

74. $\dfrac{2}{3} + \dfrac{1}{3x}$ $\dfrac{2x+1}{3x}$

75. $\dfrac{1}{a-2} + \dfrac{1}{a+2}$ $\dfrac{2a}{a^2-4}$

76. $\dfrac{1}{x+2} + \dfrac{1}{x+3}$

76. $\dfrac{2x+5}{(x+2)(x+3)}$

77. $\dfrac{1}{3} + \dfrac{1}{3a}$ $\dfrac{a+1}{3a}$

78. $\dfrac{1}{4} + \dfrac{1}{4x}$ $\dfrac{x+1}{4x}$

5.5 Complex Fractions

IN THIS SECTION:

- **An Example**
- **Simplifying Complex Fractions**

In this section we will use the idea of least common denominator to simplify complex fractions. We will first give an example of a problem that can be solved using a complex fraction.

An Example

A survey of college students found that 1/2 of the female students had jobs and 2/3 of the male students had jobs. It was also found that 1/4 of the female students worked in fast-food restaurants and 1/6 of the male students worked in fast-food restaurants. If an equal number of male and female students were surveyed, then what percentage of the working students work in fast-food restaurants?

Unfortunately, there are no simple examples of complex fractions. This example is given so that students can see that a complex fraction can occur in a real situation.

To find the required percentage, we will divide the number of students working in fast-food restaurants by the total number of working students. Let x represent the number of males in the survey. Since the number of females surveyed is also x, we have

$$\frac{1}{4}x + \frac{1}{6}x = \text{the total number of students working in fast-food restaurants}$$

and

$$\frac{1}{2}x + \frac{2}{3}x = \text{the total number of working students.}$$

We can express the percentage of working students who work in fast-food restaurants as the ratio of the above expressions:

$$\frac{\dfrac{1}{4}x + \dfrac{1}{6}x}{\dfrac{1}{2}x + \dfrac{2}{3}x} = \text{the percentage of students who work in fast-food restaurants}$$

This expression is called a complex fraction. We will simplify it in Example 1.

Simplifying Complex Fractions

A **complex fraction** is a fraction having rational expressions in the numerator, denominator, or both. Examples of complex fractions are given in the next three examples. To simplify a complex fraction, we multiply the numerator and denominator by the LCD of all of the denominators. We will illustrate this procedure in the following examples.

EXAMPLE 1 Simplify the complex fraction:

$$\frac{\dfrac{1}{4}x + \dfrac{1}{6}x}{\dfrac{1}{2}x + \dfrac{2}{3}x}$$

Solution The LCD of the denominators 2, 3, 4, and 6 is 12. Multiply numerator and denominator by 12 to eliminate the fractions as follows:

$$\frac{\dfrac{1}{4}x + \dfrac{1}{6}x}{\dfrac{1}{2}x + \dfrac{2}{3}x} = \frac{\left(\dfrac{1}{4}x + \dfrac{1}{6}x\right)12}{\left(\dfrac{1}{2}x + \dfrac{2}{3}x\right)12} = \frac{3x + 2x}{6x + 8x} = \frac{5x}{14x} = \frac{5}{14}$$

$(1/4)12 = 3$
$(1/6)12 = 2$
$(1/2)12 = 6$
$(2/3)12 = 8$

Note that we did not multiply the numerator and denominator of each fraction by 12, but we multiplied the numerator and denominator of the complex fraction by 12. This eliminates all of the fractions except the one in the final answer.   ◀

The complex fraction that we found in the example of the working students was simplified in Example 1. We can answer the question posed in the example of the working students by converting 5/14 to a percent. Approximately 36% of the working students work in fast-food restaurants.

EXAMPLE 2   Simplify the complex fraction:

$$\frac{2 - \dfrac{1}{x}}{\dfrac{1}{x^2} - \dfrac{1}{2}}$$

Solution   The LCD for the denominators x, x^2, and 2 is $2x^2$.

$$\frac{2 - \dfrac{1}{x}}{\dfrac{1}{x^2} - \dfrac{1}{2}} = \frac{\left(2 - \dfrac{1}{x}\right)2x^2}{\left(\dfrac{1}{x^2} - \dfrac{1}{2}\right)2x^2} \qquad \textbf{Multiply the numerator and denominator by } 2x^2.$$

$$= \frac{2(2x^2) - \dfrac{1}{x}(2x^2)}{\dfrac{1}{x^2}(2x^2) - \dfrac{1}{2}(2x^2)} \qquad \textbf{Distributive property}$$

$$= \frac{4x^2 - 2x}{2 - x^2} \qquad \textbf{Simplify.}$$

The numerator of this answer can be factored, but the rational expression cannot be reduced.   ◀

EXAMPLE 3   Simplify the complex fraction:

$$\frac{\dfrac{1}{x - 2} - \dfrac{2}{x + 2}}{\dfrac{3}{2 - x} + \dfrac{4}{x + 2}}$$

Solution   Since $x - 2$ and $2 - x$ are opposites, we can use $(x - 2)(x + 2)$ as the

LCD. Multiply the numerator and denominator by $(x - 2)(x + 2)$:

$$\frac{\dfrac{1}{x-2} - \dfrac{2}{x+2}}{\dfrac{3}{2-x} + \dfrac{4}{x+2}} = \frac{\dfrac{1}{x-2}(x-2)(x+2) - \dfrac{2}{x+2}(x-2)(x+2)}{\dfrac{3}{2-x}(x-2)(x+2) + \dfrac{4}{x+2}(x-2)(x+2)}$$

$$= \frac{x + 2 - 2(x - 2)}{3(-1)(x+2) + 4(x-2)} \qquad \frac{x-2}{2-x} = -1$$

$$= \frac{x + 2 - 2x + 4}{-3x - 6 + 4x - 8} \qquad \textbf{Distributive property}$$

$$= \frac{-x + 6}{x - 14} \qquad \textbf{Combine like terms.} \qquad \blacktriangleleft$$

Warm-ups

True or false?

1. The LCD for the denominators 4, x, 6, and x^2 is $12x^3$. F

2. The LCD for the denominators $a - b$, $2b - 2a$, and 6 is $6a - 6b$. T

3. The fraction $\dfrac{4117}{7983}$ is a complex fraction. F

4. The LCD for the denominators $a - 3$ and $3 - a$ is $a^2 - 9$. F

5. The largest common denominator for the fractions $\dfrac{1}{2}$, $\dfrac{1}{3}$, and $\dfrac{1}{4}$ is 24. F

6. To simplify the complex fraction $\dfrac{\dfrac{1}{2} + \dfrac{x}{3}}{\dfrac{1}{4} + \dfrac{1}{5}}$, we should multiply the numerator and denominator by $60x$. F

7. To simplify the complex fraction $\dfrac{1 + \dfrac{2}{b}}{\dfrac{2}{a} + 5}$, we should multiply the numerator and denominator by $10ab$. F

8. $\dfrac{x - \dfrac{1}{2}}{x + \dfrac{3}{2}} = \dfrac{2x - 1}{2x + 3}$ for any value of x except $-3/2$. T

9. To simplify the complex fraction $\dfrac{\dfrac{1}{a} + \dfrac{2}{b}}{\dfrac{2}{a} + \dfrac{3}{b}}$, we should multiply the numerator and denominator by $\dfrac{ab}{ab}$. F

10. $\dfrac{\dfrac{1}{2} + \dfrac{1}{3}}{1 + \dfrac{1}{2}} = \dfrac{5}{6} \div \dfrac{3}{2}$ T

5.5 EXERCISES

Simplify the complex fractions. See Example 1.

1. $\dfrac{\dfrac{1}{2} + \dfrac{1}{3}}{\dfrac{1}{4} - \dfrac{1}{2}}$ $-\dfrac{10}{3}$

2. $\dfrac{\dfrac{1}{3} - \dfrac{1}{4}}{\dfrac{1}{3} + \dfrac{1}{6}}$ $\dfrac{1}{6}$

3. $\dfrac{\dfrac{2}{5} + \dfrac{5}{6} - \dfrac{1}{2}}{\dfrac{1}{2} - \dfrac{1}{3} + \dfrac{1}{15}}$ $\dfrac{22}{7}$

4. $\dfrac{\dfrac{2}{5} - \dfrac{2}{9} - \dfrac{1}{3}}{\dfrac{1}{3} + \dfrac{1}{5} + \dfrac{2}{15}}$ $-\dfrac{7}{30}$

5. $\dfrac{3 + \dfrac{1}{2}}{5 - \dfrac{3}{4}}$ $\dfrac{14}{17}$

6. $\dfrac{1 + \dfrac{1}{12}}{1 - \dfrac{1}{12}}$ $\dfrac{13}{11}$

7. $\dfrac{1 - \dfrac{1}{17}}{1 + \dfrac{1}{17}}$ $\dfrac{8}{9}$

8. $\dfrac{\dfrac{1}{15}}{\dfrac{2}{35}}$ $\dfrac{7}{6}$

Simplify the complex fractions. See Example 2.

9. $\dfrac{\dfrac{1}{a} + \dfrac{3}{b}}{\dfrac{1}{b} - \dfrac{3}{a}}$ $\dfrac{3a + b}{a - 3b}$

10. $\dfrac{\dfrac{1}{x} - \dfrac{3}{2}}{\dfrac{3}{4} + \dfrac{1}{x}}$ $\dfrac{4 - 6x}{3x + 4}$

11. $\dfrac{5 - \dfrac{3}{a}}{3 + \dfrac{1}{a}}$ $\dfrac{5a - 3}{3a + 1}$

12. $\dfrac{4 + \dfrac{3}{y}}{1 - \dfrac{2}{y}}$ $\dfrac{4y + 3}{y - 2}$

13. $\dfrac{\dfrac{1}{2} - \dfrac{2}{x}}{3 - \dfrac{1}{x^2}}$ $\dfrac{x^2 - 4x}{6x^2 - 2}$

14. $\dfrac{\dfrac{2}{a} + \dfrac{5}{3}}{\dfrac{3}{a} - \dfrac{3}{a^2}}$ $\dfrac{6a + 5a^2}{9a - 9}$

15. $\dfrac{\dfrac{3}{2b} + \dfrac{1}{b}}{\dfrac{3}{4} - \dfrac{1}{b^2}}$ $\dfrac{10b}{3b^2 - 4}$

16. $\dfrac{\dfrac{3}{2w} + \dfrac{4}{3w}}{\dfrac{1}{4w} - \dfrac{5}{9w}}$ $-\dfrac{102}{11}$

Simplify the following. See Example 3.

17. $\dfrac{x + \dfrac{4}{x - 2}}{x - \dfrac{x + 1}{x - 2}}$ $\dfrac{x^2 - 2x + 4}{x^2 - 3x - 1}$

18. $\dfrac{x - \dfrac{x - 6}{x - 1}}{x - \dfrac{x + 15}{x - 1}}$ $\dfrac{x^2 - 2x + 6}{x^2 - 2x - 15}$

19. $\dfrac{1 - \dfrac{3}{y + 1}}{3 + \dfrac{1}{y - 1}}$ $\dfrac{y^2 - 3y + 2}{3y^2 + y - 2}$

20. $\dfrac{2 - \dfrac{1}{a-3}}{3 - \dfrac{1}{a-3}}$ $\dfrac{2a-7}{3a-10}$

21. $\dfrac{\dfrac{1}{3-x} - 5}{\dfrac{1}{x-3} - 2}$ $\dfrac{5x-14}{2x-7}$

22. $\dfrac{\dfrac{2}{x-5} - x}{\dfrac{3x}{5-x} - 1}$ $\dfrac{x^2 - 5x - 2}{4x - 5}$

23. $\dfrac{\dfrac{2}{w-1} - \dfrac{3}{w}}{\dfrac{4}{w} + \dfrac{5}{w-1}}$ $\dfrac{-w+3}{9w-4}$

24. $\dfrac{\dfrac{1}{x+2} - \dfrac{3}{x+3}}{\dfrac{2}{x+3} + \dfrac{3}{x+2}}$ $\dfrac{-2x-3}{5x+13}$

25. $\dfrac{\dfrac{1}{a-b} - \dfrac{1}{a+b}}{\dfrac{1}{b-a} + \dfrac{1}{b+a}}$ -1

26. $\dfrac{\dfrac{1}{2+x} - \dfrac{1}{2-x}}{\dfrac{1}{x+2} - \dfrac{1}{x-2}}$ $-\dfrac{x}{2}$

27. $\dfrac{1 - \dfrac{5}{a-1}}{3 - \dfrac{2}{1-a}}$ $\dfrac{a-6}{3a-1}$

28. $\dfrac{\dfrac{1}{3} - \dfrac{2}{9-x}}{\dfrac{1}{6} - \dfrac{1}{x-9}}$ $\dfrac{2x-6}{x-15}$

29. $\dfrac{\dfrac{1}{m-3} - \dfrac{4}{m}}{\dfrac{3}{m-3} + \dfrac{1}{m}}$ $\dfrac{-3m+12}{4m-3}$

30. $\dfrac{\dfrac{1}{y+3} - \dfrac{4}{3}}{\dfrac{1}{3} - \dfrac{2}{y+3}}$ $\dfrac{4y+9}{3-y}$

5.6 Solving Equations

IN THIS SECTION:

- **Multiplying by the LCD**
- **Extraneous Roots**

Many problems in algebra can be solved by using equations involving rational expressions. In this section we will learn how to solve equations that involve rational expressions, and in Sections 5.7 and 5.8 we will solve problems using these equations.

Multiplying by the LCD

To solve an equation involving rational expressions, we usually multiply each side of the equation by the LCD for all the denominators involved.

EXAMPLE 1 Solve:

$$\frac{1}{x} + \frac{1}{6} = \frac{1}{4}$$

Solution We multiply each side of the equation by $12x$, the LCD for the denominators 4, 6, and x:

$$12x\left(\frac{1}{x} + \frac{1}{6}\right) = 12x\left(\frac{1}{4}\right) \qquad \text{Multiply each side by } 12x.$$

$$12x\left(\frac{1}{x}\right) + \overset{2}{12x}\left(\frac{1}{6}\right) = \overset{3}{12x}\left(\frac{1}{4}\right) \qquad \text{Apply the distributive property.}$$

$$12 + 2x = 3x \qquad \text{Note that all denominators are eliminated.}$$

$$12 = x \qquad \text{Subtract } 2x \text{ from each side.}$$

Check that 12 satisfies the original equation:

$$\frac{1}{12} + \frac{1}{6} = \frac{1}{12} + \frac{2}{12} = \frac{3}{12} = \frac{1}{4} \qquad\qquad \blacktriangleleft$$

EXAMPLE 2   Solve the equation:

$$\frac{100}{x} + \frac{100}{x + 5} = 9$$

Solution   The LCD for the denominators x and $x + 5$ is $x(x + 5)$.

$$x(x + 5)\frac{100}{x} + x(x + 5)\frac{100}{x+5} = x(x + 5)9 \qquad \text{Multiply each side by } x(x + 5).$$

$$(x + 5)100 + x(100) = (x^2 + 5x)9 \qquad \text{All denominators are eliminated.}$$

$$100x + 500 + 100x = 9x^2 + 45x \qquad \text{Simplify.}$$

$$500 + 200x = 9x^2 + 45x$$

$$0 = 9x^2 - 155x - 500$$

$$0 = (9x + 25)(x - 20) \qquad \text{Factor.}$$

$$9x + 25 = 0 \qquad \text{or} \qquad x - 20 = 0 \qquad \text{Apply the zero factor property.}$$

$$x = -\frac{25}{9} \qquad \text{or} \qquad x = 20$$

There are two solutions to the equation, $-25/9$ or 20. Check that each of these satisfies the original equation. $\blacktriangleleft$

Extraneous Roots

In a rational expression we can replace the variable only by real numbers that do not cause the denominator to be 0. When solving equations involving rational expressions, we must check every solution to see whether it causes 0 to appear in a denominator. If a number causes the denominator to be 0, then it cannot be a solution to the equation. A number that appears to be a solution (or root) but causes 0 in a denominator is called an **extraneous root.**

EXAMPLE 3   Solve the equation:

$$\frac{1}{x-2} = \frac{x}{2x-4} + 1$$

Solution   Since $2x - 4 = 2(x - 2)$, the LCD is $2(x - 2)$.

$$2(x-2)\frac{1}{x-2} = 2(x-2)\frac{x}{2x-4} + 2(x-2)\cdot 1 \qquad \textbf{Multiply each side by } 2(x-2).$$
$$2 = x + 2x - 4$$
$$2 = 3x - 4$$
$$6 = 3x$$
$$2 = x$$

Math at Work

A good photographer can tell a complete story in a fraction of a second. This story might show a politician realizing that the election has been lost or an athlete frowning in concentration. In the photograph, the subject's face is out of focus. What has gone wrong?

The focal length of a lens is the distance between the lens and the film inside the camera. For a 50-mm lens the image of an object at infinity (a distance such as the horizon) is in focus on the film if the film is 50 mm behind the lens. We say that the focal length of the lens is 50 mm. For a 250-mm telephoto lens (a long lens) an object at infinity is in focus on the film if the film is 250 mm behind the lens. For any lens, when the object is not at infinity (see the drawing), the focal length f for a lens is related to the object distance o and the image distance i by the formula

$$\frac{1}{f} = \frac{1}{o} + \frac{1}{i}.$$

If a photograph is taken of an object that is 600 mm from a 50-mm lens, then we can use the formula to find that the image distance is approximately 54.5 mm. The image comes into focus 4.5 mm behind the film and is out of focus on the film. Focusing a lens alters the focal length of the lens and allows the photographer to photograph nearby objects in focus.

What is the image distance for an object that is 1000 mm from a 250-mm telephoto lens? What is the image distance if the object is 250 mm from a 250-mm lens?   333.3 mm, infinity

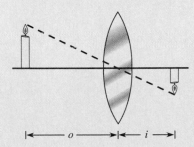

Check 2 in the original equation:

$$\frac{1}{2-2} = \frac{x}{2\cdot 2 - 4} + 1$$

Since the denominator $2-2$ has a value of 0, 2 is not a solution to the equation; it is an extraneous root. The equation has no solutions. ◄

EXAMPLE 4 Solve the equation:

$$\frac{1}{x} + \frac{1}{x-3} = \frac{x-2}{x-3}$$

Solution The LCD for the denominators x and $x-3$ is $x(x-3)$.

Ask students to make up an equation with an extraneous root. For example, start with $x = 5$, then write $x + 1 = 6$, then $\frac{x+1}{x-5} = \frac{6}{x-5}$.

$$x(x-3)\cdot\frac{1}{x} + x(x-3)\cdot\frac{1}{x-3} = x(x-3)\cdot\frac{x-2}{x-3} \qquad \text{Multiply each side by } x(x-3).$$

$$x - 3 + x = x(x-2)$$
$$2x - 3 = x^2 - 2x$$
$$0 = x^2 - 4x + 3$$
$$0 = (x-3)(x-1)$$
$$x - 3 = 0 \quad \text{or} \quad x - 1 = 0$$
$$x = 3 \quad \text{or} \quad x = 1$$

If $x = 3$, the denominator $x - 3$ has a value of 0. If $x = 1$, the original equation is satisfied. The only solution to the equation is 1. ◄

Warm-ups

True or false?

1. The idea of LCD has nothing to do with solving equations involving rational expressions. F

2. To solve $\frac{1}{x} + \frac{1}{3x} = \frac{3}{4}$, we must first change $\frac{1}{x}$ to $\frac{3}{3x}$. F

3. A solution to the equation $\frac{3}{x} - \frac{4}{x-5} = \frac{2}{x+3}$ is 5. F

4. To solve $\frac{1}{x-1} + 2 = \frac{1}{x+1}$, we multiply each side by $x^2 - 1$. T

5. Both 7/8 and -1 are solutions to the equation $\frac{3}{x+4} - \frac{6}{x+1} = \frac{7}{2}$. F

6. A solution to the equation $\frac{4}{x} = \frac{2}{5}$ is 10. T

7. To solve the equation $\dfrac{3}{x} + \dfrac{5}{x-2} = \dfrac{2}{3}$, we must first add the rational expressions on the left-hand side.   F

8. To solve the equation $x^2 = 8x$, we divide each side by x.   F

9. A solution to the equation $\dfrac{1}{x} + \dfrac{1}{2} = \dfrac{3}{4}$ is 4.   T

10. Both -1 and 2 satisfy the equation $\dfrac{1}{x-2} + \dfrac{3}{x+1} = 0$.   F

5.6 EXERCISES

Solve each equation. See Example 1.

1. $\dfrac{1}{x} + \dfrac{1}{2} = \dfrac{3}{4}$ 4

2. $\dfrac{3}{x} + \dfrac{1}{4} = \dfrac{5}{8}$ 8

3. $\dfrac{2}{3x} + \dfrac{1}{2x} - \dfrac{7}{24}$ 4

4. $\dfrac{x}{3} \quad 5 = \dfrac{x}{2} - 7$ 12

5. $\dfrac{x}{3} - \dfrac{x}{2} = \dfrac{x}{3} - 11$ 30

6. $\dfrac{x}{5} - \dfrac{2}{3} = \dfrac{x}{6} + \dfrac{1}{3}$ 30

7. $\dfrac{1}{6x} - \dfrac{1}{8x} = \dfrac{1}{72}$ 3

8. $\dfrac{2}{3x} - \dfrac{1}{6x} = \dfrac{1}{2}$ 1

Solve each equation. Watch for extraneous roots. See Examples 2–4.

9. $\dfrac{x}{2} = \dfrac{5}{x+3}$ $-5, 2$

10. $\dfrac{x}{3} = \dfrac{4}{x+1}$ $-4, 3$

11. $\dfrac{2}{x+1} = \dfrac{1}{x} + \dfrac{1}{6}$ 2, 3

12. $\dfrac{1}{x+1} - \dfrac{1}{2x} = \dfrac{1}{x}$ -3

13. $\dfrac{1}{x-1} + \dfrac{2}{x} = \dfrac{x}{x-1}$ 2

14. $\dfrac{4}{x} + \dfrac{3}{x-3} = \dfrac{x}{x-3} - \dfrac{1}{3}$ 6

15. $\dfrac{5}{x+2} + \dfrac{2}{x} \quad \dfrac{3}{x-3} \quad \dfrac{x-1}{x-3}$ No solution

16. $\dfrac{6}{x-2} + \dfrac{7}{x-8} - \dfrac{x-1}{x-8}$ No solution

17. $1 + \dfrac{3x}{x-2} = \dfrac{6}{x-2}$ No solution

18. $\dfrac{5}{x-3} = \dfrac{x+7}{2x-6} + 1$ No solution

Solve each equation.

19. $\dfrac{a}{4} = \dfrac{5}{2}$ 10

20. $\dfrac{y}{3} = \dfrac{6}{5}$ $\dfrac{18}{5}$

21. $\dfrac{1}{2x-4} + \dfrac{1}{x-2} = \dfrac{3}{2}$ 3

22. $\dfrac{7}{3x-9} - \dfrac{1}{x-3} = \dfrac{4}{3}$ 4

23. $\dfrac{x-3}{5} = \dfrac{x-3}{x}$ 3, 5

24. $\dfrac{a+4}{2} = \dfrac{a+4}{a}$ $-4, 2$

25. $\dfrac{3}{2x+4} - \dfrac{1}{x+2} = \dfrac{1}{3x+1}$ 3

26. $\dfrac{5}{2m+6} - \dfrac{1}{m+1} = \dfrac{1}{m+3}$ 3

27. $\dfrac{1}{x+2} = \dfrac{x}{x+2}$ 1

28. $\dfrac{-3}{w+2} = \dfrac{w}{w+2}$ -3

29. $\dfrac{w}{6} = \dfrac{3w}{11}$ 0

30. $\dfrac{2m}{3} = \dfrac{3m}{2}$ 0

31. $\dfrac{4}{x-2} - \dfrac{1}{2-x} = \dfrac{25}{x+6}$ 4

32. $\dfrac{3}{x+1} - \dfrac{1}{1-x} = \dfrac{10}{x^2-1}$ 3

33. $\dfrac{5}{x} = \dfrac{x}{5}$ $-5, 5$

34. $\dfrac{-3}{x} = \dfrac{x}{-3}$ $-3, 3$

35. $\dfrac{1}{x^2 - 9} + \dfrac{3}{x + 3} = \dfrac{4}{x - 3}$ -20

36. $\dfrac{3}{x - 2} - \dfrac{5}{x + 3} = \dfrac{1}{x^2 + x - 6}$ 9

37. $\dfrac{3}{x^2 - x - 6} = \dfrac{2}{x^2 - 4}$ 0

38. $\dfrac{8}{x^2 + x - 6} = \dfrac{6}{x^2 - 9}$ 6

5.7 Ratio and Proportion

IN THIS SECTION:

- Ratio
- Proportion

In this section we will use the ideas of rational expressions in ratio and proportion problems. We will solve proportions just as we solved equations in Section 5.6.

Ratio

In Chapter 1 we defined a rational number as the ratio of two integers. We will now give a more general definition of ratio. If a and b are any real numbers, with $b \neq 0$, we define the **ratio** of a and b to be

$$\dfrac{a}{b}.$$ (also written $a:b$ and read as "a to b")

A ratio is a comparison of two numbers. Some examples of ratios are

$$\dfrac{3}{4}, \quad \dfrac{4.2}{2.1}, \quad \dfrac{1/4}{1/2}, \quad \dfrac{3.6}{5}, \quad \text{and} \quad \dfrac{100}{1}.$$

Ratios are treated just like fractions. We can reduce ratios, and we can build them up. We generally express ratios as ratios of integers. When possible, we will convert a ratio into an equivalent ratio of integers in lowest terms. The following example will illustrate this.

EXAMPLE 1 Find an equivalent ratio of integers in lowest terms for each of the following ratios:

a) $\dfrac{4.2}{2.1}$

b) $\dfrac{\dfrac{1}{4}}{\dfrac{1}{2}}$

c) $\dfrac{3.6}{5}$

Solution

a) Since both the numerator and denominator have one decimal place, we multiply the numerator and denominator by 10 to eliminate the decimals:

$$\frac{4.2}{2.1} = \frac{4.2(10)}{2.1(10)} = \frac{42}{21} = \frac{21 \cdot 2}{21 \cdot 1} = \frac{2}{1} \qquad \text{Do not omit the 1 in a ratio.}$$

Thus the ratio of 4.2 to 2.1 is the same as the ratio 2 to 1.

b) We can simplify this expression like the complex fractions of Section 5.5. Multiply the numerator and denominator of this ratio by 4:

$$\frac{\dfrac{1}{4}}{\dfrac{1}{2}} = \frac{\dfrac{1}{4}(4)}{\dfrac{1}{2}(4)} = \frac{1}{2}$$

c) We can get a ratio of whole numbers if we multiply the numerator and denominator by 10:

$$\frac{3.6}{5} = \frac{3.6(10)}{5(10)} = \frac{36}{50}$$

$$= \frac{18}{25} \qquad \text{Reduce to lowest terms.} \qquad \blacktriangleleft$$

Ratios are used in many situations to compare quantities.

EXAMPLE 2 In a 50-pound bag of lawn fertilizer there are 8 pounds of nitrogen and 12 pounds of potash. What is the ratio of nitrogen to potash?

Solution The nitrogen and potash occur in this fertilizer in the ratio of 8 pounds to 12 pounds. Reducing this, we get

$$\frac{8}{12} = \frac{2 \cdot 4}{3 \cdot 4} = \frac{2}{3}.$$

Thus the ratio of nitrogen to potash is 2/3. $\qquad \blacktriangleleft$

EXAMPLE 3 What is the ratio of length to width for an 8 1/2-inch by 11-inch sheet of typing paper?

Solution The length is 11 inches, and the width is 8 1/2 inches. Expressing the width as the decimal number 8.5, we can write the ratio

$$\frac{11}{8.5} = \frac{11(2)}{8.5(2)} = \frac{22}{17}.$$

The ratio of length to width is 22 to 17. $\qquad \blacktriangleleft$

Ratios give us a means of comparing the size of two quantities. For this reason *the numbers compared in a ratio should be expressed in the same units*. For example, if one dog is 24 inches high and another is 1 foot high, then the ratio of their heights is 2 to 1, not 24 to 1.

Proportion

A **proportion** is any statement expressing the equality of two ratios. The statement

$$\frac{a}{b} = \frac{c}{d} \qquad \text{or} \qquad a:b = c:d$$

is a proportion. In any proportion the numbers in the positions of a and d above are called the **extremes.** The numbers in the positions of b and c above are called the **means.** In the proportion

$$\frac{11}{8.5} = \frac{22}{17},$$

the means are 8.5 and 22, and the extremes are 11 and 17.

If we multiply each side of the proportion

$$\frac{a}{b} = \frac{c}{d}$$

by the LCD, bd, we get

$$\frac{a}{b} \cdot bd = \frac{c}{d} \cdot bd,$$

or

$$a \cdot d = b \cdot c.$$

We can express this result by saying that *the product of the extremes is equal to the product of the means*. We call this fact the extremes-means property.

PROPERTY
Extremes-Means Property

Suppose a, b, c, and d are real numbers with $b \neq 0$ and $d \neq 0$. If

$$\frac{a}{b} = \frac{c}{d}, \text{ then } ad = bc.$$

We can use this property when solving proportions.

EXAMPLE 4 Solve the proportion for x:

$$\frac{3}{x} = \frac{5}{x + 5}$$

Solution

$$\frac{3}{x} = \frac{5}{x + 5}$$

$$3(x + 5) = 5x \qquad \text{Apply the extremes-means property.}$$

$$3x + 15 = 5x$$

$$15 = 2x$$

$$\frac{15}{2} = x \qquad \blacktriangleleft$$

EXAMPLE 5   The ratio of men to women at Brighton City College is 2 to 3. If there are 894 men, then how many women are there?

Solution   Let x represent the number of women. Since the ratio of men to women is 2 to 3, the ratio $894/x$ should be equal to the ratio $2/3$. This gives us the proportion:

$$\frac{2}{3} = \frac{894}{x}$$

$$2x = 2682 \qquad \text{Apply the extremes-means property.}$$

$$x = 1341$$

The number of women is 1341. $\qquad \blacktriangleleft$

Note that the proportion in Example 5 could be solved by multiplying each side by the LCD as we did when we solved other equations involving rational expressions. The extremes-means property gives us a shortcut for solving proportions.

EXAMPLE 6   The ratio of injuries to deaths in traffic accidents on a typical Memorial Day weekend is 20 to 1. This past Memorial Day weekend the number of injuries was 2850 more than the number of deaths. How many deaths were there?

Solution   If x represents the number of deaths, then $x + 2850$ represents the number of injuries. The ratio of $x + 2850$ to x must be the same as the ratio of 20 to 1. We can write this as a proportion and solve it.

$$\frac{x + 2850}{x} = \frac{20}{1}$$

$$x + 2850 = 20x \qquad \text{Apply the extremes-means property.}$$

$$2850 = 19x$$

$$150 = x$$

There were 150 deaths. $\qquad \blacktriangleleft$

The next example shows how conversions from one unit of measurement to another can be done with proportions.

EXAMPLE 7 There are 3 feet in 1 yard. How many feet are there in 12 yards?

Solution Let x represent the number of feet in 12 yards. There are two proportions that we can write to solve the problem:

$$\frac{3 \text{ feet}}{x \text{ feet}} = \frac{1 \text{ yard}}{12 \text{ yards}} \qquad \frac{3 \text{ feet}}{1 \text{ yard}} = \frac{x \text{ feet}}{12 \text{ yards}}$$

The ratios in the second proportion violate the rule of only comparing measurements that are expressed in the same units. Note that each side of the second proportion is actually the ratio 1 to 1, since 3 feet = 1 yard and x feet = 12 yards. For doing conversions we can use ratios like this to compare measurements in different units. Applying the extremes-means property to either proportion gives

$$3 \cdot 12 = x \cdot 1$$

or

$$x = 36.$$

So there are 36 feet in 12 yards. ◀

Warm-ups

True or false?

1. The ratio of 40 men to 30 women can be expressed as the ratio 4 to 3. T

2. The ratio of 3 feet to 2 yards can be expressed as the ratio 3 to 2. F

3. If the ratio of men to women in the Chamber of Commerce is 3 to 2 and there are 20 men, then there must be 30 women. F

4. The ratio of 1.5 to 2 is equal to the ratio of 3 to 4. T

5. A statement that two ratios are equal is called a proportion. T

6. In any proportion the product of the extremes is equal to the product of the means. T

7. If $\dfrac{2}{x} = \dfrac{3}{5}$, then $5x = 6$. F

8. The ratio of the heights of a 12-inch cactus to a 3-foot cactus is the same as the ratio 4 to 1. F

9. If 30 out of 100 lawyers prefer aspirin and the rest do not, then the ratio of lawyers who prefer aspirin to those who do not is 30 to 100. F

10. If $\dfrac{x+5}{x} = \dfrac{2}{3}$, then $3x + 15 = 2x$. T

5.7 EXERCISES

For each of the following ratios, find an equivalent ratio of integers in lowest terms. See Example 1.

1. $\dfrac{2.5}{3.5}$ $\dfrac{5}{7}$

2. $\dfrac{4.8}{1.2}$ $\dfrac{4}{1}$

3. $\dfrac{.32}{.6}$ $\dfrac{8}{15}$

4. $\dfrac{.05}{.8}$ $\dfrac{1}{16}$

5. $\dfrac{35}{10}$ $\dfrac{7}{2}$

6. $\dfrac{88}{33}$ $\dfrac{8}{3}$

7. $\dfrac{4.5}{7}$ $\dfrac{9}{14}$

8. $\dfrac{3}{2.5}$ $\dfrac{6}{5}$

9. $\dfrac{\frac{1}{2}}{\frac{1}{5}}$ $\dfrac{5}{2}$

10. $\dfrac{\frac{2}{3}}{\frac{3}{4}}$ $\dfrac{8}{9}$

11. $\dfrac{\frac{5}{1}}{\frac{1}{3}}$ $\dfrac{15}{1}$

12. $\dfrac{\frac{4}{1}}{\frac{1}{4}}$ $\dfrac{16}{1}$

Find a ratio for each of the following and write it as a ratio of integers in lowest terms. See Examples 2 and 3.

13. Find the ratio of men to women in a bowling league containing 12 men and 8 women. $\dfrac{3}{2}$

14. Among 100 coffee drinkers, 36 said they preferred their coffee black, and the rest did not prefer their coffee black. Find the ratio of those who drink black coffee to those who drink nonblack coffee. $\dfrac{9}{16}$

15. A life insurance company found that among their last 100 claims there were three dozen smokers. What is the ratio of smokers to nonsmokers in this group of claimants? $\dfrac{9}{16}$

16. A hunter shot at 60 rabbits and hit only a dozen. What is her ratio of hits to misses? $\dfrac{1}{4}$

17. While watching television for one week, a consumer group counted 1240 acts of violence and 40 acts of kindness. What is the violence to kindness ratio for television according to this group. $\dfrac{31}{1}$

18. If the rise is 3/2 and the run is 5, then what is the ratio of the rise to the run. $\dfrac{3}{10}$

Solve each proportion. See Example 4.

19. $\dfrac{4}{r} = \dfrac{2}{3}$ 6

20. $\dfrac{1}{x} = \dfrac{3}{8}$ $\dfrac{8}{3}$

21. $\dfrac{a}{2} = \dfrac{-1}{5}$ $-\dfrac{2}{5}$

22. $\dfrac{b}{3} = \dfrac{-3}{4}$ $\dfrac{9}{4}$

23. $-\dfrac{5}{9} = \dfrac{3}{x}$ $-\dfrac{27}{5}$

24. $-\dfrac{3}{4} = \dfrac{5}{x}$ $-\dfrac{20}{3}$

25. $\dfrac{10}{x} = \dfrac{34}{x + 12}$ 5

26. $\dfrac{x}{3} = \dfrac{x + 1}{2}$ -3

27. $\dfrac{x}{x + 1} = \dfrac{x + 3}{x}$ $\dfrac{3}{4}$

28. $\dfrac{x + 3}{x - 1} = \dfrac{x + 2}{x - 3}$ -7

29. $\dfrac{x - 1}{x - 2} = \dfrac{x - 3}{x + 4}$ $\dfrac{5}{4}$

30. $\dfrac{x}{x - 3} = \dfrac{x}{x + 9}$ 0

Use a proportion to solve each problem. See Examples 5, 6, and 7.

34. 4,200,000

31. The ratio of new shows to reruns on cable TV is 2 to 27. If Frank counted only eight new shows one evening, then how many reruns were there?   108

32. If four out of five doctors prefer fast food, then at a convention of 445 doctors, how many prefer fast food?   356

33. If 220 out of 500 voters surveyed said that they would vote for the incumbent, then how many votes could the incumbent expect out of the 400,000 voters in the state?   176,000

34. A taste test with 200 randomly selected people found that only three of them said they would buy a box of new Sweet Wheats cereal. How many boxes could the cereal makers expect to sell in a country of 280 million people?

35. The ratio of sports cars to luxury cars sold in Wentworth one month was 3 to 2. If there are 20 more sports cars sold than luxury cars, then how many of each were sold that month?   60 sport, 40 luxury

36. The ratio of foxes to rabbits in the Deerfield Forest Preserve is 2 to 9. If there are 35 fewer foxes than rabbits, then how many of each are there? 45 rabbits, 10 foxes

37. If there are 12 inches in one foot, then how many inches are there in 7 feet? 84 inches

38. If there are 3 feet in one yard, then how many yards are there in 28 feet? $\frac{28}{3}$ yards

39. If Alonzo travels 230 miles in 3 hours, then how many miles does he travel in 7 hours? 536.7 miles

40. If Evangelia can hike 19 miles in two days on the Appalachian Trail, then how many days will it take her to hike 63 miles? $\frac{126}{19}$ days

5.8 Applications

IN THIS SECTION:

- **Formulas**
- **Uniform Motion Problems**
- **Work Problems**
- **Miscellaneous Problems**

In this section we will study additional applications of rational expressions.

Formulas

Many formulas involve rational expressions.

EXAMPLE 1 The formula

$$\frac{y-4}{x+2} = \frac{3}{2}$$

is the equation for a certain straight line. We will study equations of this type further in Chapter 8. Solve it for y.

Solution To isolate y on the left-hand side of the equation, we multiply each side by $x + 2$:

$$(x+2) \cdot \frac{y-4}{x+2} = (x+2)\frac{3}{2} \qquad \text{Multiply by } x + 2.$$

$$y - 4 = \frac{3}{2}x + 3 \qquad \text{Simplify.}$$

$$y = \frac{3}{2}x + 7 \qquad \text{Add 4 to each side.}$$

Since the original equation is a proportion, we could have used the extremes-means property to solve it for y.   ◄

EXAMPLE 2 Solve the formula $\dfrac{D}{T} = R$ for T.

Solution

$$T \cdot \frac{D}{T} = T \cdot R \qquad \text{Multiply each side by } T.$$

$$D = TR$$

$$\frac{D}{R} = \frac{TR}{R} \qquad \text{Divide each side by } R.$$

$$\frac{D}{R} = T \qquad \text{Simplify.}$$

The formula solved for T is $T = \dfrac{D}{R}$.   ◄

In the next example different subscripts are used on a variable to indicate that they are different variables. Think of R_1 as the first resistance, R_2 as the second resistance, and R as a combined resistance.

EXAMPLE 3 The formula

$$\frac{1}{R} = \frac{1}{R_1} + \frac{1}{R_2}$$

(from physics) expresses the relationship between different amounts of resistance. Solve it for R_2.

Solution

$$RR_1R_2 \cdot \frac{1}{R} = RR_1R_2 \cdot \frac{1}{R_1} + RR_1R_2 \cdot \frac{1}{R_2} \qquad \text{Multiply each side by the LCD: } RR_1R_2.$$

$$R_1R_2 = RR_2 + RR_1 \qquad \text{All denominators are eliminated.}$$

$$R_1R_2 - RR_2 = RR_1 \qquad \text{Get all terms involving } R_2 \text{ onto the left-hand side.}$$

$$(R_1 - R)R_2 = RR_1 \qquad \text{Factor out } R_2.$$

$$R_2 = \frac{RR_1}{R_1 - R} \qquad \text{Divide each side by } R_1 - R.$$ ◄

EXAMPLE 4 In the formula of Example 1, find x if $y = -3$.

Solution Substitute $y = -3$ into the formula, then solve for x:

$$\frac{-3 - 4}{x + 2} = \frac{3}{2}$$

$$\frac{-7}{x + 2} = \frac{3}{2}$$

$$3x + 6 = -14 \qquad \textbf{Apply the extremes-means property.}$$

$$3x = -20$$

$$x = -\frac{20}{3}$$

◀

Uniform Motion Problems

EXAMPLE 5 Susan drove 1500 miles to Daytona Beach for spring break. On the way back she averaged 10 miles per hour less, and the drive back took her 5 hours longer. Find Susan's average speed on the way to Daytona Beach.

Solution If x represents her average speed going there, then $x - 10$ is her average speed for the return trip. See Figure 5.1. We use the formula $T = D/R$ to make the following table:

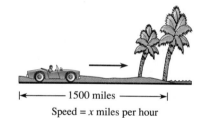

1500 miles
Speed = x miles per hour

Speed = $x - 10$ miles per hour

Figure 5.1

	D	R	T	
Going	1500	x	$\dfrac{1500}{x}$	⟵ Smaller time
Returning	1500	$x - 10$	$\dfrac{1500}{x - 10}$	⟵ Larger time

We can write an equation expressing the relationship between the two times. Subtract 5 hours from the larger time to get an expression equal to the smaller time:

$$\frac{1500}{x} = \frac{1500}{x - 10} - 5$$

$$x(x - 10)\frac{1500}{x} = x(x - 10)\frac{1500}{x - 10} - x(x - 10)5 \qquad \textbf{Multiply by } x(x - 10).$$

$$1500x - 15{,}000 = 1500x - 5x^2 + 50x$$

$$-15{,}000 = -5x^2 + 50x \qquad \textbf{Simplify.}$$

$$-3000 = -x^2 + 10x$$

$$x^2 - 10x - 3000 = 0 \qquad \textbf{Divide each side by 5.}$$

$$(x + 50)(x - 60) = 0 \qquad \textbf{Factor.}$$

$$x + 50 = 0 \qquad \text{or} \qquad x - 60 = 0$$

$$x = -50 \qquad \text{or} \qquad x = 60$$

The answer $x = -50$ is a solution to the equation but does not indicate the average speed of the car. Her average speed going to Daytona Beach was 60 mph.   ◄

Work Problems

If you can perform a task in 3 hours, then you are working at the rate of 1/3 of the task per hour. For problems involving work, we will always assume that the work is done at a constant rate. Thus if a job takes x hours to complete, we know that $1/x$ of it is done per hour.

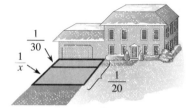

Figure 5.2

EXAMPLE 6   After a heavy snowfall, Brian can shovel all of the driveway in 30 minutes. If his brother Allen helps, the job takes only 20 minutes. How long would it take Allen to do the job by himself?

Solution   Let x represent the number of minutes it would take Allen to do the job by himself. Brian shovels 1/30 of the driveway per minute, and Allen shovels $1/x$ of the driveway per minute. When they work together, 1/20 of the driveway gets done per minute. Assuming that they do not interfere with one another, the portions they each do in one minute must total the portion they do together in one minute. See Fig. 5.2.

$$\frac{1}{30} + \frac{1}{x} = \frac{1}{20}$$

$$60x \cdot \frac{1}{30} + 60x \cdot \frac{1}{x} = 60x \cdot \frac{1}{20} \qquad \text{Multiply each side by } 60x.$$

$$2x + 60 = 3x$$

$$60 - x$$

When checking this answer, point out that in twenty minutes Allen does 20/60 = 1/3 of the driveway, while Brian does 20/30 = 2/3 of the driveway.

It would take Allen 60 minutes to shovel the driveway by himself.   ◄

Miscellaneous Problems

EXAMPLE 7   Tamara bought 50 pounds of fruit consisting of Florida oranges and Texas grapefruit. She paid twice as much per pound for the grapefruit as she did for the oranges. If Tamara bought $12 worth of oranges and $16 worth of grapefruit, then how many pounds of each did she buy?

Solution   If x represents the number of pounds of oranges, then $50 - x$ is the number of pounds of grapefruit. See Fig. 5.3. Thus

$$\frac{\$12}{x} = \text{the price per pound of the oranges,}$$

and

$$\frac{\$16}{50 - x} = \text{the price per pound of the grapefruit.}$$

x lb

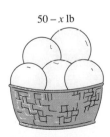

Oranges

$50 - x$ lb

Grapefruit

Figure 5.3

Point out the similarity between this problem and a uniform motion problem. Price per pound corresponds to rate, the number of pounds corresponds to time, and total cost corresponds to distance.

The equation expresses the fact that the price per pound of the grapefruit is twice that of the oranges.

$$2\left(\frac{12}{x}\right) = \frac{16}{50 - x}$$

$$\frac{24}{x} = \frac{16}{50 - x}$$

$$16x = 1200 - 24x \qquad \text{Apply the extremes-means property.}$$

$$40x = 1200$$

$$x = 30$$

$$50 - x = 20$$

Tamara purchased 20 pounds of grapefruit and 30 pounds of oranges. Check this answer. ◄

Warm-ups

True or false?

1. The formula $t = \dfrac{1 - t}{m}$, solved for m, is $m = \dfrac{1 - t}{t}$. T

2. To solve $\dfrac{1}{m} + \dfrac{1}{n} = \dfrac{1}{2}$ for m, we multiply each side by $2mn$. T

3. If $D = RT$, then $T = D/R$. T

4. If Fiona drives 300 miles in x hours, then her average speed is $x/300$ miles per hour. F

5. If Miguel drives 20 hard bargains in $x + 3$ hours, then he is driving $20/(x + 3)$ hard bargains per hour. T

6. If Jared can paint a house in y days, then he paints $1/y$ of the house per day. T

7. If $1/x$ is one less than $2/(x + 3)$, then the equation that expresses this relationship is $\dfrac{1}{x} - 1 = \dfrac{2}{x + 3}$. F

8. If a and b are nonzero and $a = \dfrac{m}{b}$, then $b = am$. F

9. Solving $P + Prt = I$ for P gives $P = I - Prt$. F

10. To solve $3R + yR = m$ for R, we must first factor the left-hand side. T

5.8 EXERCISES

Solve each equation for y. See Example 1.

1. $\dfrac{y-1}{x-3} = 2$ $y = 2x - 5$

2. $\dfrac{y-2}{x-4} = -2$ $y = -2x + 10$

3. $\dfrac{y-1}{x+6} = -\dfrac{1}{2}$ $y = -\dfrac{1}{2}x - 2$

4. $\dfrac{y+5}{x-2} = -\dfrac{1}{2}$ $y = -\dfrac{1}{2}x - 4$

5. $\dfrac{y+a}{x-b} = m$ $y = mx - mb - a$

6. $\dfrac{y-h}{x+k} = a$ $y = ax + ak + h$

7. $\dfrac{y-1}{x+4} = -\dfrac{1}{3}$ $y = -\dfrac{1}{3}x - \dfrac{1}{3}$

8. $\dfrac{y-1}{x+3} = -\dfrac{3}{4}$ $y = -\dfrac{3}{4}x - \dfrac{5}{4}$

Solve each formula for the indicated variable. See Examples 2 and 3.

9. $A = \dfrac{B}{C}$ for C $C = \dfrac{B}{A}$

10. $P = \dfrac{A}{C+D}$ for A $A = PC + PD$

11. $\dfrac{1}{a} + m = \dfrac{1}{p}$ for p $p = \dfrac{a}{1+am}$

12. $\dfrac{2}{f} + t = \dfrac{3}{m}$ for m $m = \dfrac{3f}{2+ft}$

13. $F = k\dfrac{m_1 m_2}{r^2}$ for m_1 $m_1 = \dfrac{r^2 F}{k m_2}$

14. $F = \dfrac{mv^2}{r}$ for r $r = \dfrac{mv^2}{F}$

15. $\dfrac{1}{a} + \dfrac{1}{b} = \dfrac{1}{f}$ for a $a = \dfrac{bf}{b-f}$

16. $\dfrac{1}{R} = \dfrac{1}{R_1} + \dfrac{1}{R_2}$ for R $R = \dfrac{R_1 R_2}{R_1 + R_2}$

17. $S = \dfrac{a}{1-r}$ for r $r = \dfrac{S-a}{S}$

18. $I = \dfrac{E}{R+r}$ for R $R = \dfrac{E - Ir}{I}$

19. $\dfrac{P_1 V_1}{T_1} = \dfrac{P_2 V_2}{T_2}$ for P_2 $P_2 = \dfrac{P_1 V_1 T_2}{T_1 V_2}$

20. $\dfrac{P_1 V_1}{T_1} = \dfrac{P_2 V_2}{T_2}$ for T_1 $T_1 = \dfrac{P_1 V_1 T_2}{P_2 V_2}$

21. $V = \dfrac{4}{3}\pi r^2 h$ for h $h = \dfrac{3V}{4\pi r^2}$

22. $h = \dfrac{S - 2\pi r^2}{2\pi r}$ for S $S = 2\pi rh + 2\pi r^2$

Find the value of the indicated variable. See Example 4.

23. In the formula of Exercise 9, if $A = 12$ and $B = 5$, find C. $C = \dfrac{5}{12}$

24. In the formula of Exercise 10, if $A = 500$, $P = 100$, and $C = 2$, find D. $D = 3$

25. In the formula of Exercise 11, if $p = 6$ and $m = 4$, find a. $a = \dfrac{6}{23}$

26. In the formula of Exercise 12, if $m = 4$ and $t = 3$, find f. $f = -\dfrac{8}{9}$

27. In the formula of Exercise 13, if $F = 32$, $r = 4$, $m_1 = 2$, and $m_2 = 6$, find k. $k = \dfrac{128}{3}$

28. In the formula of Exercise 14, if $F = 10$, $v = 8$, and $r = 6$, find m. $m = \dfrac{15}{16}$

29. In the formula of Exercise 15, if $f = 3$ and $a = 2$, find b. $b = -6$

30. In the formula of Exercise 16, if $R = 3$ and $R_1 = 5$, find R_2. $R_2 = \dfrac{15}{2}$

31. In the formula of Exercise 17, if $S = 3/2$ and $r = 1/5$, find a. $a = \dfrac{6}{5}$

32. In the formula of Exercise 18, if $I = 15$, $E = 3$, and $R = 2$, find r. $r = \dfrac{9}{5}$

Show a complete solution to each of the following problems. See Example 5.

33. Marcie can walk 8 miles in the same time as it takes Frank to walk 6 miles. If Marcie walks 1 mile per hour faster than Frank, then how fast does each of them walk?

34. Junior's boat will go 15 miles per hour in still water. If he can go 12 miles downstream in the same amount of time as it takes to go 9 miles upstream, then what is the speed of the current?

35. Pat travels 70 miles on her milk route, and Bob travels 75 miles on his route. Pat travels 5 miles per hour slower than Bob, and her route takes 1/2 hour longer than Bob's. How fast is each one traveling?

36. Smith bicycled 45 miles, and Jones bicycled 70 miles. Jones averaged 5 miles per hour more than Smith, and his

33. Marcie: 4 mph, Frank: 3 mph   **34.** $\dfrac{15}{7}$ mph   **35.** Bob: 25 mph, Pat: 20 mph

trip took one-half hour longer than Smith's. How fast was each one traveling?

37. Raffaele ran 8 miles and then walked 6 miles. If he ran 5 miles per hour faster than he walked and the total time was 2 hours, then how fast did he walk? 5 mph

38. Luisa participated in a triathalon in which she swam 3 miles, ran 5 miles, and then bicycled 10 miles. Luisa ran twice as fast as she swam, and she cycled three times as fast as she swam. If her total time for the triathalon was 1 hour and 46 minutes, then how fast did she swim? 5 mph

36. Smith: 15 mph, Jones: 20 mph or Smith: 30 mph, Jones: 35 mph

Photo for Exercise 38

Show a complete solution to each of the following problems. See Example 6.

39. Kiyoshi can paint a certain fence in 3 hours by himself. If Red helps, the job takes only 2 hours. How long would it take Red to paint the fence by himself? 6 hours

40. Every week Linda must stuff 1000 envelopes. She can do the job by herself in 6 hours. If Laura helps, they get the job done in 5 1/2 hours. How long would it take Laura to do the job by herself? 66 hours

41. Mr. McGregor has discovered that a large dog can destroy his entire garden in 2 hours and a small boy can do the same job in 1 hour. How long would it take the large dog and the small boy working together to destroy Mr. McGregor's garden? 40 minutes

42. With only the small valve open, all of the liquid can be drained from a large vat in 4 hours. With only the large valve open, all of the liquid can be drained from the same

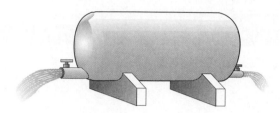

Figure for Exercise 42

vat in 2 hours. How long would it take to drain the vat with both valves open?

43. Edgar can blow the leaves off of the sidewalks around the capitol using a gasoline-powered blower in 2 hours. Ellen can do the same job in 8 hours using a broom. How long would it take them working together?

44. It takes a computer 8 days to print all of the personalized letters for a national sweepstakes. A new computer is purchased that can do the same job in 5 days. How long would it take to do the job with both computers working on it?

42. $\frac{4}{3}$ hours   43. 1 hour 36 minutes   44. $\frac{40}{13}$ days

Show a complete solution to each of the following problems. See Example 7.

45. Bertha bought 18 pounds of fruit consisting of apples and bananas. She paid $9 for the apples and $2.40 for the bananas. If the price per pound of the apples was 3 times that of the bananas, then how many pounds of each type of fruit did she buy? 8 pounds bananas, 10 pounds apples

46. In the play-off game the ball was carried by either Anderson or Brown on 21 plays. Anderson gained 36 yards, and Brown gained 54 yards. If Brown averaged twice as many yards per carry as Anderson, then on how many plays did Anderson carry the ball? 12

47. Last week, Kim's Electric Service used 110 gallons of gasoline in their two trucks. The large truck was driven 800 miles, and the small truck was driven 600 miles. If the small truck gets twice as many miles per gallon as the large truck, then how many gallons of gasoline did the large truck use? 80 gallons

48. Sally received a bill for a total of 8 hours labor on the repair of her bulldozer. She paid $50 to the master mechanic and $90 to his apprentice. If the master mechanic gets $10 more per hour than his apprentice, then how many hours did each work on the bulldozer?

48. Master: 2 hours, apprentice: 6 hours

Wrap-up

CHAPTER 5

SUMMARY

	Concepts	Examples
Rational expression	The ratio of two polynomials with the denominator not equal to 0	$\dfrac{x-1}{2x-3}$ $x \neq 3/2$
Rule for reducing rational expressions	If $a \neq 0$ and $c \neq 0$, then $\dfrac{ab}{ac} = \dfrac{b}{c}.$ Divide out the common factors.	$\dfrac{8x+2}{4x} = \dfrac{2(4x+1)}{2(2x)} = \dfrac{4x+1}{2x}$

Quotient Rule for exponents	Suppose $a \neq 0$ and m and n are positive integers.
	If $m \geq n$, then
	$\dfrac{a^m}{a^n} = a^{m-n}.$
	
	If $n > m$, then
	$\dfrac{a^m}{a^n} = \dfrac{1}{a^{n-m}}.$
		$\dfrac{x^7}{x^5} = x^2$
		
		$\dfrac{x^2}{x^5} = \dfrac{1}{x^3}$

Multiplication of rational expressions	If $b \neq 0$ and $d \neq 0$, then $\dfrac{a}{b} \cdot \dfrac{c}{d} = \dfrac{ac}{bd}$	$\dfrac{3}{x} \cdot \dfrac{6}{x^2} = \dfrac{18}{x^3}$
Division of rational expressions	If $b \neq 0$, $c \neq 0$, and $d \neq 0$, then $\dfrac{a}{b} \div \dfrac{c}{d} = \dfrac{a}{b} \cdot \dfrac{d}{c}$ (Invert and multiply.)	$\dfrac{a}{x} \div \dfrac{5}{4x} = \dfrac{a}{x} \cdot \dfrac{4x}{5} = \dfrac{4a}{5}$
Least common denominators	The LCD of a group of denominators is the smallest number that is a multiple of all of them.	8, 12 LCD = 24
Finding the least common denominator	1. Factor each denominator completely. Use exponent notation for repeated factors. 2. Write the product of all of the different factors that appear in the denominators. 3. On each factor, use the highest power that appears on that factor in any of the denominators.	$6a^2b$, $4ab^3$ $4 = 2^2$ $6 = 2 \cdot 3$ LCD $= 12a^2b^3$

Addition and subtraction of rational expressions	If $b \neq 0$, then $$\frac{a}{b} + \frac{c}{b} = \frac{a+c}{b} \text{ and } \frac{a}{b} - \frac{c}{b} = \frac{a-c}{b}.$$ If the denominators are not identical, we change each fraction to an equivalent fraction so that all denominators are identical.	$\dfrac{2x}{x-3} + \dfrac{7x}{x-3} = \dfrac{9x}{x-3}$ $\dfrac{2}{x} + \dfrac{1}{3x} = \dfrac{6}{3x} + \dfrac{1}{3x} = \dfrac{7}{3x}$
Rules for adding or subtracting simple fractions	If $b \neq 0$ and $d \neq 0$, then $$\frac{a}{b} + \frac{c}{d} = \frac{ad+bc}{bd}$$ and $$\frac{a}{b} - \frac{c}{d} = \frac{ad-bc}{bd}.$$	$\dfrac{1}{2} + \dfrac{1}{3} = \dfrac{5}{6}$ $\dfrac{2}{5} - \dfrac{3}{7} = \dfrac{-1}{35}$
To simplify complex fractions	Multiply the numerator and denominator by the LCD.	$\dfrac{\left(\dfrac{1}{2} + \dfrac{1}{3}\right)12}{\left(\dfrac{1}{3} - \dfrac{3}{4}\right)12} = \dfrac{6+4}{4-9} = -2$
Equations with rational expressions	Multiply each side by the LCD.	$\dfrac{1}{x} - \dfrac{1}{3} = \dfrac{1}{2x} - \dfrac{1}{6}$ $6x\left(\dfrac{1}{x} - \dfrac{1}{3}\right) = 6x\left(\dfrac{1}{2x} - \dfrac{1}{6}\right)$ $6 - 2x = 3 - x$
Proportion	Equation expressing the equality of two ratios	$\dfrac{1}{x} = \dfrac{3}{5}$
Extremes-means property	If $b \neq 0$ and $d \neq 0$, then $\dfrac{a}{b} = \dfrac{c}{d}$ is equivalent to $ad = bc$. Shortcut for solving proportions	$\dfrac{2}{x-3} = \dfrac{5}{6}$ $12 = 5x - 15$

REVIEW EXERCISES

5.1 *Reduce each rational expression to lowest terms.*

1. $\dfrac{a^3c^3}{a^5c}$ $\dfrac{c^2}{a^2}$

2. $\dfrac{x^2-1}{3x-3}$ $\dfrac{x+1}{3}$

3. $\dfrac{39x^6}{15x}$ $\dfrac{13x^5}{5}$

4. $\dfrac{3x^2-9x+6}{10-5x}$ $\dfrac{3-3x}{5}$

5.2 *Perform the indicated operations.*

5. $\dfrac{12b^6}{15a^5} \cdot \dfrac{25a^2}{20b^3}$ $\dfrac{b^3}{a^3}$

6. $\dfrac{x^2-1}{3x} \cdot \dfrac{6x}{2x-2}$ $x+1$

7. $\dfrac{w-2}{3w} \div \dfrac{4w-8}{6w}$ $\dfrac{1}{2}$

8. $\dfrac{2y+2x}{x-xy} \div \dfrac{x^2+2xy+xy^2}{y^2-y}$ $\dfrac{-2xy-2y^2}{x^2(x+2y+y^2)}$

9. $\dfrac{2xy}{3} \div y$ $\dfrac{2x}{3}$

10. $4ab \div \dfrac{1}{2}$ $8ab$

11. $\dfrac{1}{k} \cdot 3k^2$ $3k$

12. $\dfrac{1}{abc} \cdot 5a^3b^5c^2$ $5a^2b^4c$

5.3 *Find the least common denominator for each pair of denominators.*

13. $4x$, $6x - 6$ $12x(x - 1)$

14. $x^2 - 4$, $x^2 - x - 2$ $(x + 1)(x - 2)(x + 2)$

15. $6ab^3$, $8a^7b^2$ $24a^7b^3$

16. $x^2 - 9$, $x^2 + 6x + 9$ $(x - 3)(x + 3)^2$

Convert each rational expression into an equivalent rational expression with the indicated denominator.

17. $\dfrac{2}{3xy} = \dfrac{?}{15x^2y}$ $\dfrac{10x}{15x^2y}$

18. $\dfrac{x}{x - 1} = \dfrac{?}{x^2 - 1}$ $\dfrac{x^2 + x}{x^2 - 1}$

19. $\dfrac{5}{y - 6} = \dfrac{?}{12 - 2y}$ $\dfrac{-10}{12 - 2y}$

20. $\dfrac{-3}{2 - t} = \dfrac{?}{2t - 4}$ $\dfrac{6}{2t - 4}$

5.4 *Perform the indicated operations.*

21. $\dfrac{3}{2x - 4} + \dfrac{1}{x^2 - 4}$ $\dfrac{3x + 8}{2(x + 2)(x - 2)}$

22. $\dfrac{3}{x - 2} - \dfrac{5}{x + 3}$ $\dfrac{-2x + 19}{(x - 2)(x + 3)}$

23. $\dfrac{2}{ab^2} - \dfrac{1}{a^2b}$ $\dfrac{2a - b}{a^2b^2}$

24. $\dfrac{x}{x - 3} - \dfrac{3x}{x^2 - 9}$ $\dfrac{x^2}{x^2 - 9}$

25. $3 + \dfrac{4}{x}$ $\dfrac{3x + 4}{x}$

26. $1 + \dfrac{3a}{2b}$ $\dfrac{2b + 3a}{2b}$

27. $a - \dfrac{1}{a}$ $\dfrac{a^2 - 1}{a}$

28. $\dfrac{4}{b} - b$ $\dfrac{4 - b^2}{b}$

29. $\dfrac{1}{a - 8} - \dfrac{2}{8 - a}$ $\dfrac{3}{a - 8}$

30. $\dfrac{5}{x - 14} + \dfrac{4}{14 - x}$ $\dfrac{1}{x - 14}$

5.5 *Simplify each complex fraction.*

31. $\dfrac{\dfrac{1}{2} - \dfrac{3}{4}}{\dfrac{2}{3} + \dfrac{1}{2}}$ $-\dfrac{3}{14}$

32. $\dfrac{\dfrac{2}{3} + \dfrac{5}{8}}{\dfrac{1}{2} - \dfrac{3}{8}}$ $\dfrac{31}{3}$

33. $\dfrac{\dfrac{1}{a} + \dfrac{2}{3b}}{\dfrac{1}{2b} - \dfrac{3}{a}}$ $\dfrac{6b + 4a}{3a - 18b}$

34. $\dfrac{\dfrac{3}{xy} - \dfrac{1}{3y}}{\dfrac{1}{6x} - \dfrac{3}{5y}}$ $\dfrac{90 - 10x}{5y - 18x}$

5.6 *Solve each equation.*

35. $\dfrac{-2}{5} = \dfrac{3}{x}$ $-\dfrac{15}{2}$

36. $\dfrac{3}{x} + \dfrac{5}{3x} = 1$ $\dfrac{14}{3}$

37. $\dfrac{14}{a^2 - 1} + \dfrac{1}{a - 1} = \dfrac{3}{a + 1}$ 9

38. $2 + \dfrac{3}{x - 5} = \dfrac{2x}{x - 5}$ No solution

39. $x - \dfrac{3x}{2 - x} = \dfrac{6}{x - 2}$ -3

40. $\dfrac{1}{x} + \dfrac{1}{3} = \dfrac{1}{2}$ 6

5.7 *Solve each proportion.*

41. $\dfrac{3}{x} = \dfrac{2}{7}$ $\dfrac{21}{2}$

42. $\dfrac{4}{x} = \dfrac{x}{4}$ $-4, 4$

43. $\dfrac{2}{x - 3} = \dfrac{5}{x}$ 5

44. $\dfrac{3}{x - 3} = \dfrac{5}{x + 4}$ $\dfrac{27}{2}$

Solve each problem by using a proportion.

45. The ratio of taxis to private automobiles in Times Square on New Year's Eve was estimated to be 15 to 2. If there were 60 taxis, then how many private automobiles were there?   8

46. The student-teacher ratio at Washington High was reported to be 27.5 to 1. If there are 42 teachers, then how many students are there?   1155

47. At Wong's Chinese Restaurant the secret recipe for white rice calls for a 2 to 1 ratio of water to rice. In one batch the chef used 28 more cups of water than rice. How many cups of each did he use?   56 cups water, 28 cups rice

48. An outboard motor calls for a fuel mixture that has a gasoline-to-oil ratio of 50 to 1. How many pints of oil should be added to 6 gallons of gasoline? $\dfrac{24}{25}$ pints

5.8 *Solve each formula for the indicated variable.*

49. $\dfrac{y - b}{m} = x$ for y $y = mx + b$

50. $\dfrac{A}{h} = \dfrac{a + b}{2}$ for a $a = \dfrac{2A - hb}{h}$

51. $F = \dfrac{mv + 1}{m}$ for m $m = \dfrac{1}{F - v}$

52. $m = \dfrac{r}{1 + rt}$ for r $r = \dfrac{m}{1 - mt}$

53. $\dfrac{y + 1}{x - 3} = 4$ for y $y = 4x - 13$

54. $\dfrac{y - 3}{x + 2} = \dfrac{-1}{3}$ for y $y = -\dfrac{1}{3}x + \dfrac{7}{3}$

Solve each problem.

55. 200 hours

55. Tracy, Stacy, and Fred assembled a very large puzzle together in 40 hours. If Stacy worked twice as fast as Fred, and Tracy worked just as fast as Stacy, then how long would it take Fred to assemble the puzzle alone?

56. Leon drove 270 miles in the same time as Pat drove 330 miles. If Pat drove 10 miles per hour faster than Leon, then how fast did each of them drive?

57. When Bert and Ernie merged their automobile dealerships, Bert had 10 more cars than Ernie. While 36% of Ernie's stock consisted of new cars, only 25% of Bert's stock consisted of new cars. If they had 33 new cars on the lot after the merger, then how many cars did each one have before the merger?   Bert: 60, Ernie: 50

58. A company specializing in magazine sales over the telephone found that of 2500 phone calls, 360 resulted in sales and were made by male callers, and 480 resulted in sales and were made by female callers. If the company makes twice as many sales per call with a woman's voice than with a man's voice, then how many of the 2500 calls were made by females?   1000

56. Leon: 45 mph, Pat: 55 mph

Miscellaneous

For each of the following, either perform the indicated operation or solve the equation, whichever is appropriate.

59. $\dfrac{1}{x} + \dfrac{1}{2x}$ $\dfrac{3}{2x}$

60. $\dfrac{1}{y} + \dfrac{1}{3y} = 2$ $\dfrac{2}{3}$

61. $\dfrac{2}{3xy} + \dfrac{1}{6x}$ $\dfrac{4 + y}{6xy}$

62. $\dfrac{3}{x - 1} - \dfrac{3}{x}$ $\dfrac{3}{x(x - 1)}$

63. $\dfrac{5}{a - 5} - \dfrac{3}{5 - a}$ $\dfrac{8}{a - 5}$

64. $\dfrac{2}{x - 2} - \dfrac{3}{x} = \dfrac{-1}{x}$ No solution

65. $\dfrac{2}{x - 1} - \dfrac{2}{x} = 1$ $-1, 2$

66. $\dfrac{2}{x - 2} \cdot \dfrac{6x - 12}{14}$ $\dfrac{6}{7}$

67. $\dfrac{-3}{x + 2} \cdot \dfrac{5x + 10}{9}$ $-\dfrac{5}{3}$

68. $\dfrac{3}{10} = \dfrac{5}{x}$ $\dfrac{50}{3}$

69. $\dfrac{1}{-3} = \dfrac{-2}{x}$ 6

70. $\dfrac{x^2 - 4}{x} \div \dfrac{4x - 8}{x}$ $\dfrac{x + 2}{4}$

71. $\dfrac{ax + am + 3x + 3m}{a^2 - 9} \div \dfrac{2x + 2m}{a - 3}$ $\dfrac{1}{2}$

72. $\dfrac{-2}{x} = \dfrac{3}{x + 2}$ $-\dfrac{4}{5}$

73. $\dfrac{2}{x^2 - 25} + \dfrac{1}{x^2 - 4x - 5}$ $\dfrac{3x + 7}{(x - 5)(x + 5)(x + 1)}$

74. $\dfrac{4}{a^2 - 1} + \dfrac{1}{2a + 2}$ $\dfrac{a + 7}{2(a + 1)(a - 1)}$

75. $\dfrac{-3}{a^2 - 9} - \dfrac{2}{a^2 + 5a + 6}$ $\dfrac{-5a}{(a - 3)(a + 3)(a + 2)}$

76. $\dfrac{-5}{a^2 - 4} - \dfrac{2}{a^2 - 3a + 2}$ $\dfrac{-7a + 1}{(a + 2)(a - 2)(a - 1)}$

77. $\dfrac{1}{a^2 - 1} + \dfrac{2}{1 - a} = \dfrac{3}{a + 1}$ $\dfrac{2}{5}$

78. $3 + \dfrac{1}{x - 2} = \dfrac{2x - 3}{x - 2}$ No solution

In place of the question mark put an expression that makes each equation an identity. Do these mentally.

79. $\dfrac{5}{x} = \dfrac{?}{2x}$ 10

80. $\dfrac{?}{a} = \dfrac{6}{3a}$ 2

81. $\dfrac{2}{a-5} = \dfrac{?}{5-a}$ -2

82. $\dfrac{-1}{a-7} = \dfrac{1}{?}$ $7-a$

83. $3 = \dfrac{?}{x}$ $3x$

84. $2a = \dfrac{?}{b}$ $2ab$

85. $m \div \dfrac{1}{2} = ?$ $2m$

86. $5x \div \dfrac{1}{x} = ?$ $5x^2$

87. $2a \div ? = 12a$ $\dfrac{1}{6}$

88. $10x \div ? = 20x^2$ $\dfrac{1}{2x}$

89. $\dfrac{a-1}{a^2-1} = \dfrac{1}{?}$ $a+1$

90. $\dfrac{?}{x^2-9} = \dfrac{1}{x-3}$ $x+3$

91. $\dfrac{1}{2} - \dfrac{1}{7} = ?$ $\dfrac{5}{14}$

92. $\dfrac{1}{2} - \dfrac{1}{5} = ?$ $\dfrac{3}{10}$

93. $\dfrac{a}{3} + \dfrac{a}{3} = ?$ $\dfrac{2a}{3}$

94. $\dfrac{x}{2} + \dfrac{x}{3} = ?$ $\dfrac{5x}{6}$

95. $\dfrac{1}{a} - \dfrac{1}{5} = ?$ $\dfrac{5-a}{5a}$

96. $\dfrac{3}{7} - \dfrac{2}{b} = ?$ $\dfrac{3b-14}{7b}$

97. $\dfrac{a}{2} - 1 = ?$ $\dfrac{a-2}{2}$

98. $\dfrac{1}{a} - 1 = ?$ $\dfrac{1-a}{a}$

99. $1 + \dfrac{1}{a} = ?$ $\dfrac{a+1}{a}$

100. $a - \dfrac{1}{x} = ?$ $\dfrac{ax-1}{x}$

101. $\dfrac{a}{2} - ? = \dfrac{a-2}{2}$ 1

102. $? - \dfrac{1}{y} = \dfrac{y-1}{y}$ 1

103. $(a-b) \div (-1) = ?$ $b-a$ **104.** $(a-7) \div (7-a) = ?$ -1 **105.** $(y-3) \div (3-y) = ?$ -1 **106.** $(2x-2) \div (x-1) = ?$

106. 2

107. $\dfrac{\frac{1}{5a}}{2} = ?$ $\dfrac{1}{10a}$

108. $\dfrac{3a}{\frac{1}{2}} = ?$ $6a$

109. $\dfrac{a-6}{5} \cdot \dfrac{1}{6-a} = ?$ $-\dfrac{1}{5}$ **110.** $\dfrac{1}{w-2} + \dfrac{1}{2-w} = ?$ 0

CHAPTER 5 TEST

What numbers cannot be used for x in each rational expression?

1. $\dfrac{2x-1}{x^2-1}$ $-1, 1$

2. $\dfrac{5}{2-3x}$ $\dfrac{2}{3}$

3. $\dfrac{1}{x}$ 0

Perform the indicated operations. Write answers in lowest terms.

4. $\dfrac{2}{15} - \dfrac{4}{9}$ $-\dfrac{14}{45}$

5. $\dfrac{1}{y} + 3$ $\dfrac{1+3y}{y}$

6. $\dfrac{3}{a-2} - \dfrac{1}{2-a}$ $\dfrac{4}{a-2}$

7. $\dfrac{2}{x^2-4} - \dfrac{3}{x^2+x-2}$ $\dfrac{-x+4}{(x+2)(x-2)(x-1)}$

8. $\dfrac{m^2-1}{(m-1)^2} \cdot \dfrac{2m-2}{3m+3}$ $\dfrac{2}{3}$

9. $\dfrac{a-b}{3} \div \dfrac{b^2-a^2}{6}$ $\dfrac{-2}{a+b}$

10. $\dfrac{5a^5b}{12a} \cdot \dfrac{2a^3b}{15ab^6}$ $\dfrac{a^6}{18b^4}$

Simplify each complex fraction.

11. $\dfrac{\frac{2}{3} + \frac{4}{5}}{\frac{2}{5} - \frac{3}{2}}$ $-\dfrac{4}{3}$

12. $\dfrac{\frac{2}{x} + \frac{1}{x-2}}{\frac{1}{x-2} - \frac{3}{x}}$ $\dfrac{3x-4}{-2x+6}$

Solve each equation.

13. $\dfrac{3}{x} = \dfrac{7}{5}$ $\dfrac{15}{7}$

14. $\dfrac{x}{x-1} - \dfrac{3}{x} = \dfrac{1}{2}$ $2, 3$

15. $\dfrac{1}{x} + \dfrac{1}{6} = \dfrac{1}{4}$ 12

Solve each formula for the indicated variable.

16. $\dfrac{y-3}{x+2} = \dfrac{-1}{5}$ for y $y = -\dfrac{1}{5}x + \dfrac{13}{5}$

17. $M = \dfrac{1}{3}b(c+d)$ for c $c = \dfrac{3M-bd}{b}$

Solve each problem.

19. Brenda: 15 mph, Randy: 20 mph or Brenda: 10 mph, Randy: 15 mph

18. When all of the grocery carts escape from the supermarket, it takes Reginald 12 minutes to round them up and bring them back. Since Norman doesn't make as much per hour as Reginald, it takes Norman 18 minutes to do the same job. How long would it take them working together to complete the roundup? 7.2 minutes

19. Brenda and her husband, Randy, bicycled cross-country together. One morning Brenda rode 30 miles. By traveling only 5 miles per hour faster and putting in one more hour, Randy covered twice the distance Brenda covered. What was the speed of each cyclist?

20. For a certain time period the ratio of the dollar value of exports to the dollar value of imports for this country was 2 to 3. If the value of exports during that time period was 48 billion dollars, then what was the value of imports? 72 billion dollars

Tying It All Together

CHAPTERS 1–5

Solve each equation.

1. $\dfrac{3}{x} = \dfrac{-2}{5}$ $-\dfrac{15}{2}$

2. $\dfrac{3}{x} = \dfrac{x}{12}$ $-6, 6$

3. $\dfrac{x}{3} = \dfrac{-2}{5}$ $-\dfrac{6}{5}$

4. $\dfrac{x}{2} = \dfrac{4}{x-2}$ $-2, 4$

5. $\dfrac{3}{5}x = -2$ $-\dfrac{10}{3}$

6. $3x - 2 = 5$ $\dfrac{7}{3}$

7. $2(x-2) = 4x$ -2

8. $2(x-2) = 2x$ No solution

9. $2(x+3) = 6x + 6$ 0

10. $2(3x+4) + x^2 = 0$ $-4, -2$

11. $4x - 4x^3 = 0$ $-1, 0, 1$

12. $x(x+4) = -3$ $-3, -1$

Solve each equation for y.

13. $2x + 3y = c$ $y = \dfrac{c-2x}{3}$

14. $\dfrac{y-3}{x-5} = \dfrac{1}{2}$ $y = \dfrac{1}{2}x + \dfrac{1}{2}$

15. $2y = ay + c$ $y = \dfrac{c}{2-a}$

16. $\dfrac{A}{y} = \dfrac{C}{B}$ $y = \dfrac{AB}{C}$

17. $\dfrac{A}{y} + \dfrac{1}{3} = \dfrac{B}{y}$ $y = 3B - 3A$

18. $\dfrac{A}{y} - \dfrac{1}{2} = \dfrac{1}{3}$ $y = \dfrac{6A}{5}$

19. $3y - 5ay = 8$ $y = \dfrac{8}{3-5a}$

20. $y^2 - By = 0$ $y = 0$ or $y = B$

21. $A = \dfrac{1}{2}h(b+y)$ $y = \dfrac{2A-hb}{h}$

22. $2(b+y) = b$ $y = -\dfrac{b}{2}$

Calculate the value of $b^2 - 4ac$ for each of the following choices of a, b, and c.

23. $a = 1, b = 2, c = -15$ 64

24. $a = 1, b = 8, c = 12$ 16

25. $a = 2, b = 5, c = -3$ 49

26. $a = 6, b = 7, c = -3$ 121

Perform each indicated operation mentally.

27. $(3x - 5) - (5x - 3)$ $-2x - 2$

28. $(2a - 5)(a - 3)$ $2a^2 - 11a + 15$

29. $x^7 \div x^3$ x^4

30. $\dfrac{x - 3}{5} + \dfrac{x + 4}{5}$ $\dfrac{2x + 1}{5}$

31. $\dfrac{1}{2} \cdot \dfrac{1}{x}$ $\dfrac{1}{2x}$

32. $\dfrac{1}{2} + \dfrac{1}{x}$ $\dfrac{x + 2}{2x}$

33. $\dfrac{1}{2} \div \dfrac{1}{x}$ $\dfrac{x}{2}$

34. $\dfrac{1}{2} - \dfrac{1}{x}$ $\dfrac{x - 2}{2x}$

35. $\dfrac{x - 3}{5} - \dfrac{x + 4}{5}$ $-\dfrac{7}{5}$

36. $\dfrac{3a}{2} \div 2$ $\dfrac{3a}{4}$

37. $(x - 8)(x + 8)$ $x^2 - 64$

38. $3x(x^2 - 7)$ $3x^3 - 21x$

39. $2a^5 \cdot 5a^9$ $10a^{14}$

40. $x^2 \cdot x^8$ x^{10}

41. $(k - 6)^2$ $k^2 - 12k + 36$

42. $(j + 5)^2$ $j^2 + 10j + 25$

43. $(g - 3) \div (3 - g)$ -1

44. $(6x^3 - 8x^2) \div 2x$ $3x^2 - 4x$

45. $\dfrac{1}{3} + \dfrac{2}{5}$ $\dfrac{11}{15}$

46. $\dfrac{2}{5} - \dfrac{1}{6}$ $\dfrac{7}{30}$

Powers and Roots

The manufacturer of a radio transmission tower recommends that guy wires be attached from the top of the tower to points on the ground. For maximum efficiency the points on the ground should be located at a distance from the base of the tower equal to 2/3 of the height of the tower. The wire used for this purpose is very expensive, and the installer must know exactly how much is needed. If the tower is 200 feet high, then how long should each guy wire be?

This problem can be solved by using the Pythagorean Theorem and square roots. The exact answer to this question is an irrational number. However, the installer cannot order an irrational length of wire. Measurements are generally given as rational numbers. When we solve this problem as Exercise 51 of Section 6.5, we will find the exact length and also a rational approximation to the exact length.

6.1 Positive Integral Exponents

IN THIS SECTION:

- **Review of Exponents**
- **Raising an Exponential Expression to a Power**
- **Power of a Product**
- **Power of a Quotient**
- **Summary of Rules**

We defined exponential expressions in Chapter 1. The product and quotient rules for positive integral exponents were given in Chapter 3, and the quotient rule was extended in Chapter 4. In this section we will continue to study the properties of exponential expressions.

Review of Exponents

In Chapter 1, exponents were defined to provide a convenient way of writing repeated factors.

DEFINITION
Positive Integral Exponents

For any positive integer n,

$$a^n = \underbrace{a \cdot a \cdot a \cdots a}_{n \text{ factors of } a}.$$

We call a the **base** and n the **exponent.**

The quantity a^n is called an **exponential expression.** With this notation we can write

$$3 \cdot 3 = 3^2, \qquad x \cdot x \cdot x \cdot x \cdot x = x^5, \qquad \text{or} \qquad 1000 = 10^3.$$

We also refer to the exponent as a **power.** For example, we say that 3^2 is 3 to the second power, x^5 is x to the fifth power, and 2^4 is the fourth power of 2.

The product rule from Chapter 3 allows us to write the product of exponential expressions with the same base as a single exponential expression. The exponent of the new expression is the sum of the exponents of the factors.

Product Rule

> If m and n are positive integers, then
> $$a^m \cdot a^n = a^{m+n}.$$

For example, we can write

$$w^5 \cdot w^3 = w^8.$$

The quotient rule from Chapter 4 allows us to combine the exponential expressions in a quotient into a single exponential expression. The exponent in the new expression is the difference of the exponents in the quotient.

Quotient Rule

> Suppose $a \neq 0$, and m and n are positive integers.
>   If $m \geq n$, then
> $$\frac{a^m}{a^n} = a^{m-n}.$$
>   If $n > m$, then
> $$\frac{a^m}{a^n} = \frac{1}{a^{n-m}}.$$

For example, we can write

$$\frac{x^9}{x^5} = x^4 \qquad \text{or} \qquad \frac{y^4}{y^7} = \frac{1}{y^3}.$$

We defined 0 as an exponent so that the quotient rule could be used even when the two exponents were equal.

DEFINITION
Zero Exponent

> If a is any nonzero real number, then
> $$a^0 = 1.$$

For example,

$$\frac{x^3}{x^3} = x^{3-3} = x^0 = 1.$$

The next example shows how we use these rules to simplify expressions involving exponents.

EXAMPLE 1 Use the rules of exponents to simplify each expression. Assume that all variables represent nonzero real numbers.

a) $2^3 \cdot 3^2$

b) $\dfrac{(3a^2b)b^9}{(6a^5)b^3}$

c) $(3x)^0(5x^2)(4x)$

Solution

a) $2^3 \cdot 3^2 = 8 \cdot 9$ $2^3 = 2 \cdot 2 \cdot 2 = 8, \; 3^2 = 3 \cdot 3 = 9$

$ = 72$

Note that since the bases are different (2 and 3), we cannot use the product rule to add the exponents.

b) $\dfrac{(3a^2b)b^9}{(6a^5)b^3} = \dfrac{3a^2b^{10}}{6a^5b^3}$ **Apply the product rule.**

$\phantom{\dfrac{(3a^2b)b^9}{(6a^5)b^3}} = \dfrac{b^7}{2a^3}$ **Apply the quotient rule.**

c) $(3x)^0(5x^2)(4x) = 1 \cdot 5x^2 \cdot 4x$ **Definition of zero exponent**

$ = 20x^3$ **Product rule**   ◀

Raising an Exponential Expression to a Power

Consider the following example of simplifying the square of x^5:

$(x^5)^2 = x^5 \cdot x^5$ **Two factors of x^5 because of the exponent 2**

$ = x^{10}$ **Use the product rule to add the exponents: $5 + 5 = 10$**

Note that 10 could also be obtained by multiplying 2 and 5.
        Consider the cube of w^4:

$(w^4)^3 = w^4 \cdot w^4 \cdot w^4$ **Three factors of w^4 because of the exponent 3**

$ = w^{12}$ **Use the product rule to add exponents: $4 + 4 + 4 = 12$**

Note that we could have obtained the exponent 12 by multiplying 4 and 3.
        In each case above, we used known rules to simplify the expression. However, these examples illustrate a new rule.

Power Rule

If m and n are nonnegative integers and $a \neq 0$, then

$$(a^m)^n = a^{mn}.$$

We can use this new rule in conjunction with the other rules.

EXAMPLE 2 Use the rules of exponents to simplify the expressions. Assume that all variables represent nonzero real numbers.

a) $3x^2(x^3)^5$ **b)** $\dfrac{(2^3)^4 \cdot 2^7}{(2^5)(2^9)}$ **c)** $\dfrac{3(x^5)^4}{15x^{22}}$

Solution

a) $3x^2(x^3)^5 = 3x^2 \cdot x^{15}$ **Apply the power rule.**

               $= 3x^{17}$ **Apply the product rule.**

b) $\dfrac{(2^3)^4 \cdot 2^7}{(2^5)(2^9)} = \dfrac{2^{12} \cdot 2^7}{2^{14}}$ **Power rule and product rule**

            $= \dfrac{2^{19}}{2^{14}}$ **Product rule**

            $= 2^5$ **Quotient rule**

            $= 32$ **Evaluate 2^5.**

c) $\dfrac{3(x^5)^4}{15x^{22}} = \dfrac{3x^{20}}{15x^{22}} = \dfrac{1}{5x^2}$ ◄

Power of a Product

Consider an example of raising a product to a power. We will use known rules to rewrite the expression:

$$(2x)^3 = 2x \cdot 2x \cdot 2x$$ **Definition of exponent 3**

$$= 2 \cdot 2 \cdot 2 \cdot x \cdot x \cdot x$$ **Commutative and associative properties**

$$= 2^3 \cdot x^3$$ **Definition of exponents**

Note that the power is distributed over each factor of the product. This example illustrates the **power of a product** rule.

RULE
Power of a Product

If a and b are real numbers and n is a positive integer, then

$$(ab)^n = a^n b^n.$$

EXAMPLE 3 Use the rules of exponents to simplify each expression. Assume that the variables are nonzero.

a) $(xy^3)^5$ **b)** $(-3m)^3$

Solution

a) $(xy^3)^5 = x^5 \cdot (y^3)^5$ **Apply the power of a product rule.**

          $= x^5 y^{15}$ **Power rule**

b) $(-3m)^3 = (-3)^3 m^3$ **Apply the power of a product rule.**

           $= -27m^3$ **Evaluate: $(-3)(-3)(-3) = -27$**   ◄

Power of a Quotient

Raising a quotient to a power is similar to raising a product to a power:

$$\left(\frac{x}{5}\right)^3 = \frac{x}{5} \cdot \frac{x}{5} \cdot \frac{x}{5} \qquad \text{Definition of exponent 3}$$

$$= \frac{x \cdot x \cdot x}{5 \cdot 5 \cdot 5} \qquad \text{Multiply the rational expressions.}$$

$$= \frac{x^3}{5^3} \qquad \text{Definition of exponents}$$

The power is distributed over each term of the quotient. This example illustrates the **power of a quotient** rule.

RULE
Power of a Quotient

If a and b are real numbers, $b \neq 0$, and n is a positive integer, then

$$\left(\frac{a}{b}\right)^n = \frac{a^n}{b^n}.$$

EXAMPLE 4 Use the rules of exponents to simplify each expression. Assume that the variables are nonzero.

a) $\left(\dfrac{2}{5x^3}\right)^2$

b) $\left(\dfrac{3x^4}{2y^3}\right)^3$

Solution

a) $\left(\dfrac{2}{5x^3}\right)^2 = \dfrac{2^2}{(5x^3)^2}$ Apply the power of a quotient rule.

$$= \frac{4}{25x^6} \qquad (5x^3)^2 = 5^2(x^3)^2 = 25x^6$$

b) $\left(\dfrac{3x^4}{2y^3}\right)^3 = \dfrac{3^3 x^{12}}{2^3 y^9}$ Apply the power of a quotient and power of a product rules.

$$= \frac{27x^{12}}{8y^9} \qquad \text{Simplify.} \qquad \blacktriangleleft$$

Summary of Rules

The rules we have discussed in this section are summarized in the following box.

RULES
Nonnegative Integral Exponents

If a and b are nonzero real numbers and m and n are nonnegative integers, then

1. $a^0 = 1$ Zero exponent

2. $a^m a^n = a^{m+n}$ Product rule

3. $\dfrac{a^m}{a^n} = a^{m-n}$ if $m \geq n$ Quotient rule

     $\dfrac{a^m}{a^n} = \dfrac{1}{a^{n-m}}$ if $n > m$

4. $(a^m)^n = a^{mn}$ Power rule

5. $(ab)^n = a^n b^n$ Power of a product

6. $\left(\dfrac{a}{b}\right)^n = \dfrac{a^n}{b^n}$ Power of a quotient

Warm-ups

True or false?

Assume that all variables represent nonzero real numbers. A statement involving variables is to be marked true only if it is an identity.

1. $2^3 = 6$ F   **2.** $2^5 \cdot 2^8 = 4^{13}$ F   **3.** $(2x)^4 = 2x^4$ F

4. $(-3x^2)^3 = 27x^6$ F   **5.** $(ab^3)^4 = a^4 b^{12}$ T   **6.** $2^3 \cdot 3^2 = 6^5$ F

7. $(-3)^0 = 1$ T   **8.** $-5^0 = 1$ F   **9.** $x^3 \cdot x^3 = x^9$ F

10. $\dfrac{a^{12}}{a^4} = a^3$ F

6.1 EXERCISES

Simplify the exponential expressions. For all exercises in this section, assume that the variables represent nonzero real numbers. See Example 1.

1. 2^4 16   **2.** 3^4 81   **3.** 5^3 125   **4.** 5^2 25

5. 0^2 0   **6.** 0^3 0   **7.** $(-1)^4$ 1   **8.** $(-1)^5$ -1

9. 1^8 1   **10.** 1^9 1   **11.** 5^4 625   **12.** $(-2)^6$ 64

13. 2^6 64   **14.** $(-4)^3$ -64 **15.** 4^3 64   **16.** 11^2 121

17. $3 \cdot 10^2$ 300   **18.** $2^3 \cdot 2^5$ 256   **19.** $2^5 \cdot 2^{10}$ 2^{15} **20.** $2x^2 \cdot 3x^3$ $6x^5$

21. $(7 - 5)^0$ 1   **22.** $(8 - 19)^0$ 1   **23.** -4^2 -16 **24.** -3^4 -81

25. $(-2)^2$ 4   **26.** $(-5)^2$ 25   **27.** -6^0 -1 **28.** -9^0 -1

29. $\dfrac{2a^3}{4a^7}$ $\dfrac{1}{2a^4}$ **30.** $\dfrac{3t^{18}}{6t^9}$ $\dfrac{t^9}{2}$ **31.** $\dfrac{2a^5 \cdot 3a^7}{15a^6}$ $\dfrac{2a^6}{5}$ **32.** $\dfrac{3y^3 \cdot 5y^4}{20y^{14}}$ $\dfrac{3}{4y^7}$

Simplify. See Example 2.

33. $(x^2)^3$ x^6 **34.** $(y^2)^4$ y^8 **35.** $2x^2 \cdot (x^2)^5$ $2x^{12}$ **36.** $(y^2)^6 \cdot 3y^5$ $3y^{17}$

37. $\dfrac{(t^2)^5}{(t^3)^4}$ $\dfrac{1}{t^2}$

38. $\dfrac{(r^4)^2}{(r^5)^3}$ $\dfrac{1}{r^7}$

39. $\dfrac{3x(x^5)^2}{6x^3(x^2)^4}$ $\dfrac{1}{2}$

40. $\dfrac{5y^3(y^5)^2}{10y^5(y^2)^6}$ $\dfrac{1}{2y^4}$

41. $8x^2y^2 \cdot (x^3)^5(y^4)^3$ $8x^{17}y^{14}$ **42.** $5r^3t \cdot (r^5)^2(t^3)^2$ $5r^{13}t^7$

Simplify. See Example 3.

43. $(2x^2)^3$ $8x^6$

44. $(3y^2)^2$ $9y^4$

45. $(-2t^5)^3$ $-8t^{15}$

46. $(-3r^3)^3$ $-27r^9$

47. $(x^2y^5)^3$ x^6y^{15}

48. $(-y^2z^3)^3$ $-y^6z^9$

49. $\dfrac{(a^4b^2)^3}{a^3b^4}$ a^9b^2

50. $\dfrac{(2ab^2)^5}{(2a^3)^4}$ $\dfrac{2b^{10}}{a^7}$

Simplify. See Example 4.

51. $\left(\dfrac{x^4}{4}\right)^3$ $\dfrac{x^{12}}{64}$

52. $\left(\dfrac{y^2}{2}\right)^3$ $\dfrac{y^6}{8}$

53. $\left(\dfrac{a^2}{b^3}\right)^4$ $\dfrac{a^8}{b^{12}}$

54. $\left(\dfrac{r^3}{t^5}\right)^2$ $\dfrac{r^6}{t^{10}}$

55. $\left(\dfrac{2x^2}{y^2}\right)^3$ $\dfrac{8x^6}{y^6}$

56. $\left(\dfrac{3y^3}{2z}\right)^4$ $\dfrac{81y^{12}}{16z^4}$

57. $\left(\dfrac{x^2y^4}{z^3}\right)^2$ $\dfrac{x^4y^8}{z^6}$

58. $\left(\dfrac{rt^4}{s^2}\right)^3$ $\dfrac{r^3t^{12}}{s^6}$

Simplify the following expressions.

59. $2^3 + 3^3$ 35

60. $5^2 - 3^2$ 16

61. $(2 + 3)^3$ 125

62. $(5 - 3)^2$ 4

63. $\left(\dfrac{2}{3}\right)^3$ $\dfrac{8}{27}$

64. $\left(\dfrac{3}{4}\right)^2$ $\dfrac{9}{16}$

65. $5^2 2^3$ 200

66. $10^3 \cdot 10^2$ 100,000

67. $10^2 \cdot 2^3$ 800

68. $3^2 \cdot 10^3$ 9000

69. $(3 - 2)^5$ 1

70. $(8 - 7)^4$ 1

71. $(8 - 2 \cdot 4)^5$ 0

72. $(2 \cdot 3 - 6)^7$ 0

73. $(2 \cdot 10^3)(3 \cdot 10^2)$ 600,000

74. $(-4 \cdot 10^4)(2 \cdot 10^3)$

75. $\dfrac{8^2}{4^3}$ 1

76. $\dfrac{4^2}{2^4}$ 1

77. $\left(\dfrac{2x^2}{x^4}\right)^3$ $\dfrac{8}{x^6}$

78. $\left(\dfrac{3y^8}{y^5}\right)^2$ $9y^6$

79. $(xt^2)^3(x^2t)^4$ $x^{11}t^{10}$

80. $(ab)^3(ba^2)^4$ $a^{11}b^7$

74. $-80,000,000$

6.2 Roots

IN THIS SECTION:

- **Fundamentals**
- **Roots and Variables**
- **Product Rule for Radicals**
- **Quotient Rule for Radicals**

In Section 6.1 we learned the basic facts about powers. In this section we will study roots and see how powers and roots are related.

Fundamentals

We know that

$$3^2 = 9, \qquad (-4)^2 = 16, \qquad \text{and} \qquad 2^3 = 8.$$

We use the idea of roots to reverse powers. We say that 3 is a square root of 9, -4 is a square root of 16, and 2 is the cube root of 8.

DEFINITION
nth Roots

> For any positive integer n, we say that a is an nth root of b if
>
> $$a^n = b.$$

Since $3^2 = 9$ and $(-3)^2 = 9$, we say that 9 has two square roots, 3 and -3. Note that 16 has two fourth roots, 2 and -2. Since

$$2^3 = 8 \qquad \text{and} \qquad (-2)^3 = -8,$$

the cube root of 8 is 2, and the cube root of -8 is -2. *There are two even roots of any positive number. There is only one real odd root of a number whether the number is positive or negative.*

We use the **radical symbol,** $\sqrt{}$, to signify roots. The symbol $\sqrt[n]{}$ is used to signify the nth root. When $n = 2$, the 2 is usually omitted. Note that 4 has two square roots, 2 and -2. We use the radical notation $\sqrt{4}$ to represent only the positive square root. The two fourth roots of 16 are 2 and -2, but we use the radical notation $\sqrt[4]{16}$ to represent only the positive fourth root. For example, we can write

$$\sqrt{4} = 2 \quad \text{since} \quad 2^2 = 4, \qquad \text{``}\sqrt{\phantom{x}}\text{'' is read as ``square root.''}$$
$$\sqrt[3]{27} = 3 \quad \text{since} \quad 3^3 = 27, \qquad \text{``}\sqrt[3]{\phantom{x}}\text{'' is read as ``cube root.''}$$

and

$$\sqrt[4]{16} = 2 \quad \text{since} \quad 2^4 = 16. \qquad \text{``}\sqrt[4]{\phantom{x}}\text{'' is read as ``fourth root.''}$$

The expression $\sqrt{3}$ represents the unique positive real number whose square is 3. The number $\sqrt{3}$ is an irrational number. If we use the square root table of Appendix A or a calculator, we find that $\sqrt{3}$ is approximately equal to 1.732. The number 1.732 is a rational number that approximates $\sqrt{3}$. Since $\sqrt{3}$ is not a rational number, the simplest representation for the exact value of the square root of 3 is $\sqrt{3}$.

DEFINITION
$\sqrt[n]{a}$

> If n is a positive *even* integer and a is positive, then the symbol $\sqrt[n]{a}$ denotes the **positive nth root of a,** and is called the **principal nth root of a.**
>
> If n is a positive *odd* integer, then the symbol $\sqrt[n]{a}$ denotes the nth root of a.
>
> If n is any positive integer, then $\sqrt[n]{0} = 0$.

In the notation $\sqrt[n]{a}$, n is called the **index of the radical,** and a is called the **radicand.** Note that even roots of negative numbers were omitted from the definition of nth roots. This is because *even powers of real numbers are never negative.* So no real number can be an even root of a negative number. Expressions such as

$$\sqrt{-9}, \quad \sqrt[4]{-81}, \quad \text{and} \quad \sqrt[6]{-64}$$

are not real numbers.

EXAMPLE 1   Find the following roots:

a) $\sqrt{25}$ **b)** $\sqrt[3]{-27}$ **c)** $\sqrt[6]{64}$

d) $-\sqrt{4}$ **e)** $\sqrt{-16}$

Solution

a) Since $5^2 = 25$, $\sqrt{25} = 5$.
b) Since $(-3)^3 = -27$, $\sqrt[3]{-27} = -3$.
c) Since $2^6 = 64$, $\sqrt[6]{64} = 2$.
d) Since $\sqrt{4} = 2$, $-\sqrt{4} = -2$.
e) This is an even root of a negative number and is not a real number. ◀

Roots and Variables

Note that the radical symbol is a grouping symbol like parentheses. All operations within the radical are done before the root is found. So $\sqrt{x^2}$ is a real number even when x is negative.

Consider the expression $\sqrt{x}$. Since this expression is not a real number if x is negative, we will assume that *the values of any variables in the radicand are nonnegative.*

If x is a nonnegative number, then $\sqrt{x^2}$ represents a nonnegative number whose square is x^2. So

$$\sqrt{x^2} = x.$$

Now consider $\sqrt{x^6}$. Since $(x^3)^2 = x^6$, we must have

$$\sqrt{x^6} = x^3$$

for any nonnegative number x. In fact, any even power of a variable is a perfect square.

Perfect Squares

The following expressions are perfect squares:

$$x^2, x^4, x^6, x^8, x^{10}, x^{12}, \ldots$$

To find the square roots of any of these perfect squares, we just use the same variable with one-half of the exponent. For example,

$$\sqrt{x^{48}} = x^{24}.$$

We have a similar situation for cube roots. Any power of a variable in which the exponent is divisible by 3 is a perfect cube.

Perfect Cubes

The following expressions are perfect cubes:

$$x^3, x^6, x^9, x^{12}, x^{15}, \ldots$$

To find the cube root of any of these expressions, we use the same variable with one-third of the exponent. Thus

$$\sqrt[3]{x^{15}} = x^5,$$

since

$$(x^5)^3 = x^{15}.$$

If the exponent is divisible by 4, we have a perfect fourth power, and so on.

EXAMPLE 2 Find the following roots. Assume that all variables represent non-negative real numbers.

a) $\sqrt{x^{22}}$ **b)** $\sqrt[3]{t^{18}}$ **c)** $\sqrt[5]{s^{30}}$

Solution

a) $\sqrt{x^{22}} = x^{11}$ since $(x^{11})^2 = x^{22}$.
b) $\sqrt[3]{t^{18}} = t^6$ since $(t^6)^3 = t^{18}$.
c) $\sqrt[5]{s^{30}} = s^6$ since $1/5$ of 30 is 6.   ◄

Product Rule for Radicals

Consider the product of two square roots, $\sqrt{2} \cdot \sqrt{3}$. If we square this product, we get

$$(\sqrt{2} \cdot \sqrt{3})^2 = (\sqrt{2})^2(\sqrt{3})^2 \qquad \text{Power of a product}$$
$$= 2 \cdot 3 \qquad (\sqrt{2})^2 = 2 \text{ and } (\sqrt{3})^2 = 3$$
$$= 6$$

The number $\sqrt{6}$ is the unique positive number whose square is 6. Since we squared $\sqrt{2} \cdot \sqrt{3}$ and obtained 6, we must have

$$\sqrt{2} \cdot \sqrt{3} = \sqrt{6}.$$

This example illustrates the product rule for radicals.

Product Rule for Radicals

The nth root of a product is equal to the product of the nth roots. In symbols,

$$\sqrt[n]{ab} = \sqrt[n]{a} \cdot \sqrt[n]{b},$$

provided that all of these roots are real numbers.

EXAMPLE 3 Simplify the following. Assume that all variables represent positive real numbers.

a) $\sqrt{4y}$ b) $\sqrt{3y^8}$

Solution

a) $\sqrt{4y} = \sqrt{4} \cdot \sqrt{y}$ **Apply the product rule.**
$\qquad\quad = 2\sqrt{y}$ **Simplify.**

b) $\sqrt{3y^8} = \sqrt{3} \cdot \sqrt{y^8}$ **Apply the product rule.**
$\qquad\quad = \sqrt{3} \cdot y^4$ $\sqrt{y^8} = y^4$
$\qquad\quad = y^4\sqrt{3}$ **A radical is usually written last in a product.**   ◄

Quotient Rule for Radicals

Since $\sqrt{2} \cdot \sqrt{3} = \sqrt{6}$, we can write

$$\frac{\sqrt{6}}{\sqrt{3}} = \sqrt{2} \qquad \text{or} \qquad \frac{\sqrt{6}}{\sqrt{3}} = \sqrt{\frac{6}{3}}.$$

This example illustrates the quotient rule for radicals.

Quotient Rule for Radicals

The nth root of a quotient is equal to the quotient of the nth roots. In symbols,

$$\sqrt[n]{\frac{a}{b}} = \frac{\sqrt[n]{a}}{\sqrt[n]{b}},$$

provided that all of these roots are real numbers and $b \neq 0$.

The next example illustrates the use of the quotient rule.

EXAMPLE 4 Simplify the following. Assume that all variables represent positive real numbers.

a) $\sqrt{\dfrac{t}{9}}$ b) $\sqrt[3]{\dfrac{x^{21}}{y^6}}$

Solution

a) $\sqrt{\dfrac{t}{9}} = \dfrac{\sqrt{t}}{\sqrt{9}}$ Apply the quotient rule.

$= \dfrac{\sqrt{t}}{3}$

b) $\sqrt[3]{\dfrac{x^{21}}{y^6}} = \dfrac{\sqrt[3]{x^{21}}}{\sqrt[3]{y^6}}$ Apply the quotient rule.

$= \dfrac{x^7}{y^2}$

Warm-ups

True or false?

1. $\sqrt{2} \cdot \sqrt{2} = 2$ T   2. $\sqrt[3]{2} \cdot \sqrt[3]{2} = 2$ F   3. $\sqrt[3]{-27} = -3$ T
4. $\sqrt{-25} = -5$ F   5. $\sqrt[5]{16} = 2$ T   6. $\sqrt{9} = 3$ T
7. $\sqrt{2^9} = 2^3$ F   8. $\sqrt{17} \cdot \sqrt{17} = 289$ F
9. If w is a positive number, then $\sqrt{w^2} = w$. T
10. If t is a positive number, then $\sqrt[4]{t^{12}} = t^3$. T

6.2 EXERCISES

Find the following roots or powers. See Example 1.

1. 6^2 36
2. 7^2 49
3. $\sqrt{36}$ 6
4. $\sqrt{49}$ 7
5. 2^5 32
6. 3^4 81
7. $\sqrt[5]{32}$ 2
8. $\sqrt[4]{81}$ 3
9. 10^3 1000
10. $(-2)^4$ 16
11. $\sqrt[3]{1000}$ 10
12. $\sqrt[4]{16}$ 2
13. $\sqrt[4]{-16}$ Not a real number
14. $\sqrt{1}$ 1
15. $\sqrt{0}$ 0
16. $\sqrt{-1}$ Not a real number
17. $\sqrt[3]{-1}$ -1
18. $\sqrt[3]{0}$ 0
19. $\sqrt[3]{1}$ 1
20. $\sqrt[4]{81}$ 3
21. $\sqrt{-81}$ Not a real number
22. $\sqrt[6]{-64}$ Not a real number
23. $\sqrt[6]{64}$ 2
24. $\sqrt[7]{128}$ 2
25. $\sqrt[3]{125}$ 5
26. $\sqrt[3]{-125}$ -5
27. $-\sqrt{100}$ -10
28. $\sqrt[4]{-50}$ Not a real number
29. $-\sqrt{36}$ -6
30. $-\sqrt{144}$ -12

Find the following roots or powers. Assume that all variables represent nonnegative real numbers. See Example 2.

31. $\sqrt{m^2}$ m
32. $\sqrt{m^6}$ m^3
33. $(y^3)^5$ y^{15}
34. $(m^2)^4$ m^8
35. $\sqrt[5]{y^{15}}$ y^3
36. $\sqrt[4]{m^8}$ m^2
37. $\sqrt[3]{y^{15}}$ y^5
38. $\sqrt{m^8}$ m^4
39. $\sqrt[3]{m^3}$ m
40. $\sqrt[4]{x^4}$ x
41. $\sqrt{3^6}$ 27
42. $\sqrt{4^2}$ 4
43. $\sqrt{2^{10}}$ 32
44. $\sqrt[3]{2^{99}}$ 2^{33}
45. $\sqrt[3]{5^9}$ 125
46. $\sqrt{10^{20}}$ 10^{10}
47. $\sqrt{10^{18}}$ 10^9
48. $\sqrt[3]{10^{18}}$ 10^6

Use the product rule for radicals to simplify each expression. See Example 3. Assume that all variables represent nonnegative real numbers.

49. $\sqrt{9y}$ $3\sqrt{y}$
50. $\sqrt{16mn}$ $4\sqrt{mn}$
51. $\sqrt{4a^2}$ $2a$
52. $\sqrt{36n^2}$ $6n$
53. $\sqrt{x^4y^2}$ x^2y

54. $\sqrt{w^6 t^2}$ $w^3 t$ **55.** $\sqrt{9m^2}$ $3m$ **56.** $\sqrt{25z^{16}}$ $5z^8$ **57.** $\sqrt{81w^{10}}$ $9w^5$ **58.** $\sqrt[3]{8y}$ $2 \cdot \sqrt[3]{y}$

59. $\sqrt[3]{27z^2}$ $3 \cdot \sqrt[3]{z^2}$ **60.** $\sqrt[3]{8y^2}$ $2 \cdot \sqrt[3]{y^2}$ **61.** $\sqrt[3]{8z^3}$ $2z$ **62.** $\sqrt[3]{-27w^3}$ $-3w$ **63.** $\sqrt[3]{-125m^6}$ $-5m^2$

64. $\sqrt[4]{16s}$ $2 \cdot \sqrt[4]{s}$ **65.** $\sqrt[4]{81w}$ $3 \cdot \sqrt[4]{w}$ **66.** $\sqrt[4]{10,000z^4}$ $10z$

Use the quotient rule for radicals to simplify each expression. See Example 4. Assume that all variables represent positive real numbers.

67. $\sqrt{\dfrac{t}{4}}$ $\dfrac{\sqrt{t}}{2}$ **68.** $\sqrt{\dfrac{9}{4}}$ $\dfrac{3}{2}$ **69.** $\sqrt{\dfrac{625}{16}}$ $\dfrac{25}{4}$ **70.** $\sqrt{\dfrac{3}{144}}$ $\dfrac{\sqrt{3}}{12}$ **71.** $\sqrt{\dfrac{2}{25}}$ $\dfrac{\sqrt{2}}{5}$

72. $\sqrt[3]{\dfrac{t}{8}}$ $\dfrac{\sqrt[3]{t}}{2}$ **73.** $\sqrt[3]{\dfrac{a}{27}}$ $\dfrac{\sqrt[3]{a}}{3}$ **74.** $\sqrt[3]{\dfrac{x^6}{y^3}}$ $\dfrac{x^2}{y}$ **75.** $\sqrt{\dfrac{a^6}{9}}$ $\dfrac{a^3}{3}$ **76.** $\sqrt{\dfrac{a^2}{b^4}}$ $\dfrac{a}{b^2}$

Use a calculator to find the approximate value of each expression to three decimal places.

77. $\sqrt{3}$ 1.732 **78.** $\sqrt{7}$ 2.646 **79.** $\sqrt{5}$ 2.236 **80.** $\sqrt{2}$ 1.414 **81.** $\sqrt{3} \cdot \sqrt{5}$ 3.873

82. $\dfrac{\sqrt{7}}{\sqrt{2}}$ 1.871 **83.** $\sqrt{15}$ 3.873 **84.** $\sqrt{3.5}$ 1.871 **85.** $\sqrt{3} \cdot \sqrt{7}$ 4.583 **86.** $\sqrt{21}$ 4.583

6.3 Simplifying Square Roots

IN THIS SECTION:

- Using the Product Rule
- Rationalizing the Denominator
- Simplifying Square Roots Involving Variables

In Section 6.2 we learned to simplify some radical expressions using the product rule. In this section we will learn three basic rules to follow for writing expressions involving square roots in simplest form.

Using the Product Rule

Since 9 is a perfect square and a factor of 45, we can write

$$\sqrt{45} = \sqrt{9 \cdot 5} \qquad \text{Factor 45 as } 9 \cdot 5.$$
$$= \sqrt{9} \cdot \sqrt{5} \qquad \text{Apply the product rule.}$$
$$= 3\sqrt{5}. \qquad \sqrt{9} = 3$$

Note that $\sqrt{45}$ is an irrational number, and $3\sqrt{5}$ is considered a simpler expression that represents the exact value of $\sqrt{45}$. When simplifying square roots, we can factor the perfect squares out of the radical and replace them with their square roots. Look for the factors.

$$4, \quad 9, \quad 16, \quad 25, \quad 36, \quad 49, \quad \text{etc.}$$

EXAMPLE 1   Simplify the following square roots:

a) $\sqrt{12}$ **b)** $\sqrt{50}$ **c)** $\sqrt{72}$

Solution

a) Since $12 = 4 \cdot 3$, we can use the product rule to write

$$\sqrt{12} = \sqrt{4} \cdot \sqrt{3} = 2\sqrt{3}.$$

b) $\sqrt{50} = \sqrt{25} \cdot \sqrt{2} = 5\sqrt{2}$

c) Note that 4, 9, and 36 are perfect squares and are factors of 72. In factoring out a perfect square it is most efficient to use the largest perfect square. Therefore

$$\sqrt{72} = \sqrt{36} \cdot \sqrt{2} = 6\sqrt{2}.$$

If we factored out 9, we could still get the correct answer as follows:

$$\sqrt{72} = \sqrt{9} \cdot \sqrt{8} = 3 \cdot \sqrt{8} = 3 \cdot \sqrt{4} \cdot \sqrt{2} = 3 \cdot 2 \cdot \sqrt{2} = 6\sqrt{2} \quad \blacktriangleleft$$

Rationalizing the Denominator

The quotient rule is used to change the root of a quotient into the quotient of the roots. For example,

$$\sqrt{\frac{3}{4}} = \frac{\sqrt{3}}{\sqrt{4}} = \frac{\sqrt{3}}{2}.$$

In this case the denominator is a rational number. Numbers such as $\sqrt{2}$, $\sqrt{3}$, and $\sqrt{5}$ are irrational numbers. If an irrational number appears in the denominator of a fraction, we can write an equivalent expression with a rational denominator. This is called **rationalizing the denominator.** In the expression below, $\sqrt{3}$ appears in the denominator. Since $\sqrt{3} \cdot \sqrt{3} = 3$, we multiply the numerator and denominator by $\sqrt{3}$:

$$\frac{\sqrt{5}}{\sqrt{3}} = \frac{\sqrt{5}}{\sqrt{3}} \cdot \frac{\sqrt{3}}{\sqrt{3}}$$

$$= \frac{\sqrt{15}}{3} \quad \text{Apply the product rule to write } \sqrt{5} \cdot \sqrt{3} \text{ as } \sqrt{15}.$$

Note that the denominator is a rational number.

EXAMPLE 2   Rationalize the denominator in each expression:

a) $\dfrac{3}{\sqrt{5}}$ **b)** $\dfrac{\sqrt{3}}{\sqrt{7}}$

Solution

a) Since $\sqrt{5} \cdot \sqrt{5} = 5$, we multiply numerator and denominator by $\sqrt{5}$:

$$\frac{3}{\sqrt{5}} = \frac{3}{\sqrt{5}} \cdot \frac{\sqrt{5}}{\sqrt{5}} = \frac{3\sqrt{5}}{5}$$

b) Since $\sqrt{7}$ is in the denominator, we multiply numerator and denominator by $\sqrt{7}$:

$$\frac{\sqrt{3}}{\sqrt{7}} = \frac{\sqrt{3}}{\sqrt{7}} \cdot \frac{\sqrt{7}}{\sqrt{7}} = \frac{\sqrt{21}}{7} \qquad \sqrt{7} \cdot \sqrt{7} = 7 \qquad \blacktriangleleft$$

In simplifying square roots we always obey the following three rules.

RULES
Simplifying Square Roots

A simplified square root expression has

1. *no* perfect square factors inside the radical,
2. *no* fractions inside the radical, and
3. *no* radicals in the denominator.

EXAMPLE 3 Simplify the expressions:

a) $\sqrt{300}$ **b)** $\sqrt{\dfrac{2}{3}}$ **c)** $\dfrac{\sqrt{10}}{\sqrt{6}}$

Solution

a) We must remove the perfect square factor of 100 from inside the radical:

$$\sqrt{300} = \sqrt{100 \cdot 3} = \sqrt{100} \cdot \sqrt{3} = 10\sqrt{3}$$

b) We first use the quotient rule to remove the fraction 2/3 from inside the radical:

$$\sqrt{\frac{2}{3}} = \frac{\sqrt{2}}{\sqrt{3}} \qquad \text{Apply the quotient rule.}$$

$$= \frac{\sqrt{2}}{\sqrt{3}} \cdot \frac{\sqrt{3}}{\sqrt{3}} \qquad \text{Multiply both numerator and denominator by } \sqrt{3} \text{ to rationalize the denominator.}$$

$$= \frac{\sqrt{6}}{3}$$

c) We first rationalize the denominator:

$$\frac{\sqrt{10}}{\sqrt{6}} = \frac{\sqrt{10}}{\sqrt{6}} \cdot \frac{\sqrt{6}}{\sqrt{6}} = \frac{\sqrt{60}}{6} \qquad \text{First rationalize the denominator.}$$

$$= \frac{\sqrt{4} \cdot \sqrt{15}}{6} \qquad \text{Factor 60 as } 4 \cdot 15.$$

$$= \frac{2 \cdot \sqrt{15}}{6} \qquad \sqrt{4} = 2$$

$$= \frac{\sqrt{15}}{3} \qquad \text{Reduce } \frac{2}{6} \text{ to } \frac{1}{3}. \qquad \blacktriangleleft$$

Simplifying Square Roots Involving Variables

In Section 6.2 we simplified some square roots involving variables. For example, if x represents a nonnegative real number, then

$$\sqrt{x^6} = x^3.$$

Recall that *any exponential expression with an even exponent is a perfect square.* With this in mind we will simplify square roots involving variables and follow the three rules stated above.

EXAMPLE 4   Simplify the expressions. Assume that all variables represent nonnegative real numbers.

a) $\sqrt{x^3}$ b) $\sqrt{8a^9}$ c) $\sqrt{18a^4b^7}$

Solution

a) $\sqrt{x^3} = \sqrt{x^2} \cdot \sqrt{x}$ Since x^2 is a perfect square factor of x^3

$\quad\quad = x\sqrt{x}$ For any nonnegative x, $\sqrt{x^2} = x$.

b) $\sqrt{8a^9} = \sqrt{4a^8} \cdot \sqrt{2a}$ The largest perfect square factor of $8a^9$ is $4a^8$.

$\quad\quad = 2a^4\sqrt{2a}$ $\sqrt{4a^8} = 2a^4$

c) $\sqrt{18a^4b^7} = \sqrt{9a^4b^6} \cdot \sqrt{2b}$ Factor out a perfect square.

$\quad\quad = 3a^2b^3\sqrt{2b}$ $\sqrt{9a^4b^6} = 3a^2b^3$ ◀

If square roots of variables appear in the denominator, then we rationalize the denominator.

EXAMPLE 5   Simplify each expression. Assume that all variables represent positive real numbers.

a) $\dfrac{5}{\sqrt{a}}$ b) $\sqrt{\dfrac{a}{b}}$ c) $\dfrac{\sqrt{2}}{\sqrt{6a}}$

Solution

a) $\dfrac{5}{\sqrt{a}} = \dfrac{5}{\sqrt{a}} \cdot \dfrac{\sqrt{a}}{\sqrt{a}}$

$\quad\quad = \dfrac{5\sqrt{a}}{a}$ Note that $\sqrt{a} \cdot \sqrt{a} = a$.

b) $\sqrt{\dfrac{a}{b}} = \dfrac{\sqrt{a}}{\sqrt{b}}$ Apply the quotient rule.

$\quad\quad = \dfrac{\sqrt{a}}{\sqrt{b}} \cdot \dfrac{\sqrt{b}}{\sqrt{b}}$ Rationalize the denominator.

$\quad\quad = \dfrac{\sqrt{ab}}{b}$ Apply the product rule: $\sqrt{a} \cdot \sqrt{b} = \sqrt{ab}$.

c) $\dfrac{\sqrt{2}}{\sqrt{6a}} = \dfrac{\sqrt{2}}{\sqrt{6a}} \cdot \dfrac{\sqrt{6a}}{\sqrt{6a}}$ **Rationalize the denominator.**

   $= \dfrac{\sqrt{12a}}{6a}$ **Simplify.**

   $= \dfrac{\sqrt{4} \cdot \sqrt{3a}}{6a}$ **Factor out the perfect square.**

   $= \dfrac{2 \cdot \sqrt{3a}}{6a}$ $\sqrt{4} = 2$

   $= \dfrac{\sqrt{3a}}{3a}$ **Reduce** $\dfrac{2}{6}$ **to** $\dfrac{1}{3}$.

Note that we cannot divide out the 3's or the a's in the last expression.   ◄

Warm-ups

True or false?

1. $\sqrt{20} = 2\sqrt{5}$ T   2. $\sqrt{18} = 9\sqrt{2}$ F   3. $\dfrac{1}{\sqrt{3}} = \dfrac{\sqrt{3}}{3}$ T   4. $\dfrac{9}{4} = \dfrac{3}{2}$ F

5. $\sqrt{a^3} = a\sqrt{a}$ for any positive value of a.   T

6. $\sqrt{a^9} = a^3$ for any positive value of a.   F

7. $\sqrt{y^{17}} = y^8\sqrt{y}$ for any positive value of y.   T

8. $\dfrac{\sqrt{6}}{2} = \sqrt{3}$ F   9. $\sqrt{4} = \sqrt{2}$ F   10. $\sqrt{283} = 17$ F

6.3 EXERCISES

Assume that all variables in the exercises represent positive real numbers. Simplify the following square roots. See Example 1.

1. $\sqrt{8}$ $2\sqrt{2}$

2. $\sqrt{20}$ $2\sqrt{5}$

3. $\sqrt{24}$ $2\sqrt{6}$

4. $\sqrt{75}$ $5\sqrt{3}$

5. $\sqrt{28}$ $2\sqrt{7}$

6. $\sqrt{40}$ $2\sqrt{10}$

7. $\sqrt{90}$ $3\sqrt{10}$

8. $\sqrt{200}$ $10\sqrt{2}$

9. $\sqrt{500}$ $10\sqrt{5}$

10. $\sqrt{98}$ $7\sqrt{2}$

11. $\sqrt{150}$ $5\sqrt{6}$

12. $\sqrt{120}$ $2\sqrt{30}$

Simplify each of the following by rationalizing the denominator. See Example 2.

13. $\dfrac{1}{\sqrt{5}}$ $\dfrac{\sqrt{5}}{5}$

14. $\dfrac{1}{\sqrt{6}}$ $\dfrac{\sqrt{6}}{6}$

15. $\dfrac{3}{\sqrt{2}}$ $\dfrac{3\sqrt{2}}{2}$

16. $\dfrac{4}{\sqrt{3}}$ $\dfrac{4\sqrt{3}}{3}$

17. $\dfrac{\sqrt{3}}{\sqrt{2}}$ $\dfrac{\sqrt{6}}{2}$

18. $\dfrac{\sqrt{7}}{\sqrt{6}}$ $\dfrac{\sqrt{42}}{6}$

19. $\dfrac{-3}{\sqrt{10}}$ $\dfrac{-3\sqrt{10}}{10}$

20. $\dfrac{-4}{\sqrt{5}}$ $-\dfrac{4\sqrt{5}}{5}$

21. $\dfrac{-10}{\sqrt{17}}$ $\dfrac{-10\sqrt{17}}{17}$

22. $\dfrac{-3}{\sqrt{19}}$ $-\dfrac{3\sqrt{19}}{19}$

23. $\dfrac{\sqrt{11}}{\sqrt{7}}$ $\dfrac{\sqrt{77}}{7}$

24. $\dfrac{\sqrt{10}}{\sqrt{3}}$ $\dfrac{\sqrt{30}}{3}$

Simplify each expression. See Example 3.

25. $\sqrt{63}$ $3\sqrt{7}$

26. $\sqrt{48}$ $4\sqrt{3}$

27. $\sqrt{\dfrac{3}{2}}$ $\dfrac{\sqrt{6}}{2}$

28. $\sqrt{\dfrac{3}{5}}$ $\dfrac{\sqrt{15}}{5}$

29. $\sqrt{\dfrac{5}{8}}$ $\dfrac{\sqrt{10}}{4}$

30. $\sqrt{\dfrac{5}{18}}$ $\dfrac{\sqrt{10}}{6}$

31. $\dfrac{\sqrt{6}}{\sqrt{10}}$ $\dfrac{\sqrt{15}}{5}$

32. $\dfrac{\sqrt{12}}{\sqrt{20}}$ $\dfrac{\sqrt{15}}{5}$

33. $\dfrac{\sqrt{75}}{\sqrt{3}}$ 5

34. $\dfrac{\sqrt{45}}{\sqrt{5}}$ 3

35. $\dfrac{\sqrt{15}}{\sqrt{10}}$ $\dfrac{\sqrt{6}}{2}$

36. $\dfrac{\sqrt{30}}{\sqrt{21}}$ $\dfrac{\sqrt{70}}{7}$

Simplify each expression. See Example 4.

37. $\sqrt{a^8}$ a^4

38. $\sqrt{t^{10}}$ t^5

39. $\sqrt{a^9}$ $a^4 \cdot \sqrt{a}$

40. $\sqrt{t^{11}}$ $t^5 \cdot \sqrt{t}$

41. $\sqrt{8a^6}$ $2a^3 \cdot \sqrt{2}$

42. $\sqrt{18w^9}$ $3w^4 \cdot \sqrt{2w}$

43. $\sqrt{20a^4b^9}$ $2a^2b^4 \cdot \sqrt{5b}$

44. $\sqrt{12x^2y^3}$ $2xy\sqrt{3y}$

45. $\sqrt{27x^3y^3}$ $3xy\sqrt{3xy}$

46. $\sqrt{45x^5y^3}$ $3x^2y\sqrt{5xy}$

Simplify each expression. See Example 5.

47. $\dfrac{1}{\sqrt{x}}$ $\dfrac{\sqrt{x}}{x}$

48. $\dfrac{1}{\sqrt{2x}}$ $\dfrac{\sqrt{2x}}{2x}$

49. $\dfrac{\sqrt{2}}{\sqrt{3a}}$ $\dfrac{\sqrt{6a}}{3a}$

50. $\dfrac{\sqrt{5}}{\sqrt{2b}}$ $\dfrac{\sqrt{10b}}{2b}$

51. $\dfrac{\sqrt{3}}{\sqrt{15y}}$ $\dfrac{\sqrt{5y}}{5y}$

52. $\dfrac{\sqrt{5}}{\sqrt{10x}}$ $\dfrac{\sqrt{2x}}{2x}$

53. $\sqrt{\dfrac{3x}{2y}}$ $\dfrac{\sqrt{6xy}}{2y}$

54. $\sqrt{\dfrac{6}{5w}}$ $\dfrac{\sqrt{30w}}{5w}$

55. $\sqrt{\dfrac{10y}{15x}}$ $\dfrac{\sqrt{6xy}}{3x}$

56. $\sqrt{\dfrac{6x}{4y}}$ $\dfrac{\sqrt{6xy}}{2y}$

57. $\sqrt{\dfrac{8x^3}{y}}$ $\dfrac{2x\sqrt{2xy}}{y}$

58. $\sqrt{\dfrac{8s^5}{t}}$ $\dfrac{2s^2 \cdot \sqrt{2st}}{t}$

Simplify the following.

59. $\sqrt{80}$ $4\sqrt{5}$

60. $\sqrt{y^{80}}$ y^{40}

61. $\sqrt{9y^9}$ $3y^4 \cdot \sqrt{y}$

62. $\sqrt{48x^2y^7}$ $4xy^3 \cdot \sqrt{3y}$

63. $\sqrt{x^2y^2}$ xy

64. $\dfrac{5}{\sqrt{5}}$ $\sqrt{5}$

65. $\dfrac{2}{\sqrt{6}}$ $\dfrac{\sqrt{6}}{3}$

66. $\dfrac{3}{\sqrt{3t}}$ $\dfrac{\sqrt{3t}}{t}$

67. $\dfrac{a}{\sqrt{a}}$ $\sqrt{a}$

68. $\dfrac{3w}{\sqrt{w}}$ $3\sqrt{w}$

Use a calculator to find a decimal approximation for each expression. Round your answer to three decimal places.

69. $\dfrac{1}{\sqrt{2}}$ .707

70. $\dfrac{\sqrt{6}}{\sqrt{2}}$ 1.732

71. $\dfrac{\sqrt{2}}{2}$ .707

72. $\sqrt{3}$ 1.732

73. $\dfrac{\sqrt{20}}{\sqrt{5}}$ 2

74. $\dfrac{\sqrt{28}}{\sqrt{7}}$ 2

6.4 Operations with Radicals

IN THIS SECTION:

- Adding and Subtracting Radicals
- Multiplying Radicals
- Dividing Radicals

In this section we will learn how to perform the basic operations of arithmetic with radical expressions.

Adding and Subtracting Radicals

Consider the expression

$$2\sqrt{3} + 5\sqrt{3}.$$

When we studied like terms, we learned that

$$2x + 5x = 7x$$

is true for any value of x. If we let $x = \sqrt{3}$, then we get

$$2\sqrt{3} + 5\sqrt{3} = 7\sqrt{3}.$$

This shows that terms involving radicals can be combined just as we combine like terms. *When we add or subtract radicals, they must have the same index and the same radicand.* For example,

$$8\sqrt[3]{7} - 6\sqrt[3]{7} = 2\sqrt[3]{7},$$

but we cannot combine the terms in the expressions

$$\sqrt{2} + \sqrt{5}, \quad \sqrt[3]{5} - \sqrt{5}, \quad \text{or} \quad 3\sqrt{2} + \sqrt[4]{6}.$$

EXAMPLE 1 Simplify the following expressions by combining like radicals. Assume that the variables represent nonnegative numbers.

a) $2\sqrt{5} + 7\sqrt{5}$
b) $3\sqrt{2} - 9\sqrt{2}$
c) $\sqrt{2} - 5\sqrt{a} + 4\sqrt{2} - 3\sqrt{a}$

Solution

a) $2\sqrt{5} + 7\sqrt{5} = 9\sqrt{5}$
b) $3\sqrt{2} - 9\sqrt{2} = -6\sqrt{2}$
c) $\sqrt{2} - 5\sqrt{a} + 4\sqrt{2} - 3\sqrt{a} = \sqrt{2} + 4\sqrt{2} - 5\sqrt{a} - 3\sqrt{a}$
   $\qquad\qquad\qquad\qquad\qquad = 5\sqrt{2} - 8\sqrt{a}$ Combine like radicals only.   ◄

We may have to simplify radicals before we add or subtract. The next example will illustrate this.

EXAMPLE 2 Simplify the expressions by combining like terms. Assume that all variables represent nonnegative real numbers.

a) $\sqrt{12} + \sqrt{75}$ b) $\dfrac{4}{\sqrt{2}} - \dfrac{\sqrt{3}}{\sqrt{6}}$ c) $\sqrt{8x^3} + x\sqrt{18x}$

Solution

a) $\sqrt{12} + \sqrt{75} = 2\sqrt{3} + 5\sqrt{3}$ $\begin{aligned}\sqrt{12} &= \sqrt{4}\sqrt{3} = 2\sqrt{3}\\ \sqrt{75} &= \sqrt{25}\sqrt{3} = 5\sqrt{3}\end{aligned}$

$\qquad\qquad\qquad = 7\sqrt{3}$

b) $\dfrac{4}{\sqrt{2}} - \dfrac{\sqrt{3}}{\sqrt{6}} = \dfrac{4}{\sqrt{2}} \cdot \dfrac{\sqrt{2}}{\sqrt{2}} - \dfrac{\sqrt{3}}{\sqrt{6}} \cdot \dfrac{\sqrt{6}}{\sqrt{6}}$ **Rationalize the denominators.**

$\qquad = \dfrac{4\sqrt{2}}{2} - \dfrac{\sqrt{18}}{6}$ **Simplify.**

$\qquad = \dfrac{4\sqrt{2}}{2} - \dfrac{3\sqrt{2}}{6}$ $\sqrt{18} = \sqrt{9}\sqrt{2} = 3\sqrt{2}$

$\qquad = \dfrac{4\sqrt{2}}{2} - \dfrac{\sqrt{2}}{2}$ **Reduce** $\dfrac{3}{6}$ **to** $\dfrac{1}{2}$.

$\qquad = \dfrac{3\sqrt{2}}{2}$ $4\sqrt{2} - \sqrt{2} = 3\sqrt{2}$

c) $\sqrt{8x^3} + x\sqrt{18x} = 2x\sqrt{2x} + 3x\sqrt{2x}$ $\sqrt{8x^3} = \sqrt{4x^2}\sqrt{2x} = 2x\sqrt{2x}$
$\qquad\qquad\qquad\qquad\qquad\qquad\qquad\qquad x\sqrt{18x} = x\sqrt{9}\sqrt{2x} = 3x\sqrt{2x}$

$\qquad\qquad\qquad = 5x\sqrt{2x}$ ◄

Multiplying Radicals

We have been using the product rule for radicals

$$\sqrt[n]{a} \cdot \sqrt[n]{b} = \sqrt[n]{ab}$$

to express a root of a product as a product of the roots of the factors. When we rationalized denominators in Section 6.3, we used the product rule to multiply radicals. We will now study multiplication of radicals in more detail.

EXAMPLE 3 Multiply and simplify. Assume that variables represent positive numbers.

a) $\sqrt{2} \cdot \sqrt{5}$ b) $2\sqrt{5} \cdot 3\sqrt{6}$
c) $\sqrt{2a^2} \cdot \sqrt{6a}$ d) $\sqrt[3]{4} \cdot \sqrt[3]{2}$

Solution

a) $\sqrt{2} \cdot \sqrt{5} = \sqrt{10}$ **Product rule for radicals**

b) $2\sqrt{5} \cdot 3\sqrt{6} = 2 \cdot 3\sqrt{5}\,\sqrt{6}$

$\qquad\qquad\quad = 6\sqrt{30}$ **Product rule for radicals**

c) $\sqrt{2a^2} \cdot \sqrt{6a} = \sqrt{12a^3}$ **Product rule for radicals**

$\qquad\qquad\quad = \sqrt{4a^2} \cdot \sqrt{3a}$ **Factor out the perfect square.**

$\qquad\qquad\quad = 2a\sqrt{3a}$ **Simplify.**

d) $\sqrt[3]{4} \cdot \sqrt[3]{2} = \sqrt[3]{8} = 2$ **Product rule for radicals**   ◄

A sum such as $\sqrt{6} + \sqrt{2}$ is in its simplest form, and so it is treated like a binomial when it occurs in a product.

EXAMPLE 4 Find the product: $3\sqrt{3}(\sqrt{6} + \sqrt{2})$.

Solution

$$3\sqrt{3}(\sqrt{6} + \sqrt{2}) = 3\sqrt{3} \cdot \sqrt{6} + 3\sqrt{3} \cdot \sqrt{2} \qquad \textbf{Distributive property}$$
$$= 3\sqrt{18} + 3 \cdot \sqrt{6}$$
$$= 3 \cdot 3\sqrt{2} + 3\sqrt{6} \qquad \sqrt{18} = \sqrt{9} \cdot \sqrt{2} = 3\sqrt{2}$$
$$= 9\sqrt{2} + 3\sqrt{6} \qquad \blacktriangleleft$$

We find the product $(\sqrt{3} + 5)(\sqrt{3} - 2)$ just as we find the product of two binomials. We can use the FOIL method.

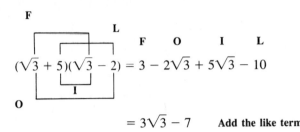

$$\begin{array}{cccc} & \textbf{F} & \textbf{O} & \textbf{I} & \textbf{L} \\ (\sqrt{3} + 5)(\sqrt{3} - 2) = & 3 & - 2\sqrt{3} & + 5\sqrt{3} & - 10 \end{array}$$

$$= 3\sqrt{3} - 7 \qquad \textbf{Add the like terms.}$$

EXAMPLE 5   Multiply and simplify:

a) $(\sqrt{5} - 2)(\sqrt{5} + 2)$
b) $(2\sqrt{3} + \sqrt{5})(\sqrt{3} - 4\sqrt{5})$

Solution

a) Note that this is the product of a sum and a difference, and recall that $(a - b)(a + b) = a^2 - b^2$.

$$(\sqrt{5} - 2)(\sqrt{5} + 2) = \sqrt{5} \cdot \sqrt{5} - 2 \cdot 2$$
$$= 5 - 4 = 1$$

b)
$$\begin{array}{cccc} & \textbf{F} & \textbf{O} & \textbf{I} & \textbf{L} \\ (2\sqrt{3} + \sqrt{5})(\sqrt{3} - 4\sqrt{5}) = & 2\sqrt{3}\,\sqrt{3} & - 2\sqrt{3} \cdot 4\sqrt{5} & + \sqrt{5}\,\sqrt{3} & - 4\sqrt{5} \cdot \sqrt{5} \end{array}$$
$$= 6 - 8\sqrt{15} + \sqrt{15} - 20$$
$$= -14 - 7\sqrt{15} \qquad \blacktriangleleft$$

Dividing Radicals

Make sure students do not overlook this idea. You do not always multiply by the denominator when the denominator is irrational.

In Section 6.2 we used the quotient rule for radicals to write a radical of a quotient as a quotient of radicals. We can also use the quotient rule for radicals to divide radicals of the same index. For example,

$$\frac{\sqrt{10}}{\sqrt{2}} = \sqrt{\frac{10}{2}} = \sqrt{5}.$$

Division of this type is very easy as long as the quotient of the two radicands is a whole number. For example,

$$\sqrt{18} \div \sqrt{3} = \sqrt{6} \quad \text{or} \quad \frac{\sqrt{30}}{\sqrt{2}} = \sqrt{15}.$$

We know that $\sqrt{6} \div \sqrt{2} = \sqrt{3}$, but now consider $\sqrt{2} \div \sqrt{6}$. To do this division, we write the quotient as a fraction and then rationalize the denominator:

$$\frac{\sqrt{2}}{\sqrt{6}} = \frac{\sqrt{2}}{\sqrt{6}} \cdot \frac{\sqrt{6}}{\sqrt{6}}$$

$$= \frac{\sqrt{12}}{6}$$

$$= \frac{2\sqrt{3}}{6} \qquad \sqrt{12} = \sqrt{4} \cdot \sqrt{3} = 2\sqrt{3}$$

$$= \frac{\sqrt{3}}{3} \qquad \text{Reduce.}$$

Note that we get the same answer if we start with $\dfrac{\sqrt{2}}{\sqrt{6}} = \sqrt{\dfrac{1}{3}}$.

EXAMPLE 6   Divide and simplify:

a) $\sqrt{30} \div \sqrt{3}$
b) $(5\sqrt{2}) \div (2\sqrt{5})$
c) $15\sqrt{6} \div 3\sqrt{2}$

Solution
a) $\sqrt{30} \div \sqrt{3} = \sqrt{10}$
b) $5\sqrt{2} \div 2\sqrt{5} = \dfrac{5\sqrt{2}}{2\sqrt{5}} = \dfrac{5\sqrt{2} \cdot \sqrt{5}}{2\sqrt{5} \cdot \sqrt{5}}$ Rationalize the denominator.

$$= \frac{5\sqrt{10}}{2 \cdot 5} \qquad \text{Product rule for radicals}$$

$$= \frac{\sqrt{10}}{2} \qquad \text{Reduce.}$$

Note that $\sqrt{10} \div 2 \neq \sqrt{5}$.

c) $15\sqrt{6} \div 3\sqrt{2} = \dfrac{15\sqrt{6}}{3\sqrt{2}} = 5\sqrt{3}$ $\sqrt{6} \div \sqrt{2} = \sqrt{3}$ ◄

Note that the product rule and the quotient rule allow us to combine roots of the same index only. We did not combine square roots with cube roots or with roots of any other index.

Be very careful not to abuse the product and quotient rules in expressions involving whole numbers and roots. For example,

$$\sqrt{2} \cdot \sqrt{3} = \sqrt{6} \quad \text{but} \quad 2 \cdot \sqrt{3} \neq \sqrt{6}.$$

Similarly, we have

$$\frac{\sqrt{14}}{\sqrt{2}} = \sqrt{7} \quad \text{but} \quad \frac{\sqrt{14}}{2} \neq \sqrt{7}.$$

In the next example we simplify expressions with radicals in the numerator and whole numbers in the denominator.

EXAMPLE 7 Simplify the radical expressions:

a) $\dfrac{4 - \sqrt{20}}{4}$

b) $\dfrac{-6 + \sqrt{27}}{3}$

Solution

a) $\dfrac{4 - \sqrt{20}}{4} = \dfrac{4 - 2\sqrt{5}}{4}$ $\sqrt{20} = \sqrt{4} \cdot \sqrt{5} = 2\sqrt{5}$

$= \dfrac{2(2 - \sqrt{5})}{2 \cdot 2}$ Factor out the GCF 2.

$= \dfrac{2 - \sqrt{5}}{2}$ Reduce.

Do not divide out the remaining 2's.

b) $\dfrac{-6 + \sqrt{27}}{3} = \dfrac{-6 + 3\sqrt{3}}{3} = \dfrac{3(-2 + \sqrt{3})}{3} = -2 + \sqrt{3}$ ◄

In Example 5a we used the rule for the product of a sum and a difference to get $(\sqrt{5} - 2)(\sqrt{5} + 2) = 1$. If we apply the same rule to other products of this type, we will get a rational number as the result. For example,

$$(\sqrt{7} + \sqrt{2})(\sqrt{7} - \sqrt{2}) = 7 - 2 = 5.$$

Expressions such as $\sqrt{5} + 2$ and $\sqrt{5} - 2$ are called **conjugates** of each other. The conjugate of $\sqrt{7} + \sqrt{2}$ is $\sqrt{7} - \sqrt{2}$. We can use conjugates to simplify a radical expression that has a sum or a difference in its denominator.

EXAMPLE 8 Simplify each expression:

a) $\dfrac{\sqrt{3}}{\sqrt{7} - \sqrt{2}}$

b) $\dfrac{4}{6 + \sqrt{2}}$

Solution

a) $\dfrac{\sqrt{3}}{\sqrt{7} - \sqrt{2}} = \dfrac{\sqrt{3}}{(\sqrt{7} - \sqrt{2})} \cdot \dfrac{(\sqrt{7} + \sqrt{2})}{(\sqrt{7} + \sqrt{2})}$ Multiply by $\sqrt{7} + \sqrt{2}$, the conjugate of $\sqrt{7} - \sqrt{2}$.

$= \dfrac{\sqrt{21} + \sqrt{6}}{7 - 2}$

$= \dfrac{\sqrt{21} + \sqrt{6}}{5}$

b) $\dfrac{4}{6+\sqrt{2}} = \dfrac{4}{(6+\sqrt{2})} \cdot \dfrac{(6-\sqrt{2})}{(6-\sqrt{2})}$ **Multiply by $6-\sqrt{2}$, the conjugate of $6+\sqrt{2}$.**

$\qquad\qquad = \dfrac{24-4\sqrt{2}}{36-2}$

$\qquad\qquad = \dfrac{24-4\sqrt{2}}{34}$

$\qquad\qquad = \dfrac{\cancel{2}(12-2\sqrt{2})}{\cancel{2}\cdot 17}$

$\qquad\qquad = \dfrac{12-2\sqrt{2}}{17}$ ◀

Warm-ups

True or false?

1. $\sqrt{9}+\sqrt{16}=\sqrt{25}$ F   2. $\dfrac{5}{\sqrt{5}}=\sqrt{5}$ T   3. $\sqrt{10}\div 2=\sqrt{5}$ F

4. $3\sqrt{2}\cdot 3\sqrt{2}=9\sqrt{2}$ F   5. $3\sqrt{5}\cdot 3\sqrt{2}=9\sqrt{10}$ T

6. $\sqrt{5}+3\sqrt{5}=4\sqrt{10}$ F   7. $\dfrac{\sqrt{15}}{3}=\sqrt{5}$ F   8. $\sqrt{2}\div\sqrt{6}=\sqrt{3}$ F

9. $\dfrac{\sqrt{27}}{\sqrt{3}}=3$ T   10. $(\sqrt{3}-1)(\sqrt{3}+1)=2$ T

6.4 EXERCISES

Assume that all variables in these exercises represent only positive real numbers. Simplify the following expressions by combining like radicals. See Example 1.

1. $4\sqrt{5}+3\sqrt{5}$ $7\sqrt{5}$ 2. $\sqrt{2}+\sqrt{2}$ $2\sqrt{2}$ 3. $3\sqrt{7}-\sqrt{7}$ $2\sqrt{7}$ 4. $\sqrt{3}-5\sqrt{3}$ $-4\sqrt{3}$

5. $\sqrt{14}+\sqrt{14}$ $2\sqrt{14}$ 6. $\sqrt{5}-\sqrt{5}$ 0 7. $\sqrt[3]{2}+\sqrt[3]{2}$ $2\cdot\sqrt[3]{2}$ 8. $4\sqrt[3]{6}-7\sqrt[3]{6}$ $-3\cdot\sqrt[3]{6}$

9. $\sqrt{2}+\sqrt{3}-5\sqrt{2}+3\sqrt{3}$ $4\sqrt{3}-4\sqrt{2}$ 10. $8\sqrt{6}-\sqrt{2}-3\sqrt{6}+5\sqrt{2}$ $5\sqrt{6}+4\sqrt{2}$

11. $3\sqrt{y}-\sqrt{x}-4\sqrt{y}-3\sqrt{x}$ $-4\sqrt{x}-\sqrt{y}$ 12. $5\sqrt{7}-\sqrt{a}+3\sqrt{7}-5\sqrt{a}$ $8\sqrt{7}-6\sqrt{a}$

13. $3x\sqrt{y}-\sqrt{a}+2x\sqrt{y}+3\sqrt{a}$ $5x\sqrt{y}+2\sqrt{a}$ 14. $ab\sqrt{b}+5ab\sqrt{b}-2\sqrt{a}+3\sqrt{a}$ $6ab\sqrt{b}+\sqrt{a}$

Simplify each expression. See Example 2.

15. $\sqrt{24}+\sqrt{54}$ $5\sqrt{6}$ 16. $\sqrt{12}+\sqrt{27}$ $5\sqrt{3}$ 17. $2\sqrt{27}-4\sqrt{75}$ $-14\sqrt{3}$

18. $\sqrt{2}-\sqrt{18}$ $-2\sqrt{2}$ 19. $\sqrt{3}-\sqrt{12}$ $-\sqrt{3}$ 20. $\sqrt{5}-\sqrt{45}$ $-2\sqrt{5}$

21. $\sqrt{x^3}+2x\sqrt{x}$ $3x\sqrt{x}$ 22. $\sqrt{8x^3}+\sqrt{2x}$ $(2x+1)\sqrt{2x}$ 23. $\dfrac{1}{\sqrt{3}}+\dfrac{\sqrt{2}}{\sqrt{6}}$ $\dfrac{2\sqrt{3}}{3}$

24. $\dfrac{3}{\sqrt{5}}+\dfrac{\sqrt{2}}{\sqrt{10}}$ $\dfrac{4\sqrt{5}}{5}$ 25. $\dfrac{1}{\sqrt{3}}+\sqrt{12}$ $\dfrac{7\sqrt{3}}{3}$ 26. $\dfrac{1}{\sqrt{2}}+3\sqrt{8}$ $\dfrac{13\sqrt{2}}{2}$

Multiply and simplify. See Example 3.

27. $2\sqrt{6} \cdot 3\sqrt{6}$ 36

28. $4\sqrt{2} \cdot 3\sqrt{2}$ 24

29. $3\sqrt{5} \cdot 4\sqrt{2}$ $12\sqrt{10}$

30. $8\sqrt{3} \cdot 3\sqrt{2}$ $24\sqrt{6}$

31. $5\sqrt{2} \cdot 3\sqrt{6}$ $30\sqrt{3}$

32. $2\sqrt{3} \cdot 6\sqrt{6}$ $36\sqrt{2}$

33. $5\sqrt{12} \cdot 3\sqrt{2}$ $30\sqrt{6}$

34. $2\sqrt{10} \cdot 3\sqrt{5}$ $30\sqrt{2}$

35. $\sqrt{2a^3} \cdot \sqrt{6a}$ $2a^2 \cdot \sqrt{3}$

36. $\sqrt{3a^3} \cdot \sqrt{a^5}$ $a^4 \cdot \sqrt{3}$

37. $\sqrt{6x^3} \cdot \sqrt{3x^2}$ $3x^2 \cdot \sqrt{2x}$

38. $\sqrt{2x^2} \cdot \sqrt{10x}$ $2x\sqrt{5x}$

Multiply and simplify. See Example 4.

39. $\sqrt{2}(\sqrt{2} + \sqrt{3})$ $2 + \sqrt{6}$

40. $\sqrt{3}(\sqrt{3} - \sqrt{2})$ $3 - \sqrt{6}$

41. $3\sqrt{2}(2\sqrt{6} + \sqrt{10})$ $12\sqrt{3} + 6\sqrt{5}$

42. $2\sqrt{3}(\sqrt{6} + 2\sqrt{15})$ $6\sqrt{2} + 12\sqrt{5}$

43. $2\sqrt{5}(\sqrt{5} - 3\sqrt{10})$ $10 - 30\sqrt{2}$

44. $\sqrt{6}(\sqrt{24} - 6)$ $12 - 6\sqrt{6}$

Multiply and simplify. See Example 5.

45. $(\sqrt{3} - 1)(\sqrt{3} + 1)$ 2

46. $(\sqrt{6} + 2)(\sqrt{6} - 2)$ 2

47. $(\sqrt{5} - 3)(\sqrt{5} + 4)$ $-7 + \sqrt{5}$

48. $(\sqrt{3} - 5)(\sqrt{3} + 2)$ $-7 - 3\sqrt{3}$

49. $(2\sqrt{5} + 1)(3\sqrt{5} - 2)$ $28 - \sqrt{5}$

50. $(2\sqrt{2} + 3)(4\sqrt{2} + 4)$ $28 + 20\sqrt{2}$

51. $(2\sqrt{3} - 3\sqrt{5})(3\sqrt{3} + 4\sqrt{5})$ $-42 - \sqrt{15}$

52. $(4\sqrt{3} + 3\sqrt{7})(2\sqrt{3} + 4\sqrt{7})$ $108 + 22\sqrt{21}$

53. $(4\sqrt{6} - 3\sqrt{2})(2\sqrt{6} + 5\sqrt{2})$ $18 + 28\sqrt{3}$

54. $(2\sqrt{15} + 4\sqrt{3})(3\sqrt{15} - 2\sqrt{3})$ $66 + 24\sqrt{5}$

55. $(2\sqrt{3} + 5)^2$ $37 + 20\sqrt{3}$

56. $(3\sqrt{2} + 1)^2$ $19 + 6\sqrt{2}$

57. $(\sqrt{3} - \sqrt{2})^2$ $5 - 2\sqrt{6}$

58. $(\sqrt{5} - \sqrt{3})^2$ $8 - 2\sqrt{15}$

Divide and simplify. See Example 6.

59. $\sqrt{10} \div \sqrt{5}$ $\sqrt{2}$

60. $\sqrt{14} \div \sqrt{2}$ $\sqrt{7}$

61. $\sqrt{5} \div \sqrt{3}$ $\dfrac{\sqrt{15}}{3}$

62. $\sqrt{3} \div \sqrt{2}$ $\dfrac{\sqrt{6}}{2}$

63. $2\sqrt{3} \div \sqrt{5}$ $\dfrac{2\sqrt{15}}{5}$

64. $3\sqrt{2} \div \sqrt{5}$ $\dfrac{3\sqrt{10}}{5}$

65. $4\sqrt{5} \div 3\sqrt{6}$ $\dfrac{2\sqrt{30}}{9}$

66. $3\sqrt{7} \div 4\sqrt{3}$ $\dfrac{\sqrt{21}}{4}$

67. $5\sqrt{14} \div 3\sqrt{2}$ $\dfrac{5\sqrt{7}}{3}$

68. $4\sqrt{15} \div 5\sqrt{2}$ $\dfrac{2\sqrt{30}}{5}$

69. $8 \div \sqrt{2}$ $4\sqrt{2}$

70. $6 \div \sqrt{3}$ $2\sqrt{3}$

Simplify each expression. See Example 7.

71. $\dfrac{2 + \sqrt{8}}{2}$ $1 + \sqrt{2}$

72. $\dfrac{3 + \sqrt{18}}{3}$ $1 + \sqrt{2}$

73. $\dfrac{-4 + \sqrt{20}}{2}$ $-2 + \sqrt{5}$

74. $\dfrac{-6 + \sqrt{45}}{3}$ $-2 + \sqrt{5}$

75. $\dfrac{4 - \sqrt{20}}{6}$ $\dfrac{2 - \sqrt{5}}{3}$

76. $\dfrac{-6 - \sqrt{27}}{6}$ $\dfrac{-2 - \sqrt{3}}{2}$

77. $\dfrac{-4 - \sqrt{24}}{-6}$ $\dfrac{2 + \sqrt{6}}{3}$

78. $\dfrac{-3 - \sqrt{27}}{-3}$ $1 + \sqrt{3}$

Simplify each expression. See Example 8.

79. $\dfrac{5}{\sqrt{3} - \sqrt{2}}$ $5\sqrt{3} + 5\sqrt{2}$

80. $\dfrac{3}{\sqrt{6} + \sqrt{2}}$ $\dfrac{3\sqrt{6} - 3\sqrt{2}}{4}$

81. $\dfrac{\sqrt{3}}{\sqrt{5} - \sqrt{3}}$ $\dfrac{\sqrt{15} + 3}{2}$

82. $\dfrac{\sqrt{2}}{\sqrt{2} + 4}$ $\dfrac{-1 + 2\sqrt{2}}{7}$

83. $\dfrac{2 + \sqrt{3}}{5 - \sqrt{3}}$ $\dfrac{13 + 7\sqrt{3}}{22}$

84. $\dfrac{\sqrt{2} - \sqrt{3}}{\sqrt{3} - 1}$ $\dfrac{\sqrt{6} - 3 + \sqrt{2} - \sqrt{3}}{2}$

85. $\dfrac{2}{2\sqrt{3} + 1}$ $\dfrac{4\sqrt{3} - 2}{11}$

86. $\dfrac{\sqrt{5} + 4}{3\sqrt{2} - \sqrt{5}}$ $\dfrac{3\sqrt{10} + 12\sqrt{2} + 4\sqrt{5} + 5}{13}$

Simplify.

87. $\sqrt{5} + \sqrt{20}$ $3\sqrt{5}$

88. $\sqrt{5} \cdot \sqrt{20}$ 10

89. $\sqrt{20} \div \sqrt{5}$ 2

90. $\sqrt{20} - \sqrt{5}$ $\sqrt{5}$

91. $2\sqrt{5} \cdot 3\sqrt{20}$ 60   **92.** $(20 + \sqrt{5})^2$ $405 + 40\sqrt{5}$ **93.** $(20 - \sqrt{5})(20 + \sqrt{5})$ 395

94. $\sqrt{5} + \dfrac{\sqrt{20}}{3}$ $\dfrac{5\sqrt{5}}{3}$ **95.** $\dfrac{4 - \sqrt{20}}{2}$ $2 - \sqrt{5}$ **96.** $\dfrac{\sqrt{20}}{\sqrt{3}}$ $\dfrac{2\sqrt{15}}{3}$ **97.** $\dfrac{\sqrt{18}}{\sqrt{2}}$ 3

98. $\dfrac{\sqrt[3]{6}}{\sqrt[3]{2}}$ $\sqrt[3]{3}$ **99.** $\dfrac{5}{\sqrt{8} - \sqrt{3}}$ $2\sqrt{2} + \sqrt{3}$ **100.** $\dfrac{1 - \sqrt{3}}{3 + \sqrt{3}}$ $\dfrac{3 - 2\sqrt{3}}{3}$

Use a calculator to find the approximate value of each expression to three decimal places.

101. $\dfrac{2 + \sqrt{3}}{2}$ 1.866   **102.** $\dfrac{3 + \sqrt{6}}{2}$ 2.725   **103.** $\dfrac{-3 + \sqrt{7}}{-4}$ .089   **104.** $\dfrac{-2 + \sqrt{3}}{-6}$ .045

105. $\dfrac{-4 - \sqrt{6}}{5 - \sqrt{3}}$ -1.974 **106.** $\dfrac{-5 - \sqrt{2}}{\sqrt{3} + \sqrt{7}}$ -1.465 **107.** $\dfrac{7 - \sqrt{8}}{\sqrt{5}}$ 1.866   **108.** $\dfrac{9 - \sqrt{5}}{\sqrt{2}}$ 4.783

6.5 Solving Equations with Radicals and Exponents

IN THIS SECTION:

- **The Square Root Property**
- **Obtaining Equivalent Equations**
- **Squaring Each Side of an Equation**
- **Solving for the Indicated Variable**
- **Applications**

Equations involving radicals and exponents occur in many applications. In this section we will learn to solve equations of this type, and we will see how these equations occur in some geometric problems.

The Square Root Property

Consider the equation

$$x^2 = 4.$$

We know that both 2 and -2 have a square of 4. So this equation is equivalent to the compound equation

$$x = 2 \quad \text{or} \quad x = -2.$$

The sentence $x = 2$ or $x = -2$ is written simply as

$$x = \pm 2.$$

We read this as "x equals positive or negative 2." Both 2 and -2 are solutions to the equation $x^2 = 4$.

Consider the equation

$$x^2 = -4.$$

Since the square of every real number is greater than or equal to 0, this equation has no real solution.

The equation $x^2 = 0$ has only one solution. Only 0 has a square of 0.

These examples illustrate the square root property.

PROPERTY
Square Root Property (How to Solve $x^2 = k$)

> If $k > 0$, the equation $x^2 = k$ is equivalent to the sentence
>
> $$x = \pm\sqrt{k} \qquad (x = \sqrt{k} \text{ or } x = -\sqrt{k}).$$
>
> If $k = 0$, the equation $x^2 = k$ is equivalent to $x = 0$.
> If $k < 0$, the equation $x^2 = k$ has no real solution.

Note that the expression $\sqrt{9}$ has a value of 3 only, but the equation $x^2 = 9$ has two solutions, 3 and -3.

EXAMPLE 1 Solve the equations:

a) $x^2 = 12$

b) $2(x + 1)^2 - 18 = 0$

c) $x^2 = -9$

d) $(x - 16)^2 = 0$

Solution

a) $x^2 = 12$

$\quad x = \pm\sqrt{12}$ Apply the square root property.

$\quad x = \pm2\sqrt{3}$ $\sqrt{12} = \sqrt{4}\,\sqrt{3} = 2\sqrt{3}$

Check that $2\sqrt{3}$ and $-2\sqrt{3}$ both satisfy the equation.

b) $2(x + 1)^2 - 18 = 0$

$\quad\quad 2(x + 1)^2 = 18$ Add 18 to each side.

$\quad\quad (x + 1)^2 = 9$ Divide each side by 2.

$\quad\quad x + 1 = \pm3$ Apply the square root property.

$\quad x + 1 = 3$ or $x + 1 = -3$

$\quad\quad x = 2$ or $x = -4$

Both -4 and 2 are solutions to the equation. Check in the original equation.

c) The equation $x^2 = -9$ has no real solution, because no real number has a square that is negative.

d) $(x - 16)^2 = 0$

$\quad x - 16 = 0$ Apply the square root property.

$\quad\quad x = 16$

The equation has only one solution, 16. Check.   ◀

Obtaining Equivalent Equations

When solving equations, we usually write down a sequence of equivalent equations in which each equation is simpler than the preceding one. In Chapter 2 we learned that we get an equivalent equation by performing the same operation on each side of an equation. To get an equivalent equation, we can

1. add the same number to each side,
2. subtract the same number from each side,
3. multiply each side by the same nonzero number, and
4. divide each side by the same nonzero number.

However, "doing the same thing to each side" is not the only way to obtain an equivalent equation. In Chapter 4 we used the zero factor property to obtain equivalent equations. For example, by the zero factor property the equation

$$(x - 3)(x + 2) = 0$$

is equivalent to the compound equation

$$x - 3 = 0 \quad \text{or} \quad x + 2 = 0.$$

In this section we just learned how to obtain equivalent equations by the square root property. This property tells us how to write an equation that is equivalent to the equation $x^2 = k$. Note that the square root property does *not* tell us to "take the square root of each side." To become proficient at solving equations, we must understand these methods. One of our main goals in algebra is to keep expanding our skills for solving equations.

Squaring Each Side of an Equation

It is often necessary to square each side of an equation to solve it. However, this is *not* a method for obtaining equivalent equations. Consider the equation

$$x = 3.$$

If we square each side, we get

$$x^2 = 9.$$

Since 3 is the only number that satisfies $x = 3$, and both -3 and 3 are solutions to $x^2 = 9$, these are *not equivalent equations*. The equation $x^2 = 9$ has an extra solution (or root). This extra solution is also called an **extraneous root**.

Now consider the equation $\sqrt{x} = 5$:

$$\sqrt{x} = 5$$
$$(\sqrt{x})^2 = 5^2 \qquad \text{Square each side.}$$
$$x = 25$$

If we check 25 in the original equation, we see that it does satisfy the original equation. In this case we squared each side, and we did not obtain an extraneous root. All of these equations are equivalent. Squaring each side of an equation may or may not result in equivalent equations.

These two examples illustrate the squaring property of equality.

PROPERTY
Squaring Property of
Equality

> When we square each side of an equation, the solutions to the new equation include all of the solutions to the original equation. However, the new equation may have extraneous roots.

This property says that *we may square each side of an equation, but we must check all of our solutions for extraneous roots.*

EXAMPLE 2 Solve each equation:

a) $\sqrt{x^2 - 16} = 3$ b) $x = \sqrt{2x + 3}$ c) $\sqrt{x^2 - 4x} = \sqrt{2 - 3x}$

Solution

a)
$$\sqrt{x^2 - 16} = 3$$
$$(\sqrt{x^2 - 16})^2 = 3^2 \qquad \text{Square each side.}$$
$$x^2 - 16 = 9$$
$$x^2 = 25$$
$$x = \pm 5 \qquad \text{Apply the square root property.}$$

Check each solution:

Check $x = 5$:	Check $x = -5$:
$\sqrt{5^2 - 16} = 3$	$\sqrt{(-5)^2 - 16} = 3$
$\sqrt{25 - 16} = 3$	$\sqrt{25 - 16} = 3$
$\sqrt{9} = 3$	$\sqrt{9} = 3$

Since both of the solutions check in the original equation, both -5 and 5 are solutions to the equation.

Point out that no negative number could be a solution to this equation.

b)
$$x = \sqrt{2x + 3}$$
$$x^2 = (\sqrt{2x + 3})^2 \qquad \text{Square each side.}$$
$$x^2 = 2x + 3$$
$$x^2 - 2x - 3 = 0$$
$$(x - 3)(x + 1) = 0 \qquad \text{Factor.}$$

$$x - 3 = 0 \quad \text{or} \quad x + 1 = 0 \qquad \text{Apply the zero factor property.}$$
$$x = 3 \quad \text{or} \quad x = -1$$

Check each of these solutions in the original equation:

Check $x = 3$:	Check $x = -1$:
$3 = \sqrt{2 \cdot 3 + 3}$	$-1 = \sqrt{2(-1) + 3}$
$3 = \sqrt{9}$ Correct	$-1 = \sqrt{1}$ Incorrect

The first of these is correct, but the second has a positive square root equal to a negative number. So -1 is an extraneous root. The only solution is 3.

c)

$$\sqrt{x^2 - 4x} = \sqrt{2 - 3x}$$

$$x^2 - 4x = 2 - 3x \qquad \text{Square each side.}$$

$$x^2 - x - 2 = 0$$

$$(x - 2)(x + 1) = 0 \qquad \text{Factor.}$$

$$x - 2 = 0 \qquad \text{or} \qquad x + 1 = 0 \qquad \text{Apply the zero factor property.}$$

$$x = 2 \qquad \text{or} \qquad x = -1$$

Check each of these solutions in the original equation:

$$\text{Check } x = 2: \qquad\qquad \text{Check } x = -1:$$

$$\sqrt{2^2 - 4 \cdot 2} = \sqrt{2 - 3 \cdot 2} \qquad \sqrt{(-1)^2 - 4(-1)} = \sqrt{2 - 3(-1)}$$

$$\sqrt{-4} = \sqrt{-4} \qquad\qquad \sqrt{5} = \sqrt{5}$$

Since $\sqrt{-4}$ is not a real number, 2 is an extraneous root. The only solution to the equation is -1.   ◄

In the next example, one of the sides of the equation is a binomial. When we square each side, we must be sure to square the binomial properly.

EXAMPLE 3   Solve the equation $x + 2 = \sqrt{-2 - 3x}$.

Solution

$$x + 2 = \sqrt{-2 - 3x}$$

$$(x + 2)^2 = (\sqrt{-2 - 3x})^2 \qquad \text{Square each side.}$$

$$x^2 + 4x + 4 = -2 - 3x \qquad \text{Square the binomial on the left. Square the square root on the right.}$$

$$x^2 + 7x + 6 = 0$$

$$(x + 6)(x + 1) = 0 \qquad \text{Factor.}$$

$$x + 6 = 0 \qquad \text{or} \quad x + 1 = 0$$

$$x = -6 \quad \text{or} \qquad x = -1$$

Check these solutions in the original equation:

$$\text{Check } x = -6: \qquad\qquad \text{Check } x = -1:$$

$$-6 + 2 = \sqrt{-2 - 3(-6)} \qquad -1 + 2 = \sqrt{-2 - 3(-1)}$$

$$-4 = \sqrt{16} \quad \text{Incorrect} \qquad 1 = \sqrt{1} \quad \text{Correct}$$

The solution -6 does not check. The only solution to the equation is -1.   ◄

Solving for the Indicated Variable

We can use the new techniques we have learned to solve formulas for one variable in terms of another variable.

EXAMPLE 4   Solve the formula $A = \pi r^2$ for r.

Solution

$$A = \pi r^2$$

$$\frac{A}{\pi} = r^2 \qquad \text{Divide each side by } \pi.$$

$$\pm \sqrt{\frac{A}{\pi}} = r \qquad \text{Apply the square root property.}$$

The formula solved for r is

$$r = \pm \sqrt{\frac{A}{\pi}}.$$

If we are thinking of r as the radius of a circle, then we use only the positive square root,

$$r = \sqrt{\frac{A}{\pi}}. \qquad\qquad \blacktriangleleft$$

Applications

Equations involving exponents can be used to solve problems in geometry. The exact answer to a problem may be an irrational number in radical notation. In this case it is usually helpful to also find a decimal approximation for the answer.

EXAMPLE 5   If the diagonal of a square window is 10 feet long, then what are the exact and approximate lengths of a side?

Solution   First make a sketch as in Fig. 6.1. Let x be the length of a side. The Pythagorean Theorem tells us that the sum of the squares of the sides is equal to the diagonal squared.

$$x^2 + x^2 = 10^2$$

$$2x^2 = 100$$

$$x^2 = 50$$

$$x = \pm\sqrt{50} = \pm 5\sqrt{2}$$

Since the length of a side must be positive, we disregard the negative solution. The exact length of a side is $5\sqrt{2}$ feet. Use a calculator to obtain $\sqrt{2} \approx 1.414$ and $5\sqrt{2} \approx 7.07$. The symbol "$\approx$" means "is approximately equal to." The approximate length of a side is 7.07 feet. $\blacktriangleleft$

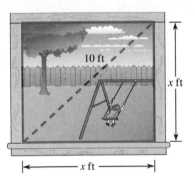

Figure 6.1

Warm-ups

True or false?

1. The equation $x^2 = 9$ is equivalent to the equation $x = 3$.   F

2. The equation $x^2 = -16$ has no real solution.   T

3. The equation $a^2 = 0$ has no solution. F

4. Both $-\sqrt{5}$ and $\sqrt{5}$ are solutions to $x^2 + 5 = 0$. F

5. The equation $-x^2 = 9$ has no solution. T

6. To solve $\sqrt{x + 4} = \sqrt{2x - 9}$, first take the square root of each side. F

7. All extraneous roots give us a denominator of zero. F

8. Squaring both sides of $\sqrt{x} = -1$ will produce an extraneous root. T

9. The equation $x^2 - 3 = 0$ is equivalent to $x = \pm\sqrt{3}$. T

10. The equation $-2 = \sqrt{6x^2 - x - 8}$ has no solution. T

6.5 EXERCISES

Solve each equation. See Example 1.

1. $x^2 = 16$ $-4, 4$

2. $x^2 = 49$ $-7, 7$

3. $x^2 = 12$ $-2\sqrt{3}, 2\sqrt{3}$

4. $x^2 = 24$ $-2\sqrt{6}, 2\sqrt{6}$

5. $3x^2 = 2$ $-\dfrac{\sqrt{6}}{3}, \dfrac{\sqrt{6}}{3}$

6. $2x^2 = 3$ $-\dfrac{\sqrt{6}}{2}, \dfrac{\sqrt{6}}{2}$

7. $9x^2 = -4$ No solution

8. $25x^2 - 1 = 0$ $-\dfrac{1}{5}, \dfrac{1}{5}$

9. $(x - 1)^2 = 4$ $-1, 3$

10. $(x + 3)^2 = -9$ No solution

11. $(x - 5)^2 = 3$ $5 - \sqrt{3}, 5 + \sqrt{3}$

12. $(x - 6)^2 = 5$ $6 - \sqrt{5}, 6 + \sqrt{5}$

13. $(x + 19)^2 = 0$ -19

14. $3x^2 = 0$ 0

Solve each equation. See Example 2.

15. $\sqrt{x - 9} = 9$ 90

16. $\sqrt{x + 3} = 4$ 13

17. $\sqrt{2x - 3} = -4$ No solution

18. $\sqrt{3x - 5} = -9$ No solution

19. $4 = \sqrt{x^2 - 9}$ $-5, 5$

20. $1 = \sqrt{x^2 - 1}$ $-\sqrt{2}, \sqrt{2}$

21. $x = \sqrt{18 - 3x}$ 3

22. $x = \sqrt{6x + 27}$ 9

23. $x = \sqrt{x}$ $0, 1$

24. $x = \sqrt{2x}$ $0, 2$

25. $\sqrt{x + 1} = \sqrt{2x - 5}$ 6

26. $\sqrt{1 - 3x} = \sqrt{x + 5}$ -1

Solve each equation. See Example 3.

27. $x - 3 = \sqrt{2x - 6}$ $3, 5$

28. $x - 1 = \sqrt{3x - 5}$ $2, 3$

29. $\sqrt{x + 13} = x + 1$ 3

30. $x + 1 = \sqrt{22 - 2x}$ 3

31. $\sqrt{10x - 44} = x - 2$ $6, 8$

32. $\sqrt{8x - 7} = x + 1$ $2, 4$

Solve each formula for the indicated variable. See Example 4.

33. $V = \pi r^2 h$ for r $r = \pm\sqrt{\dfrac{V}{\pi h}}$

34. $V = \dfrac{4}{3}\pi r^2 h$ for r $r = \pm\sqrt{\dfrac{3V}{4\pi h}}$

35. $a^2 + b^2 = c^2$ for b $b = \pm\sqrt{c^2 - a^2}$

36. $y = ax^2 + c$ for x $x = \pm\sqrt{\dfrac{y - c}{a}}$

37. $b^2 - 4ac = 0$ for b $b = \pm2\sqrt{ac}$

38. $s = \dfrac{1}{2}gt^2 + v$ for t $t = \pm\sqrt{\dfrac{2s - 2v}{g}}$

39. $v = \sqrt{2pt}$ for t $t = \dfrac{v^2}{2p}$

40. $y = \sqrt{2x}$ for x $x = \dfrac{y^2}{2}$

Find the exact answer to each problem. If the answer is irrational, then find a decimal approximation to the answer using three decimal places. See Example 5.

41. Find the length of the side of a square whose area is 18 square feet. $3\sqrt{2}$ feet or 4.243 feet

42. Find the length of the side of a square wheat field whose area is 75 square miles. $5\sqrt{3}$ miles or 8.660 miles

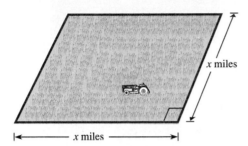

x miles

x miles

Figure for Exercise 42

43. Find the length of the side of a square coffee table whose diagonal is 6 feet. $3\sqrt{2}$ foot or 4.243 feet

44. Find the length of the side of a square whose diagonal measures 1 yard. $\dfrac{\sqrt{2}}{2}$ yards or .707 yard

45. Find the length of the diagonal of a square floor tile whose sides measure 1 foot each. $\sqrt{2}$ feet or 1.414 feet

46. The sandbox at Totland is shaped like a square with an area of 20 square meters. Find the length of the diagonal.

47. Find the length of the diagonal of a rectangular bathtub with sides of 3 feet and 4 feet.   5 feet

46. $2\sqrt{10}$ meters or 6.325 meters

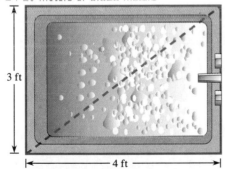

3 ft

4 ft

Figure for Exercise 47

48. What is the length of the diagonal of a rectangular office whose sides are 6 feet and 8 feet?   10 feet

49. If we neglect air resistance, the number of feet that a body falls from rest during t seconds is given by $s = 16t^2$. How long would it take a pine cone to fall from the top of a 100-foot pine tree?   2.5 seconds

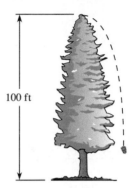

100 ft

Figure for Exercise 49

50. If a baseball diamond is actually a square, 90 feet on each side, then how far is it from home plate to second base?

$90\sqrt{2}$ feet or 127.279 feet

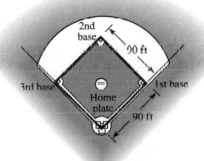

2nd base

90 ft

3rd base

1st base

Home plate

90 ft

Figure for Exercise 50

51. A guy wire is to be attached from the top of a 200-foot tower to a point on the ground whose distance from the base of the tower is 2/3 of the height of the tower. Find the length of the guy wire.

$\dfrac{200\sqrt{13}}{3}$ feet or 240.37 feet

52. The size of a rectangular television screen is commonly given by the manufacturer as the length of the diagonal of the rectangle. If a television screen measures 10 inches wide and 8 inches high, then what is the length of the diagonal of the rectangle? What is the approximate size of this television screen to the nearest inch? $2\sqrt{41}$ inches or 13 inches

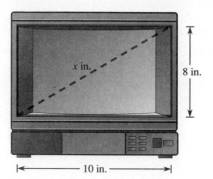

Figure for Exercise 52

Solve each equation.

53. $3x^2 - 6 = 0$ $-\sqrt{2}, \sqrt{2}$

54. $5x^2 + 3 = 0$ No solution

55. $\sqrt{2x - 3} = \sqrt{3x + 1}$ No solution

56. $\sqrt{2x - 4} = \sqrt{x - 9}$ No solution

57. $(2x - 1)^2 = 8$ $\dfrac{1 + 2\sqrt{2}}{2}, \dfrac{1 - 2\sqrt{2}}{2}$

58. $(3x - 2)^2 = 18$ $\dfrac{2 - 3\sqrt{2}}{3}, \dfrac{2 + 3\sqrt{2}}{3}$

59. $\sqrt{2x - 9} = 0$ $\dfrac{9}{2}$

60. $\sqrt{5 - 3x} = 0$ $\dfrac{5}{3}$

61. $x + 1 = \sqrt{2x + 10}$ 3

62. $x - 3 = \sqrt{2x + 18}$ 9

63. $3(x + 1)^2 - 27 = 0$ $-4, 2$

64. $2(x - 3)^2 - 50 = 0$ $-2, 8$

65. $(2x - 5)^2 = 0$ $\dfrac{5}{2}$

66. $(3x - 1)^2 = 0$ $\dfrac{1}{3}$

Use a calculator to find approximate solutions to the following equations. Round your answers to three decimal places.

67. $x^2 = 3.25$ $-1.803, 1.803$

68. $(x + 1)^2 = 20.3$ $-5.506, 3.506$

69. $\sqrt{x + 2} = 1.73$ $.993$

70. $\sqrt{2.3x - 1.4} = 3.3$ 5.343

71. $1.3(x - 2.4)^2 = 5.4$ $.362, 4.438$

72. $-2.4x^2 = -9.55$ $-1.995, 1.995$

6.6 Negative and Rational Exponents

IN THIS SECTION:

- **Negative Exponents**
- **The Rules**
- **Rational Exponents**

In this section we will learn that exponents can be used to indicate multiplicative inverses and roots.

Negative Exponents

The product of any nonzero number x and its reciprocal $1/x$ is 1. Consider the product of 2^5 and its reciprocal:

$$2^5 \cdot \frac{1}{2^5} = 1$$

If we use 2^{-5} to represent the reciprocal of 2^5, then we can use the product rule for exponents to evaluate the above product:

$$2^5 \cdot 2^{-5} = 2^{5+(-5)} = 2^0 = 1$$

In general we have the following definition.

DEFINITION
Negative Integral Exponents

If a is a nonzero real number and n is a positive integer, then

$$a^{-n} = \frac{1}{a^n}.$$

Note that 2^5 and 2^{-5} are reciprocals of each other. The fact that 2^{-5} is the reciprocal of 2^5 is expressed as

$$2^{-5} = \frac{1}{2^5}.$$

The fact that 2^5 is the reciprocal of 2^{-5} is expressed as

$$2^5 = \frac{1}{2^{-5}}.$$

It is convenient to think of this fact as a rule for changing the sign of an exponent.

Sign Change Rule

If a is a nonzero real number and n is *any integer*, then

$$a^n = \frac{1}{a^{-n}}.$$

The sign change rule allows us to move an exponential expression from numerator to denominator or from denominator to numerator by merely changing the sign of its exponent. For example,

$$a^6 = \frac{1}{a^{-6}}, \qquad 3^{-5} = \frac{1}{3^5}, \qquad \text{and} \qquad \frac{1}{2} = 2^{-1}.$$

EXAMPLE 1 Evaluate the following expressions:

a) 2^{-3}

b) $(-2)^3$

c) $(-2)^{-3}$

d) $\dfrac{2^{-3}}{3^{-2}}$

Solution

a) $2^{-3} = \dfrac{1}{2^3} = \dfrac{1}{8}$ Note that 2^{-3} is a positive number.

b) $(-2)^3 = (-2)(-2)(-2) = -8$

c) $(-2)^{-3} = \dfrac{1}{(-2)^3} = \dfrac{1}{-8} = -\dfrac{1}{8}$

d) $\dfrac{2^{-3}}{3^{-2}} = \dfrac{1}{2^3} \cdot 3^2$ Sign change rule

   $= \dfrac{1}{8} \cdot 9$

   $= \dfrac{9}{8}$ ◀

The Rules

All of the rules that we learned in Section 6.1 for positive integral exponents are true if we allow the exponents to be any integers. We restate the product rule as follows.

Product Rule

If m and n are *any integers* and $a \neq 0$, then

$$a^m \cdot a^n = a^{m+n}$$

The quotient rule is simpler when we allow any integers as exponents.

Quotient Rule

If m and n are *any integers* and $a \neq 0$, then

$$\frac{a^m}{a^n} = a^{m-n}.$$

EXAMPLE 2 Use the rules of exponents to simplify the following. Write your answers with positive exponents only. Assume that the variables represent nonzero real numbers.

a) $b^{-3}b^5$

b) $\dfrac{m^{-6}}{m^{-2}}$

c) $x^{-3}x^2$

d) $\dfrac{y^5}{y^{-3}}$

Solution

a) $b^{-3}b^5 = b^{-3+5}$ Product rule

       $= b^2$

b) $\dfrac{m^{-6}}{m^{-2}} = m^{-6-(-2)}$ Quotient rule

       $= m^{-4}$ $-6 - (-2) = -4$

       $= \dfrac{1}{m^4}$ Apply the sign change rule to write the answer with a positive exponent.

c) $x^{-3}x^2 = x^{-1}$ Product rule

       $= \dfrac{1}{x}$ Sign change rule

d) $\dfrac{y^5}{y^{-3}} = y^8$ $5 - (-3) = 8$ ◀

We now restate the power rule, the power of a product rule, and the power of a quotient rule to include any integers as exponents.

The Power Rules

> If m and n are any integers, $a \neq 0$, and $b \neq 0$, then
>
> $$(a^m)^n = a^{mn}. \qquad \text{Power rule}$$
> $$(ab)^n = a^n b^n. \qquad \text{Power of a product}$$
> $$\left(\frac{a}{b}\right)^n = \frac{a^n}{b^n}. \qquad \text{Power of a quotient}$$

The use of these rules is illustrated in the following example.

EXAMPLE 3 Use the rules of exponents to simplify the following. Write your answers with positive exponents only. Assume that all variables represent nonzero real numbers.

a) $(a^{-3})^2$ **b)** $(2x^{-3})^{-2}$ **c)** $\left(\dfrac{2x^{-3}}{y^2}\right)^{-2}$

Solution

a) $(a^{-3})^2 = a^{-6}$ Power rule: $(-3)(2) = -6$

       $= \dfrac{1}{a^6}$ Sign change rule

b) $(2x^{-3})^{-2} = 2^{-2}(x^{-3})^{-2}$ Power of a product rule

       $= 2^{-2}x^6$ Power rule: $(-3)(-2) = 6$

       $= \dfrac{x^6}{2^2}$ Sign change rule

       $= \dfrac{x^6}{4}$

c) $\left(\dfrac{2x^{-3}}{y^2}\right)^{-2} = \dfrac{2^{-2}x^6}{y^{-4}}$ **Power of a quotient rule**

$\qquad\qquad = \dfrac{x^6 y^4}{2^2}$ **Sign change rule**

$\qquad\qquad = \dfrac{x^6 y^4}{4}$ ◀

Rational Exponents

Since an exponent of 3 indicates the cube of a number, it seems reasonable to use an exponent of 1/3 to indicate the cube root of a number. For example, we write

$$2^3 = 8$$

and

$$8^{1/3} = 2.$$

DEFINITION
$a^{1/n}$

> If n is any positive integer, then
> $$a^{1/n} = \sqrt[n]{a},$$
> provided that $\sqrt[n]{a}$ is a real number.

The expression $4^{1/2}$ indicates the positive square root of 4, so

$$4^{1/2} = \sqrt{4} = 2.$$

The expression $(-8)^{1/3}$ indicates the cube root of -8, so

$$(-8)^{1/3} = \sqrt[3]{-8} = -2.$$

The expression $(-9)^{1/2}$ is not a real number, since an even root of any negative number is not a real number.

We can extend the definition of exponent $1/n$ to include any rational number as an exponent. The numerator of the rational number indicates power, and the denominator indicates root. For example, the expression

$$8^{2/3}$$

is used to represent the square of the cube root of 8. So we write

┌ **The square**
The cube root ┐ │
           ↓ ↓

$$8^{2/3} = (8^{1/3})^2 = (2)^2 = 4.$$

DEFINITION
Rational Exponents

If m and n are positive integers, then $a^{m/n} = (a^{1/n})^m$ provided that $a^{1/n}$ is a real number.

We define negative rational exponents just like negative integral exponents.

DEFINITION
Negative Rational Exponents

If m and n are positive integers and $a \neq 0$, then

$$a^{-m/n} = \frac{1}{a^{m/n}},$$

provided that $a^{1/n}$ is a real number.

The operations indicated by the exponent can be performed in any order, provided that we are working with a defined expression. For example, $(-3)^{4/2}$ is undefined, but if we do the power and then the root, we get a real number.

Remember that in a negative rational exponent the denominator indicates root, the numerator indicates power, and the negative sign indicates reciprocal. Consider the expression $8^{-2/3}$:

$$8^{-2/3} \quad \begin{array}{l} \text{Cube root} \\ \text{Second power} \\ \text{Reciprocal} \end{array}$$

To evaluate $8^{-2/3}$, we find that the cube root of 8 is 2; then we square 2 to get 4 and take the reciprocal of 4 to get 1/4. Thus

$$8^{-2/3} = \frac{1}{4}.$$

We get the same value regardless of the order in which we perform the three operations.

EXAMPLE 4 Evaluate the following expressions:

a) $27^{1/3}$ b) $4^{3/2}$ c) $27^{-2/3}$

Note that the exercises corresponding to this example do not contain any undefined expressions.

Solution
a) $27^{1/3} = 3$ The cube root of 27 is 3.
b) $4^{3/2} = 8$ The square root of 4 is 2, and 2 cubed is 8.

c) $27^{-2/3} = \dfrac{1}{27^{2/3}}$ Negative exponent indicates reciprocal.

$\qquad\quad = \dfrac{1}{9}$ The cube root of 27 is 3, and 3 squared is 9.   ◀

Fortunately, all of the rules for exponents that we learned for integral exponents also hold for rational exponents. We will illustrate the use of those rules in the next example.

EXAMPLE 5   Use the rules of exponents to simplify each expression. Write answers with positive exponents. Assume that all variables represent positive real numbers.

a) $2^{1/2} \cdot 2^{3/2}$ **b)** $\dfrac{x}{x^{2/3}}$

c) $(b^{1/2})^{1/3}$ **d)** $(x^4 y^{-6})^{1/2}$

Solution

a) $2^{1/2} \cdot 2^{3/2} = 2^2$ $\dfrac{1}{2} + \dfrac{3}{2} = \dfrac{4}{2} = 2$

$\qquad\qquad\quad = 4$

b) $\dfrac{x}{x^{2/3}} = x^{1/3}$ $1 - \dfrac{2}{3} = \dfrac{1}{3}$

c) $(b^{1/2})^{1/3} = b^{1/6}$ $\dfrac{1}{2} \cdot \dfrac{1}{3} = \dfrac{1}{6}$

d) $(x^4 y^{-6})^{1/2} = x^2 y^{-3}$ $\dfrac{1}{2} \cdot 4 = 2, \; \dfrac{1}{2} \cdot (-6) = -3$

$\qquad\qquad\qquad = \dfrac{x^2}{y^3}$ ◀

Warm-ups

True or false?

1. $3^{-2} = \dfrac{1}{9}$ T   **2.** $\dfrac{3^{-2}}{3^{-1}} = \dfrac{1}{3}$ T   **3.** $10^{-1} = .1$ T

4. $10^{-2} = .01$ T   **5.** $9^{1/2} = 3$ T   **6.** $2^{1/2} \cdot 2^{1/2} = 4^{1/2}$ T

7. $\dfrac{2}{2^{1/2}} = 2^{1/2}$ T   **8.** $6^{-1/2} = \dfrac{1}{3}$ F   **9.** $16^{-1/4} = -2$ F

10. $9^{-3/2} = \dfrac{1}{27}$ T

6.6 EXERCISES

Variables in all exercises represent positive real numbers. Evaluate each expression. See Example 1.

1. 3^{-1} $\dfrac{1}{3}$

2. 4^{-2} $\dfrac{1}{16}$

3. 2^{-4} $\dfrac{1}{16}$

4. 3^{-3} $\dfrac{1}{27}$

5. -2^{-2} $-\dfrac{1}{4}$

6. -3^{-2} $-\dfrac{1}{9}$

7. $(-2)^{-2}$ $\dfrac{1}{4}$

8. $(-3)^{-2}$ $\dfrac{1}{9}$

9. $\dfrac{4^{-3}}{8^{-2}}$ 1

10. $\dfrac{5^{-2}}{10^{-1}}$ $\dfrac{2}{5}$

Simplify each expression. Write your answers with positive exponents only. See Example 2.

11. $x^{-1}x^2$ x

12. $y^{-3}y^5$ y^2

13. x^2x^{-6} $\dfrac{1}{x^4}$

14. y^5y^{-7} $\dfrac{1}{y^2}$

15. $a^{-2}a^{-3}$ $\dfrac{1}{a^5}$

16. $b^{-3}b^{-5}$ $\dfrac{1}{b^8}$

17. $(t^{-2})^{-5}$ t^{10}

18. $w^{-4}w^{-6}$ $\dfrac{1}{w^{10}}$

19. $\dfrac{t^{-3}}{t^5}$ $\dfrac{1}{t^8}$

20. $\dfrac{w^{-4}}{w^3}$ $\dfrac{1}{w^7}$

21. $\dfrac{x^{-5}}{x^{-6}}$ x

22. $\dfrac{y^{-6}}{y^{-9}}$ y^3

Simplify each expression. Write your answers with positive exponents only. See Example 3.

23. $(x^2)^{-5}$ $\dfrac{1}{x^{10}}$

24. $(y^{-2})^4$ $\dfrac{1}{y^8}$

25. $(a^{-3})^{-3}$ a^9

26. $(b^{-5})^{-2}$ b^{10}

27. $(2x^{-3})^{-4}$ $\dfrac{x^{12}}{16}$

28. $(3y^{-1})^{-2}$ $\dfrac{y^2}{9}$

29. $(x^2y^{-3})^{-2}$ $\dfrac{y^6}{x^4}$

30. $(s^{-2}t^4)^{-1}$ $\dfrac{s^2}{t^4}$

31. $\left(\dfrac{x^{-1}}{y^{-3}}\right)^{-2}$ $\dfrac{x^2}{y^6}$

32. $\left(\dfrac{a^{-2}}{b^3}\right)^{-3}$ a^6b^9

33. $\left(\dfrac{2a^{-3}}{a^{-2}}\right)^{-4}$ $\dfrac{a^4}{16}$

34. $\left(\dfrac{3w^2}{w^4}\right)^{-2}$ $\dfrac{w^4}{9}$

Evaluate each expression. See Example 4.

35. $25^{1/2}$ 5

36. $16^{1/2}$ 4

37. $125^{1/3}$ 5

38. $81^{1/4}$ 3

39. $125^{2/3}$ 25

40. $1000^{2/3}$ 100

41. $25^{3/2}$ 125

42. $16^{3/2}$ 64

43. $27^{-4/3}$ $\dfrac{1}{81}$

44. $16^{-3/4}$ $\dfrac{1}{8}$

45. $4^{-3/2}$ $\dfrac{1}{8}$

46. $25^{-3/2}$ $\dfrac{1}{125}$

Use the rules of exponents to simplify each expression. Write your answers with positive exponents only. See Example 5.

47. $x^{1/4}x^{1/4}$ $x^{1/2}$

48. $y^{1/3}y^{2/3}$ y

49. $(x^6)^{1/3}$ x^2

50. $(y^{-4})^{1/2}$ $\dfrac{1}{y^2}$

51. $\dfrac{x^2}{x^{1/2}}$ $x^{3/2}$

52. $\dfrac{a^{1/2}}{a^{1/3}}$ $a^{1/6}$

53. $\dfrac{t^{1/2}}{t^{1/4}}$ $t^{1/4}$

54. $\dfrac{w^{1/4}}{w^{1/3}}$ $\dfrac{1}{w^{1/12}}$

55. $(x^2y^6)^{1/2}$ xy^3

56. $(t^3w^6)^{1/3}$ tw^2

57. $(x^{-2}y^8)^{-1/2}$ $\dfrac{x}{y^4}$

58. $(4w^{-2}t^{-4})^{-1/2}$ $\dfrac{wt^2}{2}$

59. $\left(\dfrac{x^{-4}}{y^{-6}}\right)^{1/2}$ $\dfrac{y^3}{x^2}$

60. $\left(\dfrac{8}{y^3}\right)^{-1/3}$ $\dfrac{y}{2}$

61. $\left(\dfrac{4}{x^6}\right)^{-1/2}$ $\dfrac{x^3}{2}$

62. $\left(\dfrac{t^{-3}}{w^{-9}}\right)^{1/3}$ $\dfrac{w^3}{t}$

Evaluate each expression.

63. $2^{-1} \cdot 3^{-1}$ $\dfrac{1}{6}$

64. $16^{-1/2}$ $\dfrac{1}{4}$

65. $2^{-1} + 3^{-1}$ $\dfrac{5}{6}$

66. 10^{-2} $\dfrac{1}{100}$

67. $(2 + 3)^{-1}$ $\dfrac{1}{5}$

68. $(-1)^{-1}$ -1

69. -1^{-1} -1

70. $5 \cdot 5^{-1}$ 1

71. $2 \cdot 2 \cdot 2 \cdot 2^{-1}$ 4

72. $(2 \cdot 3^{-1})^{-1}$ $\dfrac{3}{2}$

73. $4^{-1/2}$ $\dfrac{1}{2}$

74. $27^{-1/3}$ $\dfrac{1}{3}$

75. $25^{-3/2}$ $\dfrac{1}{125}$

76. $81^{3/4}$ 27

77. $9^{-1/2}$ $\dfrac{1}{3}$

78. $100^{-1/2}$ $\dfrac{1}{10}$

6.7 Scientific Notation

IN THIS SECTION:

- Basic Ideas
- Converting from Scientific Notation
- Converting into Scientific Notation
- Computations with Scientific Notation

Many of the numbers occurring in science are either very large or very small. The speed of light is 983,569,000 feet per second. One millimeter is equal to .000001 kilometer. In scientific notation, positive and negative exponents are used to get a convenient way of writing very large and very small numbers.

Basic Ideas

To understand scientific notation, we need to first review the integral powers of 10:

$$10^1 = 10 \qquad 10^{-1} = \frac{1}{10} \qquad\qquad = .1$$

$$10^2 = 100 \qquad 10^{-2} = \frac{1}{10^2} = \frac{1}{100} \quad = .01$$

$$10^3 = 1000 \qquad 10^{-3} = \frac{1}{10^3} = \frac{1}{1000} \quad = .001$$

$$10^4 = 10,000 \qquad 10^{-4} = \frac{1}{10^4} = \frac{1}{10,000} = .0001$$

Consider the results of multiplying by a positive power of 10. Multiplying a number by 10, 100, or 1000 just moves the decimal point either one, two, or three places to the right. For example,

$$
\begin{array}{ccc}
5.3 & 5.3 & 5.3 \\
\underline{\times\ 10} & \underline{\times\ 100} & \underline{\times\ 1000} \\
53 & 530 & 5300 \\
\end{array}
$$

      One place   Two places   Three places

Multiplying a number by a negative power of 10 moves the decimal point to the left. For example,

$$
\begin{array}{ccc}
5.3 & 5.3 & 5.3 \\
\underline{\times\ .1} & \underline{\times\ .01} & \underline{\times\ .001} \\
.53 & .053 & .0053 \\
\end{array}
$$

      One place   Two places   Three places

Notice that $1000 = 10^3$, and the decimal moved three places to the right when we multiplied by 10^3. When we multiplied by 10^{-3} or .001, we moved the decimal point three places to the left.

The numbers 6×10^3 and 2.5×10^{-4} are examples of numbers written in scientific notation. In scientific notation the times symbol, $\times$, is used to indicate multiplication. A number in scientific notation is written as a product of a number between 1 and 10 and a power of 10. There is one digit to the left of the decimal point.

Converting from Scientific Notation

In scientific notation a number larger than 10 is written with a positive power of 10, and any number smaller than 1 is written with a negative power of 10.

To convert a number in scientific notation to one in standard notation, we merely move the decimal point, since we are multiplying by a power of 10. For example,

$$3.27 \times 10^9 = 3,270,000,000.$$

Positive 9 indicates a large number, so move to the right.

9 places to right

Of course, it is not necessary to put the decimal point in when writing a whole number.

If the number in scientific notation has a negative power of 10, the decimal point is moved to the left:

$$7.213 \times 10^{-5} = .00007213$$

Negative 5 indicates a small number, so move to the left.

5 places to left

In general, we use the following strategy to convert from scientific notation to standard notation.

STRATEGY
Converting from Scientific Notation

1. Determine the number of places to move the decimal point by examining the exponent on the 10.
2. Move to the right for a positive exponent and to the left for a negative exponent.

EXAMPLE 1 Write these numbers in standard notation:

a) 7.02×10^6 b) 8.13×10^{-5}
c) -2.347×10^5 d) -1.6×10^{-4}

Solution

a) Since the exponent is positive, this number is larger than 10. Move the decimal point six places to the right:

$$7.02 \times 10^6 = 7020000. = 7,020,000$$

b) Since the exponent is negative, this number is smaller than 1. Move the decimal point five places to the left:

$$8.13 \times 10^{-5} = .0000813$$

c) Convert any negative number just as if it was a positive number and then put a negative sign on the result:

$$-2.347 \times 10^5 = -234,700$$

d) $-1.6 \times 10^{-4} = -.00016$ ◄

Math at Work

In science fiction movies it happens all the time: The hero and heroine jump into a spacecraft and zoom past the stars like a car passing fenceposts along a highway. Humans have orbited the earth and landed on the moon. It is conceivable that astronauts will soon roam the planets of our own solar system. Why isn't the high-speed star travel depicted in science fiction stories possible yet?

The reason we cannot expect star travel in the near future has to do with the amount of time, energy, and speed it would take to reach even the nearest stars. Astronomers measure distance in space in light years, or the distance light travels in a year. No object in space can travel at or exceed the speed of light, which is 300,000 kilometers per second. Astronomers have found that the nearest star to our sun, Proxima Centauri, is located 4.2 light years away. At the speed of a Boeing 747 aircraft, a trip to Proxima Centauri would take about 4 million years.

For us to be able to travel at speeds approaching the speed of light would entail vast amounts of energy. It has been estimated that to take a spacecraft to Proxima Centauri in ten years would require 40,000 times the energy that the United States uses in one year. The energy for that trip would have to be stored in a vast amount of fuel. Until we have the ability to generate that much energy, star travel will be impossible.

Galaxies as far as 10 billion light years away are visible to astronomers. Using the fact that one light year equals approximately 10^{13} kilometers, find the distance in kilometers to these galaxies. At 80,000 kilometers per hour, how many years would it take to get to Proxima Centauri? 10^{22} km, 1.4×10^{13} km

Converting to Scientific Notation

To convert a positive number to scientific notation, we just reverse the strategy for converting from scientific notation.

STRATEGY
Converting to Scientific Notation

1. Count the number of places (n) that the decimal point must be moved so that it will follow the first nonzero digit of the number.
2. If the original number was larger than 10 use 10^n.
3. If the original number was smaller than 1 use 10^{-n}.

To convert a negative number to scientific notation, ignore the negative sign, convert the number, and then attach the negative sign to the result.

EXAMPLE 2 Convert the following numbers into scientific notation.

a) 7,346,200   b) .0000348
c) −14,000   d) −.00000002

Solution
a) $7,346,200 = 7.3462 \times 10^6$ Positive 6, since 7,346,200 is larger than 10
b) $.0000348 = 3.48 \times 10^{-5}$ Negative 5, since .0000348 is smaller than 1
c) $-14,000 = -1.4 \times 10^4$
d) $-.00000002 = -2 \times 10^{-8}$ ◄

Computations with Scientific Notation

An important use of scientific notation is in computations. Numbers in scientific notation are nothing more than exponential expressions, and we have already studied operations with exponential expressions in this chapter. We use the same rules of exponents on numbers in scientific notation that we use on any other exponential expressions.

EXAMPLE 3 Perform the indicated computations. Write the answers in scientific notation.

a) $(3 \times 10^6)(2 \times 10^8)$ b) $\dfrac{4 \times 10^5}{8 \times 10^{-2}}$ c) $(5 \times 10^{-7})^3$

Solution
a) $(3 \times 10^6)(2 \times 10^8) = 3 \cdot 2 \cdot 10^6 \cdot 10^8 = 6 \times 10^{14}$

b) $\dfrac{4 \times 10^5}{8 \times 10^{-2}} = \dfrac{4}{8} \cdot \dfrac{10^5}{10^{-2}} = \dfrac{1}{2} \cdot 10^7$ Quotient rule: $5 - (-2) = 7$

$= (.5)10^7$ $1/2 = .5$
$= 5 \times 10^{-1} \cdot 10^7$ $.5 = 5 \times 10^{-1}$
$= 5 \times 10^6$ Product rule

c) $(5 \times 10^{-7})^3 = 5^3 \cdot (10^{-7})^3$ **Power of a product rule**
$= 125 \cdot 10^{-21}$ **Power rule**
$= 1.25 \times 10^2 \cdot 10^{-21}$ $125 = 1.25 \times 10^2$
$= 1.25 \times 10^{-19}$ **Product rule: 2 + (−21) = −19** ◀

EXAMPLE 4 Perform these computations by first converting each number into scientific notation. Give your answer in scientific notation.

a) $(3,000,000)(.0002)$ **b)** $(20,000,000)^3(.0000003)$

Solution

a) $(3,000,000)(.0002) = 3 \times 10^6 \cdot 2 \times 10^{-4}$ **Convert to scientific notation.**
$= 6 \times 10^2$ **Apply the product rule: 6 + (−4) = 2.**
b) $(20,000,000)^3(.0000003) = (2 \times 10^7)^3 \cdot 3 \times 10^{-7}$ **Scientific notation**
$= 8 \times 10^{21} \cdot 3 \times 10^{-7}$ **Power of a product rule**
$= 24 \cdot 10^{14}$ **Product rule**
$= 2.4 \times 10^1 \cdot 10^{14}$ $24 = 2.4 \times 10^1$
$= 2.4 \times 10^{15}$ **Product rule** ◀

Scientific calculators are very useful for performing computations involving scientific notation. If a number is too large to be displayed in standard notation, then a scientific calculator automatically displays the number in scientific notation. Most calculators are limited to powers of 10 from −99 to 99. Use a calculator with scientific notation to do Exercises 45–60. Refer to your calculator manual to see how to enter numbers in scientific notation.

Warm-ups

True or false?

1. $(5.34)(1000) = 5340$ T
2. $(432)(.01) = 4.32$ T
3. $10^{-2} = .001$ F
4. $23.7 = 2.37 \times 10^{-1}$ F
5. $.000036 = 3.6 \times 10^{-5}$ T
6. $25 \cdot 10^7 = 2.5 \times 10^8$ T
7. $.442 \times 10^{-3} = 4.42 \times 10^{-4}$ T
8. $(3 \times 10^{-9})^2 = 9 \times 10^{-18}$ T
9. $(2 \times 10^{-5})(4 \times 10^4) = 8 \times 10^{-20}$ F
10. $\dfrac{8 \times 10^{-9}}{2 \times 10^{-5}} = 4 \times 10^{-4}$ T

6.7 EXERCISES

Write each number in standard notation. See Example 1.

1. 9.86×10^9 9,860,000,000
2. 1.37×10^{-3} .00137
3. -1.55×10^{-5} −.0000155
4. -3.77×10^{-6} −.00000377
5. 4.007×10^4 40,070
6. 9×10^{-5} .00009

7. 1×10^5 100,000

8. 1×10^{-6} .000001

9. 1×10^{-1} .1

10. 1×10^6 1,000,000

11. -5.55×10^2 -555

12. -8.9×10^{-3} $-.0089$

Write each number in scientific notation. See Example 2.

13. 9000 9×10^3

14. 5,298,000 5.298×10^6

15. .00078 7.8×10^{-4}

16. .000214 2.14×10^{-4}

17. $-3,450,000$ -3.45×10^6

18. $-.007$ -7×10^{-3}

19. .0000085 8.5×10^{-6}

20. 5,670,000,000 5.67×10^9

21. $-.0000255$ -2.55×10^{-5}

22. -4500 -4.5×10^3

23. 50 5×10^1

24. 80 8×10^1

Perform the computations. Write each number in scientific notation. See Example 3.

25. $(3 \times 10^5)(2 \times 10^{-15})$ 6×10^{-10}

26. $(2 \times 10^{-9})(4 \times 10^{23})$ 8×10^{14}

27. $\dfrac{4 \times 10^{-8}}{2 \times 10^{30}}$ 2×10^{-38}

28. $\dfrac{9 \times 10^{-4}}{3 \times 10^{-6}}$ 3×10^2

29. $\dfrac{3 \times 10^{20}}{6 \times 10^{-8}}$ 5×10^{27}

30. $\dfrac{1 \times 10^{-8}}{4 \times 10^7}$ 2.5×10^{-16}

31. $(3 \times 10^{12})^2$ 9×10^{24}

32. $(2 \times 10^{-5})^3$ 8×10^{-15}

33. $(5 \times 10^4)^3$ 1.25×10^{14}

34. $(5 \times 10^{14})^{-1}$ 2×10^{-15}

35. $(4 \times 10^{32})^{-1}$ 2.5×10^{-33}

36. $(6 \times 10^{11})^2$ 3.6×10^{23}

Perform the following computations by first converting each number into scientific notation. Write your answer in scientific notation. See Example 4.

37. $(4300)(2,000,000)$ 8.6×10^9

38. $(40,000)(4,000,000,000)$ 1.6×10^{14}

39. $(4,200,000)(.00005)$ 2.1×10^2

40. $(.00075)(4,000,000)$ 3×10^3

41. $(300)^3(.000001)^5$ 2.7×10^{-23}

42. $(200)^4(.0005)^3$ 2×10^{-1}

43. $\dfrac{(4000)(90,000)}{(.00000012)}$ 3×10^{15}

44. $\dfrac{(30,000)(80,000)}{(.000006)(.002)}$ 2×10^{17}

Perform the following computations with the aid of a scientific calculator. Write answers in scientific notation. Round the decimal part to three decimal places.

▦ **45.** $(6.3 \times 10^6)(1.45 \times 10^{-4})$ 9.135×10^2

▦ **46.** $(8.35 \times 10^9)(4.5 \times 10^3)$ 3.758×10^{13}

▦ **47.** $(5.36 \times 10^{-4}) + (3.55 \times 10^{-5})$ 5.715×10^{-4}

▦ **48.** $(8.79 \times 10^8) + (6.48 \times 10^9)$ 7.359×10^9

▦ **49.** $(3.56 \times 10^{85})(4.43 \times 10^{96})$ 1.577×10^{182}

▦ **50.** $(4.36 \times 10^{55})(7.7 \times 10^{88})$ 3.357×10^{144}

▦ **51.** $(8 \times 10^{99}) + (3 \times 10^{99})$ 1.1×10^{100}

▦ **52.** $(8 \times 10^{-99}) + (9 \times 10^{-99})$ 1.7×10^{-98}

▦ **53.** $\dfrac{(3.5 \times 10^5)(4.3 \times 10^{-6})}{3.4 \times 10^{-8}}$ 4.426×10^7

▦ **54.** $\dfrac{(3.5 \times 10^{-8})(4.4 \times 10^{-4})}{2.43 \times 10^{45}}$ 6.337×10^{-57}

▦ **55.** The distance from the earth to the sun is 93 million miles. Express this distance in feet. (1 mile = 5280 feet.) 4.910×10^{11} feet

▦ **56.** The speed of light is 9.83569×10^8 feet per second. How long does it take light to get from the sun to the earth?

▦ **57.** How long does it take a spacecraft traveling at 2×10^{35} miles per hour (warp factor 4) to travel 93 million miles. 4.65×10^{-28} hours

▦ **58.** If the radius of a very small circle is 2.35×10^{-8} centimeters, then what is the circle's area?

▦ **59.** If the circumference of a circle is 5.68×10^9 feet, then what is its radius? 9.040×10^8 feet

▦ **60.** If the diameter of a circle is 1.3×10^{-12} meters, then what is its radius? 6.5×10^{-13} meters

56. 499.2 seconds or 8.3 minutes   **58.** 1.735×10^{-15} square centimeters

Wrap-up

CHAPTER 6

SUMMARY

Concepts	Examples

Definition of positive integral exponents

For any positive integer n,

$$a^n = \underbrace{a \cdot a \cdot a \cdots \cdot a}_{n \text{ factors of } a}$$

The base is a, and the exponent is n.

$2^3 = 8$

$3^3 = 27$

Definition of negative integral exponents

If a is a nonzero real number and n is a positive integer, then

$$a^{-n} = \frac{1}{a^n}.$$

$2^{-3} = \dfrac{1}{2^3}$

Definition of nth roots

Let n be a natural number. We say that a is an nth root of b, if

$$a^n = b.$$

$2^4 = 16$, $(-2)^4 = 16$: 2 and -2 are fourth roots of 16.

Radical notation

If n is a positive *even* integer and a is positive, then the symbol $\sqrt[n]{a}$ denotes the positive nth root of a, and is called the principal nth root of a.

$\sqrt[4]{16} = 2$

If n is a positive *odd* integer, then the symbol $\sqrt[n]{a}$ denotes the nth root of a.

$\sqrt[3]{-8} = -2$

If n is any positive integer, then $\sqrt[n]{0} = 0$.

$\sqrt[3]{8} = 2$

Definition of $a^{1/n}$

If n is any positive integer, then $a^{1/n} = \sqrt[n]{a}$, provided that $\sqrt[n]{a}$ is a real number.

$8^{1/3} = \sqrt[3]{8}$

$(-4)^{1/2}$ is not real.

Definition of rational exponents

If m and n are positive integers, then $a^{m/n} = (a^{1/n})^m$, provided that $a^{1/n}$ is a real number.

$8^{2/3} = (8^{1/3})^2 = 2^2 = 4$

$(-16)^{3/4}$ is not real.

| Definition of negative rational exponents | If m and n are positive integers and $a \neq 0$, then

$a^{-m/n} = \dfrac{1}{a^{m/n}}$, provided that $a^{1/n}$ is a real number. | $8^{-2/3} = \dfrac{1}{8^{2/3}}$ |

Rules of Exponents

Suppose a and b are nonzero real numbers, and m and n are rational numbers. The following rules hold, provided that all expressions represent real numbers.

Examples

Sign change	$a^n = \dfrac{1}{a^{-n}}$	$3^2 = \dfrac{1}{3^{-2}}$, $4^{-3} = \dfrac{1}{4^3}$
Zero exponent	$a^0 = 1$	$9^0 = 1$, $-9^0 = -1$, $(-9)^0 = 1$
Product rule	$a^m a^n = a^{m+n}$	$3^2 \cdot 3^4 = 3^6$, $x^5 x^{-2} = x^3$
Quotient rule	$\dfrac{a^m}{a^n} = a^{m-n}$	$\dfrac{3^5}{3^7} = 3^{-2}$, $\dfrac{x^{10}}{x^7} = x^3$
Power rule	$(a^m)^n = a^{mn}$	$(2^2)^3 = 2^6$ $(w^{3/4})^4 = w^3$
Power of a product	$(ab)^n = a^n b^n$	$(2t)^3 = 8t^3$
Power of a quotient	$\left(\dfrac{a}{b}\right)^n = \dfrac{a^n}{b^n}$	$\left(\dfrac{x}{3}\right)^3 = \dfrac{x^3}{27}$

Rules for Radicals

The following rules hold, provided that all roots are real numbers and n is a positive integer.

Examples

Product Rule for radicals	$\sqrt[n]{ab} = \sqrt[n]{a} \cdot \sqrt[n]{b}$	$\sqrt{2} \cdot \sqrt{3} = \sqrt{6}$ $\sqrt{9y} = 3\sqrt{y}$
Quotient rule for radicals	$\sqrt[n]{\dfrac{a}{b}} = \dfrac{\sqrt[n]{a}}{\sqrt[n]{b}}$	$\sqrt{\dfrac{5}{4}} = \dfrac{\sqrt{5}}{2}$ $\sqrt{15} \div \sqrt{5} - \sqrt{3}$
Rules for simplifying square roots	A simplified square root expression has 1. *no* perfect square factors inside the radical, 2. *no* fractions inside the radical, and 3. *no* radicals in the denominator.	$\sqrt{12} = \sqrt{4 \cdot 3} = 2\sqrt{3}$ $\sqrt{\dfrac{5}{2}} = \dfrac{\sqrt{5}}{\sqrt{2}}$ $\dfrac{\sqrt{5}}{\sqrt{2}} = \dfrac{\sqrt{5}}{\sqrt{2}}\dfrac{\sqrt{2}}{\sqrt{2}} = \dfrac{\sqrt{10}}{2}$

| **Solving Equations Involving Squares and Square Roots** | **Examples** |

Square root property (how to solve $x^2 = k$)

If $k > 0$, the equation $x^2 = k$ is equivalent to the sentence $x = \pm\sqrt{k}$ ($x = \sqrt{k}$ or $x = -\sqrt{k}$).

If $k = 0$, the equation $x^2 = k$ is equivalent to $x = 0$.

If $k < 0$, the equation $x^2 = k$ has no real solution.

$x^2 = 6$
$x = \pm\sqrt{6}$
$t^2 = 0$
$t = 0$
$x^2 = -8$
No solution

Squaring property of equality

If we square each side of an equation, then we must check all of our solutions for extraneous roots.

$\sqrt{x} = -3$

$(\sqrt{x})^2 = (-3)^2$
$x = 9$
Extraneous root

| **Scientific Notation** | **Examples** |

Converting from scientific notation

1. Determine the number of places to move the decimal point by examining the exponent on the 10.
2. Move to the right for a positive exponent and to the left for a negative exponent.

$5.6 \times 10^3 = 5600$

$9 \times 10^{-4} = .0009$

Converting into scientific notation

1. Count the number of places (n) that the decimal point must be moved so that it will follow the first nonzero digit of the number.
2. If the original number was larger than 10 use 10^n.
3. If the original number was smaller than 1 use 10^{-n}.

$304.6 = 3.046 \times 10^2$
$.0035 = 3.5 \times 10^{-3}$

REVIEW EXERCISES

For the following exercises, assume that all of the variables represent positive real numbers.

6.1 *Simplify the following. Write your answers with positive exponents.*

1. 2^5 32
2. -2^4 -16
3. 10^3 1000
4. $5 \cdot 5^0$ 5
5. $x^5 x^8$ x^{13}
6. $a^3 a^9$ a^{12}
7. $\dfrac{a^8}{a^3}$ a^5
8. $\dfrac{a^{10}}{a^4}$ a^6
9. $\dfrac{a^3}{a^7}$ $\dfrac{1}{a^4}$
10. $\dfrac{b^2}{b^6}$ $\dfrac{1}{b^4}$
11. $(x^3)^4$ x^{12}
12. $(x^5)^{10}$ x^{50}
13. $(2x^3)^3$ $8x^9$
14. $(3y^5)^2$ $9y^{10}$
15. $\left(\dfrac{a}{3b^3}\right)^2$ $\dfrac{a^2}{9b^6}$
16. $\left(\dfrac{a^2}{5b}\right)^3$ $\dfrac{a^6}{125b^3}$

6.2 *Find the following roots or powers.*

17. $\sqrt[5]{32}$ 2
18. $\sqrt[3]{-27}$ -3
19. 10^4 10,000
20. 10^5 100,000
21. $\sqrt[3]{1000}$ 10
22. $\sqrt{100}$ 10
23. $(x^6)^2$ x^{12}
24. $(a^5)^2$ a^{10}

25. $\sqrt{x^{12}}$ x^6
26. $\sqrt{a^{10}}$ a^5
27. $\sqrt[3]{x^6}$ x^2
28. $\sqrt[3]{a^9}$ a^3

29. $\sqrt{4x^2}$ $2x$
30. $\sqrt{9y^4}$ $3y^2$
31. $\sqrt[3]{125x^6}$ $5x^2$
32. $\sqrt[3]{8y^{12}}$ $2y^4$

33. $\sqrt{\dfrac{4x^{16}}{y^{14}}}$ $\dfrac{2x^8}{y^7}$
34. $\sqrt{\dfrac{9y^8}{t^{10}}}$ $\dfrac{3y^4}{t^5}$
35. $\sqrt{\dfrac{w^2}{16}}$ $\dfrac{w}{4}$
36. $\sqrt{\dfrac{a^4}{25}}$ $\dfrac{a^2}{5}$

6.3 *Simplify each expression.*

37. $\sqrt{72}$ $6\sqrt{2}$
38. $\sqrt{48}$ $4\sqrt{3}$
39. $\dfrac{1}{\sqrt{3}}$ $\dfrac{\sqrt{3}}{3}$
40. $\dfrac{2}{\sqrt{5}}$ $\dfrac{2\sqrt{5}}{5}$

41. $\sqrt{\dfrac{3}{5}}$ $\dfrac{\sqrt{15}}{5}$
42. $\sqrt{\dfrac{5}{6}}$ $\dfrac{\sqrt{30}}{6}$
43. $\dfrac{\sqrt{33}}{\sqrt{3}}$ $\sqrt{11}$
44. $\dfrac{\sqrt{50}}{\sqrt{5}}$ $\sqrt{10}$

45. $\dfrac{\sqrt{3}}{\sqrt{8}}$ $\dfrac{\sqrt{6}}{4}$
46. $\dfrac{\sqrt{2}}{\sqrt{18}}$ $\dfrac{1}{3}$
47. $\sqrt{y^6}$ y^3
48. $\sqrt{z^{10}}$ z^5

49. $\sqrt{24t^8}$ $2t^4 \cdot \sqrt{6}$
50. $\sqrt{8p^6}$ $2p^3 \cdot \sqrt{2}$
51. $\sqrt{12t^3}$ $2t\sqrt{3t}$
52. $\sqrt{18q^7}$ $3q^3 \cdot \sqrt{2q}$

53. $\dfrac{\sqrt{2}}{\sqrt{x}}$ $\dfrac{\sqrt{2x}}{x}$
54. $\dfrac{\sqrt{5}}{\sqrt{y}}$ $\dfrac{\sqrt{5y}}{y}$
55. $\sqrt{\dfrac{3a}{2s}}$ $\dfrac{\sqrt{6as}}{2s}$
56. $\sqrt{\dfrac{5x}{3w}}$ $\dfrac{\sqrt{15xw}}{3w}$

6.4 *Perform the computations and simplify.*

57. $2\sqrt{7} + 8\sqrt{7}$ $10\sqrt{7}$
58. $3\sqrt{6} - 5\sqrt{6}$ $-2\sqrt{6}$
59. $\sqrt{12} - \sqrt{27}$ $-\sqrt{3}$

60. $\sqrt{18} + \sqrt{50}$ $8\sqrt{2}$
61. $2\sqrt{3} \cdot 5\sqrt{3}$ 30
62. $-3\sqrt{6} \cdot 2\sqrt{6}$ -36

63. $-3 \cdot 5\sqrt{3}$ $-15\sqrt{3}$
64. $4 \cdot 6\sqrt{8}$ $48\sqrt{2}$
65. $-3(5 + \sqrt{3})$ $-15 - 3\sqrt{3}$

66. $4(6 + \sqrt{8})$ $24 + 8\sqrt{2}$
67. $\sqrt{3}(\sqrt{6} - \sqrt{15})$ $3\sqrt{2} - 3\sqrt{5}$
68. $\sqrt{2}(\sqrt{6} - \sqrt{2})$ $2\sqrt{3} - 2$

69. $(\sqrt{3} - 5)(\sqrt{3} + 5)$ -22
70. $(\sqrt{2} + \sqrt{7})(\sqrt{2} - \sqrt{7})$ -5
71. $(2\sqrt{5} - \sqrt{6})^2$ $26 - 4\sqrt{30}$

72. $(3\sqrt{2} + \sqrt{6})^2$ $24 + 12\sqrt{3}$
73. $3\sqrt{5} \div 6\sqrt{2}$ $\dfrac{\sqrt{10}}{4}$
74. $6\sqrt{5} \div 4\sqrt{3}$ $\dfrac{\sqrt{15}}{2}$

75. $\dfrac{4 - \sqrt{20}}{10}$ $\dfrac{2 - \sqrt{5}}{5}$
76. $\dfrac{6 - \sqrt{12}}{-2}$ $-3 + \sqrt{3}$
77. $\dfrac{3}{1 - \sqrt{5}}$ $-\dfrac{3 + 3\sqrt{5}}{4}$

78. $\dfrac{\sqrt{2}}{\sqrt{6} + \sqrt{3}}$ $\dfrac{2\sqrt{3} - \sqrt{6}}{3}$

6.5 *Solve each equation.*

84. $-1 - 2\sqrt{5},\ -1 + 2\sqrt{5}$

79. $x^2 = 400$ $-20, 20$
80. $x^2 = 121$ $-11, 11$
81. $5x^2 = 2$ $-\dfrac{\sqrt{10}}{5}, \dfrac{\sqrt{10}}{5}$

82. $3x^2 = 7$ $-\dfrac{\sqrt{21}}{3}, \dfrac{\sqrt{21}}{3}$
83. $(x - 4)^2 = 18$ $4 - 3\sqrt{2}, 4 + 3\sqrt{2}$
84. $(x + 1)^2 = 20$

85. $\sqrt{x} = 9$ 81
86. $\sqrt{x} = 20$ 400
87. $x = \sqrt{36 - 5x}$ 4

88. $x = \sqrt{2 - x}$ 1
89. $x + 2 = \sqrt{52 + 2x}$ 6
90. $x - 4 = \sqrt{x - 4}$ $4, 5$

Solve each formula for t.

91. $t^2 - 8sw = 0$ $t = \pm 2\sqrt{2sw}$
92. $(t + b)^2 = b^2 - 4ac$ $t = -b \pm \sqrt{b^2 - 4ac}$

93. $3a = \sqrt{bt}$ $t = \dfrac{9a^2}{b}$
94. $a - \sqrt{t} = w$ $t = (a - w)^2$

6.6 *Simplify each expression. Answers with exponents should have positive exponents only.*

95. 5^{-3} $\dfrac{1}{125}$
96. 6^{-2} $\dfrac{1}{36}$
97. $25^{1/2}$ 5
98. $9^{3/2}$ 27

99. $64^{-1/2}$ $\dfrac{1}{8}$

100. $125^{-2/3}$ $\dfrac{1}{25}$

101. $x^{-3}x^{-5}$ $\dfrac{1}{x^8}$

102. $t^{-4}t^9$ t^5

103. $(-2x^{-1})^{-2}$ $\dfrac{x^2}{4}$

104. $(-3x^2)^{-2}$ $\dfrac{1}{9x^4}$

105. $w^{-3} \div w^{-7}$ w^4

106. $m^3 \div m^{-8}$ m^{11}

107. $\left(\dfrac{2x^{-2}}{y^3}\right)^{-3}$ $\dfrac{x^6y^9}{8}$

108. $\left(\dfrac{-3y^{-4}}{t^2}\right)^{-2}$ $\dfrac{y^8t^4}{9}$

109. $\left(\dfrac{9t^{-6}}{s^{-4}}\right)^{-1/2}$ $\dfrac{t^3}{3s^2}$

110. $\left(\dfrac{8y^{-3}}{x^6}\right)^{-2/3}$ $\dfrac{y^2x^4}{4}$

111. $(x^2y^2)^{-1/2}$ $\dfrac{1}{xy}$

112. $(t^3u^{-6})^{1/3}$ $\dfrac{t}{u^2}$

113. $u^{1/4}u^{1/2}$ $u^{3/4}$

114. $t^{-1/2}t$ $t^{1/2}$

6.7 *Convert each number in scientific notation to a number in standard notation, and convert each number in standard notation to a number in scientific notation.*

115. 5000 5×10^3

116. .00009 9×10^{-5}

117. 3.4×10^5 340,000

118. 5.7×10^{-8} .000000057

119. .0000461 4.61×10^{-5}

120. 44,000 4.4×10^4

121. 5.69×10^{-6} .00000569

122. 5.5×10^9 5,500,000,000

Perform each computation. Write each answer in scientific notation.

123. $(3.5 \times 10^8)(2.0 \times 10^{-12})$ 7×10^{-4}

124. $(9 \times 10^{12})(2 \times 10^{17})$ 1.8×10^{30}

125. $(2 \times 10^{-4})^4$ 1.6×10^{-15}

126. $(-3 \times 10^5)^3$ -2.7×10^{16}

127. $(.00000004)(2,000,000,000)$ 8×10^1

128. $(3,000,000,000) \div (.000002)$ 1.5×10^{15}

129. $(.0000002)^5$ 3.2×10^{-34}

130. $(50,000,000,000)^3$ 1.25×10^{32}

CHAPTER 6 TEST

Simplify each expression.

1. 2^5 32

2. $\sqrt{144}$ 12

3. $\sqrt[3]{-27}$ -3

4. 2^{-2} $\dfrac{1}{4}$

5. $16^{1/4}$ 2

6. $\sqrt{24}$ $2\sqrt{6}$

7. $\sqrt{\dfrac{3}{8}}$ $\dfrac{\sqrt{6}}{4}$

8. $(-3)^0$ 1

9. $\sqrt{8} + \sqrt{2}$ $3\sqrt{2}$

10. $(2 + \sqrt{3})^2$ $7 + 4\sqrt{3}$

11. $(3\sqrt{2} - \sqrt{7})(3\sqrt{2} + \sqrt{7})$ 11

12. $\sqrt{21} \div \sqrt{3}$ $\sqrt{7}$

13. $\sqrt{20} \div \sqrt{3}$ $\dfrac{2\sqrt{15}}{3}$

14. $\dfrac{2 + \sqrt{8}}{2}$ $1 + \sqrt{2}$

15. $27^{4/3}$ 81

16. $\sqrt{3}(\sqrt{6} - \sqrt{3})$ $3\sqrt{2} - 3$

Simplify. Assume that all variables represent positive real numbers and write answers with positive exponents only.

17. $3x^2 \cdot 5x^7$ $15x^9$

18. $(2x^6)^3$ $8x^{18}$

19. $(-3x^5y)^3$ $-27x^{15}y^3$

20. $\dfrac{2y^{-5}}{8y^9}$ $\dfrac{1}{4y^{14}}$

21. $\dfrac{t^{-7}}{t^{-3}}$ $\dfrac{1}{t^4}$

22. $(x^3y^9)^{1/3}$ xy^3

23. $(-2s^{-3}t^2)^{-2}$ $\dfrac{s^6}{4t^4}$

24. $\left(\dfrac{2w^3}{u^4}\right)^3$ $\dfrac{8w^9}{u^{12}}$

25. $\sqrt{\dfrac{3}{t}}$ $\dfrac{\sqrt{3t}}{t}$

26. $\sqrt{4y^6}$ $2y^3$

27. $\sqrt[3]{8y^{12}}$ $2y^4$

28. $\sqrt{18t^7}$ $3t^3 \cdot \sqrt{2t}$

Solve each equation.

29. $(x + 3)^2 = 36$ −9, 3   **30.** $\sqrt{x + 7} = 5$ 18   **31.** $5x^2 = 2$ $-\dfrac{\sqrt{10}}{5}, \dfrac{\sqrt{10}}{5}$ **32.** $(3x − 4)^2 = 0$ $\dfrac{4}{3}$

Show a complete solution to the problem.

33. Find the exact length of the side of a square whose diagonal is 5 meters. $\dfrac{5\sqrt{2}}{2}$ meters

Convert to scientific notation.

34. 5,433,000 5.433×10^6 **35.** .0000065 6.5×10^{-6}

Perform each computation by converting to scientific notation. Give the answer in scientific notation.

36. $(80,000)(.000006)$ 4.8×10^{-1} **37.** $(.0000003)^4$ 8.1×10^{-27}

Tying It All Together

CHAPTERS 1−6

Solve each equation or inequality. For the inequalities, also sketch the graph of the inequality. Graphs in Answers sections.

1. $2x + 3 = 0$ $-\dfrac{3}{2}$ **2.** $2x = 3$ $\dfrac{3}{2}$ **3.** $2x + 3 > 0$ $x > -\dfrac{3}{2}$ **4.** $-2x + 3 > 0$ $x < \dfrac{3}{2}$

5. $2(x + 3) = 0$ −3   **6.** $2x^2 = 3$ $-\dfrac{\sqrt{6}}{2}, \dfrac{\sqrt{6}}{2}$ **7.** $\dfrac{x}{3} = \dfrac{2}{x}$ $-\sqrt{6}, \sqrt{6}$ **8.** $\dfrac{x - 1}{x} = \dfrac{x}{x - 2}$ $\dfrac{2}{3}$

9. $(2x + 3)^2 = 0$ $-\dfrac{3}{2}$ **10.** $(2x + 3)(x - 3) = 0$ $-\dfrac{3}{2}, 3$ **11.** $2x^2 + 3 = 0$ No solution

12. $(2x + 3)^2 = 1$ −2, −1   **13.** $(2x + 3)^2 = -1$ No solution   **14.** $\sqrt{2x^2 - 14} = x - 1$ 3

Let $a = 2$, $b = -3$, and $c = -9$. Calculate the value of each of the following algebraic expressions.

15. b^2 9   **16.** $-4ac$ 72   **17.** $b^2 - 4ac$ 81   **18.** $\sqrt{b^2 - 4ac}$ 9

19. $-b + \sqrt{b^2 - 4ac}$ 12   **20.** $-b - \sqrt{b^2 - 4ac}$ −6   **21.** $\dfrac{-b + \sqrt{b^2 - 4ac}}{2a}$ 3   **22.** $\dfrac{-b - \sqrt{b^2 - 4ac}}{2a}$ $-\dfrac{3}{2}$

Factor each trinomial completely.

23. $x^2 - 6x + 9$ $(x - 3)^2$ **24.** $x^2 + 10x + 25$ $(x + 5)^2$ **25.** $x^2 + 12x + 36$ $(x + 6)^2$

26. $x^2 - 20x + 100$ $(x - 10)^2$ **27.** $2x^2 - 8x + 8$ $2(x - 2)^2$ **28.** $3x^2 + 6x + 3$ $3(x + 1)^2$

Perform the indicated operations with the binomials.

29. $(3 + 2x) - (6 - 5x)$ $7x - 3$

30. $(5 + 3t)(4 - 5t)$ $-15t^2 - 13t + 20$

31. $(8 - 6j)(3 + 4j)$ $-24j^2 + 14j + 24$

32. $(1 - j) + (5 + 7j)$ $6j + 6$

33. $(3 - 4j) - (2 - 5j)$ $j + 1$

34. $(2 + j)^2$ $j^2 + 4j + 4$

35. $(j - 7)(j + 7)$ $j^2 - 49$

36. $(3 - 2j)(3 + 2j)$ $-4j^2 + 9$

37. $(1 - j)^2$ $j^2 - 2j + 1$

38. $(-4 - 6t) - (-3 - 8t)$ $2t - 1$

39. $(1 + j)(3 - 4j)$ $-4j^2 - j + 3$

40. $(2 - 6j)(1 + 3j)$ $-18j^2 + 2$

41. $(1 - 2j) + (-6 + 5j)$ $3j - 5$

42. $(-2 - j) + (4 - 5j)$ $-6j + 2$

43. $\dfrac{4 - 6x}{2}$ $2 - 3x$

44. $\dfrac{-3 - 9j}{3}$ $-1 - 3j$

45. $\dfrac{8 - 12j}{-4}$ $-2 + 3j$

46. $\dfrac{20 - 5j}{-5}$ $-4 + j$

Quadratic Equations

Alberta and Ernie have planted about an acre of strawberries every year since their retirement. On Monday morning, Alberta picked the strawberries and Ernie sold the berries. On Tuesday morning, Alberta's back was killing her, so Ernie picked and Alberta sold the berries. Unfortunately, it took Ernie two hours longer to get the berries picked than it took Alberta, and they missed some of their best customers. So on Wednesday they worked together and got the entire patch picked in two hours. How long did it take Ernie to pick the berries by himself?

This problem is similar to some of the work problems that we solved in Chapter 5. However, the solution to this problem is not a rational number and cannot be obtained by the techniques of Chapter 5. To solve this problem, we must be able to solve a quadratic equation that cannot be solved by factoring. In this chapter we will learn techniques that allow us to solve *any* quadratic equation. In Exercise 7 of Section 7.4 we will find out how long it takes Ernie to pick the berries.

7.1 Familiar Quadratic Equations

IN THIS SECTION:

- Definition
- The Simplest Quadratic Equation
- Solving by Factoring

We solved some quadratic equations in Chapters 4, 5, and 6. However, we did not call them quadratic equations. In this section we define a quadratic equation, and review the types that we have already learned to solve.

Definition

An equation such as

$$3x^2 - 5x + 7 = 0$$

is called a **quadratic equation.** A quadratic equation has a term involving the second power of the variable. More formally, we make the following definition.

DEFINITION
Quadratic Equation

> A quadratic equation is an equation that can be written in the form
>
> $$ax^2 + bx + c = 0,$$
>
> where a, b, and c are real numbers and $a \neq 0$.

A quadratic equation is not always written in the form of the definition. Some of the quadratic equations that we solved in Chapters 4, 5, and 6 were in the form of the definition, and some were not. The equations

$$x^2 = 10, \qquad 5(x - 2)^2 = 20, \qquad \text{and} \qquad x^2 - 5x + 6 = 0$$

are quadratic equations that we solved in previous chapters.

The Simplest Quadratic Equation

If $b = 0$ in the general form of the quadratic equation, then the quadratic equation can be solved by the square root property.

EXAMPLE 1 Solve the equations:

a) $x^2 - 9 = 0$ **b)** $2x^2 - 3 = 0$ **c)** $-3(x + 1)^2 = -6$

Solution

We discuss the square root property first because once students learn the quadratic formula they often overlook this method. There is no need for the quadratic formula or factoring when $b = 0$.

a) Solve the equation for x^2, and then use the square root property:

$$x^2 - 9 = 0$$
$$x^2 = 9 \quad \text{Add 9 to each side.}$$
$$x = \pm 3 \quad \text{Apply the square root property.}$$

Both -3 and 3 are solutions to this equation. Check these solutions in the original equation.

b)
$$2x^2 - 3 = 0$$
$$2x^2 = 3$$
$$x^2 = \frac{3}{2}$$
$$x = \pm\sqrt{\frac{3}{2}} \quad \text{Apply the square root property.}$$
$$x = \pm\frac{\sqrt{3}}{\sqrt{2}} \cdot \frac{\sqrt{2}}{\sqrt{2}} \quad \text{Rationalize the denominator.}$$
$$x = \pm\frac{\sqrt{6}}{2}$$

The solutions to this equation are $-\dfrac{\sqrt{6}}{2}$ and $\dfrac{\sqrt{6}}{2}$. Check these solutions in the original equation.

Note that in part (c) of Example 1 we have some of the key steps in completing the square.

c) This equation is not quite like the other two parts of this example. If we actually squared the quantity $(x + 1)$, we would get a term involving x. Then b would not be equal to zero, as it is in the other two equations. However, it is much simpler not to square $(x + 1)$ and to solve this equation like the others here:

$$-3(x + 1)^2 = -6$$
$$(x + 1)^2 = 2 \quad \text{Divide each side by } -3.$$
$$x + 1 = \pm\sqrt{2} \quad \text{Use the square root property.}$$
$$x = -1 \pm \sqrt{2} \quad \text{Subtract 1 from each side.}$$

The solutions to this equation are $-1 + \sqrt{2}$ and $-1 - \sqrt{2}$. Be sure to check. ◄

Solving by Factoring

In Chapter 4 we learned to factor trinomials and to use factoring to solve quadratic equations. Quadratic equations are solved by factoring as follows.

STRATEGY

Solving Quadratic Equations by Factoring

1. Write the equation with 0 on the right-hand side.
2. Factor the left-hand side.
3. Use the zero factor property to get two simpler linear equations. (Set each factor equal to 0.)
4. Solve the two linear equations.
5. Check the answers in the original quadratic equation.

EXAMPLE 2   Solve each equation by factoring:

a) $x^2 + 2x = 8$

b) $3x^2 + 13x - 10 = 0$

c) $\dfrac{1}{6}x^2 - \dfrac{1}{2}x = 3$

Solution

a)
$$x^2 + 2x = 8$$
$$x^2 + 2x - 8 = 0 \qquad \text{Get 0 on the right side.}$$
$$(x + 4)(x - 2) = 0 \qquad \text{Factor the trinomial.}$$
$$x + 4 = 0 \qquad \text{or} \qquad x - 2 = 0 \qquad \text{Apply the zero factor property.}$$
$$x = -4 \qquad \text{or} \qquad x = 2$$

Both -4 and 2 are solutions to the equation. Check the solutions in the original equation.

b)
$$3x^2 + 13x - 10 = 0$$
$$(3x - 2)(x + 5) = 0 \qquad \text{Factor the trinomial.}$$
$$3x - 2 = 0 \qquad \text{or} \qquad x + 5 = 0 \qquad \text{Apply the zero factor property.}$$
$$3x = 2 \qquad \text{or} \qquad x = -5$$
$$x = \frac{2}{3} \qquad \text{or} \qquad x = -5$$

Both -5 and 2/3 are solutions to the equation. We leave the check to the reader.

c)
$$\frac{1}{6}x^2 - \frac{1}{2}x = 3$$
$$x^2 - 3x = 18 \qquad \text{Multiply each side by 6.}$$
$$x^2 - 3x - 18 = 0 \qquad \text{Get 0 on the right side.}$$
$$(x - 6)(x + 3) = 0 \qquad \text{Factor the trinomial.}$$
$$x - 6 = 0 \qquad \text{or} \qquad x + 3 = 0 \qquad \text{Apply the zero factor property.}$$
$$x = 6 \qquad \text{or} \qquad x = -3$$

The solutions to the equation are -3 and 6. Be sure to check the solutions in the original equation.   ◀

Warm-ups

True or false?

1. Both -4 and 4 satisfy the equation $x^2 - 16 = 0$. T
2. The equation $(x - 3)^2 = 8$ is equivalent to $x - 3 = 2\sqrt{2}$. F
3. Every quadratic equation can be solved by factoring. F
4. Both -5 and 4 are solutions to the equation $(x - 4)(x + 5) = 0$. T
5. The quadratic equation $x^2 = -3$ has no solutions. T
6. The equation $x^2 = 0$ has no solutions. F
7. The equation $(2x + 3)(4x - 5) = 0$ is equivalent to $x = 3/2$ or $x = 5/4$. F
8. The only solution to the equation $(x + 2)^2 = 0$ is -2. T
9. The equation $(x - 3)(x - 5) = 4$ is equivalent to $x - 3 = 2$ or $x - 5 = 2$. F
10. All quadratic equations have two distinct solutions. F

7.1 EXERCISES

Solve. See Example 1.

1. $x^2 - 36 = 0$ $-6, 6$
2. $x^2 - 81 = 0$ $-9, 9$
3. $x^2 + 10 = 0$ No solution
4. $x^2 + 4 = 0$ No solution
5. $5x^2 = 50$ $-\sqrt{10}, \sqrt{10}$
6. $7x^2 = 14$ $-\sqrt{2}, \sqrt{2}$
7. $3t^2 - 5 = 0$ $-\dfrac{\sqrt{15}}{3}, \dfrac{\sqrt{15}}{3}$
8. $5y^2 - 7 = 0$ $-\dfrac{\sqrt{35}}{5}, \dfrac{\sqrt{35}}{5}$
9. $-3y^2 + 8 = 0$ $-\dfrac{2\sqrt{6}}{3}, \dfrac{2\sqrt{6}}{3}$
10. $-5w^2 + 12 = 0$ $-\dfrac{2\sqrt{15}}{5}, \dfrac{2\sqrt{15}}{5}$
11. $(x - 3)^2 - 4$ $1, 5$
12. $(x + 5)^2 = 9$ $-8, -2$
13. $(y - 2)^2 - 18$ $2 - 3\sqrt{2}, 2 + 3\sqrt{2}$
14. $(m - 5)^2 = 20$ $5 - 2\sqrt{5}, 5 + 2\sqrt{5}$
15. $2(x + 1)^2 = \dfrac{1}{2}$ $-\dfrac{3}{2}, -\dfrac{1}{2}$
16. $-3(x - 1)^2 = -\dfrac{3}{4}$ $\dfrac{3}{2}, \dfrac{1}{2}$
17. $(x - 1)^2 = \dfrac{1}{2}$ $\dfrac{2 - \sqrt{2}}{2}, \dfrac{2 + \sqrt{2}}{2}$
18. $(y + 2)^2 = \dfrac{1}{2}$ $\dfrac{-4 - \sqrt{2}}{2}, \dfrac{-4 + \sqrt{2}}{2}$
19. $\left(x + \dfrac{1}{2}\right)^2 = \dfrac{1}{2}$ $\dfrac{-1 - \sqrt{2}}{2}, \dfrac{-1 + \sqrt{2}}{2}$
20. $\left(x - \dfrac{1}{2}\right)^2 = \dfrac{3}{2}$ $\dfrac{1 - \sqrt{6}}{2}, \dfrac{1 + \sqrt{6}}{2}$
21. $(x - 11)^2 = 0$ 11
22. $(x + 45)^2 = 0$ -45

Solve by factoring. See Example 2.

23. $x^2 - 2x - 15 = 0$ $-3, 5$
24. $x^2 - x - 12 = 0$ $-3, 4$
25. $x^2 + 6x + 9 = 0$ -3
26. $x^2 + 10x + 25 = 0$ -5
27. $4x^2 - 4x = 8$ $-1, 2$
28. $3x^2 + 3x = 90$ $-6, 5$
29. $3x^2 - 6x = 0$ $0, 2$
30. $-5x^2 + 10x = 0$ $0, 2$
31. $-4t^2 + 6t = 0$ $0, \dfrac{3}{2}$
32. $-6w^2 + 15w = 0$ $0, \dfrac{5}{2}$
33. $2x^2 + 11x - 21 = 0$ $-7, \dfrac{3}{2}$
34. $2x^2 - 5x + 2 = 0$ $\dfrac{1}{2}, 2$
35. $x^2 - 10x + 25 = 0$ 5
36. $x^2 - 4x + 4 = 0$ 2
37. $2x^2 - 7x = 30$ $-\dfrac{5}{2}, 6$
38. $15x^2 - 2x - 1 = 0$ $-\dfrac{1}{5}, \dfrac{1}{3}$

Solve each equation.

39. $x^2 - 2x = 2(3 - x)$ $-\sqrt{6}, \sqrt{6}$

40. $x^2 + 2x = \dfrac{1 + 4x}{2}$ $-\dfrac{\sqrt{2}}{2}, \dfrac{\sqrt{2}}{2}$

41. $x = \dfrac{27}{12 - x}$ 3, 9

42. $x = \dfrac{6}{x + 1}$ $-3, 2$

43. $\sqrt{3x - 8} = x - 2$ 3, 4

44. $\sqrt{3x - 14} = x - 4$ 5, 6

Solve each problem.

45. If the diagonal of a square is 5 meters, then what is the length of a side? $\dfrac{5\sqrt{2}}{2}$ meters

46. If the side of a square is 5 meters, then what is the length of the diagonal? $5\sqrt{2}$ meters

47. The formula $S = \dfrac{n^2 + n}{2}$ gives the sum of the first n positive integers. For what value of n is this sum equal to 45? 9

48. From the art museum, Howard walked eight blocks east and then four blocks north to reach the public library. How far was he then from where he started? (The answer is not 12 blocks.) $4\sqrt{5}$ blocks

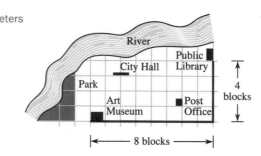

Figure for Exercise 48

7.2 Solving Any Quadratic Equation

IN THIS SECTION:

- Perfect Square Trinomials
- Solving a Quadratic Equation by Completing the Square

The quadratic equations solved in Section 7.1 were solved by factoring or the square root property. Some quadratic equations cannot be solved by either of those methods. In this section we will learn a method that works on any quadratic equation.

Perfect Square Trinomials

The new method for solving any quadratic equation depends on perfect square trinomials. Recall that a perfect square trinomial is the square of a binomial. Just as we recognize the numbers

$$1, 4, 9, 16, 25, 36, \ldots$$

as being the squares of the positive integers, we must recognize a perfect square trinomial. The following is a list of some perfect square trinomials with a leading coefficient of 1:

$$x^2 + 2x + 1 = (x + 1)^2 \qquad x^2 - 2x + 1 = (x - 1)^2$$
$$x^2 + 4x + 4 = (x + 2)^2 \qquad x^2 - 4x + 4 = (x - 2)^2$$
$$x^2 + 6x + 9 = (x + 3)^2 \qquad x^2 - 6x + 9 = (x - 3)^2$$
$$x^2 + 8x + 16 = (x + 4)^2 \qquad x^2 - 8x + 16 = (x - 4)^2$$

To solve quadratic equations using perfect square trinomials, we must be able to determine the last term of a perfect square trinomial when given the first two terms. For example, the perfect square trinomial whose first two terms are $x^2 + 6x$ is $x^2 + 6x + 9$. This process is called **completing the square.**

If the coefficient of x^2 is 1, there is a simple rule for identifying the last term in a perfect square trinomial.

RULE
Finding the Last Term

> The last term of a perfect square trinomial is the square of one-half of the coefficient of the middle term. In symbols, the perfect square trinomial whose first two terms are $x^2 + bx$ is $x^2 + bx + \left(\dfrac{b}{2}\right)^2$.

Consider finding the perfect square trinomial whose first two terms are

$$x^2 + 12x.$$

One-half of 12 is 6, and 6 squared is 36. Thus

$$x^2 + 12x + 36$$

is the perfect square trinomial that we are seeking, and

$$x^2 + 12x + 36 = (x + 6)^2.$$

EXAMPLE 1 Find the perfect square trinomial whose first two terms are given:

a) $x^2 + 10x$ **b)** $x^2 - 20x$ **c)** $x^2 + 3x$ **d)** $x^2 - x$

Solution

a) One-half of 10 is 5, and 5 squared is 25, so the perfect square trinomial is $x^2 + 10x + 25$.

b) One-half of -20 is -10, and -10 squared is 100, so the perfect square trinomial is $x^2 - 20x + 100$.

c) One-half of 3 is 3/2, and 3/2 squared is 9/4, so the perfect square trinomial is $x^2 + 3x + 9/4$.

d) One-half of -1 is $-1/2$, and $(-1/2)^2 = 1/4$, so the perfect square is $x^2 - x + 1/4$.   ◄

EXAMPLE 2   Factor each of the perfect square trinomials of Example 1 as the square of a binomial.

Solution

a) $x^2 + 10x + 25 = (x + 5)^2$ **b)** $x^2 - 20x + 100 = (x - 10)^2$

c) $x^2 + 3x + \dfrac{9}{4} = \left(x + \dfrac{3}{2}\right)^2$ **d)** $x^2 - x + \dfrac{1}{4} = \left(x - \dfrac{1}{2}\right)^2$ ◀

Solving a Quadratic Equation by Completing the Square

We can use completing the square and the square root property to solve any quadratic equation. The following examples show how this is done.

EXAMPLE 3   Solve by completing the square: $x^2 + 6x - 7 = 0$.

Solution   We first add 7 to each side of the equation:

$$x^2 + 6x = 7$$

Now we complete the square on the left-hand side. One-half of 6 is 3, and 3 squared is 9.

$$x^2 + 6x + 9 = 7 + 9 \qquad \text{Add 9 to each side.}$$
$$(x + 3)^2 = 16 \qquad \text{Factor the left-hand side.}$$
$$x + 3 = \pm 4 \qquad \text{Use the square root property.}$$
$$x + 3 = 4 \quad \text{or} \quad x + 3 = -4$$
$$x = 1 \quad \text{or} \qquad x = -7$$

The solutions to the equation are -7 and 1. Check.   ◀

Note that all of the perfect square trinomials that we have used had a leading coefficient of 1. If the leading coefficient is not 1, then we must divide each side of the equation by the leading coefficient to get an equation with a leading coefficient of 1. The steps to follow in completing the square are summarized as follows.

Note that now we are not just identifying a perfect square trinomial. The perfect square trinomial is part of an equation. We obtain the perfect square trinomial by adding the same number to each side.

STRATEGY
Completing the Square

To solve a quadratic equation by completing the square,
1. The coefficient of x^2 must be 1.
2. Get only the x^2 and the x terms on the left-hand side.
3. Add to each side the square of 1/2 the coefficient of x.
4. Factor the left-hand side as the square of a binomial.
5. Apply the square root property.
6. Solve for x.
7. Simplify.

EXAMPLE 4   Solve by completing the square: $2x^2 - 5x - 3 = 0$.

Solution   Our perfect square trinomial must begin with x^2 and not $2x^2$.

$$\frac{2x^2 - 5x - 3}{2} = \frac{0}{2} \qquad \text{Divide each side by 2.}$$

$$x^2 - \frac{5}{2}x - \frac{3}{2} = 0 \qquad \text{Simplify.}$$

$$x^2 - \frac{5}{2}x = \frac{3}{2} \qquad \text{Add 3/2 to each side.}$$

$$x^2 - \frac{5}{2}x + \frac{25}{16} = \frac{3}{2} + \frac{25}{16} \qquad \begin{array}{l}\text{Complete the square.}\\ \frac{1}{2}\cdot\frac{5}{2} = \frac{5}{4}, \left(\frac{5}{4}\right)^2 = \frac{25}{16}\end{array}$$

$$\left(x - \frac{5}{4}\right)^2 = \frac{49}{16} \qquad \text{Factor the left-hand side.}$$

$$x - \frac{5}{4} = \pm\frac{7}{4} \qquad \text{Use the square root property.}$$

$$x - \frac{5}{4} = \frac{7}{4} \quad \text{or} \quad x - \frac{5}{4} = \frac{7}{4}$$

$$x = \frac{12}{4} \quad \text{or} \qquad x = -\frac{2}{4}$$

$$x = 3 \quad \text{or} \qquad x = -\frac{1}{2}$$

The solutions to the equation are $-1/2$ and 3. Check.   ◀

The equations in Examples 3 and 4 could have been solved by factoring. The next example illustrates a quadratic equation that cannot be solved by factoring but can be solved by completing the square.

EXAMPLE 5   Solve by completing the square: $x^2 + 4x - 3 = 0$.

Note that most students do not like to check this kind of answer, but it is instructive to do so.

Solution

$$x^2 + 4x - 3 = 0$$

$$x^2 + 4x = 3 \qquad \text{Add 3 to each side.}$$

$$x^2 + 4x + 4 = 3 + 4 \qquad \text{Complete the square.}$$

$$(x + 2)^2 = 7 \qquad \text{Factor the left-hand side.}$$

$$x + 2 = \pm\sqrt{7} \qquad \text{Apply the square root property.}$$

$$x + 2 = \sqrt{7} \quad \text{or} \quad x + 2 = -\sqrt{7}$$

$$x = -2 + \sqrt{7} \quad \text{or} \qquad x = -2 - \sqrt{7}$$

Both $-2 + \sqrt{7}$ and $-2 - \sqrt{7}$ are solutions to this equation. Check these answers in the original equation.   ◀

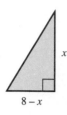

x

$8 - x$

Figure 7.1

In the next example we use completing the square to solve a geometric problem.

EXAMPLE 6   The sum of the lengths of the two legs of a right triangle is 8 feet. If the area of this right triangle is 5 square feet, then what are the lengths of the legs?

Solution   If x represents the length of one leg, then $8 - x$ represents the length of the other. See Fig. 7.1.

The area of any triangle is given by the formula

$$A = \frac{1}{2}bh.$$

Math at Work

Every year, as winter turns into spring, Americans of all ages join in the national pastime of baseball. Recreational leagues form in schools and businesses across the country. For every league, someone must schedule the games.

One popular way for league managers to schedule games is to have each team in the league play every other team once. How many games would this involve?

Consider a small league of four baseball teams: Antelopes, Bears, Cats, and Dodgers. Since A must play each of the others, we must have the pairs AB, AC, and AD. Since B has already played A, only the games BC and BD must be scheduled. Team C has already played A and B, so only a game with D must be scheduled. Listing the matches, we can observe a pattern:

AB	BC	CD
AC	BD	
AD		

The number of games to be played is $3 + 2 + 1$, a total of 6. If n represents the number of teams in a league, then the expression $\frac{1}{2}n^2 - \frac{1}{2}n$ gives the number of games it takes for each team to play every other team once. For $n = 4$,

$$\frac{1}{2} \cdot 4^2 - \frac{1}{2} \cdot 4 = 6.$$

The Middletown Baseball League plays 120 games in a season. If the league organizer wants to arrange a schedule in which each team plays every other team once, then how many teams should be invited to participate in the league?   16

In this case we have

$$5 = \frac{1}{2}x(8 - x).$$

Rearrange the equation to use completing the square:

$$2 \cdot 5 = \not{2} \cdot \frac{1}{\not{2}}x(8 - x) \qquad \text{Multiply each side by 2.}$$

$$10 = 8x - x^2 \qquad \text{Simplify.}$$

$$x^2 - 8x + 10 = 0$$

$$x^2 - 8x \quad\quad = -10$$

$$x^2 - 8x + 16 = -10 + 16 \qquad \text{Add 16 to each side to complete} \\ \text{the square.}$$

$$(x - 4)^2 = 6 \qquad \text{Factor the left-hand side.}$$

$$x - 4 = \pm\sqrt{6} \qquad \text{Use the square root property.}$$

$$x - 4 = \sqrt{6} \qquad \text{or} \quad x - 4 = -\sqrt{6}$$

$$x = 4 + \sqrt{6} \quad \text{or} \qquad x = 4 - \sqrt{6}$$

If $x = 4 + \sqrt{6}$,   then $8 - x = 8 - (4 + \sqrt{6}) = 4 - \sqrt{6}$.

If $x = 4 - \sqrt{6}$,   then $8 - x = 8 - (4 - \sqrt{6}) = 4 + \sqrt{6}$.

So there is only one solution. The lengths of the legs are $4 + \sqrt{6}$ feet and $4 - \sqrt{6}$ feet. Check.   ◀

Warm-ups

True or false?

1. Completing the square is used for finding the area of a square.   F

2. The polynomial $x^2 + \frac{2}{3}x + \frac{4}{9}$ is a perfect square trinomial.   F

3. Every quadratic equation can be solved by factoring.   F

4. The polynomial $x^2 - x + 1$ is a perfect square trinomial.   F

5. Every quadratic equation can be solved by completing the square.   T

6. The solutions to the equation $x - 2 = \pm\sqrt{3}$ are $2 + \sqrt{3}$ and $2 - \sqrt{3}$.   T

7. There are no real numbers that satisfy $(x + 7)^2 = -5$.   T

8. In completing the square for $x^2 - 5x = 4$, we should add 25/4 to each side.   T

9. One-half of four-fifths is two-fifths.   T

10. One-half of three-fourths is three-eighths.   T

7.2 EXERCISES

Find the perfect square trinomial whose first two terms are given. See Example 1.

1. $x^2 + 6x$ $x^2 + 6x + 9$

2. $x^2 - 4x$ $x^2 - 4x + 4$

3. $x^2 + 14x$ $x^2 + 14x + 49$

4. $x^2 + 16x$ $x^2 + 16x + 64$

5. $x^2 - 16x$ $x^2 - 16x + 64$

6. $x^2 - 14x$ $x^2 - 14x + 49$

7. $t^2 - 18t$ $t^2 - 18t + 81$

8. $w^2 + 18w$ $w^2 + 18w + 81$

9. $m^2 + 3m$ $m^2 + 3m + \dfrac{9}{4}$

10. $n^2 - 5n$ $n^2 - 5n + \dfrac{25}{4}$

11. $z^2 + z$ $z^2 + z + \dfrac{1}{4}$

12. $v^2 - v$ $v^2 - v + \dfrac{1}{4}$

13. $x^2 - \dfrac{1}{2}x$ $x^2 - \dfrac{1}{2}x + \dfrac{1}{16}$

14. $y^2 + \dfrac{1}{3}y$ $y^2 + \dfrac{1}{3}y + \dfrac{1}{36}$

15. $y^2 + \dfrac{1}{4}y$ $y^2 + \dfrac{1}{4}y + \dfrac{1}{64}$

16. $z^2 - \dfrac{4}{3}z$ $z^2 - \dfrac{4}{3}z + \dfrac{4}{9}$

Factor each perfect square trinomial as the square of a binomial. See Example 2.

17. $x^2 + 10x + 25$ $(x + 5)^2$

18. $x^2 - 6x + 9$ $(x - 3)^2$

19. $m^2 - 2m + 1$ $(m - 1)^2$

20. $n^2 + 4n + 4$ $(n + 2)^2$

21. $x^2 + x + \dfrac{1}{4}$ $\left(x + \dfrac{1}{2}\right)^2$

22. $y^2 - y + \dfrac{1}{4}$ $\left(y - \dfrac{1}{2}\right)^2$

23. $t^2 + \dfrac{1}{3}t + \dfrac{1}{36}$ $\left(t + \dfrac{1}{6}\right)^2$

24. $v^2 - \dfrac{2}{3}v + \dfrac{1}{9}$ $\left(v - \dfrac{1}{3}\right)^2$

25. $x^2 + \dfrac{2}{5}x + \dfrac{1}{25}$ $\left(x + \dfrac{1}{5}\right)^2$

26. $y^2 - \dfrac{1}{4}y + \dfrac{1}{64}$ $\left(y - \dfrac{1}{8}\right)^2$

Solve each quadratic equation by completing the square. See Examples 3 *and* 4.

27. $x^2 + 2x - 15 = 0$ $-5, 3$

28. $x^2 + 2x - 24 = 0$ $-6, 4$

29. $x^2 - 4x - 21 = 0$ $-3, 7$

30. $x^2 - 4x - 12 = 0$ $-2, 6$

31. $x^2 + 6x + 9 = 0$ -3

32. $x^2 - 10x + 25 = 0$ 5

33. $2t^2 - 3t + 1 = 0$ $\dfrac{1}{2}, 1$

34. $2t^2 - 3t - 2 = 0$ $-\dfrac{1}{2}, 2$

35. $2w^2 - 7w + 6 = 0$ $\dfrac{3}{2}, 2$

36. $4t^2 + 5t - 6 = 0$ $-2, \dfrac{3}{4}$

37. $3x^2 + 2x - 1 = 0$ $-1, \dfrac{1}{3}$

38. $3x^2 - 8x - 3 = 0$ $-\dfrac{1}{3}, 3$

Solve each quadratic equation by completing the square. See Example 5.

39. $x^2 + 2x - 6 = 0$ $-1 - \sqrt{7}, -1 + \sqrt{7}$

40. $x^2 + 4x - 4 = 0$ $-2 - 2\sqrt{2}, -2 + 2\sqrt{2}$

41. $x^2 + 6x + 1 = 0$ $-3 - 2\sqrt{2}, -3 + 2\sqrt{2}$

42. $x^2 - 6x - 3 = 0$ $3 - 2\sqrt{3}, 3 + 2\sqrt{3}$

43. $x^2 - x - 3 = 0$ $\dfrac{1 - \sqrt{13}}{2}, \dfrac{1 + \sqrt{13}}{2}$

44. $x^2 + x - 1 = 0$ $\dfrac{-1 - \sqrt{5}}{2}, \dfrac{-1 + \sqrt{5}}{2}$

45. $x^2 + 3x - 3 = 0$ $\dfrac{-3 - \sqrt{21}}{2}, \dfrac{-3 + \sqrt{21}}{2}$

46. $x^2 - 3x + 1 = 0$ $\dfrac{3 - \sqrt{5}}{2}, \dfrac{3 + \sqrt{5}}{2}$

47. $2x^2 - x - 4 = 0$ $\dfrac{1 - \sqrt{33}}{4}, \dfrac{1 + \sqrt{33}}{4}$

48. $4x^2 + 2x - 1 = 0$ $\dfrac{-1 - \sqrt{5}}{4}, \dfrac{-1 + \sqrt{5}}{4}$

Solve each equation by whichever method is appropriate.

49. $(x - 5)^2 = 7$ $5 - \sqrt{7}, 5 + \sqrt{7}$

50. $x^2 + x = 12$ $-4, 3$

51. $3x^2 - 5 = 0$ $-\dfrac{\sqrt{15}}{3}, \dfrac{\sqrt{15}}{3}$

52. $2x^2 + 16 = 0$ No solution

53. $3x^2 + 1 = 0$ No solution

54. $x^2 + 6x + 7 = 0$ $-3 - \sqrt{2}, -3 + \sqrt{2}$

55. $x^2 + 8 = 8x$ $4 - 2\sqrt{2}, 4 + 2\sqrt{2}$

56. $2x^2 + 5x = 42$ $-6, \dfrac{7}{2}$

57. $(2x - 7)^2 = 0$ $\dfrac{7}{2}$

58. $x^2 - 7 = 0$ $-\sqrt{7}, \sqrt{7}$

59. $y^2 + 6y = 11$ $-3 - 2\sqrt{5}, -3 + 2\sqrt{5}$

60. $y^2 + 6y = 0$ $-6, 0$

Use a quadratic equation and completing the square to solve each problem. See Example 6.

61. Joan has saved the candles from her birthday cake for every year of her life. If Joan has 78 candles, then how old is Joan? (See Exercise 47 of Section 7.1.)   12 years old

62. A rectangle has a perimeter of 12 inches and an area of 6 square inches. What are the length and width of the rectangle? $L = 3 + \sqrt{3}$ inches, $W = 3 - \sqrt{3}$ inches

63. The sum of two numbers is 12 and their product is 34. What are the numbers? $6 - \sqrt{2}$ and $6 + \sqrt{2}$

64. The sum of the measures of the base and height of a triangle is 10 inches. If the area of the triangle is 11 square inches, then what are the measures of the base and height? $5 - \sqrt{3}$ inches, $5 + \sqrt{3}$ inches

---

7.3 The Quadratic Formula

IN THIS SECTION:

- A Formula for Solving Any Quadratic Equation
- The Discriminant
- Which Method to Use

In Section 7.2 we learned that every quadratic equation can be solved by completing the square. We can also use completing the square to get a formula for solving any quadratic equation. In this section we will learn the quadratic formula.

A Formula for Solving Any Quadratic Equation

We prevent the development of the quadratic formula for completeness. Most students at this level will find this development difficult to follow. Remind students that the important part of all of this is the end result, the formula.

To develop a formula for solving any quadratic equation, we use completing the square on the general quadratic equation

$$ax^2 + bx + c = 0.$$

Assume that a is positive for now and divide each side by a:

$$\frac{ax^2 + bx + c}{a} = \frac{0}{a}$$

$$x^2 + \frac{b}{a}x + \frac{c}{a} = 0$$

$$x^2 + \frac{b}{a}x = -\frac{c}{a} \qquad \text{Subtract } \frac{c}{a} \text{ from each side.}$$

One-half of b/a is $b/(2a)$. To complete the square on the left-hand side, we add $\dfrac{b^2}{4a^2}$ to each side:

$$x^2 + \frac{b}{a}x + \frac{b^2}{4a^2} = \frac{b^2}{4a^2} - \frac{c}{a}$$

Factor the left-hand side and get a common denominator for the right-hand side:

$$\left(x + \frac{b}{2a}\right)^2 = \frac{b^2}{4a^2} - \frac{4ac}{4a^2}$$

$$\left(x + \frac{b}{2a}\right)^2 = \frac{b^2 - 4ac}{4a^2}$$

$$x + \frac{b}{2a} = \pm\sqrt{\frac{b^2 - 4ac}{4a^2}} \qquad \text{Apply the square root property.}$$

$$x = \frac{-b}{2a} \pm \frac{\sqrt{b^2 - 4ac}}{2a} \qquad \text{Since } a > 0,\ \sqrt{4a^2} = 2a.$$

$$x = \frac{-b \pm \sqrt{b^2 - 4ac}}{2a}$$

We assumed that a was positive so that $\sqrt{4a^2} = 2a$ would be correct. If a is negative, then $\sqrt{4a^2} = -2a$. Either way, the result is the same. It is called the **quadratic formula.** The formula gives x in terms of the coefficients a, b, and c. The quadratic formula is generally used instead of completing the square to solve a quadratic equation that cannot be factored.

Quadratic Formula

The solution to the equation $ax^2 + bx + c = 0$, where $a \neq 0$, is given by the formula

$$x = \frac{-b \pm \sqrt{b^2 - 4ac}}{2a}.$$

EXAMPLE 1 Solve by using the quadratic formula:

a) $x^2 + 2x - 3 = 0$ **b)** $2x^2 = 3 - x$ **c)** $3x^2 - 6x + 1 = 0$

Solution

a) To use the formula, we first identify a, b, and c. For the equation

$$x^2 + 2x - 3 = 0,$$

$$\underset{a}{\uparrow} \quad \underset{b}{\uparrow} \quad \underset{c}{\uparrow}$$

$a = 1$, $b = 2$, and $c = -3$. Now use these values in the quadratic formula:

$$x = \frac{-b \pm \sqrt{b^2 - 4ac}}{2a}$$

$$x = \frac{-2 \pm \sqrt{(2)^2 - 4(1)(-3)}}{2(1)}$$

$$x = \frac{-2 \pm \sqrt{16}}{2} \qquad 2^2 - 4(1)(-3) = 16$$

$$x = \frac{-2 \pm 4}{2}$$

$$x = 1 \quad \text{or} \quad x = -3 \qquad \frac{-2+4}{2} = 1, \; \frac{-2-4}{2} = -3$$

The solutions to the equation are -3 and 1. Check these solutions in the original equation.

b) To identify a, b, and c, we must have the equation written in the general form:

$$2x^2 = 3 - x$$

$$2x^2 + x - 3 = 0$$

Now we can see that $a = 2$, $b = 1$, and $c = -3$. Insert these values into the formula:

$$x - \frac{-1 \pm \sqrt{(1)^2 - 4(2)(-3)}}{2(2)} \qquad \text{Since } b = 1, \; -b = -1.$$

$$x = \frac{-1 \pm \sqrt{25}}{4} \qquad (1)^2 - 4(2)(-3) = 25$$

$$x = \frac{-1 \pm 5}{4}$$

$$x = 1 \quad \text{or} \quad x = -\frac{3}{2} \qquad \frac{-1+5}{4} = 1, \; \frac{-1-5}{4} = -\frac{3}{2}$$

The solutions to the equation are $-3/2$ and 1. Check.

c) For $3x^2 - 6x + 1 = 0$, we have $a = 3$, $b = -6$, and $c = 1$:

$$x = \frac{6 \pm \sqrt{(-6)^2 - 4(3)(1)}}{2(3)} \qquad \text{Since } b = -6, \; -b = 6.$$

$$x = \frac{6 \pm \sqrt{24}}{6} = \frac{6 \pm 2\sqrt{6}}{6} = \frac{2(3 \pm \sqrt{6})}{2(3)}$$

$$x = \frac{3 \pm \sqrt{6}}{3}$$

The two solutions to this quadratic equation are the irrational numbers

$$\frac{3 + \sqrt{6}}{3} \quad \text{and} \quad \frac{3 - \sqrt{6}}{3}.$$ ◄

The Discriminant

We have seen quadratic equations with two real solutions, one real solution, and no real solutions. The reason for this is in the square root in the numerator of the quadratic formula.

If the value of $b^2 - 4ac$ is positive, we get two solutions because the square root of this value can be added to $-b$ or subtracted from $-b$. This is the case in each part of Example 1.

If the value of $b^2 - 4ac$ is 0, we get only one solution:

$$x = \frac{-b \pm \sqrt{0}}{2a} = \frac{-b}{2a}$$

If the value of $b^2 - 4ac$ is negative, there are no real solutions because the square root of a negative number appears in the numerator of the formula.

The quantity $b^2 - 4ac$ is called the **discriminant** because its value determines the number of real solutions to the quadratic equation. Table 7.1 summarizes these facts about the discriminant.

TABLE 7.1

Value of $b^2 - 4ac$	Number of Real Solutions to $ax^2 + bx + c = 0$
Positive	2
Zero	1
Negative	0

EXAMPLE 2   Find the value of the discriminant and determine how many real solutions there are to each equation:

a) $3x^2 - 5x + 1 = 0$ b) $x^2 + 6x + 9 = 0$
c) $2x^2 + 1 = x$

Solution

a) For the equation $3x^2 - 5x + 1 = 0$, we have $a = 3$, $b = -5$, and $c = 1$. Now calculate the value of the discriminant:

$$b^2 - 4ac = (-5)^2 - 4(3)(1) = 25 - 12 = 13$$

Since the discriminant is positive, there are two real solutions to this quadratic equation.

b) For the equation $x^2 + 6x + 9 = 0$, we have $a = 1$, $b = 6$, and $c = 9$:

$$b^2 - 4ac = (6)^2 - 4(1)(9) = 36 - 36 = 0$$

Since the discriminant is zero, there is only one real solution to the equation.

c) We must first rewrite the equation:

$$2x^2 + 1 = x$$
$$2x^2 - x + 1 = 0 \qquad \text{Subtract } x \text{ from each side.}$$

The values of a, b, and c are $a = 2$, $b = -1$, and $c = 1$:

$$b^2 - 4ac = (-1)^2 - 4(2)(1) = 1 - 8 = -7$$

Since the discriminant is negative, the equation has no real solutions. ◄

Which Method to Use

Remind students that when they are requested to demonstrate a certain method, they must use the requested method.

If the quadratic equation is simple enough, we can solve it by factoring or by the square root property. These are the methods that should be tried first. *All quadratic equations can be solved by the quadratic formula.* Remember that the quadratic formula is just a shortcut to completing the square and is easier to use than completing the square. The idea of completing the square is used in other topics in algebra, but we do not need it for solving quadratic equations. The available methods are summarized as follows.

Solving the Quadratic Equation: $ax^2 + bx + c = 0$

Methods	Comments	Examples
Square root property	Use when $b = 0$.	$x^2 = 3$, $(x - 2)^2 = 8$
Factoring	Use when the polynomial can be factored.	$x^2 + 5x + 6 = 0$ $(x + 2)(x + 3) = 0$
Quadratic formula	Use when the first two methods do not apply.	$x^2 + 5x + 3 = 0$
Completing the square	Use the quadratic formula instead.	

Warm-ups

True or false?

1. Completing the square is used to develop the quadratic formula. T
2. For the equation $x^2 - x + 1 = 0$, we have $a = 1$, $b = -x$, and $c = 1$. F
3. For the equation $x^2 - 3 = 5x$, we have $a = 1$, $b = -3$, and $c = 5$. F
4. The quadratic formula can be expressed as $x = -b \pm \dfrac{\sqrt{b^2 - 4ac}}{2a}$. F
5. If $a = 2$, $b = -6$, and $c = 0$, then the quadratic equation with those coefficients has two real solutions. T
6. All quadratic equations have two distinct real solutions. F
7. It will still be necessary to use completing the square on the quadratic equations that cannot be solved by the quadratic formula. F
8. The quadratic equation $x^2 - 8x + 16 = 0$ has only one solution. T
9. For the quadratic equation $-3x^2 + 5x = 0$, we have $a = -3$, $b = 5$, and $c = 0$. T
10. The only solution to $x^2 + 6x + 9 = 0$ is -3. T

7.3 EXERCISES

Solve by using the quadratic formula. See Example 1.

1. $x^2 + 2x - 15 = 0$ $-5, 3$
2. $x^2 - 3x - 18 = 0$ $-3, 6$
3. $x^2 + 10x + 25 = 0$ -5
4. $x^2 - 12x + 36 = 0$ 6
5. $2x^2 + x - 6 = 0$ $-2, \dfrac{3}{2}$
6. $2x^2 + x - 15 = 0$ $-3, \dfrac{5}{2}$
7. $4x^2 + 4x - 3 = 0$ $-\dfrac{3}{2}, \dfrac{1}{2}$
8. $4x^2 + 8x + 3 = 0$ $-\dfrac{3}{2}, -\dfrac{1}{2}$
9. $2y^2 - 6y + 3 = 0$ $\dfrac{3 - \sqrt{3}}{2}, \dfrac{3 + \sqrt{3}}{2}$
10. $3y^2 + 6y + 2 = 0$ $\dfrac{-3 - \sqrt{3}}{3}, \dfrac{-3 + \sqrt{3}}{3}$
11. $2t^2 + 4t + 1 = 0$ $\dfrac{-2 - \sqrt{2}}{2}, \dfrac{-2 + \sqrt{2}}{2}$
12. $w^2 - 4w + 2 = 0$ $2 - \sqrt{2}, 2 + \sqrt{2}$

Find the value of the discriminant and state how many real solutions there are to each quadratic equation. See Example 2.

13. $4x^2 - 4x + 1 = 0$ 0, one
14. $9x^2 + 6x + 1 = 0$ 0, one
15. $6x^2 - 7x + 4 = 0$ -47, none
16. $-3x^2 + 5x - 7 = 0$ -59, none
17. $-5t^2 - t + 9 = 0$ 181, two
18. $-2w^2 - 6w + 5 = 0$ 76, two
19. $4x^2 - 12x + 9 = 0$ 0, one
20. $9x^2 + 12x + 4 = 0$ 0, one
21. $x^2 + x + 4 = 0$ -15, none
22. $y^2 - y + 2 = 0$ -7, none
23. $x - 5 = 3x^2$ -59, none
24. $4 - 3x = x^2$ 25, two

Use the method of your choice to solve each equation.

25. $x^2 + \dfrac{3}{2}x = 1$ $-2, \dfrac{1}{2}$
26. $x^2 - \dfrac{7}{2}x = 2$ $-\dfrac{1}{2}, 4$
27. $(x - 1)^2 + (x - 2)^2 = 5$ $0, 3$
28. $x^2 + (x - 3)^2 = 29$ $-2, 5$
29. $\dfrac{1}{x} + \dfrac{1}{x + 2} = \dfrac{5}{12}$ $-\dfrac{6}{5}, 4$
30. $\dfrac{1}{x} + \dfrac{1}{x + 1} = \dfrac{5}{6}$ $-\dfrac{3}{5}, 2$
31. $x^2 + 6x + 8 = 0$ $-4, -2$
32. $2x^2 - 5x - 3 = 0$ $-\dfrac{1}{2}, 3$
33. $x^2 - 9x = 0$ $0, 9$

34. $x^2 - 9 = 0$ $-3, 3$

35. $(x + 5)^2 = 9$ $-8, -2$

36. $(3x - 1)^2 = 0$ $\dfrac{1}{3}$

37. $x(x - 3) = 2 - 3(x + 4)$ No solution

38. $(x - 1)(x + 4) = (2x - 4)^2$ $\dfrac{4}{3}, 5$

39. $\dfrac{x}{3} = \dfrac{x + 2}{x}$ $\dfrac{3 - \sqrt{33}}{2}, \dfrac{3 + \sqrt{33}}{2}$

40. $\dfrac{x - 2}{x} = \dfrac{5}{x + 2}$ $\dfrac{5 - \sqrt{41}}{2}, \dfrac{5 + \sqrt{41}}{2}$

41. $2x^2 - 3x = 0$ $0, \dfrac{3}{2}$

42. $x^2 = 5$ $-\sqrt{5}, \sqrt{5}$

Use a calculator to find the approximate solutions to each quadratic equation. Round answers to two decimal places.

43. $x^2 - 3x - 3 = 0$ $-.791, 3.791$

44. $x^2 - 2x - 2 = 0$ $-.73, 2.73$

45. $x^2 - x - 3.2 = 0$ $-1.357, 2.357$

46. $x^2 - 4.3x + 3 = 0$ $.88, 3.42$

47. $5.29x^2 - 3.22x + .49 = 0$ $.304$

48. $2.6x^2 + 3.1x - 5 = 0$ $-2.11, .91$

7.4 Applications

IN THIS SECTION:

- ● **Geometric Applications**
- ● **Work Problems**
- ● **Formulas**

Most of the problems in this section require the use of the quadratic formula. It is helpful to have a calculator to obtain approximate answers and to aid in checking.

In this section we will solve problems that involve quadratic equations. When solving a problem, we may solve the quadratic equation by any of the methods that we have learned.

Geometric Applications

EXAMPLE 1 The length of a rectangular flower bed is 2 feet longer than the width. If the area is 6 square feet, then what are the exact values of the length and width of the flower bed. See Fig. 7.2.

Solution If x represents the width, then $x + 2$ represents the length. Since the area is 6, we can write

$$x(x + 2) = 6 \qquad \text{Since } A = L \cdot W$$
$$x^2 + 2x - 6 = 0.$$

Since the polynomial is prime, we use the quadratic formula:

$$x = \frac{-2 \pm \sqrt{(2)^2 - 4(1)(-6)}}{2(1)}$$

$$x = \frac{-2 \pm \sqrt{28}}{2} = \frac{-2 \pm 2\sqrt{7}}{2} = \frac{2(-1 \pm \sqrt{7})}{2}$$

$$x = -1 \pm \sqrt{7}$$

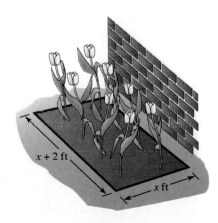

$x + 2$ ft

x ft

Figure 7.2

Since $-1 - \sqrt{7}$ is a negative number, it cannot be the width of the rectangle. If

$$x = -1 + \sqrt{7},$$

then

$$x + 2 = -1 + \sqrt{7} + 2 = 1 + \sqrt{7}.$$

The width is $-1 + \sqrt{7}$ feet, and the length is $1 + \sqrt{7}$ feet. We can check this by multiplying the length and the width:

$$(-1 + \sqrt{7})(1 + \sqrt{7}) = -1 + \sqrt{7} - \sqrt{7} + 7 = 6 \qquad \blacktriangleleft$$

Work Problems

When the solution to a problem is an irrational number, we can find the exact answer as well as a decimal approximation to the answer.

EXAMPLE 2 Lois can mow the lawn by herself in 2 hours less time than Louis takes to mow the lawn by himself. When they work together, it takes them only 6 hours to mow the lawn. How long would it take each of them to mow the lawn working alone? Find the exact and approximate answers.

Use a calculator to check the approximate answers in the original equation.

Solution If x represents the number of hours it takes Lois to mow the lawn, then $x + 2$ represents the number of hours it takes Louis to mow the lawn. We can make a table to classify all of the given information. See Table 7.2.

We write the equation for a work problem here in the same way that we wrote equations for work problems in Section 5.8:

$$\frac{1}{x} + \frac{1}{x + 2} = \frac{1}{6}$$

$$6x(x + 2)\frac{1}{x} + 6x(x + 2)\frac{1}{x + 2} = 6x(x + 2)\frac{1}{6} \qquad \text{Multiply by the LCD,} \atop 6x(x + 2).$$

$$6x + 12 + 6x = x^2 + 2x$$

$$12x + 12 = x^2 + 2x$$

$$-x^2 + 10x + 12 = 0$$

$$x^2 - 10x - 12 = 0 \qquad \text{Multiply each side by } -1 \text{ to check} \atop \text{for factoring.}$$

$$x = \frac{10 \pm \sqrt{(-10)^2 - 4(1)(-12)}}{2(1)} \qquad \text{Since the poly-} \atop \text{nomial is prime,} \atop \text{use the quadratic} \atop \text{formula.}$$

$$x = \frac{10 \pm \sqrt{148}}{2} = \frac{10 \pm 2\sqrt{37}}{2} = 5 \pm \sqrt{37}$$

Use a calculator to find that

$$x = 5 - \sqrt{37} \approx -1.08 \qquad \text{and} \qquad x = 5 + \sqrt{37} \approx 11.08.$$

TABLE 7.2

	Time	Portion done in one hour
Lois	x	$\dfrac{1}{x}$
Louis	$x + 2$	$\dfrac{1}{x + 2}$
Together	6	$\dfrac{1}{6}$

Since x must be positive, the time for Lois is $5 + \sqrt{37} \approx 11.08$ hours and the time for Louis is $7 + \sqrt{37} \approx 13.08$ hours.   ◄

Formulas

EXAMPLE 3 If an object is given an initial velocity of v_0 feet per second, from an altitude of s_0 feet, then its altitude S after t seconds is given by the formula

$$S = -16t^2 + v_0 t + s_0.$$

A soccer ball bounces straight up into the air off the head of a soccer player from an altitude of 6 feet with an initial velocity of 40 feet per second. How long does it take for the ball to reach the earth? Find the exact answer and an approximate answer.

Solution The time that it takes for the ball to reach the earth is the value of t for which S has a value of 0. To find t, we use $S = 0$, $v_0 = 40$, and $s_0 = 6$ in the given formula:

$$0 = -16t^2 + 40t + 6$$
$$16t^2 - 40t - 6 = 0$$
$$8t^2 - 20t - 3 = 0 \qquad \text{Divide each side by 2.}$$
$$t = \frac{20 \pm \sqrt{(-20)^2 - 4(8)(-3)}}{2(8)} \qquad \text{The polynomial is prime, so use the quadratic formula.}$$
$$t = \frac{20 \pm \sqrt{496}}{16} = \frac{20 \pm 4\sqrt{31}}{16} = \frac{5 \pm \sqrt{31}}{4}$$

Since the time is positive, we have

$$t = \frac{5 + \sqrt{31}}{4} \approx 2.64.$$

Use a calculator to check the approximate answer in the original equation.

It takes the ball approximately 2.64 seconds to reach the earth.   ◄

Warm-ups

True or false?

1. Two numbers that have a sum of 10 can be represented by x and $x + 10$. F

2. The area of a right triangle is one-half the product of the lengths of the legs. T

3. If the speed of a boat in still water is x mph and the current is 5 mph, then the speed of the boat moving with the current is $5x$ mph. F

4. If it takes Boudreaux x hours to eat a 50-pound bag of crawfish, then he eats $50/x$ of the bag per hour. F

5. If the Spirit of New Orleans went 900 miles in $x + 2$ hours, then its average speed was $900/(x + 2)$ miles per hour. T

6. The quantity $\dfrac{7 - \sqrt{50}}{2}$ is negative. T

7. If the length of one side of a square is $x + 9$ meters, then the area of the square is $x^2 + 81$ square meters. F

8. If Julia can mow her entire lawn in x hours, then she mows $1/x$ of the lawn per hour. T

9. If John's boat goes 20 miles per hour in still water, then against a 5-mph current, it will go 15 miles per hour. T

10. The quantity $(5 - \sqrt{27})$ is positive. F

7.4 EXERCISES

Find the exact solution to each problem. See Example 1.

1. The length of a rectangle is 2 meters longer than the width. If the area is 10 square meters, then what are the length and width? $L = 1 + \sqrt{11}$ meters, $W = -1 + \sqrt{11}$ meters

2. One leg of a right triangle is 4 centimeters longer than the other leg. If the area of this triangle is 8 square centimeters, then what are the lengths of the legs?

2. $-2 + 2\sqrt{5}$ centimeters, $2 + 2\sqrt{5}$ centimeters

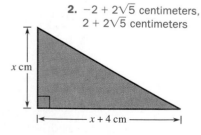

Figure for Exercise 2

3. If the diagonal of a square is 8 feet long, then what is the length of the side of the square? $4\sqrt{2}$ feet

4. If one side of a rectangle is 2 meters shorter than the other side and the diagonal is 10 meters long, then what are the dimensions of the rectangle?   6 meters by 8 meters

5. The base of a parallelogram is 6 inches longer than its height. If the area of the parallelogram is 10 square inches, then what are the base and height?

5. Base: $3 + \sqrt{19}$ inches, height: $-3 + \sqrt{19}$ inches

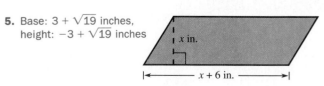

Figure for Exercise 5

6. Find two positive real numbers that have a sum of 8 and a product of 4. $4 + 2\sqrt{3}$ and $4 - 2\sqrt{3}$

Solve each problem. Give the exact answer and an approximate answer rounded to two decimal places. See Examples 2 *and* 3.

7. On Monday, Alberta picked berries from the strawberry patch, and Ernie sold the berries. On Tuesday, Ernie picked and Alberta sold, but it took him 2 hours longer to get the berries picked than it took Alberta. On Wednesday, they worked together and got all of the berries picked in 2 hours. How long did it take Ernie to pick the berries by himself? $3 + \sqrt{5}$ or 5.24 hours

8. Claude and Melvin read the water meters for the city of Ponchatoula. When Claude reads all of the meters by himself, it takes him a full day longer than it takes Melvin to read all of the meters by himself. If they can get the job done working together in 2 days, then how long does it take Claude by himself? $\dfrac{5 + \sqrt{17}}{2}$ or 4.56 days

9. Nancy traveled 6 miles upstream to do some fly fishing. It took her 20 minutes longer to get there than to return. If the current in the river is 2 miles per hour, then how fast will her boat go in still water? $2\sqrt{19}$ or 8.72 mph

10. Gladys and Bonita commute to work daily. Bonita drives 40 miles and averages 9 miles per hour more than Gladys. Gladys drives 50 miles, and she is on the road one-half hour longer than Bonita. How fast does each of them drive?   Gladys: 36 mph, Bonita: 45 mph

11. Olin's garden is currently 5 feet wide and 8 feet long. He bought enough okra seed to plant 100 square feet in okra. If he wants to increase the width and the length each by the same amount to plant all of his okra, then what should the increase be?

12. Lillian has a 5-foot-square bed of tulips. She plans to surround this bed with a crocus bed of uniform width. If she has enough crocus bulbs to plant 100 square feet of crocuses, then how wide should the crocus bed be?

13. A punter kicks a football straight up from a height of 4 feet with an initial velocity of 60 feet per second. How long will it take for the ball to reach the earth?

Photo for Exercise 9

11. $\dfrac{-13 + \sqrt{409}}{2}$
or 3.61 foot

12. $\dfrac{-5 + 5\sqrt{5}}{2}$
or 3.09 feet

13. $\dfrac{15 + \sqrt{241}}{8}$
or 3.82 seconds

14. $\dfrac{25 + \sqrt{655}}{4}$
or 12.65 seconds

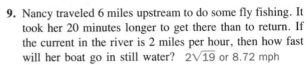

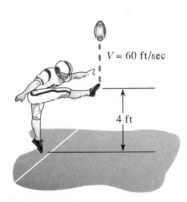

$V = 60$ ft/sec

4 ft

Figure for Exercise 13

14. Dwight accidentally fired his rifle straight into the air while sitting in his deer stand 30 feet off the ground. If the bullet left the barrel with a velocity of 200 feet per second, then how long did it take for the bullet to fall to the earth?

7.5 Complex Numbers

IN THIS SECTION:

- **Definition**
- **Operations with Complex Numbers**
- **Square Roots of Negative Numbers**
- **Complex Solutions to Quadratic Equations**
- **Summary**

In this chapter we have seen quadratic equations that have no solution in the set of real numbers. In this section we will learn that the set of real numbers is contained in the set of complex numbers. Quadratic equations that have no real solutions have solutions that are complex numbers.

Definition

Complex numbers can be used to complete the discussion of quadratic equations. They do not appear anywhere else in this text and so this section may be omitted if necessary.

The complex numbers are based on the symbol $\sqrt{-1}$. In the real number system this symbol has no meaning. In the set of complex numbers this symbol is given meaning. We call it i. We make the definition that

$$i = \sqrt{-1}$$

and

$$i^2 = -1.$$

DEFINITION
Complex Numbers

> The set of complex numbers is the set of all numbers of the form
>
> $$a + bi,$$
>
> where a and b are real numbers, $i = \sqrt{-1}$, and $i^2 = -1$.

In the complex number $a + bi$, a is called the **real part** and b is called the **imaginary part.** If $b \neq 0$, the number $a + bi$ is called an **imaginary number.**

In dealing with complex numbers, we treat $a + bi$ as if it were a binomial, i being a variable. Thus we would write $2 + (-3)i$ as $2 - 3i$. We agree that $2 + i3$, $3i + 2$, and $i3 + 2$ are just different ways of writing $2 + 3i$. Some examples of complex numbers are

$$2 + 3i, \quad -2 - 5i, \quad 0 + 4i, \quad 9 + 0i, \quad \text{and} \quad 0 + 0i.$$

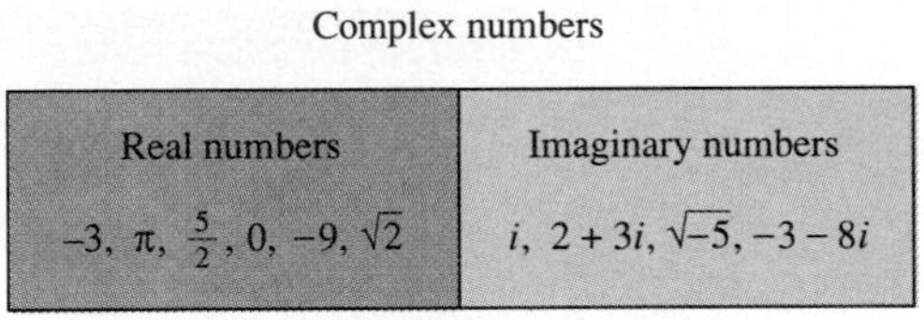

Complex numbers

Real numbers	Imaginary numbers
$-3, \pi, \frac{5}{2}, 0, -9, \sqrt{2}$	$i, \ 2 + 3i, \ \sqrt{-5}, -3 - 8i$

Figure 7.3

For simplicity we will write only $4i$ for $0 + 4i$. The complex number $9 + 0i$ is the real number 9, and $0 + 0i$ is the real number 0. Any complex number with $b = 0$ is a real number. The diagram in Fig. 7.3 shows the relationships between the complex numbers, the real numbers, and the imaginary numbers.

Operations with Complex Numbers

We perform the operations of addition and subtraction of complex numbers as if the complex numbers were binomials with i a variable. We define addition and subtraction of complex numbers as follows.

DEFINITION
Addition and Subtraction of Complex Numbers

The complex numbers $a + bi$ and $c + di$ are added and subtracted as follows:

$$(a + bi) + (c + di) = (a + c) + (b + d)i$$
$$(a + bi) - (c + di) = (a - c) + (b - d)i$$

EXAMPLE 1 Perform the indicated operations:

a) $(2 + 3i) + (4 + 5i)$ b) $(2 - 3i) + (-1 - i)$
c) $(3 + 4i) - (1 + 7i)$ d) $(2 - 3i) - (-2 - 5i)$

Solution

Remind students that to perform operations with complex numbers they must think of them as binomials with i being a variable.

a) $(2 + 3i) + (4 + 5i) = 6 + 8i$
b) $(2 - 3i) + (-1 - i) = 1 - 4i$
c) $(3 + 4i) - (1 + 7i) = 3 + 4i - 1 - 7i$
$$= 2 - 3i$$
d) $(2 - 3i) - (-2 - 5i) = 2 - 3i + 2 + 5i$
$$= 4 + 2i \qquad \blacktriangleleft$$

The formal definition of multiplication of complex numbers is stated as follows.

**DEFINITION
Multiplication of
Complex Numbers**

> The complex numbers $a + bi$ and $c + di$ are multiplied as follows:
>
> $$(a + bi)(c + di) = (ac - bd) + (ad + bc)i$$

Multiplication of complex numbers is actually much easier to perform than it appears in the definition. We multiply complex numbers just as we multiply polynomials. We can use the FOIL method, and we must remember that i times i is -1.

EXAMPLE 2   Perform the indicated operations:

a) $(-2 - 5i)(6 - 7i)$ **b)** $(5i)^2$ **c)** $(-5i)^2$
d) $(3 - 2i)(3 + 2i)$ **e)** $2i(1 - 3i)$

Solution

a) Use FOIL to multiply these complex numbers:

$$\begin{aligned}
(-2 - 5i)(6 - 7i) &= -12 + 14i - 30i + 35i^2 \\
&= -12 - 16i + 35(-1) \qquad i^2 = -1 \\
&= -12 - 16i - 35 \\
&= -47 - 16i
\end{aligned}$$

b) $(5i)^2 = 25i^2 = 25(-1) = -25$
c) $(-5i)^2 = (-5)^2 i^2 = 25(-1) = -25$
d) This is the product of a sum and a difference:

$$\begin{aligned}
(3 - 2i)(3 + 2i) &= 9 - 4i^2 \\
&= 9 - 4(-1) \\
&= 9 + 4 \\
&= 13
\end{aligned}$$

e) $2i(1 - 3i) = 2i - 6i^2$ **Distributive property**
$= 2i - 6(-1)$
$= 6 + 2i$ ◄

To divide a complex number by a real number, we divide each term by the real number. For example,

$$\frac{4 - 6i}{2} = 2 - 3i.$$

To understand division by a complex number, first notice that the product of the two imaginary numbers in Example 2(d) is a real number:

$$(3 - 2i)(3 + 2i) = 13$$

We say that $3 - 2i$ and $3 + 2i$ are **complex conjugates** of one another. Their product is a real number.

DEFINITION
Complex Conjugates

> The complex numbers $a + bi$ and $a - bi$ are called **complex conjugates** of one another. Their product is the real number $a^2 + b^2$.

We use the idea of complex conjugates to divide complex numbers.

DEFINITION
Division of
Complex Numbers

> We divide the complex number $a + bi$ by the complex number $c + di$ as follows:
>
> $$\frac{a + bi}{c + di} = \frac{(a + bi)(c - di)}{(c + di)(c - di)}$$

Note that the division process is similar to rationalizing the denominator. We *multiply the numerator and denominator of the quotient by the complex conjugate of the denominator*.

EXAMPLE 3 Perform the indicated operations:

a) $\dfrac{2}{3 - 4i}$ **b)** $\dfrac{6}{2 + i}$ **c)** $\dfrac{3 - 2i}{i}$

Solution

a) Multiply the numerator and denominator by $3 + 4i$, the conjugate of $3 - 4i$:

$$\frac{2}{3 - 4i} = \frac{2(3 + 4i)}{(3 - 4i)(3 + 4i)} = \frac{6 + 8i}{9 - 16i^2} = \frac{6 + 8i}{25} = \frac{6}{25} + \frac{8}{25}i$$

b) Multiply the numerator and denominator by $2 - i$, the conjugate of $2 + i$:

$$\frac{6}{2 + i} = \frac{6(2 - i)}{(2 + i)(2 - i)} = \frac{12 - 6i}{4 - i^2} = \frac{12 - 6i}{5} = \frac{12}{5} - \frac{6}{5}i$$

c) Multiply the numerator and denominator by $-i$, the conjugate of i:

$$\frac{3 - 2i}{i} = \frac{(3 - 2i)(-i)}{i(-i)} = \frac{-3i + 2i^2}{-i^2} = \frac{-3i - 2}{1} = -2 - 3i \qquad \blacktriangleleft$$

Square Roots of Negative Numbers

In Example 1 we saw that both

$$(5i)^2 = -25 \qquad \text{and} \qquad (-5i)^2 = -25.$$

Since the square of each of these complex numbers is -25, both $5i$ and $-5i$ are square roots of -25. When we use the radical notation, we write

$$\sqrt{-25} = 5i.$$

The square root of a negative number is not a real number, it is a complex number.

**DEFINITION
Square Root of a
Negative Number**

For any positive number b, $\sqrt{-b} = i\sqrt{b}$.

For example, $\sqrt{-9} = i\sqrt{9} = 3i$ and $\sqrt{-7} = i\sqrt{7}$. Note that the expression $\sqrt{7}i$ could easily be mistaken for the expression $\sqrt{7i}$, where i is under the radical. For this reason, when the coefficient of i is a radical, we write i preceding the radical.

EXAMPLE 4 Write each expression in the form $a + bi$, where a and b are real numbers:

a) $2 + \sqrt{-4}$ b) $\dfrac{2 + \sqrt{-12}}{2}$ c) $\dfrac{-2 - \sqrt{-18}}{3}$

Solution

a) $2 + \sqrt{-4} = 2 + i\sqrt{4} = 2 + 2i$

b) $\dfrac{2 + \sqrt{-12}}{2} = \dfrac{2 + i \cdot \sqrt{12}}{2}$

$\qquad = \dfrac{2 + 2i\sqrt{3}}{2} \qquad \sqrt{12} = \sqrt{4} \cdot \sqrt{3} = 2\sqrt{3}$

$\qquad = 1 + i\sqrt{3} \qquad$ Divide by 2.

c) $\dfrac{-2 - \sqrt{-18}}{3} = \dfrac{-2 - i\sqrt{18}}{3} = \dfrac{-2 - 3i\sqrt{2}}{3} = -\dfrac{2}{3} - i\sqrt{2}$ ◀

Complex Solutions to Quadratic Equations

The equation $x^2 = -4$ has no real solutions, but it has complex solutions:

$$x^2 = -4$$
$$x = \pm\sqrt{-4}$$
$$x = \pm i\sqrt{4}$$
$$x = \pm 2i$$

Check:

$$(2i)^2 = 4i^2 = 4(-1) = -4$$
$$(-2i)^2 = 4i^2 = -4$$

Both $2i$ and $-2i$ are solutions to the equation.

Consider the general quadratic equation

$$ax^2 + bx + c = 0,$$

where a, b, and c are real numbers. If the discriminant $b^2 - 4ac$ is positive, then it has two real solutions. If the discriminant is 0, then it has one real solution. If the discriminant is negative, it has two complex solutions. In the complex number system, all quadratic equations have solutions.

EXAMPLE 5 Find the complex solutions to the quadratic equations:

a) $x^2 - 2x + 5 = 0$ **b)** $2x^2 + 3x + 5 = 0$

Solution

a) To solve $x^2 - 2x + 5 = 0$, use $a = 1$, $b = -2$, and $c = 5$ in the quadratic formula:

$$x = \frac{2 \pm \sqrt{(-2)^2 - 4(1)(5)}}{2(1)}$$

$$= \frac{2 \pm \sqrt{-16}}{2} = \frac{2 \pm 4i}{2}$$

$$- 1 \pm 2i$$

The solutions are $1 - 2i$ and $1 + 2i$.

b) To solve $2x^2 + 3x + 5 = 0$, use $a = 2$, $b = 3$, and $c = 5$ in the quadratic formula:

$$x = \frac{-3 \pm \sqrt{(3)^2 - 4(2)(5)}}{2(2)}$$

$$-\frac{-3 \pm \sqrt{-31}}{4} - \frac{-3 \pm i\sqrt{31}}{4}$$

The solutions are $\dfrac{-3 + i\sqrt{31}}{4}$ and $\dfrac{-3 - i\sqrt{31}}{4}$. ◀

Summary

The following table summarizes the basic facts about complex numbers.

Complex Number Facts

1. Definition of i: $i = \sqrt{-1}$ and $i^2 = -1$.

2. A complex number has the form $a + bi$, where a and b are real numbers.

3. If b is a positive real number, then $\sqrt{-b} = i\sqrt{b}$.

4. Definition of operations:
   a) Addition: $(a + bi) + (c + di) = (a + c) + (b + d)i$
   b) Subtraction: $(a + bi) - (c + di) = (a - c) + (b - d)i$
   c) Multiplication: $(a + bi)(c + di) = (ac - bd) + (ad + bc)i$
   d) Division: $\dfrac{a + bi}{c + di} = \dfrac{(a + bi)(c - di)}{(c + di)(c - di)}$

5. The complex number $a + 0i$ is the real number a.

Warm-ups

True or false?

1. Zero is the only real number that is also a complex number. F
2. $\sqrt{-5} = 5i$ F
3. $\sqrt{-36} = \pm 6i$ F
4. Both $2i$ and $-2i$ are solutions to the equation $x^2 = 4$. F
5. $(3 + i) + (2 - 4i) = 5 - 3i$ T
6. $(4 - 2i) - (3 - 5i) = 1 + 3i$ T
7. $(4 - i)(4 + i) = 17$ T
8. $i^4 = 1$ T
9. If we consider complex numbers, then all quadratic equations have two distinct solutions. F
10. $(4 - 8i) \div 2 = 2 - 4i$ T

7.5 EXERCISES

Perform the indicated operations. See Example 1.

1. $(3 + 5i) + (2 + 4i)$ $5 + 9i$
2. $(8 + 3i) + (1 + 2i)$ $9 + 5i$
3. $(-1 + i) + (2 - i)$ 1
4. $(-2 - i) + (-3 + 5i)$ $-5 + 4i$
5. $(4 - 5i) - (2 + 3i)$ $2 - 8i$
6. $(3 - 2i) - (7 + 6i)$ $-4 - 8i$
7. $(-3 - 5i) - (-2 - i)$ $-1 - 4i$
8. $(-4 - 8i) - (-2 - 3i)$ $-2 - 5i$
9. $(8 - 3i) - (9 - 3i)$ -1
10. $(5 + 6i) - (-3 + 6i)$ 8

Perform the indicated operations. See Example 2.

11. $3(2 - 3i)$ $6 - 9i$
12. $-4(3 - 2i)$ $-12 + 8i$
13. $(6i)^2$ -36
14. $(3i)^2$ -9
15. $(-6i)^2$ -36
16. $(-3i)^2$ -9
17. $(2 + 3i)(3 - 5i)$ $21 - i$
18. $(4 - i)(3 - 6i)$ $6 - 27i$
19. $(5 - 2i)^2$ $21 - 20i$
20. $(3 + 4i)^2$ $-7 + 24i$
21. $(4 - 3i)(4 + 3i)$ 25
22. $(-3 + 5i)(-3 - 5i)$ 34
23. $(1 - i)(1 + i)$ 2
24. $(3 - i)(3 + i)$ 10

Perform the indicated operations. See Example 3.

25. $(2 - 6i) \div 2$ $1 - 3i$

26. $(-3 + 6i) \div (-3)$ $1 - 2i$

27. $\dfrac{-2 + 8i}{2}$ $-1 + 4i$

28. $\dfrac{6 - 9i}{-3}$ $-2 + 3i$

29. $\dfrac{4 - 12i}{3 + i}$ $-4i$

30. $\dfrac{-4 + 10i}{5 - i}$ $-\dfrac{15}{13} + \dfrac{23}{13}i$

31. $\dfrac{4i}{3 + 2i}$ $\dfrac{8}{13} + \dfrac{12}{13}i$

32. $\dfrac{5}{4 - 5i}$ $\dfrac{20}{41} + \dfrac{25}{41}i$

33. $\dfrac{2 + i}{2 - i}$ $\dfrac{3}{5} + \dfrac{4}{5}i$

34. $\dfrac{i - 5}{5 - i}$ -1

Write each expression in the form a + bi, where a and b are real numbers. See Example 4.

35. $5 + \sqrt{-9}$ $5 + 3i$

36. $6 + \sqrt{-16}$ $6 + 4i$

37. $-3 - \sqrt{-7}$ $-3 - i\sqrt{7}$

38. $2 - \sqrt{-3}$ $2 - i\sqrt{3}$

39. $\dfrac{-2 + \sqrt{-12}}{2}$ $-1 + i\sqrt{3}$

40. $\dfrac{-6 - \sqrt{-18}}{3}$ $-2 - i\sqrt{2}$

41. $\dfrac{-8 - \sqrt{-20}}{-4}$ $2 + \dfrac{1}{2}i\sqrt{5}$

42. $\dfrac{6 + \sqrt{-24}}{-2}$ $-3 - i\sqrt{6}$

43. $\dfrac{-4 + \sqrt{-28}}{6}$ $-\dfrac{2}{3} + \dfrac{1}{3}i\sqrt{7}$

44. $\dfrac{6 - \sqrt{-45}}{6}$ $1 - \dfrac{1}{2}i\sqrt{5}$

45. $\dfrac{-2 + \sqrt{-100}}{-10}$ $\dfrac{1}{5} - i$

46. $\dfrac{-3 + \sqrt{-81}}{-9}$ $\dfrac{1}{3} - i$

Find the complex solutions to each quadratic equation. See Example 5.

47. $x^2 + 81 = 0$ $-9i, 9i$

48. $x^2 + 100 = 0$ $10i, 10i$

49. $x^2 + 5 = 0$ $-i\sqrt{5}, i\sqrt{5}$

50. $x^2 + 6 = 0$ $-i\sqrt{6}, i\sqrt{6}$

51. $3y^2 + 2 = 0$ $-i\dfrac{\sqrt{6}}{3}, i\dfrac{\sqrt{6}}{3}$

52. $5y^2 + 3 = 0$ $-i\dfrac{\sqrt{15}}{5}, i\dfrac{\sqrt{15}}{5}$

53. $x^2 - 4x + 5 = 0$ $2 - i, 2 + i$

54. $x^2 - 6x + 10 = 0$ $3 - i, 3 + i$

55. $y^2 + 13 = 6y$ $3 - 2i, 3 + 2i$

56. $y^2 + 29 = 4y$ $2 - 5i, 2 + 5i$

57. $x^2 - 4x + 7 = 0$ $2 - i\sqrt{3}, 2 + i\sqrt{3}$

58. $x^2 - 10x + 27 = 0$ $5 - i\sqrt{2}, 5 + i\sqrt{2}$

59. $9y^2 - 12y + 5 = 0$ $\dfrac{2 - i}{3}, \dfrac{2 + i}{3}$

60. $2y^2 - 2y + 1 = 0$ $\dfrac{1}{2} - \dfrac{1}{2}i, \dfrac{1}{2} + \dfrac{1}{2}i$

61. $x^2 - x + 1 = 0$ $\dfrac{1 - i\sqrt{3}}{2}, \dfrac{1 + i\sqrt{3}}{2}$

62. $4x^2 - 20x + 27 = 0$ $\dfrac{5}{2} - \dfrac{1}{2}i\sqrt{2}, \dfrac{5}{2} + \dfrac{1}{2}i\sqrt{2}$

63. $-4x^2 + 8x - 9 = 0$ $\dfrac{2 - i\sqrt{5}}{2}, \dfrac{2 + i\sqrt{5}}{2}$

64. $-9x^2 + 12x - 10 = 0$ $\dfrac{2}{3} - \dfrac{1}{3}i\sqrt{6}, \dfrac{2}{3} + \dfrac{1}{3}i\sqrt{6}$

Solve each problem.

65. Evaluate $(2 - 3i)^2 + 4(2 - 3i) - 9$. $-6 - 24i$

66. Evaluate $(3 + 5i)^2 - 2(3 + 5i) + 5$. $-17 + 20i$

67. What is the value of $x^2 - 8x + 17$, if $x = 4 - i$? 0

68. What is the value of $x^2 - 6x + 34$, if $x = 3 + 5i$? 0

69. Find the product $[x - (6 - i)][x - (6 + i)]$. $x^2 - 12x + 37$

70. Find the product $[x - (3 + 7i)][x - (3 - 7i)]$. $x^2 - 6x + 58$

Wrap-up

CHAPTER 7

SUMMARY

	Concepts	Examples
Quadratic equation	Any equation that can be written in the form $$ax^2 + bx + c = 0$$ where a, b, and c are real numbers and $a \neq 0$.	$x^2 = 10$ $(x + 3)^2 = 8$ $x^2 + 5x - 7 = 0$
Methods for solving quadratic equations	Factoring	$x^2 + 5x + 6 = 0$ $(x + 3)(x + 2) = 0$
	Square root property	$(x - 3)^2 = 6$ $\quad x - 3 = \pm\sqrt{6}$
	Completing the square (works on any quadratic): Take one-half of the middle term, square it, then add it to each side.	$x^2 + 6x \quad\quad = -7$ $x^2 + 6x + 9 = -7 + 9$ $\quad (x + 3)^2 = 2$
	Quadratic formula (works on any quadratic) $$x = \frac{-b \pm \sqrt{b^2 - 4ac}}{2a}$$	$2x^2 - 3x - 6 = 0$ $x = \dfrac{3 \pm \sqrt{9 - 4(2)(-6)}}{2(2)}$
Types of solutions	Determined by the discriminant $b^2 - 4ac$ $\quad b^2 - 4ac > 0$: two real solutions	$x^2 + 5x - 9 = 0$ $5^2 - 4(1)(-9) > 0$
	$\quad b^2 - 4ac = 0$: one real solution	$x^2 + 6x + 9 = 0$ $6^2 - 4(1)(9) = 0$
	$\quad b^2 - 4ac < 0$: no real solutions $\quad\quad\quad\quad\quad$ (two complex solutions)	$x^2 + 3x + 10 = 0$ $3^2 - 4(1)(10) < 0$
Complex numbers	Numbers of the form $a + bi$, where a and b are real $i = \sqrt{-1},\ i^2 = -1$	$5 + 4i$ $-3i$ $\sqrt{2} - i\sqrt{3}$
Square root of a negative number	If b is a positive real number, then $\sqrt{-b} = i\sqrt{b}$.	$\sqrt{-3} = i\sqrt{3}$ $\sqrt{-4} = i\sqrt{4} = 2i$
Complex conjugates	The complex numbers $a + bi$ and $a - bi$ are called complex conjugates of one another. Their product is real.	$(5 + 2i)(5 - 2i) = 25 + 4 = 29$

Complex number operations	Addition: Add the like terms.	$(2 + 3i) + (3 - 5i) = 5 - 2i$

Subtraction: Subtract the like terms.

$(2 - 5i) - (4 - 2i) = -2 - 3i$

Multiplication: Multiply in the same way that we multiply polynomials. Remember that $i^2 = -1$.

$(3 - 4i)(2 + 5i) = 26 + 7i$

Division: Multiply the quotient by the conjugate of the denominator and simplify.

$$\frac{4 - 6i}{5 + 2i} = \frac{(4 - 6i)(5 - 2i)}{(5 + 2i)(5 - 2i)}$$

REVIEW EXERCISES

7.1 *Solve each equation.*

1. $x^2 - 9 = 0$ $-3, 3$

2. $x^2 - 1 = 0$ $-1, 1$

3. $x^2 - 18 = 0$ $-3\sqrt{2}, 3\sqrt{2}$

4. $x^2 - 45 = 0$ $-3\sqrt{5}, 3\sqrt{5}$

5. $x^2 - 9x = 0$ $0, 9$

6. $x^2 - x = 0$ $0, 1$

7. $x^2 \quad x = 2$ $-1, 2$

8. $x^2 - 9x = 10$ $-1, 10$

9. $(x - 9)^2 = 10$ $9 \quad \sqrt{10}, 9 + \sqrt{10}$

10. $(x + 5)^2 = 14$ $-5 - \sqrt{14}, -5 + \sqrt{14}$

11. $4x^2 - 12x + 9 = 0$ $\dfrac{3}{2}$

12. $9x^2 + 6x + 1 = 0$ $-\dfrac{1}{3}$

13. $t^2 - 9t + 20 = 0$ $4, 5$

14. $s^2 - 4s + 3 = 0$ $1, 3$

15. $\dfrac{x}{2} = \dfrac{7}{x + 5}$ $-7, 2$

16. $\sqrt{x + 4} = \dfrac{2x - 1}{3}$ 5

7.2 *Solve by completing the square.*

17. $x^2 + 4x - 7 = 0$ $-2 - \sqrt{11}, -2 + \sqrt{11}$

18. $x^2 + 6x - 3 = 0$ $-3 - 2\sqrt{3}, -3 + 2\sqrt{3}$

19. $x^2 + 3x - 28 = 0$ $7, 4$

20. $x^3 - x - 6 = 0$ $-2, 3$

21. $x^2 + 3x - 5 = 0$ $\dfrac{-3 - \sqrt{29}}{2}, \dfrac{-3 + \sqrt{29}}{2}$

22. $x^2 + \dfrac{4}{3}x - \dfrac{1}{3} = 0$ $\dfrac{2 \quad \sqrt{7}}{3}, \dfrac{-2 + \sqrt{7}}{3}$

23. $2x^2 + 9x - 5 = 0$ $-5, \dfrac{1}{2}$

24. $2x^2 + 6x - 5 = 0$ $\dfrac{-3 - \sqrt{19}}{2}, \dfrac{-3 + \sqrt{19}}{2}$

7.3 *Find the value of the discriminant, and tell how many real solutions each equation has.*

25. $25t^2 - 10t + 1 = 0$ 0, one

26. $3x^2 + 2 = 0$ -24, none

27. $-3w^2 + 4w - 5 = 0$ -44, none

28. $5x^2 - 7x = 0$ 49, two

29. $-3v^2 + 4v = -5$ 76, two

30. $49u^2 + 42u + 9 = 0$ 0, one

Use the quadratic formula to solve each equation.

31. $6x^2 + x - 2 = 0$ $-\dfrac{2}{3}, \dfrac{1}{2}$

32. $-6x^2 + 11x + 10 = 0$ $-\dfrac{2}{3}, \dfrac{5}{2}$

33. $x^2 - x = 4$ $\dfrac{1 - \sqrt{17}}{2}, \dfrac{1 + \sqrt{17}}{2}$

34. $y^2 - 2y = 4$ $1 - \sqrt{5}, 1 + \sqrt{5}$

35. $5x^2 - 6x - 1 = 0$ $\dfrac{3 - \sqrt{14}}{5}, \dfrac{3 + \sqrt{14}}{5}$

36. $3x^2 - 5x = 0$ $0, \dfrac{5}{3}$

7.4 *For each problem find the exact and approximate answers. Round the decimal answers to three decimal places.*

37. Chuck is standing 12 meters from a tree, watching a bird's nest that is 5 meters above eye level. Find the distance from Chuck's eyes to the nest. 13 meters

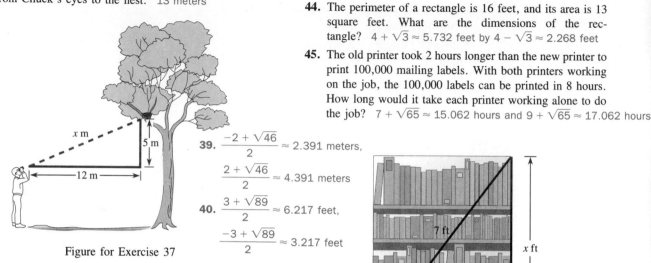

x m

5 m

12 m

Figure for Exercise 37

38. Find the diagonal of a square if the length of each side is 20 yards. $20\sqrt{2} \approx 28.284$ yards

39. The hypotenuse of a right triangle measures 5 meters, and one leg is 2 meters longer than the other. Find the lengths of the legs.

40. The width of a rectangular bookcase is 3 feet shorter than the height. If the diagonal is 7 feet, then what are the dimensions of the bookcase?

41. The base of a triangle is 4 inches longer than the height. If the area of the triangle is 20 square inches, then what are the lengths of the base and height?

42. The base of a parallelogram is 1 meter longer than the height. If the area of the parallelogram is 8 square meters, then what are the lengths of the base and height?

42. Base: $\dfrac{1 + \sqrt{33}}{2} \approx 3.372$ meters, height: $\dfrac{-1 + \sqrt{33}}{2} \approx 2.372$ meters

43. Find two positive real numbers whose sum is 6 and whose product is 7. $3 + \sqrt{2} \approx 4.414$ and $3 - \sqrt{2} \approx 1.586$

44. The perimeter of a rectangle is 16 feet, and its area is 13 square feet. What are the dimensions of the rectangle? $4 + \sqrt{3} \approx 5.732$ feet by $4 - \sqrt{3} \approx 2.268$ feet

45. The old printer took 2 hours longer than the new printer to print 100,000 mailing labels. With both printers working on the job, the 100,000 labels can be printed in 8 hours. How long would it take each printer working alone to do the job? $7 + \sqrt{65} \approx 15.062$ hours and $9 + \sqrt{65} \approx 17.062$ hours

39. $\dfrac{-2 + \sqrt{46}}{2} \approx 2.391$ meters,

$\dfrac{2 + \sqrt{46}}{2} \approx 4.391$ meters

40. $\dfrac{3 + \sqrt{89}}{2} \approx 6.217$ feet,

$\dfrac{-3 + \sqrt{89}}{2} \approx 3.217$ feet

7 ft

x ft

x − 3 ft

Figure for Exercise 40

46. When Blake uses his old tiller, it takes him 3 hours longer to till the garden than it takes Cassie using her new tiller. If Cassie will not let Blake use her new tiller, and they can till the garden together in 6 hours, then how long would it take each one working alone?

41. Base: $2 + 2\sqrt{11} \approx 8.633$ inches, height: $-2 + 2\sqrt{11} \approx 4.633$ inches

46. Cassie: $\dfrac{9 + \sqrt{153}}{2} \approx 10.685$ hours, Blake: $\dfrac{15 + \sqrt{153}}{2} \approx 13.685$ hours

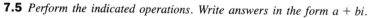

7.5 *Perform the indicated operations. Write answers in the form a + bi.*

47. $(-2 + 3i) + (5 - 6i)$ $3 - 3i$

48. $(2 - 5i) + (-9 - 4i)$ $-7 - 9i$

49. $(-5 + 4i) - (-2 - 3i)$ $-3 + 7i$

50. $(1 - i) - (1 + i)$ $-2i$

51. $(2 - 9i)(3 + i)$ $15 - 25i$

52. $2i - 3(6 - 2i)$ $8i - 18$

53. $(3 + 8i)^2$ $-55 + 48i$

54. $(-5 - 2i)(-5 + 2i)$ 29

55. $\dfrac{-2 - \sqrt{-8}}{2}$ $-1 - i\sqrt{2}$

56. $\dfrac{-6 + \sqrt{-54}}{-3}$ $2 - i\sqrt{6}$

57. $\dfrac{1 + 3i}{6 - i}$ $\dfrac{3}{37} + \dfrac{19}{37}i$

58. $\dfrac{3i}{8 + 3i}$ $\dfrac{9}{73} + \dfrac{24}{73}i$

59. $\dfrac{5 + i}{4 - i}$ $\dfrac{19}{17} + \dfrac{9}{17}i$

60. $\dfrac{3 + 2i}{i}$ $2 - 3i$

Find the complex solutions to the quadratic equations.

61. $x^2 + 121 = 0$ $-11i, 11i$

62. $x^2 + 120 = 0$ $-2i\sqrt{30}, 2i\sqrt{30}$

63. $x^2 - 16x + 65 = 0$ $8 - i, 8 + i$

64. $x^2 - 10x + 28 = 0$ $5 - i\sqrt{3}, 5 + i\sqrt{3}$

65. $2x^2 - 3x + 9 = 0$ $\dfrac{3 - 3i\sqrt{7}}{4}, \dfrac{3 + 3i\sqrt{7}}{4}$

66. $3x^2 - 6x + 4 = 0$ $\dfrac{3 - i\sqrt{3}}{3}, \dfrac{3 + i\sqrt{3}}{3}$

CHAPTER 7 TEST

Calculate the value of $b^2 - 4ac$ and state how many real solutions each equation has.

1. $9x^2 - 12x + 4 = 0$ 0, one

2. $-2x^2 + 3x - 5 = 0$ -31, none

3. $-2x^2 + 5x - 1 = 0$ 17, two

Solve by using the quadratic formula.

4. $5x^2 + 2x - 3 = 0$ $-1, \dfrac{3}{5}$

5. $2x^2 - 4x - 3 = 0$ $\dfrac{2 - \sqrt{10}}{2}, \dfrac{2 + \sqrt{10}}{2}$

Solve by completing the square.

6. $x^2 + 4x - 21 = 0$ $-7, 3$

7. $x^2 + 3x - 5 = 0$ $\dfrac{-3 - \sqrt{29}}{2}, \dfrac{-3 + \sqrt{29}}{2}$

Solve by any method.

8. $x(x + 1) = 20$ $-5, 4$

9. $x^2 - 28x + 75 = 0$ 3, 25

10. $\dfrac{x - 1}{3} = \dfrac{x + 1}{2x} - \dfrac{1}{2}$ 3

Find the exact solution to the problem.

11. Find two positive numbers that have a sum of 10 and a product of 23. $5 - \sqrt{2}, 5 + \sqrt{2}$

Perform the indicated operations. Write answers in the form $a + bi$.

12. $(2 - 3i) + (8 + 6i)$ $10 + 3i$

13. $(-2 - 5i) - (4 - 12i)$ $-6 + 7i$

14. $(-6i)^2$ -36

15. $(3 - 5i)(4 + 6i)$ $42 - 2i$

16. $(8 - 2i)(8 + 2i)$ 68

17. $(4 - 6i) \div 2$ $2 - 3i$

18. $\dfrac{-2 + \sqrt{-12}}{2}$ $-1 + i\sqrt{3}$

19. $\dfrac{6 - \sqrt{-18}}{-3}$ $-2 + i\sqrt{2}$

20. $\dfrac{5i}{4 + 3i}$ $\dfrac{3}{5} + \dfrac{4}{5}i$

Find the complex solutions to the quadratic equations.

21. $x^2 + 6x + 12 = 0$ $-3 - i\sqrt{3}, -3 + i\sqrt{3}$

22. $-5x^2 + 6x - 5 = 0$ $\dfrac{3}{5} - \dfrac{4}{5}i, \dfrac{3}{5} + \dfrac{4}{5}i$

Tying It All Together

CHAPTERS 1–7

Solve each equation.

1. $2x - 1 = 0$ $\quad \dfrac{1}{2}$

2. $2(x - 1) = 0$ $\quad 1$

3. $2x^2 - 1 = 0$ $\quad -\dfrac{\sqrt{2}}{2}, \dfrac{\sqrt{2}}{2}$

4. $(2x - 1)^2 = 8$ $\quad \dfrac{1 - 2\sqrt{2}}{2}, \dfrac{1 + 2\sqrt{2}}{2}$

5. $2x^2 - 4x - 1 = 0$ $\quad \dfrac{2 - \sqrt{6}}{2}, \dfrac{2 + \sqrt{6}}{2}$

6. $2x^2 - 4x = 0$ $\quad 0, 2$

7. $2x^2 + x = 1$ $\quad -1, \dfrac{1}{2}$

8. $x - 2 = \sqrt{2x - 1}$ $\quad 5$

9. $\dfrac{1}{x} = \dfrac{x}{2x - 15}$
$$1 - i\sqrt{14}, 1 + i\sqrt{14}$$

10. $\dfrac{1}{x} - \dfrac{1}{x - 1} = -\dfrac{1}{2}$ $\quad -1, 2$

Solve each equation for y.

11. $5x - 4y = 8$ $\quad y = \dfrac{5}{4}x - 2$

12. $3x - y = 9$ $\quad y = 3x - 9$

13. $\dfrac{y - 4}{x + 2} = \dfrac{2}{3}$ $\quad y = \dfrac{2}{3}x + \dfrac{16}{3}$

14. $ay + b = 0$ $\quad y = -\dfrac{b}{a}$

15. $ay^2 + by + c = 0$ $\quad y = \dfrac{-b \pm \sqrt{b^2 - 4ac}}{2a}$

16. $y - 1 = -\dfrac{2}{3}(x - 9)$ $\quad y = -\dfrac{2}{3}x + 7$

17. $\dfrac{2}{3}x + \dfrac{1}{2}y = \dfrac{1}{9}$ $\quad y = -\dfrac{4}{3}x + \dfrac{2}{9}$

18. $x^2 + y^2 = a^2$ $\quad y = \pm\sqrt{a^2 - x^2}$

Let $m = \dfrac{y_1 - y_2}{x_1 - x_2}$. *Find the value of m for each of the following choices of* x_1, x_2, y_1, *and* y_2.

19. $x_1 = 1$, $x_2 = 5$, $y_1 = 3$, $y_2 = 5$ $\quad \dfrac{1}{2}$

20. $x_1 = -3$, $x_2 = 5$, $y_1 = 5$, $y_2 = 7$ $\quad \dfrac{1}{4}$

21. $x_1 = 10$, $x_2 = -4$, $y_1 = 8$, $y_2 = 6$ $\quad \dfrac{1}{7}$

22. $x_1 = 3$, $x_2 = 5$, $y_1 = 8$, $y_2 = 4$ $\quad -2$

23. $x_1 = -2$, $x_2 = -4$, $y_1 = -6$, $y_2 = 8$ $\quad -7$

24. $x_1 = -3$, $x_2 = 5$, $y_1 = 9$, $y_2 = -7$ $\quad -2$

Linear Equations in Two Variables

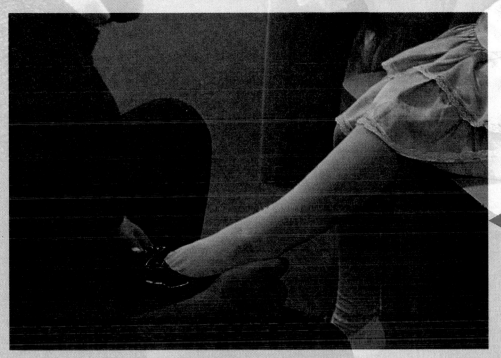

According to the Sears catalog, if a child's foot is $7\frac{3}{4}$ inches long, then the child wears a size 13 shoe. If a child has a foot that is $5\frac{3}{4}$ inches long, then the child wears a size 7 shoe. The shoe size is determined by the length of the foot. Not mentioned in the Sears catalog is the fact that there is a linear equation relating shoe size and length of the foot. This fact allows us to determine a formula relating the two variables, foot size and shoe size. What size shoe fits a child with a $6\frac{1}{4}$-inch foot?

In this chapter we will study linear equations and in Section 8.6 we will learn how to determine a linear equation expressing one variable in terms of another. This is precisely what we need to answer the shoe size question. We will solve this problem in Exercise 21 of Section 8.6.

8.1 The Rectangular Coordinate System

IN THIS SECTION:

- Ordered Pairs
- The Rectangular Coordinate System
- Plotting Points
- Graphing an Equation

In Chapter 1 we learned to graph numbers on a number line. We also used number lines to illustrate the solution to inequalities in Chapter 2. In this section we learn to graph pairs of numbers in a coordinate system made up of a pair of number lines. We will use this coordinate system to illustrate the solution to equations and inequalities in two variables.

Ordered Pairs

The equation $y = 2x - 1$ is an equation in two variables. This equation is satisfied if we choose a value for x and a value for y that make it true. If we choose $x = 2$ and $y = 3$, then

$$y = 2x - 1$$

becomes

$$\underset{\underset{y}{\uparrow}}{3} = 2(\underset{\underset{x}{\uparrow}}{2}) - 1.$$

Since this is a true statement, we say that the pair of numbers $x = 2$ and $y = 3$ satisfies the equation. The notation $(2, 3)$ is used to represent

$$x = 2 \quad \text{and} \quad y = 3.$$

We call $(2, 3)$ an **ordered pair.** The format is always to write the value for x first and the value for y second. The first number of the ordered pair is called the **x-coordinate,** and the second number is called the **y-coordinate.** Note that the ordered pair $(3, 2)$ does not satisfy the equation $y = 2x - 1$, since

$$\underset{\underset{y}{\uparrow}}{2} \neq 2(\underset{\underset{x}{\uparrow}}{3}) - 1.$$

There are infinitely many ordered pairs that satisfy the equation $y = 2x - 1$. It is easy to find some examples of them. Choose any value for x, say $x = -5$, and then calculate y by using the equation

$$y = 2(-5) - 1 = -10 - 1 = -11.$$

The ordered pair $(-5, -11)$ satisfies the equation $y = 2x - 1$.

EXAMPLE 1 Each of the ordered pairs below is missing one coordinate. Complete each ordered pair so that it satisfies the equation $y = -3x + 4$:

a) $(2, \quad)$ **b)** $(\quad , -5)$ **c)** $(0, \quad)$

Solution

a) The x-coordinate of $(2, \quad)$ is 2. Let $x = 2$ in the equation $y = -3x + 4$:

$$y = -3(2) + 4 = -6 + 4 = -2$$

The ordered pair $(2, -2)$ satisfies the equation.

b) The y-coordinate of $(\quad , -5)$ is -5. Let $y = -5$ in the equation $y = -3x + 4$:

$$-5 = -3x + 4$$
$$-9 = -3x$$
$$3 = x$$

The ordered pair $(3, -5)$ satisfies the equation.

c) If $x = 0$, then $y = -3(0) + 4 = 4$. So the ordered pair $(0, 4)$ satisfies the equation.   ◀

The Rectangular Coordinate System

To better understand ordered pairs of numbers, we will study the **rectangular (or Cartesian) coordinate system.** The rectangular coordinate system consists of two number lines drawn at a right angle to one another, intersecting at zero on each number line. On the horizontal number line the positive numbers are to the right of zero, and on the vertical number line the positive numbers are above zero. The rectangular coordinate system is shown in Fig. 8.1

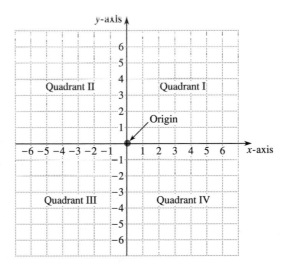

Figure 8.1

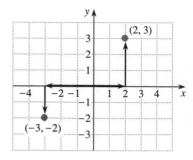

Figure 8.2

The horizontal number line is called the **x-axis,** and the vertical number line is called the **y-axis.** The point at which they intersect is called the **origin.** The two number lines divide the plane into four regions called **quadrants.** They are numbered as shown in Fig. 8.1. The quadrants do not include any points on the axes.

Plotting Points

Just as every real number corresponds to a point on the number line, *every pair of real numbers corresponds to a point in the rectangular coordinate system.* For example, the point corresponding to the pair (2, 3) is found by starting at the origin and moving two units to the right and then three units up. The point corresponding to the pair (−3, −2) is found by starting at the origin and moving three units to the left and then two units down. Both of these points are shown in Fig. 8.2.

When we locate a point in the rectangular coordinate system, we say that we are **plotting the point.** Since ordered pairs of numbers correspond to points in the coordinate plane, we frequently refer to an ordered pair as a point.

EXAMPLE 2 Locate (or plot) the points (2, 5), (−1, 4), (−3, −4), and (3, −2).

Solution To locate (2, 5), start at the origin, move two units to the right, then up five units. To locate (−1, 4), start at the origin, move one unit to the left, then up four units. All four points are shown in Fig. 8.3. ◀

Graphing an Equation

Consider again the equation

$$y = 2x - 1$$

and find several points that satisfy this equation. If $x = 1$, then

$$y = 2(1) - 1 = 1.$$

The ordered pair (1, 1) satisfies the equation. Figure 8.4 shows a table of values for x and y that satisfy the equation and a rectangular coordinate system with those points located (or graphed).

Notice how the points graphed in Fig. 8.4 appear to lie in a straight line. If we choose x to be any real number and find the point (x, y) that satisfies the equation $y = 2x - 1$, we will get another point along this line. Plotting the points (x, y) for all possible choices of x will give the complete line shown in Fig. 8.5. We say that this line is the graph of the equation $y = 2x - 1$. The graph gives us a picture of all ordered pairs that satisfy the equation.

When we draw the graph of the points that satisfy the equation, we say that we are **graphing the equation.** There are infinitely many ordered pairs that satisfy $y = 2x - 1$. All of them lie on the line shown in Fig. 8.5.

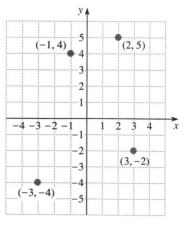

Figure 8.3

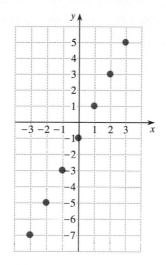

$$y = 2x - 1$$

x	y	
1	1	
2	3	
3	5	← $5 = 2(3) - 1$
0	−1	
−1	−3	← $-3 = 2(-1) - 1$
−2	−5	
−3	−7	

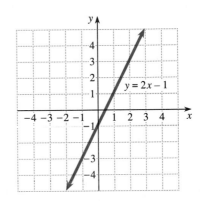

Figure 8.4   **Figure 8.5**

Of course it does not take five points to determine a line, but it is good practice.

EXAMPLE 3   Graph the equation $y + 3x = 2$. Plot at least five points.

Solution   First solve the equation for y:

$$y = -3x + 2$$

Next, we arbitrarily select values for x and calculate the corresponding value for y. The table of values and the resulting line are shown in Fig. 8.6.

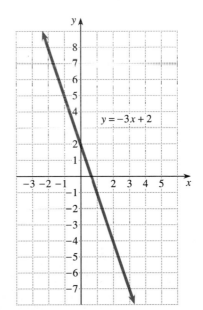

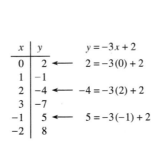

$$y = -3x + 2$$

x	y	
0	2	← $2 = -3(0) + 2$
1	−1	
2	−4	← $-4 = -3(2) + 2$
3	−7	
−1	5	← $5 = -3(-1) + 2$
−2	8	

Figure 8.6   ◄

EXAMPLE 4   Graph the equation $0 \cdot y + x = 3$. Plot at least five points.

Solution   If we choose a value of 3 for x, then we can choose any number for y, since y is multiplied by 0. A table of values and the resulting graph are shown in Fig. 8.7. The equation $0 \cdot y + x = 3$ is usually written simply as $x = 3$.

x	y
3	-4
3	-2
3	0
3	1
3	3

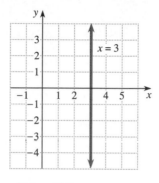

Figure 8.7   ◄

All of the equations we have considered so far have involved single-digit numbers. If an equation involves large numbers, then we must change the scale on the x-axis, the y-axis, or both to accommodate the numbers involved. Make a table of points to be graphed, then change the scale on the coordinate system so that the points can be graphed. The change of scale is arbitrary, and the graph will look different for different scales.

EXAMPLE 5   Graph the equation $y = 20x + 500$. Plot at least five points.

Solution   A table of values is shown in Fig. 8.8. To fit these points onto a graph, we change the scale on the x-axis to let each division represent ten units and change the scale on the y-axis to let each division represent 200 units. The graph is shown in Fig. 8.8.

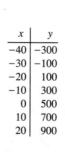

x	y
-40	-300
-30	-100
-20	100
-10	300
0	500
10	700
20	900

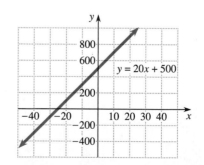

Figure 8.8   ◄

Warm-ups

True or false?

1. The point $(2, 4)$ satisfies the equation $2y - 3x = -8$.   F
2. If the point $(1, 5)$ satisfies an equation, then the point $(5, 1)$ also satisfies the equation.   F
3. The origin is in quadrant I.   F
4. The point $(4, 0)$ is on the y-axis.   F
5. The graph of $0 \cdot y + x = 9$ is the same as the graph of $x = 9$.   T
6. The graph of $x = -5$ is a vertical line.   T
7. The graph of $0 \cdot x + y = 6$ is a horizontal line.   T
8. If the point $(a, 2)$ satisfies the equation $x + 2y = 5$, then a must equal 1.   T
9. The point $(5, -3)$ is in quadrant II.   F
10. The point $(-349, 0)$ is on the x-axis.   T

8.1 EXERCISES

Complete the given ordered pairs so that each ordered pair satisfies the given equation. See Example 1.

1. $y = 3x + 9$: $(0, \ )$, $(\ , 24)$, $(2, \ )$ $(0, 9)$, $(5, 24)$, $(2, 15)$
2. $y = 2x + 5$: $(8, \ )$, $(-1, \ )$, $(\ , -1)$
3. $y = -3x - 7$: $(0, \ )$, $(-4, \ )$, $(\ , -1)$
4. $y = 5x - 3$: $(\ , 2)$, $(-3, \ )$, $(0, \ )$
5. $y = -12x + 5$: $(0, \ )$, $(10, \ )$, $(\ , 17)$
6. $y = 18x + 200$: $(1, \ )$, $(-10, \ )$, $(\ , 200)$

  2. $(8, 21)$, $(-1, 3)$, $(-3, -1)$ 3. $(0, -7)$, $(-4, 5)$, $(-2, -1)$ 4. $(-1, 2)$, $(-3, 12)$, $(0, -3)$

Plot the following points on a rectangular coordinate system. See Example 2.   Graphs in Answers sections.

7. $(1, 5)$
8. $(4, 3)$
9. $(-2, 1)$
10. $(-3, 5)$
11. $(3, -1/2)$
12. $(2, -1/3)$
13. $(-2, -4)$
14. $(-1/2, -5)$
15. $(0, 3)$
16. $(0, 2)$
17. $(-3, 0)$
18. $(5, 0)$
19. $(\pi, 1)$
20. $(-2, \pi)$
21. $(\sqrt{2}, 4)$
22. $(-3, \sqrt{2})$
23. $(0, 0)$
24. $(-2, -1/3)$
25. $(-2/3, 4)$
26. $(1, 1)$

  5. $(0, 5)$, $(10, -115)$, $(-1, 17)$ 6. $(1, 218)$, $(-10, 20)$, $(0, 200)$

Graph each of the following equations. Plot at least five points for each. See Examples 3 and 4.   Graphs in Answers sections.

27. $y = x + 1$
28. $y = x - 1$
29. $y = 2x + 1$
30. $y = 3x - 1$
31. $y = 3x - 2$
32. $y = 2x + 3$
33. $y = x$
34. $y = -x$
35. $y = 1 - x$
36. $y = 2 - x$
37. $y = -2x + 3$
38. $y = -3x + 2$
39. $y = -3$
40. $y = 2$
41. $x = 2$
42. $x = -4$
43. $2x + y = 5$
44. $3x + y = 5$
45. $x + 2y = 4$
46. $x - 2y = 6$
47. $x - 3y = 6$
48. $x + 4y = 5$
49. $y = .36x + .4$
50. $y = .27x - .42$

For each point, name the quadrant in which it lies or the axis on which it lies.

51. $(-3, 45)$ Quadrant II   **52.** $(-33, 47)$ Quadrant II   **53.** $(-3, 0)$ x-axis   **54.** $(0, -9)$ y-axis

55. $(-2.36, -5)$ Quadrant III   **56.** $(89, 0)$ x-axis   **57.** $(3.4, 8.8)$ Quadrant I   **58.** $(\sqrt{2}, 44)$ Quadrant I

59. $(-\sqrt{3}, 50)$ Quadrant II   **60.** $(-6, -1/2)$ Quadrant III   **61.** $(0, -99)$ y-axis   **62.** $(8.4, \pi)$ Quadrant I

Plot all of the following points on the same rectangular coordinate system. See Example 5. Graphs in Answers sections.

63. $(40, 900)$ **64.** $(-300, 1200)$ **65.** $(20, -390)$ **66.** $(-500, 634)$

67. $(-2, -3)$ **68.** $(40, -40)$ **69.** $(-2000, 4500)$ **70.** $(-200, -90)$

Graph each of the following equations. Plot at least five points for each equation. See Example 5. Graphs in Answers sections.

71. $y = x + 1200$ **72.** $y = 2x - 3000$ **73.** $y = 50x - 2000$

74. $y = -300x + 4500$ **75.** $y = -400x + 2000$ **76.** $y = 500x + 3$

Complete the given ordered pairs so that each ordered pair satisfies the given equation.

77. $x + y = 7$: $(0, \ )$, $(\ , 3)$, $(-2, \ )$ $(0, 7)$, $(4, 3)$, $(-2, 9)$ **78.** $x - y = 10$: $(15, \ )$, $(\ , 8)$, $(-3, \ )$

79. $y = 2x - 4$: $(5, \ )$, $(\ , 2)$, $(\ , -6)$ $(5, 6)$, $(3, 2)$, $(-1, -6)$ **80.** $y = -3x + 1$: $(4, \ )$, $(\ , 7)$, $(\ , -8)$

81. $2x - 3y = 6$: $(3, \ )$, $(\ , -2)$, $(12, \ )$ **82.** $3x + 5y = 0$: $(-5, \ )$, $(\ , -3)$, $(10, \ )$

83. $x = 5$: $(\ , -3)$, $(\ , 5)$, $(\ , 0)$ $(5, -3)$, $(5, 5)$, $(5, 0)$ **84.** $y = -6$: $(3, \ )$, $(-1, \ )$, $(4, \ )$ $(3, -6)$, $(-1, -6)$, $(4, -6)$

78. $(15, 5)$, $(18, 8)$, $(-3, -13)$ **80.** $(4, -11)$, $(-2, 7)$, $(3, -8)$ **81.** $(3, 0)$, $(0, -2)$, $(12, 6)$ **82.** $(-5, 3)$, $(5, -3)$, $(10, -6)$

8.2 Coordinate Geometry

IN THIS SECTION:

- **Distance between Two Points**
- **Midpoint Formula**
- **Geometric Figures**

In geometry we study properties of figures such as squares, rectangles, parallelograms, and triangles. They are called **plane figures,** because we think of them as lying in a plane. In this section we will study figures in a plane that has a rectangular coordinate system, a **coordinate plane.** We use the coordinates to establish properties of the geometric figures. Coordinate geometry is also called analytical geometry.

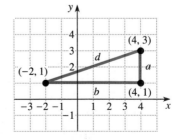

Figure 8.9

Distance between Two Points

Consider the two points $(4, 3)$ and $(-2, 1)$. As shown in Fig. 8.9, the distance between these points, d, is the length of the line segment with endpoints at $(4, 3)$ and $(-2, 1)$. The line segment is the hypotenuse of a right triangle.

If we know the lengths of the legs, then we can find the length of the hypotenuse by using the Pythagorean theorem. We can see from Fig. 8.9 that the length of side a is 2 and the length of side b is 6. The lengths of the sides can also be obtained by subtracting the y-coordinates for a and the x-coordinates for b:

$$\text{Length of side } a = 3 - 1 = 2$$
$$\text{Length of side } b = 4 - (-2) = 6$$

By the Pythagorean theorem,

$$d^2 = 2^2 + 6^2$$
$$d^2 = 4 + 36$$
$$d^2 = 40$$
$$d = \sqrt{40} \qquad \text{Use only the positive square root for distance.}$$
$$d = 2\sqrt{10} \qquad \sqrt{40} = \sqrt{4} \cdot \sqrt{10} = 2\sqrt{10}$$

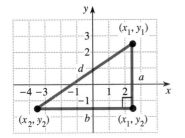

Figure 8.10

So the exact distance between the points (4, 3) and (−2, 1) is $2\sqrt{10}$.

We can easily generalize this procedure to find a formula for the distance between any two points (x_1, y_1) and (x_2, y_2). As shown in Fig. 8.10, the distance between these points is the length of the hypotenuse of a right triangle.

The length of side a is $y_1 - y_2$, and the length of side b is $x_1 - x_2$. By the Pythagorean theorem we can write

$$d^2 = (x_1 - x_2)^2 + (y_1 - y_2)^2.$$

This equation can be solved for d by using the square root property. Since distance is always greater than or equal to zero, we do not need the negative square root.

Distance Formula

The distance, d, between the points (x_1, y_1) and (x_2, y_2) is given by the formula

$$d = \sqrt{(x_1 - x_2)^2 + (y_1 - y_2)^2}.$$

EXAMPLE 1   Find the distance between (5, −3) and (−4, −1).

Solution   Let $(x_1, y_1) = (5, -3)$ and $(x_2, y_2) = (-4, -1)$. Now substitute the appropriate values into the distance formula:

$$d = \sqrt{[5 - (-4)]^2 + [-3 - (-1)]^2}$$
$$= \sqrt{(9)^2 + (-2)^2}$$
$$= \sqrt{81 + 4}$$
$$= \sqrt{85}$$

The exact distance between the points is $\sqrt{85}$.   ◀

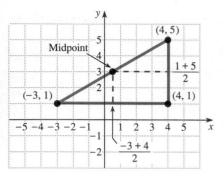

Figure 8.11

Midpoint Formula

Consider the two points $(-3, 1)$ and $(4, 5)$. We know how to find the distance between these two points. Now we will find the **midpoint** of the line segment with endpoints $(-3, 1)$ and $(4, 5)$. The two points are shown in Fig. 8.11.

Using similar triangles, we can see that a point halfway between $(-3, 1)$ and $(4, 5)$ has its x-coordinate halfway between -3 and 4 and its y-coordinate halfway between 1 and 5. To find a number halfway between two numbers, add the numbers and divide by 2. Thus

$$\text{the } x\text{-coordinate of the midpoint is } \frac{-3 + 4}{2} = \frac{1}{2},$$

and

$$\text{the } y\text{-coordinate of the midpoint is } \frac{1 + 5}{2} = 3.$$

So $(1/2, 3)$ is the midpoint of the line segment with endpoints $(-3, 1)$ and $(4, 5)$.

The procedure for finding the midpoint is the same for any line segment in the coordinate system.

Midpoint Formula

> The midpoint of the line segment with endpoints (x_1, y_1) and (x_2, y_2) is the point
>
> $$\left(\frac{x_1 + x_2}{2}, \frac{y_1 + y_2}{2} \right).$$

EXAMPLE 2 Find the midpoint of the line segment with endpoints $(-3, -5)$ and $(6, -1)$.

Solution By the midpoint formula the x-coordinate of the midpoint is

$$\frac{-3 + 6}{2} = \frac{3}{2},$$

and the y-coordinate of the midpoint is

$$\frac{-5 + (-1)}{2} = \frac{-6}{2} = -3.$$

The midpoint is $(3/2, -3)$. ◀

Geometric Figures

EXAMPLE 3 Consider the rectangle whose vertices are $(-2, -1)$, $(3, -1)$, $(3, 2)$, and $(-2, 2)$. Show that the two diagonals are equal in length.

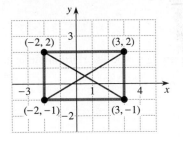

Figure 8.12

Solution First draw a diagram. The rectangle is shown in Fig. 8.12. Use the distance formula to find the length of the diagonal with endpoints $(-2, 2)$ and $(3, -1)$:

$$\sqrt{(-2-3)^2 + [2-(-1)]^2} = \sqrt{(-5)^2 + 3^2} = \sqrt{25+9} = \sqrt{34}$$

Now use the distance formula to find the length of the diagonal with endpoints $(-2, -1)$ and $(3, 2)$:

$$\sqrt{(-2-3)^2 + (-1-2)^2} = \sqrt{(-5)^2 + (-3)^2} = \sqrt{25+9} = \sqrt{34}$$

The diagonals of the rectangle are equal in length. ◄

This example illustrates a more common fact from geometry. Ask students for a general statement of this idea.

EXAMPLE 4 Consider the parallelogram with vertices $(-1, -2)$, $(2, -1)$, $(3, 1)$, and $(0, 0)$. Show that the diagonals of this parallelogram bisect each other.

Solution A sketch of the parallelogram is shown in Fig. 8.13. If the midpoint of one diagonal is the same point as the midpoint of the other diagonal, then the diagonals bisect each other. Use the midpoint formula to find the midpoint of the diagonal with endpoints $(0, 0)$ and $(2, -1)$:

$$\left(\frac{0+2}{2}, \frac{0+(-1)}{2}\right) = (1, -1/2)$$

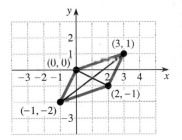

Figure 8.13

Now use the midpoint formula to find the midpoint of the diagonal with endpoints $(-1, -2)$ and $(3, 1)$:

$$\left(\frac{-1+3}{2}, \frac{-2+1}{2}\right) = (1, -1/2)$$

Since the two diagonals have the same midpoint, the diagonals bisect each other. ◄

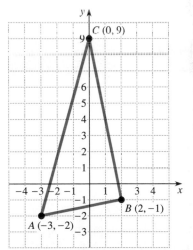

Figure 8.14

EXAMPLE 5 Show that the triangle whose vertices are $A(-3, -2)$, $B(2, -1)$, and $C(0, 9)$ is a right triangle.

Solution By the Pythagorean theorem a triangle is a right triangle if and only if the sum of the squares of the legs is equal to the hypotenuse squared. We can find the lengths of the legs and the hypotenuse with the distance formula. The triangle is shown in Fig. 8.14.

$$\text{Length of } AB = \sqrt{(-3-2)^2 + [-2-(-1)]^2} = \sqrt{25+1} = \sqrt{26}$$
$$\text{Length of } AC = \sqrt{(-3-0)^2 + (-2-9)^2} = \sqrt{9+121} = \sqrt{130}$$
$$\text{Length of } BC = \sqrt{(2-0)^2 + (-1-9)^2} = \sqrt{4+100} = \sqrt{104}$$

Since the hypotenuse is the longest side of a right triangle, side AC is the only side that could be the hypotenuse. The sum of the squares of the lengths of the legs is

$$(\sqrt{104})^2 + (\sqrt{26})^2 = 104 + 26 = 130.$$

The Warm-up Exercises illustrate some geometric facts that are important for doing the exercises of this section.

The length of the hypotenuse squared is

$$(\sqrt{130})^2 = 130.$$

By the Pythagorean theorem the triangle is a right triangle, and the right angle is at point B.   ◀

Warm-ups

True or false?

1. The distance formula gives us the distance between any two points in the rectangular coordinate system. T

2. The distance between the points (3, 5) and (3, 9) is 4. T

3. The distance between the points (−2, 7) and (4, 7) is 6. T

4. The midpoint of the line segment with endpoints (2, 5) and (6, 7) is (4, 6). T

5. The midpoint of the line segment with endpoints (−3, 0) and (3, 1) is (0, 1/2). T

6. If a quadrilateral (four-sided figure) has four sides with the same length, then it must be a square. F

7. If the opposite sides of a quadrilateral are equal in length, and it has one right angle, then the quadrilateral is a rectangle. T

8. If the opposite sides of a quadrilateral are equal in length, then the quadrilateral is a parallelogram. T

9. In any right triangle the hypotenuse is the longest side. T

10. An equilateral triangle has three sides that are equal in length. T

8.2 EXERCISES

Find the distance between each pair of points. See Example 1.

1. (1, 2), (3, 4) $2\sqrt{2}$
2. (2, 3), (6, 9) $2\sqrt{13}$
3. (−2, 1), (3, 4) $\sqrt{34}$
4. (−3, 2), (5, −6) $8\sqrt{2}$
5. (−1, −2), (−3, 4) $2\sqrt{10}$
6. (−3, −4), (−2, 3) $5\sqrt{2}$
7. (0, 3), (4, 0) 5
8. (−5, 0), (0, 12) 13
9. (0, −6), (−8, 0) 10
10. (0, 0), (−1, −1) $\sqrt{2}$
11. ($\sqrt{2}$, $\sqrt{3}$), (0, 0) $\sqrt{5}$
12. ($2\sqrt{3}$, $\sqrt{7}$), ($\sqrt{3}$, 0) $\sqrt{10}$
13. (1, 3), (.5, .7) 2.354
14. (.4, −.5), (−.3, .9) 1.565

Find the midpoint of the line segment with the given endpoints. See Example 2.

15. (3, 6), (7, 8) (5, 7)
16. (1, 4), (5, 8) (3, 6)
17. (−1, 3), (5, −9) (2, −3)
18. (−2, 4), (8, −2) (3, 1)
19. (−3, −5), (−4, 6) (−3.5, .5)
20. (−4, −3), (9, −2) $\left(\frac{5}{2}, -\frac{5}{2}\right)$
21. (0, −3), (2, 5) (1, 1)
22. (−2, 0), (3, 9) $\left(\frac{1}{2}, \frac{9}{2}\right)$
23. (0, 0), (1, 1) (.5, .5)

24. $(-1, -1)$, $(1, 1)$ $(0, 0)$

25. $(2\sqrt{2}, 3\sqrt{3})$, $(4\sqrt{2}, \sqrt{3})$ $(3\sqrt{2}, 2\sqrt{3})$

26. $(1, \sqrt{2})$, $(-3, 5\sqrt{2})$ $(-1, 3\sqrt{2})$

27. $(3.4, -5.26)$, $(-4.22, 9.44)$ $(-.41, 2.09)$

28. $(-1.23, 4.35)$, $(-3.67, -1.83)$ $(-2.45, 1.26)$

Solve each geometric problem. See Examples 3, 4, and 5.

29. Consider the quadrilateral whose vertices are $(-2, -1)$, $(-1, -4)$, $(5, -2)$, and $(4, 1)$. Show that its diagonals are equal in length. Diagonals: $5\sqrt{2}$

30. Consider the quadrilateral whose vertices are $(-3, -3)$, $(-1, -4)$, $(3, 4)$, $(1, 5)$. Show that its diagonals bisect each other. Midpoint of each diagonal: $\left(0, \dfrac{1}{2}\right)$

31. Consider the quadrilateral whose vertices are $(-1, -1)$, $(0, -3)$, $(4, 3)$, and $(3, 5)$. Show that this quadrilateral is a parallelogram. (If the opposite sides of a quadrilateral are equal in length, then the quadrilateral is a parallelogram.)

32. Determine whether the points $(-2, 0)$, $(-1, -3)$, $(6, 1)$, and $(4, 4)$ are the vertices of a parallelogram. No

33. Determine whether the points $(1, 2)$, $(-2, 5)$, and $(4, 8)$ are the vertices of an isosceles triangle. Yes

34. Determine whether the points $(2, 7)$, $(0, -3)$, and $(4, -2)$ are the vertices of an isosceles triangle. No

35. Determine whether the points $(-1, 1)$, $(-2, -4)$, and $(4, -1)$ are the vertices of an equilateral triangle. No

36. Determine whether the points $(-1, 0)$, $(3, 0)$, and $(1, 2\sqrt{3})$ are the vertices of an equilateral triangle. Yes

37. Determine whether the triangle with vertices $(-2, -3)$, $(2, -2)$, and $(1, 2)$ is a right triangle. Yes

38. Determine whether the triangle with vertices $(-4, 1)$, $(-3, -2)$, and $(3, 1)$ is a right triangle. No

39. Determine whether the points $(-2, -3)$, $(1, 2)$, and $(4, 7)$ lie on the same straight line. (If the sum of the lengths of the two shortest line segments is equal to the length of the longest line segment, then the three points lie on a line.) Yes

40. Determine whether the points $(-2, -4)$, $(0, 3)$, and $(2, 8)$ lie on the same straight line. No

41. Show that the points $(1, 1)$, $(5, 3)$, $(7, 7)$, and $(3, 5)$ are the vertices of a rhombus. (If all sides of a quadrilateral are the same length, then the quadrilateral is a rhombus.)

42. Show that the diagonals of the rhombus of Exercise 41 bisect each other. Midpoint of each diagonal: $(4, 4)$

43. Show that the diagonals of the rhombus of Exercise 41 are perpendicular to each other.

44. Consider the square with vertices $(-4, -1)$, $(2, -1)$, $(2, 5)$, and $(-4, 5)$. Show that the diagonals of this square are perpendicular.

31. Opposite sides are $2\sqrt{13}$ and $\sqrt{5}$

41. Each side: $2\sqrt{5}$

8.3 Slope

IN THIS SECTION:

- **Concepts**
- **Slope Formula**
- **Parallel Lines**
- **Perpendicular Lines**
- **Geometric Applications**

In Section 8.2 we learned to use the rectangular coordinate system in geometry. We used the distance formula to find the length of any line segment in the coordinate plane. The length of a line segment is a number that describes that line segment. In

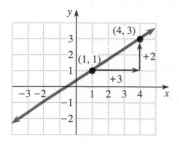

Figure 8.15

this section we will study lines in the coordinate plane. Lines extend in two directions without end. We cannot compute the length of a line, but we can calculate a number called the **slope** of the line. The slope of a line is a measurement of the "steepness" of the line.

Concepts

To calculate the slope of a line, imagine that we are to move from one point on the line to another point on the line. We cannot travel along the line, but we must move in the directions of the axes. Consider the line shown in Fig. 8.15. To go from (1, 1) to (4, 3), we move three units to the right and then two units upward. In going from (1, 1) to (4, 3) there is a change of $+2$ in the y-coordinate and a change of $+3$ in the x-coordinate. See Fig. 8.15. Slope is defined to be the ratio of these two changes.

DEFINITION
Slope

$$\text{Slope} = \frac{\text{change in } y\text{-coordinate}}{\text{change in } x\text{-coordinate}}$$

The slope of the line in Fig. 8.15 is

$$\frac{+2}{+3} = \frac{2}{3}. \quad \begin{array}{l} \longleftarrow \text{ Change in } y\text{-coordinate} \\ \longleftarrow \text{ Change in } x\text{-coordinate} \end{array}$$

If we move from the point (4, 3) to the point (1, 1), there is a change of -2 in the y-coordinate and a change of -3 in the x-coordinate. See Fig. 8.16. In this case we get

$$\text{slope} = \frac{-2}{-3} = \frac{2}{3}.$$

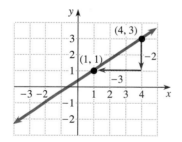

Figure 8.16

Note that going from (4, 3) to (1, 1) gives us the same slope as going from (1, 1) to (4, 3).

We call the change in y-coordinate the **rise** and the change in x-coordinate the **run.** Moving up is a positive rise, and moving down is a negative rise. Moving to the right is a positive run, and moving to the left is a negative run. We usually use the letter m to stand for slope.

DEFINITION
Another Definition of Slope

$$\text{Slope} = m = \frac{\text{change in } y\text{-coordinate}}{\text{change in } x\text{-coordinate}} = \frac{\text{rise}}{\text{run}}$$

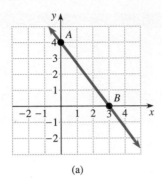
(a)

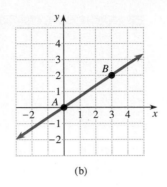
(b)

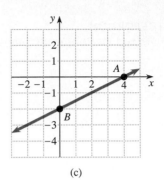
(c)

Figure 8.17

EXAMPLE 1 Find the slopes of the lines in Fig. 8.17 by going from point A to point B.

Solution

a) The coordinates of point A are $(0, 4)$, and the coordinates of point B are $(3, 0)$. Going from A to B, the change in y is -4 and the change in x is $+3$. So $m = -4/3$.

b) Going from A to B, the rise is 2 and the run is 3. Thus $m = 2/3$.

c) Going from A to B, the rise is -2 and the run is -4. Thus $m = -2/-4 = 1/2$.
◀

The ratio of rise to run is the ratio of the lengths of the two legs of a right triangle whose hypotenuse is on the line. As long as one leg is vertical and the other is horizontal, all such triangles for a certain line have the same shape. These triangles are similar triangles. The ratio of the length of the vertical side to the length of the horizontal side for any two such triangles is the same number. So we get the same value for the slope no matter which two points of the line are used to calculate it or in which order the points are used.

EXAMPLE 2 Find the slope of the line shown in Fig. 8.18, using

a) points A and B,   b) points A and C,   c) points B and C.

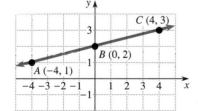

Figure 8.18

Solution

a) $m = \dfrac{\text{rise}}{\text{run}} = \dfrac{1}{4}$ b) $m = \dfrac{\text{rise}}{\text{run}} = \dfrac{2}{8} = \dfrac{1}{4}$ c) $m = \dfrac{1}{4}$ ◀

Slope Formula

One way to obtain the rise and run is from a graph. The rise and run can also be found by using the coordinates of two points on the line.

DEFINITION
Formal Definition of Slope

The slope of the line containing the points (x_1, y_1) and (x_2, y_2) is denoted by the letter m, where

$$m = \frac{y_2 - y_1}{x_2 - x_1}$$

provided that $x_2 - x_1 \neq 0$.

EXAMPLE 3   Find the slope of each of the following lines:

a) The line through $(0, 5)$ and $(6, 3)$
b) The line through $(-3, 4)$ and $(-5, -2)$
c) The line through $(-4, 2)$ and the origin

Solution

a) Let $(x_1, y_1) = (0, 5)$ and $(x_2, y_2) = (6, 3)$. Which point is called (x_1, y_1) is arbitrary.

$$m = \frac{y_2 - y_1}{x_2 - x_1} = \frac{3 - 5}{6 - 0} = \frac{-2}{6} = -\frac{1}{3}$$

b) Let $(x_1, y_1) = (-3, 4)$ and $(x_2, y_2) = (-5, -2)$:

$$m = \frac{y_2 - y_1}{x_2 - x_1} = \frac{-2 - 4}{-5 - (-3)} = \frac{-6}{-2} = 3$$

c) Let $(x_1, y_1) = (0, 0)$ and $(x_2, y_2) = (-4, 2)$:

$$m = \frac{2 - 0}{-4 - 0} = \frac{2}{-4} = -\frac{1}{2}$$ ◀

Note that slope is not defined if $x_2 - x_1 = 0$. This means that the x-coordinates of two points on the line are equal. The x-coordinates are the same only for points on a vertical line. So we say that *slope is undefined for vertical lines*.

If $y_2 - y_1 = 0$, then the points have equal y-coordinates and lie on a horizontal line. *The slope for any horizontal line is zero*. See Fig. 8.19.

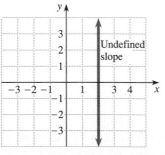
Vertical line

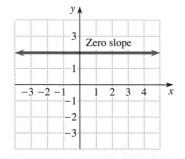
Horizontal line

Figure 8.19

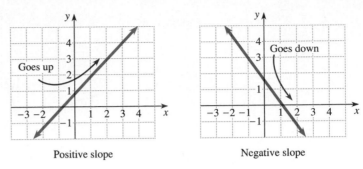

Figure 8.20

Note that if we move from left to right along a line with positive slope, we will be rising. If we move from left to right along a line with negative slope, we will be falling. See Fig. 8.20.

Parallel Lines

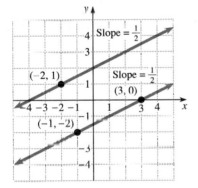

Figure 8.21

EXAMPLE 4 Draw a line through the point $(-2, 1)$ with slope $1/2$ and a line through $(3, 0)$ with slope $1/2$.

Solution Since slope is the ratio of rise to run, a slope of $1/2$ means that we can locate a second point of the line by starting at $(-2, 1)$ and going up one unit and to the right two units. For the line through $(3, 0)$ we start at $(3, 0)$ and go up one unit and to the right two units. See Fig. 8.21.   ◀

The two lines we sketched in Fig. 8.21 appear to be parallel. This is an example of a general situation. Whether two lines are parallel can be determined by their slopes.

Parallel Lines

Nonvertical lines are parallel if and only if they have equal slopes. Any two vertical lines are parallel to each other.

Perpendicular Lines

EXAMPLE 5 Draw two lines through the point $(-1, 2)$, one with slope $-1/3$ and the other with slope 3.

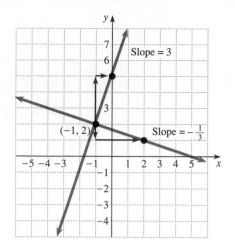

Figure 8.22

Solution   Since slope is the ratio of rise to run, a slope of $-1/3$ means that we can locate a second point on the line by starting at $(-1, 2)$ and going down one unit and to the right three units. For the line with slope 3 we start at $(-1, 2)$ and go up three units and to the right one unit. See Fig. 8.22.   ◄

The two lines we sketched in Fig. 8.22 appear to be perpendicular to one another. This is an example of a general situation. If the slope of one line is the opposite of the reciprocal of the slope of another line, the lines are perpendicular. For example, lines with slopes $3/4$ and $-4/3$ are perpendicular.

Perpendicular Lines

Two lines with slopes m_1 and m_2 are perpendicular if and only if

$$m_1 = -\frac{1}{m_2}.$$

Geometric Applications

EXAMPLE 6   Use slope to determine whether the four points $(-4, 1)$, $(-3, -3)$, $(5, -1)$, and $(4, 3)$ are the vertices of a rectangle.

Solution   First we sketch the figure determined by these points. See Fig. 8.23. It appears to be a rectangle, but we must prove that it is. Calculate the slope of

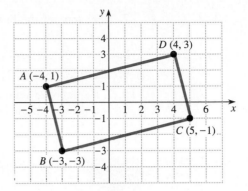

Figure 8.23

each side:

$$m_{AB} = \frac{1 - (-3)}{-4 - (-3)} = \frac{4}{-1} = -4$$

$$m_{BC} = \frac{-3 - (-1)}{-3 - 5} = \frac{-2}{-8} = \frac{1}{4}$$

$$m_{CD} = \frac{-1 - 3}{5 - 4} = \frac{-4}{1} = -4$$

$$m_{AD} = \frac{1 - 3}{-4 - 4} = \frac{-2}{-8} = \frac{1}{4}$$

Since the opposite sides have the same slope, the opposite sides are parallel. The figure is a parallelogram. Since -4 is the opposite of the reciprocal of $1/4$, the intersecting lines are perpendicular. The figure is a rectangle. ◀

Warm-ups

True or false?

1. Slope is a measurement of the steepness of a line. T
2. Slope is rise divided by run. T
3. Every line in the coordinate plane has a number corresponding to it called the slope of the line. F
4. The line through the point $(1, 1)$ and the origin has slope 1. T
5. Slope, like distance, can never be negative. F
6. A line with slope 2 is perpendicular to any line with slope -2. F

7. The slope of the line that crosses the *y*-axis at (0, 3) and the *x*-axis at (4, 0) is 3/4. F

8. Two different lines cannot have the same slope. F

9. The line through (1, 3) and (−5, 3) has zero slope. T

10. If the opposite sides of a quadrilateral are parallel, then the quadrilateral is a parallelogram. T

8.3 EXERCISES

Determine the slope of each line. See Examples 1 *and* 2.

1. 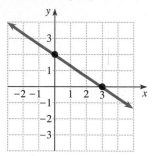 $-\dfrac{2}{3}$

2. 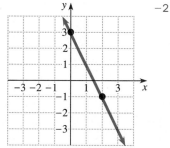 $\dfrac{2}{3}$

3. 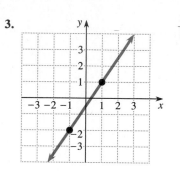 $\dfrac{3}{2}$

4. 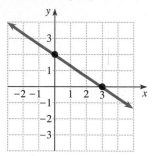 -2

5. 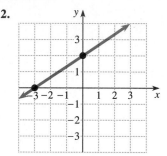 2

6. 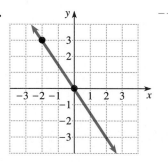 $-\dfrac{3}{2}$

7. 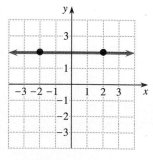 0

8. 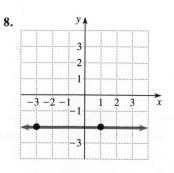 0

9. 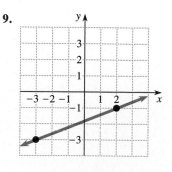 $\dfrac{2}{5}$

10. 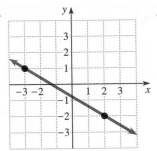 $-\dfrac{3}{5}$

11. 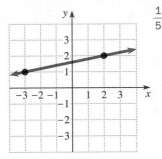 $\dfrac{1}{5}$

12. 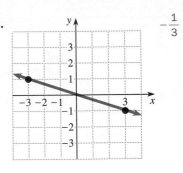 $-\dfrac{1}{3}$

Find the slope of the line that goes through each of the following pairs of points. See Example 3.

13. (1, 2), (3, 6)   2

14. (2, 5), (6, 10) $\dfrac{5}{4}$

15. (2, 4), (5, −1) $-\dfrac{5}{3}$

16. (3, 1), (6, −2)   −1

17. (−2, 4), (5, 9) $\dfrac{5}{7}$

18. (−1, 3), (3, 5) $\dfrac{1}{2}$

19. (−2, −3), (−5, 1) $-\dfrac{4}{3}$

20. (−6, −3), (−1, 1) $\dfrac{4}{5}$

21. (−3, 4), (3, −2)   −1

22. (−1, 3), (5, −2) $-\dfrac{5}{6}$

23. (1/2, 2), (−1, 1/2)   1

24. (1/3, 2), (−1/3, 1) $\dfrac{3}{2}$

25. (2, 3), (2, −9)   Undefined

26. (−3, 6), (8, 6)   0

27. (−2, −5), (9, −5)   0

28. (4, −9), (4, 6)   Undefined

29. (.3, .9), (−.1, −.3)   3

30. (−.1, .2), (.5, .8)   1

Solve the following. See Examples 4 and 5. Make a sketch for each problem.   Graphs in Answers sections.

31. Draw a graph showing a line through (1, −2) with slope 1/2 and a line through (−1, 1) with slope 1/2.

32. Draw a graph showing a line through (0, 3) with slope 1 and a line through (0, 0) with slope 1.

33. Draw a graph showing two lines through the point (1, 2), one with slope 1/2 and the other with slope −2.

34. Draw a graph showing two lines through the point (−2, 1), one with slope 2/3 and the other with slope −3/2.

35. What is the slope of a line perpendicular to a line with slope 3/4? $-\dfrac{4}{3}$

36. What is the slope of a line perpendicular to a line with slope −1?   1

37. What is the slope of a line that goes through (1, 2) and runs parallel to the line through (−2, −3) and (4, 0)? $\dfrac{1}{2}$

38. What is the slope of a line that goes through the origin and runs parallel to the line through (−4, 0) and (0, 6)? $\dfrac{3}{2}$

39. Find the slope of the line through (1, 3) that is perpendicular to the line through (−2, 4) and (3, −1).   1

40. Find the slope of the line through the origin that is perpendicular to the line through (0, −3) and (3, 0).   −1

41. Slopes of opposite sides are $\dfrac{1}{5}$ and −3. **42.** Slopes of opposite sides are −1 and $-\dfrac{1}{3}$. **43.** Slopes of opposite sides are $\dfrac{1}{2}$ and −2.

Use slope to solve each geometric figure problem. See Example 6.

41. Show that the points (−3, 2), (2, 3), (3, 0), and (−2, −1) are the vertices of a parallelogram.

42. Show that the points (−3, 2), (0, 1), (1, 0), and (−2, 1) are the vertices of a parallelogram.

43. Show that the points (0, 1), (−2, 0), (1, −1), and (−1, −2) are the vertices of a square.

44. Show that the points (−4, 1), (2, 3), (3, 0), and (−3, −2) are the vertices of a rectangle.

45. Show that the points (−4, −3), (−3, −4), and (5, 4) are the vertices of a right triangle.

46. Show that the points (−4, −3), (−3, −4), and (1, 2) are the vertices of a right triangle.

47. Determine whether the points (−3, −8), (4, 9), and (1, 2) are all on the same straight line.   No

48. Determine whether the points (2, −7), (0, −1), and (−3, 8) are all on the same straight line.   Yes

44. Slopes of opposite sides are −3 and $\dfrac{1}{3}$.   **45.** Two sides have slopes 1 and −1.   **46.** Two sides have slopes 1 and −1.

8.4 Equations of Lines

IN THIS SECTION:

- Slope-Intercept Form
- Standard Form
- Using Slope-Intercept Form for Graphing

In Section 8.1 we graphed all of the pairs of numbers that satisfied an equation in two variables. Those graphs were straight lines. Here we will reverse that process. We start with a line or a description of a line and write an equation for the line.

Slope-Intercept Form

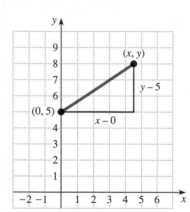

Figure 8.24

Consider the line that goes through the point (0, 5) and has slope 2/3. See Fig. 8.24. The point (0, 5) is on the y-axis, and is called the **y-intercept** of the line. We can use any two points of a line to find the slope. If we calculate the slope of this line using an arbitrary point of the line (x, y), and the y-intercept (0, 5), we get

$$m = \frac{y - 5}{x - 0} = \frac{y - 5}{x}.$$

Since the slope of this line is 2/3, any point on the line must satisfy the equation

$$\frac{y - 5}{x} = \frac{2}{3}.$$

To solve the equation for y, multiply each side by x:

$$y - 5 = \frac{2}{3}x$$

$$y = \frac{2}{3}x + 5 \qquad \text{Add 5 to each side.}$$

$$\underset{\text{Slope}}{\uparrow} \qquad \underset{\text{y-intercept (0, 5)}}{\uparrow}$$

The slope of 2/3 and the y-intercept of (0, 5) can be seen in the final form of the equation. For this reason it is called the **slope-intercept form** of the equation of the line.

Slope-Intercept Form

The equation $y = mx + b$ is the equation of a line with y-intercept (0, b) and slope m.

EXAMPLE 1 Write the equation in slope-intercept form for the line through $(0, -2)$ with slope 3.

Solution We use the form $y = mx + b$ with $b = -2$ and $m = 3$. The equation is $y = 3x - 2$. ◄

Standard Form

The graph of the equation $x = 3$ is a vertical line. Since slope is not defined for vertical lines, this line does not have an equation in slope-intercept form. Only nonvertical lines have equations in slope-intercept form. However, there is a form that includes all lines. It is called **standard form.**

Standard Form

If A, B, and C are real numbers with A and B not both zero, then

$$Ax + By = C$$

is called the **standard form** of the equation of a line. Any equation that can be written in this form is called a **linear equation in two variables.**

The equation $x = 3$ is included in this form. If we let $A = 1$, $B = 0$, and $C = 3$, we get

$$1 \cdot x + 0 \cdot y = 3,$$

which is equivalent to $x = 3$.

Any linear equation in standard form with $B \neq 0$ can be written in slope-intercept form by solving for y.

EXAMPLE 2 Find the slope and y intercept of the line $3x - 2y = 6$.

Solution Solve for y to get slope-intercept form:

$$3x - 2y = 6$$
$$-2y = -3x + 6$$
$$y = \frac{3}{2}x - \frac{6}{2}$$
$$y = \frac{3}{2}x - 3 \qquad \text{This is slope-intercept form.}$$

The slope is $3/2$, and the y-intercept is $(0, -3)$. ◄

EXAMPLE 3 Write the equation of the line $y = \frac{2}{5}x + 3$ in standard form using only integers.

Solution To get standard form, subtract $\frac{2}{5}x$ from each side:

$$y = \frac{2}{5}x + 3$$

$$-\frac{2}{5}x + y = 3$$

To eliminate the fraction multiply each side by 5:

$$5\left(-\frac{2}{5}x + y\right) = 5 \cdot 3$$

$$-2x + 5y = 15$$

$$2x - 5y = -15 \qquad \text{Multiply each side by } -1.$$

This is not the only answer with integral coefficients. Multiplying this equation by any nonzero integer would give an equivalent equation with integral coefficients. It is customary to have a positive coefficient for the first term.   ◄

Using Slope-Intercept Form for Graphing

In Section 8.1 we graphed the equation of a line by finding several points that satisfy the equation and then drawing a straight line through them. We can also graph a line from the information given in the slope-intercept form.

STRATEGY
Graphing a Line from
Slope-Intercept Form

> To graph a line using its slope and y-intercept,
> 1. start at the y-intercept,
> 2. use the rise and run to locate a second point, and
> 3. then draw a line through the two points.

EXAMPLE 4 Graph the line $2x - 3y = 3$.

Solution First write it in slope-intercept form:

$$2x - 3y = 3$$

$$-3y = -2x + 3$$

$$y = \frac{2}{3}x - 1 \qquad \text{Divide each side by } -3.$$

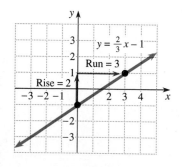

Figure 8.25

The slope is 2/3, and the y-intercept is $(0, -1)$. A slope of 2/3 means a rise of 2 and a run of 3. Start at $(0, -1)$ and go up two units and to the right three units to locate a second point on the line. Now draw a line through the two points. See Fig. 8.25 for the graph of $2x - 3y = 3$.   ◄

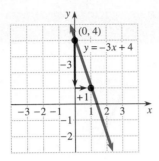

Figure 8.26

EXAMPLE 5   Graph the line $y = -3x + 4$.

Solution   The slope is -3 and the y-intercept is $(0, 4)$. Since $-3 = -3/1$, we use a rise of -3 and a run of 1. To locate a second point on the line, start at $(0, 4)$ and go down three units and to the right one unit. Draw a line through the two points. See Fig. 8.26.   ◄

EXAMPLE 6   Write the equation in slope-intercept form for the line **through** $(0, 4)$ that is perpendicular to the line $2x - 4y = 1$.

Solution   First find the slope of $2x - 4y = 1$:

$$2x - 4y = 1$$
$$-4y = -2x + 1$$
$$y = \frac{-2}{-4}x + \frac{1}{-4}$$
$$y = \frac{1}{2}x - \frac{1}{4} \qquad \text{The slope of this line is } 1/2.$$

The slope of the line in which we are interested is the opposite of the reciprocal of $1/2$. So the line has slope -2 and y-intercept $(0, 4)$. Its equation is $y = -2x + 4$.   ◄

Warm-ups

True or false?

1. There is only one line with y-intercept $(0, 3)$ and slope $-4/3$.   T
2. The equation of the line through $(1, 2)$ with slope 3 is $y = 3x + 2$.   F
3. The vertical line $x = -2$ has no y-intercept.   T
4. The equation $x = 5$ has a graph that is a vertical line.   T
5. The line $y = x - 3$ is perpendicular to the line $y = 5 - x$.   T
6. The line $y = 2x - 3$ is parallel to the line $y = 4x - 3$.   F
7. The line $2y = 3x - 8$ has a slope of 3.   F
8. Every straight line in the coordinate plane has an equation in standard form.   T
9. The line $x = 2$ is perpendicular to the line $y = 5$.   T
10. The line $y = x$ has no y-intercept.   F

8.4 EXERCISES

Write an equation in slope-intercept form (if possible) for each of the lines shown. See Example 1.

1. 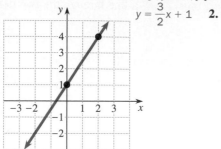 $y = \dfrac{3}{2}x + 1$

2. 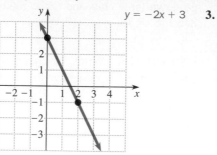 $y = -2x + 3$

3. 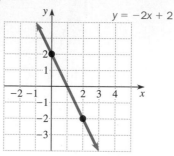 $y = -2x + 2$

4. 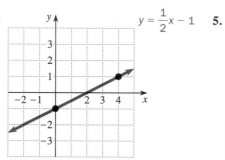 $y = \dfrac{1}{2}x - 1$

5. 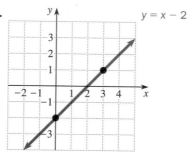 $y = x - 2$

6. 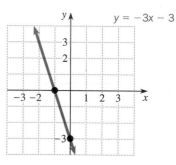 $y = -3x - 3$

7. 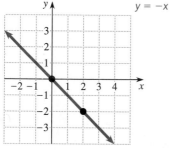 $y = -x$

8. 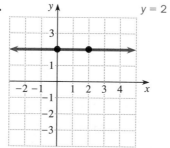 $y = 2$

9. 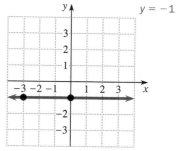 $y = -1$

10. 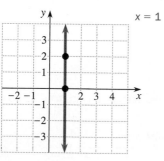 $x = 1$

11. 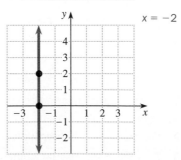 $x = -2$

12. 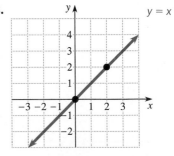 $y = x$

Find the slope and y-intercept for each of the following lines. See Example 2.

13. $y = 3x - 9$ 3, (0, −9)

14. $y = -5x + 4$ −5, (0, 4)

15. $y = 4$ 0, (0, 4)

16. $y = -5$ 0, (0, −5)

17. $y = -3x$ −3, (0, 0)

18. $y = 2x$ 2, (0, 0)

19. $x + y = 5$ −1, (0, 5)

20. $x - y = 4$ 1, (0, −4)

21. $x - 2y = 4$ $\frac{1}{2}$, (0, −2)

22. $x + 2y = 3$ $-\frac{1}{2}$, $\left(0, \frac{3}{2}\right)$

23. $2x - 5y = 10$ $\frac{2}{5}$, (0, −2)

24. $2x + 3y = 9$ $-\frac{2}{3}$, (0, 3)

25. $2x - y + 3 = 0$ 2, (0, 3)

26. $3x - 4y - 8 = 0$ $\frac{3}{4}$, (0, −2)

27. $x = -3$ Undefined, no y-intercept

28. $\frac{2}{3}x = 4$ Undefined, no y-intercept

Write each equation in standard form using only integers. See Example 3.

29. $y = -x + 2$ $x + y = 2$

30. $y = 3x - 5$ $3x - y = 5$

31. $y = \frac{1}{2}x + 3$ $x - 2y = -6$

32. $y = \frac{2}{3}x - 4$ $2x - 3y = 12$

33. $y = \frac{3}{2}x - \frac{1}{3}$ $9x - 6y = 2$

34. $y = \frac{4}{5}x + \frac{2}{3}$ $12x - 15y = -10$

35. $y = -\frac{3}{5}x + \frac{7}{10}$ $6x + 10y = 7$

36. $y = -\frac{2}{3}x - \frac{5}{6}$ $4x + 6y = -5$

37. $x - 6 = 0$ $x = 6$

38. $\frac{1}{2}x - 9 = 0$ $x = 18$

39. $\frac{3}{4}y = 5$ $3y = 20$

40. $\frac{3}{4}y = \frac{5}{9}x$ $20x - 27y = 0$

Draw the graph of each line. Use the method of Examples 4 and 5.   Graphs in Answers sections.

41. $y = 2x - 1$

42. $y = 3x - 2$

43. $y = -3x + 5$

44. $y = -4x + 1$

45. $y = \frac{3}{4}x - 2$

46. $y = \frac{3}{2}x - 4$

47. $2y + x = 0$

48. $2x + y = 0$

49. $3x - 2y = 10$

50. $4x + 3y = 9$

51. $y - 2 = 0$

52. $y + 5 = 0$

53. $x = 8$

54. $x = -1$

Graphs in Answers sections.

Write an equation in slope-intercept form for each of the lines described. See Example 6. In each case, make a sketch.

55. The line through (0, 6) that is perpendicular to the line $y = 3x - 5$

56. The line through (0, −1) that is perpendicular to the line $y = x$ $y = -x - 1$

57. The line with y-intercept (0, 3) that is parallel to the line $2x + y = 5$ $y = -2x + 3$

58. The line through the origin that is parallel to the line $2x - 5y = 8$ $y = \frac{2}{5}x$

59. The line through (2, 3) that runs parallel to the x-axis

60. The line through (−3, 5) that runs parallel to the y-axis

61. The line through (0, 4) and (5, 0) $y = -\frac{4}{5}x + 4$

62. The line through (0, −3) and (4, 0) $y = \frac{3}{4}x - 3$

55. $y = -\frac{1}{3}x + 6$ **59.** $y = 3$ **60.** $x = -3$

8.5 More on Equations of Lines

IN THIS SECTION:

- Point-Slope Form
- Parallel Lines
- Perpendicular Lines

In Section 8.4 we wrote the equation of a line given its slope and y-intercept. In this section we will learn to write the equation of a line given the slope and any other point on the line.

Point-Slope Form

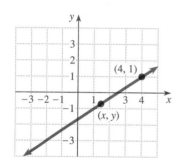

Figure 8.27

Consider a line through the point (4, 1) with slope 2/3. We can calculate the slope of this line by using any two points on the line. See Fig. 8.27. If we use the point (4, 1) and an arbitrary point (x, y) to calculate the slope, we get

$$m = \frac{y - 1}{x - 4}.$$

Since the slope is 2/3, any point on the line satisfies

$$\frac{y - 1}{x - 4} = \frac{2}{3}.$$

If we multiply each side of this equation by $(x - 4)$, we get

$$y - 1 = \frac{2}{3}(x - 4).$$

Note how the coordinates of the point (4, 1) and the slope 2/3 appear in the above equation. We can use the same procedure to get the equation of any line given one point on the line and the slope. The resulting equation is called the **point-slope form** of the equation of the line.

Point-Slope Form

The point-slope form of the equation of the line through the fixed point (x_1, y_1) with slope m is

$$y - y_1 = m(x - x_1).$$

EXAMPLE 1 Find the equation of the line through $(-2, 3)$ with slope $1/2$, and write it in slope-intercept form.

Solution Substitute $x_1 = -2$, $y_1 = 3$, and $m = 1/2$ into the point-slope form:

$$y - 3 = \frac{1}{2}[x - (-2)] \qquad \textbf{This is point-slope form.}$$

To write the equation in slope-intercept form, solve it for y:

$$y - 3 = \frac{1}{2}(x + 2)$$

$$y - 3 = \frac{1}{2}x + 1$$

$$y = \frac{1}{2}x + 4 \qquad \textbf{This is slope-intercept form.} \qquad \blacktriangleleft$$

EXAMPLE 2 Find the equation of the line that contains the points $(-3, -2)$ and $(4, -1)$, and write it in standard form.

Solution First find the slope using the two given points:

$$m = \frac{-2 - (-1)}{-3 - 4} = \frac{-1}{-7} = \frac{1}{7}$$

Now use one of the points, say $(-3, -2)$, and a slope of $1/7$ in the point-slope form. Using the other given point would give the same equation in standard form. (Try it.)

$$y - (-2) = \frac{1}{7}[x - (-3)] \qquad \textbf{This is point-slope form.}$$

$$y + 2 = \frac{1}{7}(x + 3)$$

$$7(y + 2) = 7 \cdot \frac{1}{7}(x + 3) \qquad \textbf{Multiply each side by 7.}$$

$$7y + 14 = x + 3$$

$$7y = x - 11 \qquad \textbf{Subtract 14 from each side.}$$

$$-x + 7y = -11 \qquad \textbf{Subtract } x \textbf{ from each side.}$$

$$x - 7y = 11 \qquad \textbf{Multiply each side by } -1. \qquad \blacktriangleleft$$

Parallel Lines

In Section 8.3 we learned that parallel lines have the same slope. For example, the lines

$$y = 2x - 4 \qquad \text{and} \qquad y = 2x + 9$$

are parallel, since each has slope 2. In the next example we write the equation of a line that is parallel to a given line and contains a given point.

EXAMPLE 3 Write the equation of the line that is parallel to the line $3x + y = 9$ and contains the point $(2, -1)$. Give the answer in slope-intercept form.

Solution Write the equation $3x + y = 9$ in slope-intercept form to determine its slope:

$$3x + y = 9$$
$$y = -3x + 9$$

The slope of this line and any line parallel to it is -3. To guarantee that the line contains the point $(2, -1)$, we use the point-slope form:

$$y - (-1) = -3(x - 2) \qquad \text{Point-slope form}$$
$$y + 1 = -3x + 6$$
$$y = -3x + 5 \qquad \text{Slope-intercept form}$$

The line $y = -3x + 5$ has slope -3 and contains the point $(2, -1)$. Check that $(2, -1)$ satisfies the equation $y = -3x + 5$. ◄

Perpendicular Lines

In Section 8.3 we learned that lines with slopes m and $-1/m$ (for $m \neq 0$) are perpendicular to each other. For example, the lines

$$y = -2x + 7 \qquad \text{and} \qquad y = \frac{1}{2}x - 8$$

are perpendicular to each other. In the next example we will write the equation of a line perpendicular to a given line and containing a given point.

EXAMPLE 4 Write the equation of the line that is perpendicular to the line $3x + 2y = 8$ and contains the point $(1, -3)$. Write the answer in slope-intercept form.

Solution First write $3x + 2y = 8$ in slope-intercept form to determine the slope:

$$3x + 2y = 8$$
$$2y = -3x + 8$$
$$y = -\frac{3}{2}x + 4 \qquad \text{Slope-intercept form}$$

The slope of the given line is $-3/2$. The slope of any line perpendicular to it is $2/3$.

Now we use point-slope form with the point $(1, -3)$ and the slope $2/3$:

$$y - (-3) = \frac{2}{3}(x - 1)$$

$$y + 3 = \frac{2}{3}x - \frac{2}{3}$$

$$y = \frac{2}{3}x - \frac{2}{3} - 3 \qquad \text{Subtract 3 from each side.}$$

$$y = \frac{2}{3}x - \frac{11}{3} \qquad \text{Slope-intercept form}$$

The last equation is the equation of the line perpendicular to the given line and containing the point $(1, -3)$. ◀

Warm-ups

True or false?

1. The formula $y = m(x - x_1)$ is the point-slope form for a line. F
2. It is impossible to find the equation of a line through the points $(2, 5)$ and $(-3, 1)$. F
3. The point-slope form will not work for the line through the points $(3, -4)$ and $(3, 5)$. T
4. The equation of the line through the origin with slope 1 is $y = x$. T
5. The slope of the line $5x + y = 4$ is 5. F
6. The slope of any line perpendicular to the line $y = 4x - 3$ is $-1/4$. T
7. The slope of any line parallel to the line $x + y = 1$ is -1. T
8. The line $2x - y = -1$ goes through the point $(-2, -3)$. T
9. The lines $2x + y = 4$ and $y = -2x + 7$ are parallel. T
10. The equation of the line through the origin that is perpendicular to the line $y = x$ is $y = -x$. T

8.5 EXERCISES

Find the equation of each of the lines described below. Write your answer in slope-intercept form. See Example 1.

1. The line through the point $(2, 3)$ with slope $1/3$
2. The line through the point $(1, 4)$ with slope $1/4$
3. The line through the point $(-2, 5)$ with slope $-1/2$
4. The line through the point $(-3, 1)$ with slope $-1/3$

5. The line with slope -3 that contains the point $(-1, -1)$ $y = -3x - 4$
6. The line with slope -2 that contains the point $(-1, -2)$ $y = -2x - 4$

1. $y = \frac{1}{3}x + \frac{7}{3}$ **2.** $y = \frac{1}{4}x + \frac{15}{4}$ **3.** $y = -\frac{1}{2}x + 4$ **4.** $y = -\frac{1}{3}x$

Find the equation of each of the lines described below. Write your answer in standard form using only integers. See Example 2.

7. The line through the points $(1, 2)$ and $(5, 8)$ $\quad 3x - 2y = -1$

8. The line through the points $(3, 5)$ and $(8, 15)$ $\quad 2x - y = 1$

9. The line through the points $(-2, -1)$ and $(3, -4)$

10. The line through the points $(-1, -3)$ and $(2, -1)$

11. The line through the points $(-2, 0)$ and $(0, 2)$ $\quad x - y = -2$

12. The line through the points $(0, 3)$ and $(5, 0)$ $\quad 3x + 5y = 15$

9. $3x + 5y = -11$ $\qquad$ **10.** $2x - 3y = 7$

Find the equation of each line described below. Write your answer in slope-intercept form. See Examples 3 and 4.

13. The line containing the point $(3, 4)$ and perpendicular to the line $y = 3x - 1$

14. The line containing the point $(-2, 3)$ and perpendicular to the line $y = 2x + 7$

15. The line parallel to $y = x - 9$ going through the point $(7, 10)$ $\quad y = x + 3$

16. The line parallel to $y = -x + 5$ going through the point $(-3, 6)$ $\quad y = -x + 3$

17. The line perpendicular to the line $3x - 2y = 10$ and passing through the point $(1, 1)$

18. The line perpendicular to the line $x - 5y = 4$ and passing through the point $(-1, 1)$ $\quad y = -5x - 4$

19. The line parallel to $2x + y = 8$ and containing the point $(-1, -3)$ $\quad y = -2x - 5$

20. The line parallel to $-3x + 2y = 9$ and containing the point $(-2, 1)$

13. $y = -\dfrac{1}{3}x + 5$ $\qquad$ **14.** $y = -\dfrac{1}{2}x + 2$

17. $y = -\dfrac{2}{3}x + \dfrac{5}{3}$ $\qquad$ **20.** $y = \dfrac{3}{2}x + 4$

Draw the following graphs and find the equation of the second line in each problem. Graphs in Answers sections.

21. Graph the line $3x + y = 5$ and the line through $(-1, 2)$ that lies perpendicular to it.

22. Graph the line $y = (1/2)x - 3$ and the line through $(1, 2)$ that lies perpendicular to it. $\quad y = -2x + 4$

23. Graph the line $-2x + y = 6$ and the line through $(2, 3)$ that lies parallel to it. $\quad y = 2x - 1$

24. Graph the line $x - 2y = 6$ and the line through $(1, 4)$ that lies parallel to it.

21. $y = \dfrac{1}{3}x + \dfrac{7}{3}$ $\qquad$ **24.** $y = \dfrac{1}{2}x + \dfrac{7}{2}$

Write each equation in standard form using only integers.

25. $y - 3 = 2(x - 5)$ $\quad 2x - y = 7$

26. $y + 2 = -3(x - 1)$ $\quad 3x + y = 1$

27. $y = \dfrac{1}{2}x - 3$ $\quad x - 2y = 6$

28. $y = \dfrac{1}{3}x + 5$ $\quad x - 3y = -15$

29. $y - 2 = \dfrac{2}{3}(x - 4)$ $\quad 2x - 3y = 2$

30. $y + 1 = \dfrac{3}{2}(x + 4)$ $\quad 3x - 2y = -10$

Write each equation in slope-intercept form.

31. $y - 1 = 5(x + 2)$ $\quad y = 5x + 11$

32. $y + 3 = -3(x - 6)$ $\quad y = -3x + 15$

33. $3x - 4y = 80$ $\quad y = \dfrac{3}{4}x - 20$

34. $2x + 3y = 90$ $\quad y = -\dfrac{2}{3}x + 30$

35. $y - \dfrac{1}{2} = \dfrac{2}{3}\left(x - \dfrac{1}{4}\right)$ $\quad y = \dfrac{2}{3}x + \dfrac{1}{3}$

36. $y + \dfrac{2}{3} = -\dfrac{1}{2}\left(x - \dfrac{2}{5}\right)$ $\quad y = -\dfrac{1}{2}x - \dfrac{7}{15}$

8.6 Applications of Linear Equations

IN THIS SECTION:

- **Examples**
- **Graphing**
- **Finding a Formula**

In this section we use linear equations to provide a gentle introduction to the concept of functions. However, the word "function" is not used in this section.

The linear equation $y = mx + b$ is a formula that determines a value of y for each given value of x. In this section we will study linear equations used as formulas.

Examples

The daily rental charge for renting a 1982 Buick at Wrenta-Wreck is $30 plus 25 cents per mile. The rental charge depends on the number of miles driven. If x represents the number of miles driven in one day, then $.25x + 30$ will give the rental charge in dollars for that day. If we let y represent the rental charge, then we can write the equation

$$y = .25x + 30.$$

This is a linear equation in slope-intercept form. The value of y depends on the value of x. We call x the **independent variable** and y the **dependent variable.**

In applications we generally use letters other than x and y for the variables. In the rental example it would be better to let m represent the number of miles and R represent the rental charge. We would then say that the rental fee R is determined by the number of miles m and write

$$R = .25m + 30.$$

There are many examples in which the value of one variable is determined from the value of another variable by means of a linear equation. When this is the case, we say that the variables have a **linear relationship.** For example, the formula

$$F = \frac{9}{5}C + 32$$

is a linear equation expressing the Fahrenheit temperature F in terms of the Celsius temperature C.

EXAMPLE 1 A car is traveling at a constant speed of 50 miles per hour. Write a linear equation that expresses the distance it travels in terms of the time it travels.

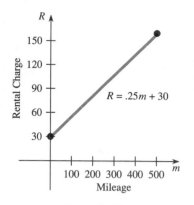

Figure 8.28

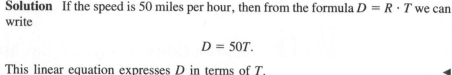

Solution If the speed is 50 miles per hour, then from the formula $D = R \cdot T$ we can write

$$D = 50T.$$

This linear equation expresses D in terms of T. ◄

Graphing

The graph of a linear equation is a straight line. A linear equation used as a formula is graphed the same way that a linear equation is graphed. If a formula is in slope-intercept form, then we can graph it using the slope and intercept as in Section 8.4. If we use letters other than x and y for the variables, then we label the axes with these letters. The horizontal axis is always the axis of the independent variable. The vertical axis is the axis of the dependent variable.

EXAMPLE 2 Graph the linear equation $R = .25m + 30$ for $0 \leq m \leq 500$. R represents the rental charge in dollars, and m represents the number of miles.

Make sure that students realize that this graph is a line segment and not a line. The restrictions on m are important.

Solution We label the x-axis with the letter m and the y-axis with the letter R. See Fig. 8.28. We adjust the scale on the m-axis to graph the values from 0 to 500. The slope of this line is .25 or 25/100. To sketch the graph, start at the R-intercept $(0, 30)$. Move 100 units to the right and up 25 units to locate a second point on the line. ◄

Finding a Formula

If two variables have a linear relationship, then there is a linear equation expressing one variable in terms of the other. In Section 8.5 we used the point-slope form to find the equation of a line given two points on the line. We can use that same procedure to translate linear relationships between two variables into equations.

EXAMPLE 3 A contractor found that his labor cost for installing 100 feet of pipe was $30. He also found that his labor cost for installing 500 feet of pipe was $120. If there is a linear relationship between labor cost C and the length L of pipe installed, then what is the formula for this relationship? What would the contractor's labor cost be for installing 240 feet of pipe?

Solution Since C is determined from L, we let C take the place of y and let L take the place of x. We can use the slope formula to find the slope of the line through the two points $(100, 30)$ and $(500, 120)$:

$$m = \frac{120 - 30}{500 - 100} = \frac{90}{400} = \frac{9}{40}$$

Point out that it is crucial that we know that there is a linear relationship between C and L. If we did not know that the relationship was linear we could not answer the question.

Now we use the point-slope form with the point $(100, 30)$ and a slope of $9/40$:

$$y - y_1 = m(x - x_1)$$

$$C - 30 = \frac{9}{40}(L - 100)$$

$$C - 30 = \frac{9}{40}L - \frac{45}{2}$$

$$C = \frac{9}{40}L - \frac{45}{2} + 30$$

$$C = \frac{9}{40}L + \frac{15}{2} \qquad \text{This linear equation expresses } C \text{ in terms of } L.$$

Now that we have a formula for C in terms of L, we can find C for any value of L. If $L = 240$ feet, then

$$C = \frac{9}{40} \cdot 240 + \frac{15}{2}$$

$$C = 54 + 7.5$$

$$C = 61.5$$

The labor cost to install 240 feet of pipe would be $61.50. ◀

Warm-ups

True or false?

1. If $z = 3r - 9$, then there is a linear relationship between z and r. T

2. The formula expressing the circumference of a circle in terms of its radius is a linear equation. T   F

3. There is a linear relationship between the area of a circle and its radius

4. There is a linear relationship between the distance that Consuela drives during her 8-hour shift at the wheel and her average speed. T

5. There is a linear relationship between Celsius temperature and Fahrenheit temperature. T

6. The slope of the line through the points $(1980, 3000)$ and $(1990, 2000)$ is 100. F

7. If your lawyer charges $90 per hour for her time, then there is a linear equation that expresses your bill for consultation in terms of the time spent with the lawyer. T

8. There is a linear equation expressing the area of a square in terms of the length of a side. F

9. There is a linear equation expressing the perimeter of a square in terms of the length of a side. T

10. There is a linear equation expressing the perimeter of a rectangle with a length of 5 meters in terms of its width. T

8.6 EXERCISES

In each case write a linear equation that expresses one variable in terms of the other. See Example 1.

1. Express length in feet in terms of length in yards. $F = 3Y$

2. Express length in yards in terms of length in feet. $Y = \frac{1}{3}F$

3. For a car averaging 65 miles per hour, express the distance it travels in terms of the time spent traveling. $D = 65T$

4. For a car traveling 6 hours, express the distance it travels in terms of its average speed. $D = 6R$

5. Express the circumference of a circle in terms of its diameter. $C = \pi D$

6. For a rectangle with a fixed width of 12 feet, express the perimeter in terms of its length. $P = 24 + 2L$

7. If Rodney makes \$7.80 per hour, then express his weekly pay in terms of the number of hours he works. $P = 7.8H$

8. If a triangle has a base of 5 feet, then express its area in terms of its height. $A = \frac{5}{2}h$

Graph each formula for the given values of the independent variable. See Example 2.   Graphs in Answers sections.

9. $P = 40n + 300, 0 \le n \le 200$

10. $C = -50r + 500, 0 \le r \le 10$

11. $R = 30t + 1000, 100 \le t \le 900$

12. $W = 3m - 4000, 1000 \le m \le 5000$

13. $C = 2\pi r, 1 \le r \le 10$

14. $P = 4s, 100 \le s \le 500$

15. $h = -7.5d + 350, 0 \le d \le 40$

16. $a = -50g + 2500, 0 \le g \le 50$

Solve each problem. See Example 2.   Graphs in Answers sections.   *Solve each problem. See Example* 3.

17. In the 1980s, People's Gas had a profit per share, P, that was determined by the equation $P = .35x + 4.60$, where x ranges from 0 to 9 corresponding to the years 1980 to 1989. What was the profit per share in 1987? Sketch the graph of this formula for x ranging from 0 to 9. \$7.05

18. For the first 6 years, the loan value of a \$30,000 automobile is determined by the formula $V = -4,000a + 30,000$ where a is the age in years of the automobile. What is the loan value of this automobile when it is 5 years old? Sketch the graph of this formula for a between 0 and 6 inclusive. \$10,000

19. When Millie called Pete's Plumbing, Pete worked 2 hours and charged Millie \$70. When her neighbor Rosalee called Pete, he worked 4 hours and charged Rosalee \$110. If there is a linear equation that determines Pete's charge from the number of hours he works, then find that equation. If Pete worked 7 hours at Gloria's house, then how much is her bill? $C = 20n + 30$, \$170

20. The sum of the measures of the interior angles of a triangle is 180°. The sum of the measures of the interior angles of a square is 360°. There is a linear relationship between the sum S of the measures of the interior angles of any n-sided polygon and the number of sides n. Express S in terms of n. What is the sum of the measures of the interior angles of an octagon? $S = 180n - 360$, 1080°

Figure for Exercise 20

Photo for Exercise 25

25. $w = -\dfrac{1}{120}t + \dfrac{3}{2}$, $\dfrac{5}{6}$ inch

width is 1.25 inches. Express w in terms of t. What is the width of the joint when the temperature is 80°F?

21. $S = 3L - \dfrac{41}{4}$, $8\dfrac{1}{2}$

21. If a child's foot is 7 3/4 inches long, then the child wears a size 13 shoe. If a child has a foot that is 5 3/4 inches long, then the child wears a size 7 shoe. There is a linear relationship between shoe size S and the length of the foot L. Write a linear equation expressing S in terms of L. What size shoe fits a child with a 6 1/4 inch foot?

22. There is a linear relationship between the Fahrenheit temperature F and the Celsius temperature C. When $C = 0$, $F = 32$. When $C = 100$, $F = 212$. Use the point-slope form to write F in terms of C. What is the Fahrenheit temperature when $C = 45$? $F = \dfrac{9}{5}C + 32$, 113°

23. There is a linear relationship between the velocity v of a falling object and the time t it has been falling. If $v = 42$ feet per second after 1 second and $v = 74$ feet per second after 2 seconds, then find the linear equation expressing v in terms of t. What is the velocity when $t = 3.5$ seconds? $V = 32t + 10$, 122 feet per second

24. There is a linear relationship between the cost C of natural gas and the number n of cubic feet of gas used. The cost of 1000 cubic feet of gas is $39, and the cost of 3000 cubic feet of gas is $99. Express C in terms of n. What is the cost of 2400 cubic feet of gas? $C = .03n + 9$, $81

25. There is a linear relationship between the width of an expansion joint on the Carl T. Hull bridge and the temperature of the roadway. When the temperature is 90°F, the width is .75 inches. When the temperature is 30°F, the

26. The perimeter P of a rectangle with a fixed width is determined by its length. If $P = 28$ when $L = 6.5$ and $P = 36$ when $L = 10.5$, then write P in terms of L. What is the perimeter when $L = 40$ feet? What is the fixed width of the rectangle? $P = 2L + 15$, 95 feet, $W = 7.5$ feet

27. There is a linear relationship between the amount A that a spring stretches beyond its natural length and the weight w placed on the spring. A weight of 3 pounds stretches a certain spring 1.8 inches, and a weight of 5 pounds stretches the same spring 3 inches. Express A in terms of w. How much will the spring stretch with a weight of 6 pounds? $A = .6w$, 3.6 inches

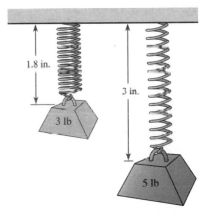

1.8 in.

3 in.

3 lb

5 lb

Figure for Exercise 27

28. $v = -32t + 100$, 4 feet per second

28. If a gun is fired upward, then there is a linear relationship between the velocity v of the bullet and the time t that has elapsed since the gun was fired. Suppose that the bullet leaves the gun at 100 feet per second (time $t = 0$) and that after 2 seconds its velocity is 36 feet per second. Express v in terms of t. What is the velocity after 3 seconds?

8.7 Functions

IN THIS SECTION:

- **Formulas**
- **Tables**
- **Definition**
- **Domain and Range**
- **The *f*-Notation**

In this section the concept of function is illustrated through numerous examples. This section could be very important to students who will see functions in a higher level course.

In Section 8.6 we studied linear relationships. The value of one variable was determined from the value of another variable by a linear equation. Whenever the value of one variable determines the value of a second variable, we say that the second variable is a function of the first. In a linear function the value of y is determined from the value of x by the formula $y = mx + b$. In this section we will discuss other types of functions. In this section, all values of the variables are real numbers.

Formulas

Your score on the final exam is a function of the number of hours you study for it. A person's blood alcohol level is a function of the number of beers consumed by that person. The area of a circle is a function of the length of the radius. These are all examples of ways in which we use the word "function." We use the word **function** to describe some process by which one variable determines the value of another variable. It is clear how the radius of a circle determines the area from the formula

$$A = \pi r^2.$$

Although it is probably true that your final exam score is a function of the number of hours you study, that is not the kind of function we will study. We must have an equation, formula, or rule by which we can determine the value of the second variable when we know the value of the first.

Any formula in which the value of one variable determines a unique value for another variable is a function. The formula

$$C = \pi d$$

expresses the circumference of a circle as a function of the length of the diameter. The formula

$$S = -16t^2 + v_0 t + s_0$$

expresses altitude S of a projectile as a function of time t, where v_0 is the initial velocity and s_0 is the initial altitude.

EXAMPLE 1 A carpet layer charges \$25 plus \$4 per square yard for installing carpet. Write the total charge C as a function of the number n of square yards of carpet installed.

Solution At \$4 per square yard, n square yards installed would cost $4n$ dollars. If we include the \$25 fee, then the total cost would be $4n + 25$ dollars. Thus the equation

$$C = 4n + 25$$

expresses C as a function of n. ◄

EXAMPLE 2 Express the volume V of a cube as a function of the length s of a side.

Solution The volume of any rectangular solid is the product of its length, width, and height. Since these are all the same for a cube, the equation

$$V = s^3$$

expresses the volume as a function of the length of a side. ◄

Tables

We do not require that a function be expressed with a formula. Consider the rate table for United Freight Service (Table 8.1). UFS will ship packages under 100 pounds anywhere in the country for the indicated cost.

TABLE 8.1

Shipping Weight in Pounds	Cost
0 to 10	\$4.60
11 to 30	\$12.75
31 to 79	\$32.90
80 to 99	\$55.82

For any *allowable* weight we can determine the shipping cost from the table. The table gives us the rule for determining the cost. We do not have a formula for determining the cost when given the weight, yet we say that the cost is a function of the weight. Note that each shipping weight determines one cost.

Table 8.2 does not look much different from Table 8.1, but there is an important difference.

Ask the students to look at this rate table and to see if anything about it bothers them. You would feel cheated if you paid $12.75 for a 10-pound package and someone else paid $4.60.

TABLE 8.2

Shipping Weight in Pounds	Cost
0 to 15	$4.60
10 to 30	$12.75
31 to 79	$32.90
80 to 99	$55.82

The cost for a 12-pound package is either $4.60 or $12.75. The weight does not determine a unique cost. In this case we would say that the cost is not a function of the weight.

EXAMPLE 3 Which of the following tables defines y as a function of x?

a)

x	y
1	3
2	6
3	9
4	12
5	15

b)

x	y
1	1
−1	1
2	2
−2	2
3	3
−3	3

c)

x	y
1988	27000
1989	27000
1990	28500
1991	29000
1992	30000
1993	30750

d)

x	y
23	48
35	27
19	28
23	37
41	56
22	34

Solution In tables (a), (b), and (c), every value of x corresponds to only one value of y. Tables (a), (b), and (c) each express y as a function of x. Notice that different values of x may correspond to the same value of y. However, in table (d) we have the value of 23 for x corresponding to two different values of y, 48 and 37, so table (d) does not express y as a function of x. ◄

Definition

So far, we have looked at functions determined by formulas and tables. We are now ready to define the concept of function.

DEFINITION
Function

A function is a rule by which any allowable value of one variable determines a *unique* value of a second variable.

Note that formulas and tables were not mentioned in the definition. Formulas and tables are just two specific ways to give a rule for determining the value of the second variable.

This idea of a first and second variable should remind us of the ordered pairs that we have been studying in this chapter. The first variable in any function is called the **independent variable,** and the second variable is called the **dependent variable.** The value of the dependent variable is determined from the value of the independent variable. In the above definition, the phrase "a unique value of a second variable" means that we will not allow a function to have any value of the independent variable paired with two different values of the dependent variable. We can give an equivalent definition of function in terms of ordered pairs.

DEFINITION
Equivalent Definition
of Function

A function is a set of ordered pairs of real numbers such that no two ordered pairs have the same first coordinates and different second coordinates.

EXAMPLE 4 Determine whether each set of ordered pairs is a function:

a) $\{(1, 2), (1, 5), (-4, 6)\}$ **b)** $\{(-1, 3), (0, 3), (6, 3), (-3, 2)\}$

Solution

a) This set of ordered pairs is not a function, since $(1, 2)$ and $(1, 5)$ have the same first coordinates but different second coordinates.
b) This set of ordered pairs is a function. Note that the same second coordinate with different first coordinates is permitted in a function. ◀

Usually, when we discuss a function, we have an equation that determines which ordered pairs are in the set. We can use the set-builder notation from Chapter 1 to describe a set of ordered pairs. For example,

$$\{(x, y) \mid y = x^2\}$$

is the set of ordered pairs such that the y-coordinate is the square of the x-coordinate. This set is a function because every value of x determines only one value of y.

EXAMPLE 5 Determine whether each set is a function:

a) $\{(x, y) \mid y = 3x^2 - 2x + 1\}$ **b)** $\{(x, y) \mid y^2 = x\}$
c) $\{(x, y) \mid x + y = 6\}$

Solution

a) This is a function because each value we select for x determines only one value for y.

b) This set is not a function because if we select $x = 9$, then we have $y^2 = 9$, or $y = \pm 3$. So both $(9, 3)$ and $(9, -3)$ belong to this set.

c) If we solve $x + y = 6$ for y, we get $y = -x + 6$. Since each value of x determines only one value for y, this is a function. In fact, this set is a linear function. ◄

We often omit the set notation and just say that the equation is a function. For example we say that the equation

$$y = 3x^2 - 2x + 1$$

is a function or defines a function, but the equation

$$y^2 = x$$

is not a function.

EXAMPLE 6 Determine which of these equations defines y as a function of x:

a) $y = |x|$ b) $y = x^3$

c) $x = |y|$ d) $y = \sqrt{x}$

Solution

a) Since every number has a unique absolute value, $y = |x|$ is a function.

b) Since every number has a unique cube, $y = x^3$ is a function.

c) The equation $x = |y|$ does not define a function, since both $(4, -4)$ and $(4, 4)$ satisfy this equation. These ordered pairs have the same first coordinate but different second coordinates.

d) We are considering only real numbers in this chapter, so we allow x to be nonnegative only. Since every nonnegative number has a unique square root, $y = \sqrt{x}$ is a function. The fact that x is not allowed to be negative in this equation has no bearing on whether this is a function. ◄

Domain and Range

The set of all possible numbers that can be used for the independent variable is called the **domain** of the function. For example, the domain of the function $y = 1/x$ is the set of all nonzero real numbers, because if $x = 0$ then $1/x$ is undefined. For some functions the domain is clearly stated when the function is given. The set of all values of the dependent variable is called the **range** of the function.

EXAMPLE 7 State the domain and range of each function.

a) $\{(3, -1), (2, 5), (1, 5)\}$ b) $y = \sqrt{x}$ c) $A = \pi r^2$ for $r > 0$

Solution

a) The domain is the set of numbers used as first coordinates, $\{1, 2, 3\}$. The range is the set of second coordinates, $\{-1, 5\}$.

b) Since $\sqrt{x}$ is a real number only for $x \geq 0$, we assume that the domain is the set of nonnegative real numbers, $\{x \mid x \geq 0\}$. The range is the set of numbers that result from taking the principal square root of every nonnegative real number. Thus the range is also the set of nonnegative real numbers, $\{y \mid y \geq 0\}$.

c) The condition $r > 0$ specifies the domain of the function. The domain is $\{r \mid r > 0\}$, the positive real numbers. Since $A = \pi r^2$, the value of A is also greater than zero. So the range is also the set of positive real numbers.   ◄

The *f*-Notation

Ask students to think of *f* as the name of the person who is pairing the *x*'s with the *y*'s. Then *f(x)* is the *y*-value calculated by *f* when given *x*.

When the variable y is a function of x, we may use the notation $f(x)$ to represent y. The symbol $f(x)$ is read as "f of x." Thus if x is the independent variable, we may use y or $f(x)$ to represent the dependent variable. For example, the function

$$y = 2x + 3$$

can also be written as

$$f(x) = 2x + 3.$$

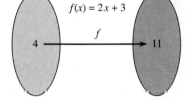

Domain   Range

Figure 8.29

We use y and $f(x)$ interchangeably. We may also use letters other than f.

Suppose $f(x) = 2x + 3$ and let $x = 4$. If we replace x by 4 in this equation, we get

$$f(4) = 2(4) + 3$$
$$f(4) = 11.$$

We read $f(4) = 11$ as "f of 4 equals 11." This means that the function f pairs 4 with 11. The ordered pair $(4, 11)$ belongs to the function f. We can use the diagram in Fig. 8.29 to picture this situation.

EXAMPLE 8 Suppose that $f(x) = x^2 - 1$ and $g(x) = -3x + 2$. Find the following:

a) $g(-2)$ **b)** $f(-1)$
c) x if $g(x) = 5$ **d)** x if $f(x) = 8$

Solution

a) Replace x by -2 in the formula $g(x) = -3x + 2$:

$$g(-2) = -3(-2) + 2 = 6 + 2 = 8$$

So $g(-2) = 8$.

b) Replace x by -1 in the formula $f(x) = x^2 - 1$:

$$f(-1) = (-1)^2 - 1$$
$$= 1 - 1$$
$$= 0$$

So $f(-1) = 0$.

c) If $g(x) = 5$, we have $5 = -3x + 2$. Solve for x:

$$-3x + 2 = 5$$
$$-3x = 3$$
$$x = -1$$

d) If $f(x) = 8$, we have

$$x^2 - 1 = 8$$
$$x^2 = 9$$
$$x = \pm 3.$$ ◄

Warm-ups

True or false?

1. Any set of ordered pairs is a function. F
2. The area of a square is a function of the length of a side. T
3. The set $\{(-1, 3), (-3, 1), (-1, -3)\}$ is a function. F
4. The set $\{(1, 5), (3, 5), (7, 5)\}$ is a function. T
5. The domain of $f(x) = \sqrt{x}$ is the set of all real numbers. F
6. The domain of $y = |x|$ is the set of nonnegative real numbers. F
7. The range of $y = |x|$ is the set of all real numbers. F
8. The set $\{(x, y) \mid x = 2y\}$ is a function. T
9. The set $\{(x, y) \mid x = y^2\}$ is a function. F
10. If $f(x) = x^2 - 5$, then $f(-2) = -1$. T

8.7 EXERCISES

Write a formula that describes the function for each of the following. See Examples 1 and 2.

1. If a small pizza costs \$5.00 plus \$.50 for each topping, express the total cost C as a function of the number of toppings t. $C = .5t + 5$

2. If the sales tax rate is 9%, express the total cost T (including tax) as a function of the total price of the groceries S. $T = 1.09S$

3. Express the circumference of a circle as a function of its radius. $C = 2\pi r$

4. Express the circumference of a circle as a function of its diameter. $C = \pi d$

5. Express the area of a circle as a function of its radius. $A = \pi r^2$

6. Express the area of a circle as a function of its diameter.

7. Express the perimeter P of a square as a function of the length s of a side. $P = 4s$

8. Express the area A of a triangle with a base of 10 as a function of its height h. $A = 5h$

9. Express the length s of a side of a square as a function of its area A. $s = \sqrt{A}$

6. $A = \dfrac{\pi d^2}{4}$

10. Express the length t of a side of a cube as a function of its volume V. $t = \sqrt[3]{V}$

11. Express the length d of the diagonal of a square as a function of the length s of a side. $d = s\sqrt{2}$

12. Express the radius of a circle as a function of its area.

12. $r = \sqrt{\dfrac{A}{\pi}}$

Determine whether each table defines the second variable as a function of the first variable. See Example 3.

13.

x	y
1	1
4	2
9	3
16	4
25	5
36	6
49	8

Yes

14.

x	y
2	4
3	9
4	16
5	25
8	36
9	49
10	100

Yes

15.

t	V
2	2
−2	2
3	3
−3	3
4	4
−4	4
5	5

Yes

16.

s	W
5	17
6	17
−1	17
−2	17
−3	17
7	17
8	17

Yes

17.

a	P
2	2
2	−2
3	3
3	−3
4	4
4	−4
5	5

No

18.

n	r
17	5
17	6
17	−1
17	−2
17	−3
17	−4
17	−5

No

19.

b	q
1970	.14
1972	.18
1974	.18
1976	.22
1978	.25
1980	.28

Yes

20.

c	h
345	.3
350	.4
355	.5
360	.6
365	.7
370	.8
375	.9

Yes

Determine whether each set of ordered pairs is a function. See Example 4.

21. $\{(1, 2), (2, 3), (3, 4)\}$ Yes

22. $\{(1, -3), (1, 3), (2, 12)\}$ No

23. $\{(-1, 4), (2, 4), (3, 4)\}$ Yes

24. $\{(1, 7), (7, 1)\}$ Yes

25. $\{(0, -1), (0, 1)\}$ No

26. $\{(50, 50)\}$ Yes

27. $\{(0, 0)\}$ Yes

28. $\{(1, 7), (-2, 7), (3, 7), (4, 7)\}$ Yes

Determine whether each set is a function. See Example 5.

29. $\{(x, y) \mid y = x - 3\}$ Yes

30. $\{(x, y) \mid y = x^2 - 2x - 1\}$ Yes

31. $\{(x, y) \mid x = |y|\}$ No

32. $\{(x, y) \mid x = y^2 + 1\}$ No

33. $\{(x, y) \mid x = y + 1\}$ Yes

34. $\{(x, y) \mid y = 1/x\}$ Yes

35. $\{(x, y) \mid x = y^2 - 1\}$ No

36. $\{(x, y) \mid x = 3y\}$ Yes

Determine whether each equation defines a function. See Example 6.

37. $x = 4y$ Yes

38. $x = \sqrt{y}$ Yes

39. $y = \sqrt{x}$ Yes

40. $y = \sqrt[4]{x}$ Yes

41. $y = \sqrt[3]{x}$ Yes

42. $y = |x - 1|$ Yes

43. $x^2 + y^2 = 25$ No

44. $x^2 - y^2 = 9$ No

Determine the domain and range of each function. See Example 7.

45. {(3, 3), (2, 5), (1, 7)}   {1, 2, 3}, {3, 5, 7}

46. {(0, 1), (2, 1), (4, 1)}   {0, 2, 4}, {1}

47. $y = |x|$ All real numbers, nonnegative real numbers

48. $y = 2x + 1$ All real numbers, all real numbers

49. $y = x^2$ All real numbers, nonnegative real numbers

50. $y = x^3$ All real numbers, all real numbers

51. $y = x + 3$ All real numbers, all real numbers

52. $y = |x - 1|$ All real numbers, nonnegative real numbers

53. $A = s^2$ for $s > 0$ {$s \mid s > 0$}, {$A \mid A > 0$}

54. $S = -16t^2$ for $t \geq 0$ {$t \mid t \geq 0$}, {$S \mid S \leq 0$}

Let $f(x) = 2x - 1$, $g(x) = x^2 - 3$, and $h(x) = |x - 1|$. Find the following. See Example 8.

55. $f(0)$ −1   **56.** $f(-1)$ −3   **57.** $g(4)$ 13   **58.** $h(3)$ 2   **59.** $g(-4)$ 13

60. $h(-1)$ 2   **61.** $h(0)$ 1   **62.** $f(1)$ 1   **63.** $h(2)$ 1   **64.** $g(0)$ −3

65. x if $f(x) = 5$ 3   **66.** x if $f(x) = 0$ $\frac{1}{2}$ **67.** x if $f(x) = -5$ −2

68. x if $f(x) = 1/2$ $\frac{3}{4}$ **69.** x if $g(x) = 22$ −5 or 5   **70.** x if $g(x) = 0$ $-\sqrt{3}$ or $\sqrt{3}$

71. x if $g(x) = -7$ The range of g does not include −7   **72.** x if $g(x) = -28$ The range of g does not include −28

Let $f(x) = \sqrt{x - 1}$ and $g(x) = x^2 - 4.2x + 2.76$. Find the following. Round answers to three decimal places.

73. $f(5.68)$ 2.163   **74.** $g(-2.7)$ 21.39   **75.** $g(3.5)$ .31

76. $f(67.2)$ 8.136   **77.** x if $f(x) = 4.8$ 24.04   **78.** x if $f(x) = 6.3$ 40.69

79. x if $g(x) = 0$ .815 or 3.385   **80.** x if $g(x) = 2$ .190 or 4.010

8.8 Variation

IN THIS SECTION:

- Direct Variation
- Finding the Constant
- Inverse Variation
- Joint Variation

This section on variation can be covered even if the section on functions is omitted.

If $y = 5x$, the value of y depends on the value of x. As x varies, so does y. Certain functions are customarily expressed in terms of variation. In this section we will learn to write formulas for those functions from verbal descriptions of the functions.

Direct Variation

Ask students for other examples of direct variation.

Suppose you drive at 60 miles per hour on the freeway. The distance D that you travel depends on the amount of time T that you travel. Using the formula $D = R \cdot T$, we can write

$$D = 60T.$$

Consider the possible values for T and D in Table 8.3.

TABLE 8.3

T (hours)	1	2	3	4	5	6
D (miles)	60	120	180	240	300	360

Note that the larger T is, the larger D will be. If T is small, D will be small. In this situation we say that D **varies directly with** T or D is **directly proportional to** T. The constant rate of 60 miles per hour is called the **variation constant** or **proportionality constant.** Notice that D is just a simple linear function of T. We are just introducing some new terms to an old idea.

DEFINITION
Direct Variation

> The statement y **varies directly as** x or y **is directly proportional to** x means that
>
> $$y = kx$$
>
> for some constant k. The constant k is a fixed nonzero real number.

Finding the Constant

EXAMPLE 1 Natasha is traveling by car, and the distance D that she travels varies directly as the rate R at which she drives. At 45 miles per hour Natasha travels 135 miles. Find the constant of variation and write D as a function of R.

Solution Since D varies directly as R, there is a constant k such that

$$D = kR.$$

Since $D = 135$ when $R = 45$, we can write

$$135 = k \cdot 45$$

or

$$3 = k.$$

Thus $D = 3R$. ◄

EXAMPLE 2 Your electric bill at Middle States Electric Co-op varies directly with the amount of electricity that you use. If the bill for 2800 kilowatts of electricity is \$196, then what would the bill be for using 4000 kilowatts of electricity?

Solution Since the amount A of the electric bill varies directly as the amount E of electricity used, we have

$$A = kE$$

for some constant k. Since 2800 kilowatts cost $196, we have

$$196 = k \cdot 2800 \qquad \text{or} \qquad .07 = k.$$

Thus $A = .07E$. If $E = 4000$, we get

$$A = .07(4000)$$
$$= 280.$$

The bill for 4000 kilowatts would be $280.   ◀

Inverse Variation

If you plan to make a 400-mile trip by car, the time it will take depends on your rate of speed. The faster you drive, the less time it will take. When you decrease your speed, the time increases. In this situation we say that the time is **inversely proportional** to the speed. Using the formula $D = RT$, we can write

$$T = \frac{400}{R}.$$

In general, we have the following definition.

DEFINITION
Inverse Variation

The statement *y* **varies inversely as** *x* or *y* **is inversely proportional to** *x* means that

$$y = \frac{k}{x}$$

for some nonzero constant k.

EXAMPLE 3 Suppose that a is inversely proportional to b and that when $b = 12$, $a = -1/3$. Find a when $b = -16$.

Solution Since a is inversely proportional to b, we have

$$a = \frac{k}{b}$$

for some constant k. Since $a = -1/3$ when $b = 12$, we can find k by substituting these values into the above formula:

$$-\frac{1}{3} = \frac{k}{12}$$

$$12\left(-\frac{1}{3}\right) = 12 \cdot \frac{k}{12}$$

$$-4 = k$$

Now to find a when $b = -16$ we can use the formula with $k = -4$:

$$a = \frac{-4}{b}$$

$$a = \frac{-4}{-16} \qquad \text{Substitute } -16 \text{ for } b.$$

$$a = \frac{1}{4} \qquad\qquad\qquad\qquad\qquad \blacktriangleleft$$

Joint Variation

If the price of carpet is $30 per square yard, then the cost C of carpeting a rectangular room depends on the width W (in yards) and the length L (in yards). As the width or length of the room increases, so does the cost. We can write the cost as a function

Math at Work

Across the country, ecologists are helping nearly extinct species to survive. Osprey hawks have been successfully reintroduced in Pennsylvania; the alligator threatens to overrun humans in Florida; and buffalo thrive in Yellowstone Park. How do ecologists determine when a species is endangered?

Ecologists use the capture-recapture method to find the size of the species population in a certain region. To count the number of trout in a lake, for example, the ecologist captures a small sample of trout, tags them for identification, and releases them back into the lake. After the tagged trout have mixed into the lake population, a second sample is taken that is large enough to include some of the tagged fish. The size of the entire population is inversely proportional to the number of tagged fish in the second sample; therefore the more tagged fish in the second sample, the lower the size of the trout population.

If the first sample includes 50 trout and the second includes 500, the constant of proportionality is $50(500) = 25{,}000$. Therefore we can estimate the population of trout P by using the formula

$$P = \frac{25{,}000}{x}$$

where x is the number of tagged fish in the second sample. If only one tagged fish is found in the second sample, the population is estimated to be 25,000. If 25 tagged fish are found in the sample, the population is estimated to be 1000.

Suppose 100 perch are tagged and released into Lake Winnebago. One month later, five tagged fish are found in a sample of 500 perch. What is the estimate of the number of perch in this lake?   10,000

of the two variables L and W:

$$C = 30LW$$

We say that C varies jointly as L and W.

DEFINITION
Joint Variation

> The statement y **varies jointly as x and z** or y **is jointly proportional to x and z** means that
>
> $$y = kxz$$
>
> for some nonzero constant k.

EXAMPLE 4   Suppose that y varies jointly with x and z and that $y = -60$ when $x = -5$ and $z = 3$. Find y when $x = 2$ and $z = 6$.

Solution   Since y varies jointly with x and z, we can write

$$y = kxz$$

for some constant k. To find the constant, substitute $y = -60$, $x = -5$, and $z = 3$ into the formula $y = kxz$:

$$-60 = k(-5)(3)$$
$$-60 = -15k$$
$$4 = k$$

To find y when $x = 2$ and $z = 6$, substitute these values into the formula $y = 4xz$:

$$y = 4(2)(6)$$
$$y = 48$$

So $y = 48$ when $x = 2$ and $z = 6$.   ◀

Notice that *these variation terms never signify addition or subtraction*. We use multiplication in the formula unless we see the word "inversely." We use division only for inverse variation.

Warm-ups

True or false?
1. If y varies directly as z, then $y = kz$ for some constant k.   T
2. If a varies inversely as b, then $a = b/k$ for some constant k.   F
3. If y varies directly as x and $y = 8$ when $x = 2$, then the variation constant is 4.   T
4. If y varies inversely as x and $y = 8$ when $x = 2$, then the variation constant is $1/4$.   F

5. If C varies jointly as h and t, then $C = ht$. F

6. The amount of sales tax on a new car varies directly with the purchase price of the car. T

7. If z varies inversely as w and $z = 10$ when $w = 2$, then $z = 20/w$. T

8. The time that it takes to travel a fixed distance varies inversely with the rate. T

9. If m varies directly as w, then $m = w + k$ for some constant k. F

10. If y varies jointly as x and z, then $y = k(x + z)$ for some constant k. F

8.8 EXERCISES

Write a formula that expresses the relationship described by each statement. Use k for the constant in each case. See Examples 1–4.

1. T varies directly as h. $T = kh$

2. m varies directly as p. $m = kp$

3. y varies inversely as r. $y = \dfrac{k}{r}$

4. u varies inversely as n. $u = \dfrac{k}{n}$

5. R is jointly proportional to t and s. $R = kts$

6. W varies jointly as u and v. $W = kuv$

7. i is directly proportional to b. $i = kb$

8. p is directly proportional to x. $p = kx$

9. A is jointly proportional to y and m. $A = kym$

10. t is inversely proportional to e. $t = \dfrac{k}{e}$

Find the variation constant and write a formula that expresses the indicated variation. See Example 1.

11. y varies directly as x, and $y = 5$ when $x = 3$. $y = \dfrac{5}{3}x$

12. m varies directly as w, and $m = 1/2$ when $w = 1/4$. $m = 2w$

13. A varies inversely as B, and $A = 3$ when $B = 2$. $A = \dfrac{6}{B}$

14. c varies inversely as d, and $c = 5$ when $d = 2$.

15. m varies inversely as p, and $m = 22$ when $p = 9$. $m = \dfrac{198}{p}$

16. s varies inversely as v, and $s = 3$ when $v = 4$. $s = \dfrac{12}{v}$

17. A varies jointly as t and u, and $A = 24$ when $t = 6$ and $u = 2$. $A = 2tu$

18. N varies jointly as p and q, and $N = 720$ when $p = 3$ and $q = 2$. $N = 120pq$

19. T varies directly as u, and $T = 9$ when $u = 2$. $T = \dfrac{9}{2}u$

20. R varies directly as p, and $R = 30$ when $p = 6$. $R = 5p$

14. $c = \dfrac{10}{d}$

Solve each variation problem. See Examples 2–4.

21. Y varies directly as x, and $Y = 100$ when $x = 20$. Find Y when $x = 5$.   25

22. n varies directly as q, and $n = 39$ when $q = 3$. Find n when $q = 8$.   104

23. a varies inversely as b, and $a = 3$, when $b = 4$. Find a when $b = 12$.   1

24. y varies inversely as w, and $y = 9$ when $w = 2$. Find y when $w = 6$.   3

25. P varies jointly as s and t, and $P = 56$ when $s = 2$ and $t = 4$. Find P when $s = 5$ and $t = 3$.   105

26. B varies jointly as u and v, and $B = 12$ when $u = 4$ and $v = 6$. Find B when $u = 5$ and $v = 8$.   20

Solve each problem.

27. The weight of an aluminum flatboat varies directly with the length of the boat. If a 12-foot boat weighs 86 pounds, then what is the weight of a 14-foot boat?   100.3 pounds

28. The price of a Christmas tree varies directly with the height. If a 5-foot tree costs \$20, then what is the price of a 6-foot tree?   \$24

Photo for Exercise 29

29. The time it takes to erect the big circus tent varies inversely as the number of elephants working on the job. If it takes four elephants 75 minutes, then how long would it take six elephants?   50 minutes

30. The volume of a gas is inversely proportional to the pressure on the gas. If the volume is 6 cubic centimeters when the pressure on the gas is 8 kilograms per square centimeter, then what is the volume when the pressure is 12 kilograms per square centimeter?   4 cubic centimeters

31. The cost of steel tubing is jointly proportional to its length and diameter. If a 10-foot tube with a 1-inch diameter costs $5.80, then what is the cost of a 15-foot tube with a 2-inch diameter? $17.40

32. The amount of sales tax varies jointly with the number of Cokes purchased and the price per Coke. If the sales tax on eight Cokes at 65 cents each is 26 cents, then what is the sales tax on six Cokes at 90 cents each?   27 cents

Wrap-up

CHAPTER 8

SUMMARY

	Concepts	Examples
Distance formula	Distance between (x_1, y_1) and (x_2, y_2) is $\sqrt{(x_1 - x_2)^2 + (y_1 - y_2)^2}$.	$(-1, 2), (2, -4)$ $\sqrt{(-1 - 2)^2 + [2 - (-4)]^2} = \sqrt{9 + 36}$ $= \sqrt{45} = 3\sqrt{5}$
Midpoint formula	The midpoint of the line segment with endpoints (x_1, y_1) and (x_2, y_2) is the point $\left(\dfrac{x_1 + x_2}{2}, \dfrac{y_1 + y_2}{2}\right)$.	$(2, -3), (4, 7)$ $\text{Midpoint} = \left(\dfrac{2 + 4}{2}, \dfrac{-3 + 7}{2}\right)$ $= (3, 2)$

Slope of a line	*(Idea)* Slope $= \dfrac{\text{change in } y}{\text{change in } x} = \dfrac{\text{rise}}{\text{run}}$	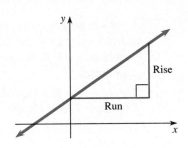

(Definition) The slope of a line through (x_1, y_1), (x_2, y_2) is

$$m = \frac{y_2 - y_1}{x_2 - x_1} \quad (x_2 - x_1 \neq 0)$$

$(3, -2)$, $(5, 6)$

$$m = \frac{6 - (-2)}{5 - 3} = \frac{8}{2} = 4$$

Types of slope

  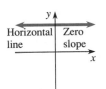

Parallel lines	Equal slopes	$y = 5x - 9$ and $y = 5x + 7$ are parallel lines.
Perpendicular lines	The slope of one line is the negative of the reciprocal of the slope of the other.	5 and $-1/5$ are slopes of perpendicular lines.
Equations of lines	*Slope-intercept form:* m is slope, $(0, b)$ is y-intercept, $y = mx + b$	Line with slope $2/3$ crossing y-axis at $(0, -1)$: $y = (2/3)x - 1$
	Point-slope form: $m = $ slope, (x_1, y_1) a point of the line $y - y_1 = m(x - x_1)$	Line through $(2, -1)$ with slope -5: $y + 1 = -5(x - 2)$
	Standard form: A and B not both zero $Ax + By = C$ Every line can be expressed in standard form, even lines for which slope is undefined.	$4x - 9y = 15$ $x = 5$ (vertical) $y = -7$ (horizontal)
Graphing a line from slope-intercept form	To graph a line using its slope and y-intercept 1. start at the y-intercept, 2. use the rise and run to locate a second point, and 3. then draw a line through the two points.	

Function	A set of ordered pairs such that no two ordered pairs have the same first coordinates and different second coordinates	$\{(1, 0), (3, 8)\}$
		$\{(x, y) \mid y = \sqrt{x}\}$
	Saying that y is a function of x means that y is determined uniquely by x.	
Domain	The set of values of the independent variable, x	$y = \sqrt{x}$
		Domain: nonnegative real numbers
Range	The set of values of the dependent variable, y	Range: nonnegative real numbers
Linear functions	If $y = mx + b$, we say that y is a linear function of x.	$F = \dfrac{9}{5}C + 32$
	The independent variable is x, and the dependent variable is y.	F is the dependent variable. C is the independent variable.
f-notation	We can use the notation $f(x)$ to represent the dependent variable.	$y = -2x + 3$ $f(x) = -2x + 3$
Variation	y varies directly as x y varies inversely as x y varies jointly as x and z	$y = kx$, where k is a constant $y = k/x$ $y = kxz$

REVIEW EXERCISES

8.1 *For each point, name the quadrant in which it lies or the axis on which it lies.*

1. $(-2, 5)$ Quadrant II   **2.** $(-3, -5)$ Quadrant III   **3.** $(3, 0)$ x-axis   **4.** $(9, 10)$ Quadrant I

5. $(0, -6)$ y-axis   **6.** $(\sqrt{2}, -3)$ Quadrant IV   **7.** $(-4, \sqrt{3})$ Quadrant II   **8.** $(0, \pi)$ y-axis

Complete the given ordered pairs so that each ordered pair satisfies the given equation.

9. $y = 3x - 5$: $(0, \ )$, $(-3, \ )$, $(4, \ )$ **10.** $y = -2x + 1$: $(9, \ )$, $(3, \ )$, $(-1, \ )$

11. $2x - 3y = 8$: $(0, \ )$, $(3, \ )$, $(-6, \ )$ **12.** $x + 2y = 1$: $(0, \ )$, $(-2, \ )$, $(2, \ )$

Sketch the graph of each equation by finding three ordered pairs that satisfy each equation.   Graphs in Answers sections.

13. $y = -3x + 4$ **14.** $y = 2x - 6$ **15.** $x + y = 7$ **16.** $x - y = 4$

8.2 *Find the length of the line segment with the given endpoints, and give the coordinates of the midpoint of that line segment.*

17. $(2, 4)$, $(-6, 5)$ $\sqrt{65}$, $(-2, 4.5)$ **18.** $(-3, 5)$, $(0, 7)$ $\sqrt{13}$, $\left(-\dfrac{3}{2}, 6\right)$ **19.** $(0, 0)$, $(3, 4)$ 5, $(1.5, 2)$

20. $(7, 9)$, $(1, 1)$ 10, $(4, 5)$ **21.** $(0, 5)$, $(12, 0)$ 13, $(6, 2.5)$ **22.** $(-3, -3)$, $(2, 1)$ $\sqrt{41}$, $\left(-\dfrac{1}{2}, -1\right)$

Solve each geometric problem.

23. Determine whether the points $(-5, 3)$, $(-1, 4)$, and $(1, -4)$ are the vertices of a right triangle. Yes

24. Determine whether the points $(-1, -1)$, $(5, 3)$, and $(0, 5)$ are the vertices of an equilateral triangle. No

9. $(0, -5)$, $(-3, -14)$, $(4, 7)$

10. $(9, -17)$, $(3, -5)$, $(-1, 3)$

11. $\left(0, -\dfrac{8}{3}\right)$, $\left(3, -\dfrac{2}{3}\right)$, $\left(-6, -\dfrac{20}{3}\right)$

12. $\left(0, \dfrac{1}{2}\right)$, $\left(-2, \dfrac{3}{2}\right)$, $\left(2, -\dfrac{1}{2}\right)$

25. Determine whether the diagonals of the quadrilateral whose vertices are $(0, -1)$, $(-1, -4)$, $(2, -3)$, and $(3, 0)$ bisect each other.   Yes

26. Use the Pythagorean theorem to determine whether the diagonals of the quadrilateral of Exercise 25 are perpendicular.   Yes

8.3 *Determine the slope of the line that goes through each of the following pairs of points.*

27. $(0, 0)$ and $(1, 1)$ 1

28. $(-1, 1)$ and $(2, -2)$ -1

29. $(-2, -3)$ and $(0, 0)$ $\dfrac{3}{2}$

30. $(-1, -2)$ and $(4, -1)$ $\dfrac{1}{5}$

31. $(-4, -2)$ and $(3, 1)$ $\dfrac{3}{7}$

32. $(0, 4)$ and $(5, 0)$ $-\dfrac{4}{5}$

Use slope to solve each geometric problem.

33. Determine whether the points $(-4, 1)$, $(-3, -3)$, and $(5, -1)$ are the vertices of a right triangle.   Yes

34. Determine whether the points $(-3, 4)$, $(-1, 2)$, $(0, 3)$, and $(2, 1)$ are the vertices of a parallelogram.   Yes

35. Slopes are 1 and -1

35. Show that the diagonals of the quadrilateral with vertices $(2, 1)$, $(6, 2)$, $(7, 6)$, and $(3, 5)$ are perpendicular.

36. Show that the opposite sides of the quadrilateral of Exercise 35 are parallel.   Slopes are 4 and $\dfrac{1}{4}$

8.4 *Find the slope and y-intercept for each line.*

37. $y = 3x - 18$ $3, (0, -18)$

38. $y = -x + 5$ $-1, (0, 5)$

39. $2x - y = 3$ $2, (0, -3)$

40. $x - 2y = 1$ $\dfrac{1}{2}, \left(0, -\dfrac{1}{2}\right)$

41. $4x - 2y - 8 = 0$ $2, (0, -4)$

42. $3x + 5y + 10 = 0$ $-\dfrac{3}{5}, (0, -2)$

Sketch the graph of each equation.   Graphs in Answers sections.

43. $y = \dfrac{2}{3}x - 5$

44. $y = \dfrac{3}{2}x + 1$

45. $-2x + y = -6$

46. $3x - y = 2$

47. $y = -4$

48. $x = 9$

Determine the equation of each line described below. Write the answer in standard form using only integers.

49. The line through $(0, 4)$ with slope $1/3$ $x - 3y = -12$

50. The line through $(-2, 0)$ with slope $-3/4$ $3x + 4y = -6$

51. The line through the origin that is perpendicular to the line $y = 2x - 1$ $x + 2y = 0$

52. The line through $(0, 9)$ that is parallel to the line $3x + 5y = 15$ $3x + 5y = 45$

53. The line through $(3, 5)$ that is parallel to the x-axis $y = 5$

54. The line through $(-2, 4)$ that is perpendicular to the x-axis $x = -2$

8.5 *Write each equation in slope-intercept form.*

55. $y - 3 = \dfrac{2}{3}(x + 6)$ $y = \dfrac{2}{3}x + 7$

56. $y + 2 = -6(x - 1)$ $y = -6x + 4$

57. $3x - 7y - 14 = 0$ $y = \dfrac{3}{7}x - 2$

58. $1 - x - y = 0$ $y = -x + 1$

59. $y - 5 = -\dfrac{3}{4}(x + 1)$ $y = -\dfrac{3}{4}x + \dfrac{17}{4}$

60. $y + 8 = \dfrac{2}{5}(x - 2)$ $y = -\dfrac{2}{5}x - \dfrac{36}{5}$

Determine the equation of each line described below. Write the answer in slope-intercept form.

61. The line through $(-4, 7)$ with slope -2 $y = -2x - 1$

62. The line through $(9, 0)$ with slope $1/2$ $y = \dfrac{1}{2}x - \dfrac{9}{2}$

63. The line through the two points $(-2, 1)$ and $(3, 7)$

64. The line through the two points $(4, 0)$ and $(-3, -5)$

65. The line through $(3, -5)$ that is parallel to the line $y = 3x - 1$ $y = 3x - 14$

66. The line through $(4, 0)$ that is perpendicular to the line $x + y = 3$ $y = x - 4$

63. $y = \dfrac{6}{5}x + \dfrac{17}{5}$ **64.** $y = \dfrac{5}{7}x - \dfrac{20}{7}$

8.6 *Graph each linear equation for the indicated values of the independent variable.*   Graphs in Answers sections.

67. $P = -3t + 400$, $10 \leq t \leq 90$

68. $R = 40w - 300$, $20 \leq w \leq 80$

69. $v = 50n + 30$, $.1 \leq n \leq .9$

70. $w = -40q + 8000$, $0 \leq q \leq 1000$

Solve each problem.

71. There is a linear relationship between the charge C for renting an air hammer from Taylor and Son Equipment Rental and the number n of days in the rental period. The charge is \$113 for two days and \$209 for five days. Write C in terms of n. What would the charge be for four days? $C = 32n + 49$, \$177

72. After two minutes on a treadmill, Jenny has a heart rate of 82. After three minutes, she has a heart rate of 86. If there is a linear relationship between Jenny's heart rate h and the time t on the treadmill, then express h in terms of t. What heart rate could be expected for Jenny after ten minutes on the treadmill? $h = 4t + 74$, 114

Photo for Exercise 72

83. All real numbers, all real numbers

8.7 *Determine whether each set of ordered pairs is a function.*

73. $\{(4, 3), (5, 3)\}$ Yes

74. $\{(0, 0), (0, 1), (0, 2)\}$ No

75. $\{(3, 4), (3, 5)\}$ No

76. $\{(1, 2), (2, 3), (3, 4)\}$ Yes

77. $\{(x, y) \mid y = 45x\}$ Yes

78. $\{(x, y) \mid x = y^3\}$ Yes

Determine whether each equation defines y as a function of x.   **84.** Nonnegative real numbers, nonpositive real numbers

79. $y = x^2 + 10$ Yes

80. $y = 2x - 7$ Yes

81. $x^2 + y^2 = 1$ No

82. $x^2 = y^2$ No

Determine the domain and range of each function.

85. $\{1, 2, 3\}$, $\{0, 2\}$

83. $f(x) = 2x - 3$

84. $y = -\sqrt{x}$

85. $\{(1, 2), (2, 0), (3, 0)\}$

86. $\{(x, y) \mid y = x^2\}$

87. $\{(x, y) \mid y = 3\}$ All real numbers, $\{3\}$

88. $g(x) = -|x|$

86. All real numbers, nonnegative real numbers

88. All real numbers, nonpositive real numbers

8.8 *Solve each variation problem.*

89. Suppose y varies directly as w. If $y = 48$ when $w = 4$, then what is y when $w = 11$?   132

90. Suppose m varies directly as t. If $m = 13$ when $t = 2$, then what is m when $t = 6$?   39

91. If y varies inversely as v, and $y = 8$ when $v = 6$, then what is y when $v = 24$?   2

92. If y varies inversely as r, and $y = 9$ when $r = 3$, then what is y when $r = 9$?   3

93. Suppose that y varies jointly as u and v and that $y = 72$ when $u = 3$ and $v = 4$. Find y when $u = 5$ and $v = 2$.   60

94. Suppose that q varies jointly as s and t and that $q = 10$ when $s = 4$ and $t = 3$. Find q when $s = 25$ and $t = 6$.   125

95. The cost of a taxi ride varies directly with the length of the ride. If a 12-minute ride costs \$9.00, then what should be the cost of a 20-minute ride?   \$15

96. The number of hours it takes to apply 296 bundles of shingles varies inversely with the number of roofers working on the job. If three roofers can complete the job in 40 hours, then how long would it take five roofers?   24 hours

Photo for Exercise 96

CHAPTER 8 TEST

For each point name the quadrant it lies in or the axis it lies on.

1. $(-2, 7)$ Quadrant II

2. $(-\pi, 0)$ x-axis

3. $(3, -6)$ Quadrant IV

4. $(0, \sqrt{2})$ y-axis

Find the distance between the given pairs of points.

5. $(1, 1)$ and $(2, 2)$ $\sqrt{2}$

6. $(-2, 3)$ and $(4, -5)$ 10

Find the midpoint of the line segment joining each pair of points.

7. $(0, 4)$ and $(3, 0)$ (1.5, 2)

8. $(3, -4)$ and $(6, 10)$ (4.5, 3)

Find the slope of the line through each pair of points.

9. $(3, 3)$ and $(4, 4)$ 1

10. $(-2, -3)$ and $(4, -8)$ $-\dfrac{5}{6}$

Write the equation of each line described below. Give the answer in slope-intercept form.

11. The line through $(0, 3)$ with slope 1/2 $y = \dfrac{1}{2}x + 3$

12. The line through $(-1, -2)$ with slope 3/7 $y = \dfrac{3}{7}x - \dfrac{11}{7}$

Write the equation of each line described below. Give the answer in standard form using only integers.

13. The line through $(2, -3)$ that is perpendicular to the line $y = -3x + 12$ $x - 3y = 11$

14. The line through $(3, 4)$ that is parallel to the line $5x + y = 7$ $y = -5x + 19$

Sketch the graph of each equation.   Graphs in Answers sections.

15. $y = \dfrac{1}{2}x - 3$

16. $2x - 3y = 6$

17. $y = 4$

18. $x = -2$

Determine the domain and range of each function.

19. $f(x) = |x| + 1$ All real numbers, $\{y \mid y \geq 1\}$

20. $y = \sqrt{x}$ $\{x \mid x \geq 0\}$, $\{y \mid y \geq 0\}$

Let $f(x) = 2x + 5$ and $g(x) = x^2 - 4$. Find the following.

21. $f(-2)$ 1

22. $g(3)$ 5

Solve each problem.

24. S = .75n + 2.50

23. Determine whether the triangle whose vertices are $(-2, -3)$, $(4, 0)$, and $(3, 2)$ is a right triangle.   Yes

24. Julie's mail-order record company charges a shipping and handling fee of $2.50 plus $.75 per record for each order shipped. Write the shipping and handling fee S as a function of the number n of records in the order.

25. The price P of a soft drink is a linear function of the volume v of the cup. A 10-ounce drink sells for 50 cents, and a 16-ounce drink sells for 68 cents. Write P as a linear

function of v. What should the price be for a 20-ounce drink? $P = 3v + 20$, 80 cents

26. The price of a watermelon varies directly with its weight. If a 30-pound watermelon sells for $4.20, then what is the price of a 20-pound watermelon? $2.80

27. The amount of time that Jason spends studying for an algebra test is inversely proportional to his score on the previous test. If Jason studied for three hours when his previous test score was 60, then how many hours would he study when his previous test score was 90?   2 hours

Tying It All Together

CHAPTERS 1–8

Simplify the following expressions.

1. 2^3 8

2. $\sqrt{12}$ $2\sqrt{3}$

3. $8^{1/3}$ 2

4. $(2^{12}) \div (2^{10})$ 4

5. $8^{-2/3}$ $\frac{1}{4}$

6. $16^{1/2}$ 4

7. $(8 - 3 \cdot 2)^2$ 4

8. $(34 \cdot 258)^0$ 1

9. $\sqrt{200}$ $10\sqrt{2}$

10. $(-1)^{-1}$ -1

11. $\sqrt{8}$ $2\sqrt{2}$

12. $\dfrac{1}{\sqrt{3}}$ $\dfrac{\sqrt{3}}{3}$

13. $\dfrac{\sqrt{30}}{\sqrt{3}}$ $\sqrt{10}$

14. $3\sqrt{2} - \sqrt{50}$ $-2\sqrt{2}$

15. $(2\sqrt{3})(5\sqrt{3})$ 30

16. $(1 - \sqrt{3})(1 + \sqrt{3})$ -2

17. $(3 + 2\sqrt{2})^2$ $17 + 12\sqrt{2}$

18. $\sqrt{20} + \sqrt{45}$ $5\sqrt{5}$

Perform the indicated operations.

19. $-3(2x - 7)$ $-3x + 21$

20. $x - 3(2x - 7)$ $-5x + 21$

21. $(x - 3)(2x - 7)$ $2x^2 - 13x + 21$

22. $(2x - 1)^2$ $4x^2 - 4x + 1$

23. $(z + 5)^2$ $z^2 + 10z + 25$

24. $(w - 7)(w + 7)$ $w^2 - 49$

Sketch a graph of each of the following equations.   Graphs in Answers sections.

25. $y = \dfrac{1}{3}x$

26. $y = 3x$

27. $y = -3x$

28. $y = -\dfrac{1}{3}x$

29. $y = 3x + 1$

30. $y = 3x - 2$

31. $y = 3$

32. $x = 3$

Solve each equation for y.

33. $3\pi y + 2 = t$ $y = \dfrac{t - 2}{3\pi}$

34. $x = \dfrac{y - b}{m}$ $y = mx + b$

35. $(y - 2)^2 = 9$ $y = -1$ or $y = 5$

36. $2y - 3 = 9$ $y = 6$

37. $\sqrt{y - 9} = 7$ $y = 58$

38. $y^2 - 3y - 40 = 0$ $y = -5$ or $y = 8$

39. $y^2 - 4y + 1 = 0$ $y = 2 \pm \sqrt{3}$

40. $\dfrac{y}{2} - \dfrac{y}{4} = \dfrac{1}{5}$ $y = \dfrac{4}{5}$

Solve each equation.

41. $5 = 4x - 7$ 3

42. $5 = 4x^2 - 7$ $-\sqrt{3}, \sqrt{3}$

43. $5 = \sqrt{x - 3}$ 28

44. $(2x - 1)^2 = 8$ $\dfrac{1 - 2\sqrt{2}}{2}, \dfrac{1 + 2\sqrt{2}}{2}$

45. $(3x - 4)(x + 9) = 0$ $-9, \dfrac{4}{3}$

46. $\dfrac{2}{3} - \dfrac{x}{6} = \dfrac{1}{2} + \dfrac{x}{4}$ $\dfrac{2}{5}$

47. $2x^2 - 7x = 0$ $0, \dfrac{7}{2}$

48. $\dfrac{3}{x} = \dfrac{x - 1}{2}$ $-2, 3$

Systems of Equations and Inequalities

Frank took care of Mr. Wilson's house while Mr. Wilson was on his spring vacation. During that time, Frank earned $50 for mowing the lawn three times and shoveling the sidewalk two times. During the last spring vacation, Frank earned $45 for mowing the lawn two times and shoveling the sidewalk three times. Frank's pay rate for mowing the lawn is the same this year as last year. His pay rate for shoveling the sidewalk also did not change. How much does Frank get paid for mowing the lawn once? How much does he get paid for shoveling the sidewalk once?

There are two unknown quantities in this problem: Frank's pay for mowing the lawn and Frank's pay for shoveling the sidewalk. We can write two equations involving these two unknowns, one for last year and one for this year. This is called a system of equations. In this chapter we will learn how to solve systems of equations. The solution to a system of equations will tell us how much Frank gets paid for mowing and for shoveling. We will solve this problem in Exercise 27 of Section 9.2.

9.1   Solving Systems of Linear Equations by Graphing

IN THIS SECTION:

- Solving a System by Graphing
- Independent, Inconsistent, and Dependent Equations

We first studied linear equations in two variables in Chapter 8. In this section we will learn to solve systems of linear equations in two variables and use systems to solve problems.

Solving a System by Graphing

Consider the linear equation

$$y = 2x - 1.$$

The graph of this equation is a straight line, and every point on the line is a solution to the equation. Now consider a second linear equation,

$$x + y = 2.$$

The graph of this equation is also a straight line, and every point on the line is a solution to this equation. A pair of equations is called a **system of equations.** A point that satisfies both equations is called a **solution to the system.**

EXAMPLE 1   Determine whether the point $(-1, 3)$ is a solution to each system of equations.

a) $3x - y = -6$
    $x + 2y = 5$

b) $y = 2x - 1$
    $x + y = 2$

Solution

a) If we let $x = -1$ and $y = 3$ in both equations of the system, we get the equations

$$3(-1) - 3 = -6 \quad \text{Correct}$$
$$-1 + 2(3) = 5 \quad \text{Correct}$$

Since both of these equations are correct, $(-1, 3)$ is a solution to the system.

b) If we let $x = -1$ and $y = 3$ in both equations of the system, we get the equations

$$3 = 2(-1) - 1 \quad \text{Incorrect}$$
$$-1 + 3 = 2 \quad \text{Correct}$$

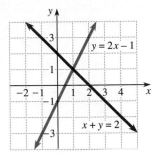

Figure 9.1

Since the first equation is not satisfied by $(-1, 3)$, the point $(-1, 3)$ is not a solution to the system. ◄

If we graph each equation of a system on the same coordinate plane, then we may be able to see the points that they have in common. Any point that is on both graphs is a solution to the system.

EXAMPLE 2 Solve the system by graphing:

$$y = 2x - 1$$
$$x + y = 2$$

Solution We first write each equation in slope-intercept form, then draw their graphs. In slope-intercept form the system becomes

$$y = 2x - 1$$
$$y = -x + 2.$$

The graph of the system is shown in Fig. 9.1. From the graph it appears that these lines intersect at $(1, 1)$. To be certain, we check that $(1, 1)$ satisfies both equations. If $x = 1$ and $y = 1$, then $y = -x + 2$ becomes $1 = -1 + 2$, and $x + y = 2$ becomes $1 + 1 = 2$. Because these equations are both true, the point $(1, 1)$ is the solution to the system. ◄

EXAMPLE 3 Solve the system by graphing:

$$3y = 2x - 6$$
$$2x - 3y = 3$$

Solution After rewriting each equation in slope-intercept form we get the following system:

$$y = \frac{2}{3}x - 2$$

$$y = \frac{2}{3}x - 1$$

Each line has slope $2/3$, but they have different y-intercepts. Their graphs are shown in Fig. 9.2. Because these two lines have the same slope, they are parallel. No point satisfies both equations. The system has no solution. ◄

Independent, Inconsistent, and Dependent Equations

Two straight lines can be placed in the coordinate plane in three different ways. We use three new terms to refer to each of these cases.

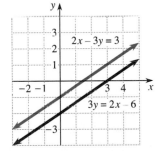

Figure 9.2

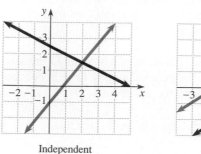
Independent

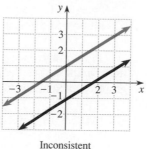
Inconsistent

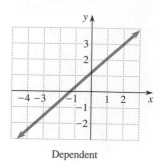
Dependent

Figure 9.3

DEFINITIONS
The Three Types of Systems

We say that the equations of the lines are

1. **independent** if the lines intersect in exactly one point,
2. **inconsistent** if the lines are parallel, and
3. **dependent** if the lines coincide.

Figure 9.3 shows each of these cases.

The system of Example 2 had a single point for the solution, so those equations are independent, or we can say that the system is independent. The graphs of the equations in Example 3 were parallel, so those equations are inconsistent, and there is no solution to that system. The next example illustrates a system of dependent equations.

EXAMPLE 4 Solve the system by graphing:

$$4x - 2y = 6$$
$$y = 2x - 3$$

Solution Rewrite the first equation in slope-intercept form for easy graphing:

$$4x - 2y = 6$$
$$-2y = -4x + 6$$
$$\frac{-2y}{-2} = \frac{-4x + 6}{-2}$$
$$y = 2x - 3$$

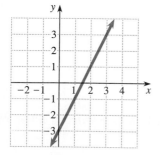

Figure 9.4

By writing the first equation in slope-intercept form, we discover that it is identical to the second equation. The graphs of the system are shown in Fig. 9.4.

Since the graphs of the two equations are identical, any point on the line satisfies both equations. The set of points on that line is written as

$$\{(x, y) \mid y = 2x - 3\}.$$

The system is dependent. ◀

Warm-ups

True or false?

1. The ordered pair $(1, 3)$ is a solution to the equation $2x + y = 5$. T

2. The ordered pair $(1, 3)$ is a solution to the system
$$2x + y = 5$$
$$x - y = 2. \quad \text{F}$$

3. Two distinct straight lines in the coordinate plane either are parallel or intersect each other in exactly one point. T

4. The system
$$y = 3x - 4$$
$$y = 3x + 5$$
has no solution. T

5. The system
$$x + y = 9$$
$$y = 9 - x$$
is a dependent system. T

6. The system
$$y = 2x - 5$$
$$y = -2x - 5$$
is an inconsistent system. F

7. The graph of $y = 3$ in the rectangular coordinate system is a horizontal straight line. T

8. Any system of two linear equations can be solved by graphing. F

9. The solution to the system
$$3x - 4y = 2$$
$$y = 1$$
is $(2, 1)$. T

10. There is only one ordered pair that satisfies the system
$$y = 5x - 7$$
$$y = 2x + 9. \quad \text{T}$$

9.1 EXERCISES

Determine which of the given points is a solution to each system of equations. See Example 1.

1. (6, 1), (3, −2), (2, 4): $2x + y = 4$ (3, −2)
$x − y = 5$

2. (−1, −1), (3, 4), (2, 3): $2x − 3y = −5$ (2, 3)
$y = x + 1$

3. (0, −2), (2, 4), (3, 7): $6x − 2y = 4$ (0, −2), (2, 4), (3, 7)
$y = 3x − 2$

4. (9, −13), (−1, 7), (0, 5): $y = −2x + 5$
$4x + 2y = 10$

5. (3, 3), (5, 7), (7, 11): $2x − y = 3$ None
$2x − y = 2$

6. (1, −2), (3, 0), (6, 3): $y = x + 5$ None
$y = x − 3$

4. (9, −13), (−1, 7), (0, 5)

Solve each system by graphing. See Example 2.

7. $y = 2x$
$y = −x + 6$ (2, 4)

8. $y = x + 1$
$y = −x + 1$ (0, 1)

9. $y = 3x − 1$
$2y − 3x = 1$ (1, 2)

10. $y = −2x + 3$
$x + y = 1$ (2, −1)

11. $y = x − 5$
$x + y = −5$ (0, −5)

12. $y + 4x = 10$
$2x − y = 2$ (2, 2)

13. $2y + x = 4$
$2x − y = −7$ (−2, 3)

14. $2x + y = −1$
$x + y = −2$ (1, −3)

15. $y = x$
$x + y = 0$ (0, 0)

16. $2y = x$
$y = 9x$ (0, 0)

17. $y = 2x − 1$
$x − 2y = −4$ (2, 3)

18. $y = x − 1$
$2x − y = 0$ (−1, −2)

Solve each system by graphing and indicate whether the system is independent, inconsistent, or dependent. See Examples 3 and 4.

19. $x − y = 3$
$3x = 3y + 12$ No solution, inconsistent

20. $x − y = 3$
$3x = 3y + 9$ {(x, y) | x − y = 3}, dependent

21. $x − y = 3$
$3x = y + 5$ (1, −2), independent

22. $3x + 2y = 6$
$2x − y = 4$ (2, 0), independent

23. $2x + y = 3$
$6x − 9 = −3y$ {(x, y) | 2x + y = 3}, dependent

24. $4y − 2x = −16$
$x − 2y = 8$ {(x, y) | x − 2y = 8}, dependent

25. $x − y = 0$
$5x = 5y$ {(x, y) | y = x}, dependent

26. $y = −3x + 1$
$2 − 2y = 6x$ {(x, y) | y = −3x + 1}, dependent

27. $x − y = −1$
$y = \frac{1}{2}x − 1$ (−4, −3), independent

28. $y = \frac{1}{3}x + 2$
$y = −\frac{1}{3}x$ (−3, 1), independent

29. $y − 4x = 4$
$y + 4x = −4$ (−1, 0), independent

30. $2y = −3x + 6$
$2y = −3x − 2$ No solution, inconsistent

9.2 Solving Systems of Linear Equations by Substitution

IN THIS SECTION:

- **Solving by Substitution**
- **Applications**

Solving a system by graphing is certainly limited by the accuracy of the graph. If the lines intersect at a point whose coordinates are not integers, then it is difficult to identify the solution from a graph. In this section we introduce a method for solving systems of linear equations in two variables that does not depend on a graph and is totally accurate.

Solving by Substitution

The next example shows how to solve a system without graphing. The method is called **substitution.**

EXAMPLE 1 Solve the system by substitution:

$$2x - 3y = 9$$
$$y - 4x = -8$$

Solution First solve $y - 4x = -8$ for y to get $y = 4x - 8$. Now use $4x - 8$ in place of y in the equation $2x - 3y = 9$:

$$2x - 3y = 9$$

$2x - 3(4x - 8) = 9$ Substitute $4x - 8$ for y.

$2x - 12x + 24 = 9$ Simplify.

$$-10x + 24 = 9$$

$$-10x = -15$$

$$x = \frac{-15}{-10} = \frac{3}{2}$$

Use the value $x = 3/2$ in $y = 4x - 8$ to find y:

$$y = 4(3/2) - 8$$
$$y = -2$$

The solution to the system is $(3/2, -2)$. Check that the point $(3/2, -2)$ satisfies both of the *original* equations.   ◀

EXAMPLE 2 Solve the system by substitution:

$$3x + 4y = 5$$
$$x = y - 1$$

Solution Since the second equation is already solved for x in terms of y, we can substitute $y - 1$ for x in the first equation and solve for y:

$$3x + 4y = 5$$
$$3(y - 1) + 4y = 5 \qquad \text{Replace } x \text{ with } y - 1.$$
$$3y - 3 + 4y = 5 \qquad \text{Simplify.}$$
$$7y - 3 = 5$$
$$7y = 8$$
$$y = \frac{8}{7}$$

Now use the value $y = 8/7$ in one of the original equations to find x. The simplest equation to use is $x = y - 1$:

$$x = \frac{8}{7} - 1$$
$$x = \frac{1}{7}$$

The solution to the system is $(1/7, 8/7)$. Check that the point $(1/7, 8/7)$ satisfies both equations. ◄

The strategy for solving by substitution can be summarized as follows.

STRATEGY
Solving a System by
Substitution

1. Solve one of the equations for one variable in terms of the others.
2. Substitute this value into the other equation to eliminate one of the variables.
3. Solve for the remaining variable.
4. Insert this value into one of the original equations to find the value of the other variable.
5. Check your solution in both equations.

The following examples illustrate how the inconsistent and dependent cases appear when we use substitution to solve the system.

EXAMPLE 3   Solve the system:

$$3x - 6y = 9$$
$$x = 2y + 5$$

Solution   Use $x = 2y + 5$ to replace x in the first equation:

$$3x - 6y = 9$$
$$3(2y + 5) - 6y = 9 \qquad \textbf{Replace } x \textbf{ by } 2y + 5.$$
$$6y + 15 - 6y = 9 \qquad \textbf{Simplify.}$$
$$15 = 9$$

Math at Work

What determines the prices of new products? Why do prices of some products go down while the prices of others go up?

Economists have determined that prices result from the collective decisions of consumers and suppliers. These decisions can be analyzed graphically by using a set of equations known as supply and demand.

An economist represents this situation on a graph, with the price of a good on one axis and the quantity sold on the other. The manufacturer's decisions form an equation of positive slope, that is, the higher the cost, the more of the good they would like to sell. Consumers, on the other hand, have the opposite demands and are willing to buy more of a good at a lower price than they are at a higher price. The point at which these two graphs cross is called the equilibrium point, and it determines prices.

A manufacturer of compact disks finds that its monthly supply equation is $P = 4 + 4x$, where P is the price for a single disk and x is in thousands of disks. This equation indicates that at a price of \$8 the manufacturer would benefit by producing only 1000 disks per month ($x = 1$). The manufacturer also finds that the demand equation for these compact disks is $P = 30 - 2.5x$. This means that consumers would buy only 2000 disks per month ($x = 2$) when the price is \$25 each. Solve the system of equations to find the equilibrium point and determine the price at which supply equals demand.   4000 CDs per month at \$20 each

Suppose that, because of lower prices of CD players, the supply equation for compact disks has changed to $P = 3 + x$ and the demand equation has changed to $P = 21 - x$. Solve this system of equations to find the new price at which supply equals demand.   9000 CDs per month at \$12 each

There are no values for x and y that will make 15 equal to 9. So there is no ordered pair that satisfies both equations. This is an inconsistent system. There is no solution to this system. These are equations for parallel lines. ◄

EXAMPLE 4 Solve the system:

$$2(y - x) = x + y - 1$$
$$y = 3x - 1$$

Solution Since the second equation is solved for y, we will eliminate the variable y in the substitution. Substitute $y = 3x - 1$ into the first equation:

$$2(3x - 1 - x) = x + (3x - 1) - 1$$
$$2(2x - 1) = 4x - 2$$
$$4x - 2 = 4x - 2$$

The last equation is an identity. It is satisfied for any value of x. Thus any value of x can be used to determine points that satisfy both equations. So these two equations are two different-appearing equations for the same straight line. This is a dependent system. The solution to the system is the set of all points on that line,

$$\{(x, y) \mid y = 3x - 1\}.$$ ◄

When solving a system by substitution, we can recognize an inconsistent system or dependent system as follows:

Inconsistent and Dependent Systems

An inconsistent system leads to a false statement. A dependent system leads to an identity.

Applications

Many of the problems that we solved in previous chapters had two unknown quantities, but we wrote only one equation to solve the problem. For problems with two unknown quantities we can use two variables and a system of equations.

EXAMPLE 5 Mrs. Robinson invested a total of $25,000 in two investments, one paying 6% and the other paying 8%. If her total income from these investments was $1790, then how much money did she invest in each?

Solution Let x represent the amount invested at 6% and let y represent the amount invested at 8%. Table 9.1 organizes the given information.

TABLE 9.1

Interest rate	Amount invested	Amount of interest
6%	x	$.06x$
8%	y	$.08y$

We write one equation describing the total of the investments, and the other equation describing the total interest:

$$x + y = 25,000 \quad \textbf{Total investments}$$

$$.06x + .08y = 1790 \quad \textbf{Total interest}$$

To solve the system, we solve the first equation for y:

$$y = 25,000 - x$$

Substitute $25,000 - x$ for y in the second equation:

$$.06x + .08(25,000 - x) = 1790$$

$$.06x + 2000 - .08x = 1790$$

$$-.02x + 2000 = 1790$$

$$-.02x = -210$$

$$x = \frac{-210}{-.02} = 10,500$$

If $x = 10,500$ and $y = 25,000 - x$, then $y = 14,500$. Mrs. Robinson invested $10,500 at 6% and $14,500 at 8%. Check. ◀

Warm-ups

True or false?

1. It is impossible to solve some systems by graphing. T
2. If $x + y = 9$ and $x = 3$, then $y = 6$. T
3. To solve the system

$$y = x - 7$$
$$2x + 3y = 4$$

by substitution, we replace y in the second equation by $x - 7$. T

4. We use the substitution method to eliminate one of the variables. T
5. We can eliminate either x or y when using the substitution method. T

6. The point $(1/2, 1/4)$ satisfies the system

$$x + 2y = 1$$
$$2x - 4y = 0. \quad \text{T}$$

7. If $2x - 3y = 7$ and $y = 1$, then $x = 5$. T

8. Solving an inconsistent system by substitution will result in a false statement. T

9. Solving a dependent system by substitution will result in an identity. T

10. Any system of two linear equations can be solved by substitution. T

9.2 EXERCISES

Solve each system by the substitution method. See Examples 1 and 2.

1. $y = x + 3$
$2x - 3y = -11$ (2, 5)

2. $y = x - 5$
$x + 2y = 8$ (6, 1)

3. $x = 2y - 4$
$2x + y = 7$ (2, 3)

4. $x = y - 2$
$-2x + y = -1$ (3, 5)

5. $2x + y = 5$
$5x + 2y = 8$ (−2, 9)

6. $5y - x = 0$
$6x - y = -2$ $\left(-\dfrac{10}{29}, -\dfrac{2}{29}\right)$

7. $x + y = 0$
$3x + 2y = -5$ (−5, 5)

8. $x - y = 6$
$3x + 4y = -3$ (3, −3)

9. $x + y = 1$
$4x - 8y = -4$ $\left(\dfrac{1}{3}, \dfrac{2}{3}\right)$

10. $x - y = 2$
$3x - 6y = 8$ $\left(\dfrac{4}{3}, -\dfrac{2}{3}\right)$

11. $2x + 3y = 2$
$4x - 9y = -1$ $\left(\dfrac{1}{2}, \dfrac{1}{3}\right)$

12. $x - 2y = 1$
$3x + 10y = -1$ $\left(\dfrac{1}{2}, -\dfrac{1}{4}\right)$

Solve each system by substitution and identify each system as independent, dependent, or inconsistent. See Examples 3 and 4.

13. $x - 2y = -2$
$x + 2y = 8$ $\left(3, \dfrac{5}{2}\right)$, independent

14. $y = -3x + 1$
$y = 2x + 4$ $\left(-\dfrac{3}{5}, \dfrac{14}{5}\right)$, independent

15. $x = 4 - 2y$
$4y + 2x = -8$ No solution, inconsistent

16. $21x - 35 = 7y$
$3x - y = 5$ $\{(x, y) \mid 3x - y = 5\}$, dependent

17. $y - 3 = 2(x - 1)$
$y = 2x + 3$ No solution, inconsistent

18. $y + 1 = 5(x + 1)$
$y = 5x - 1$ No solution, inconsistent

19. $3x - 2y = 7$
$3x + 2y = 7$ $\left(\dfrac{7}{3}, 0\right)$, independent

20. $2x + 5y = 5$
$3x - 5y = 6$ $\left(\dfrac{11}{5}, \dfrac{3}{25}\right)$, independent

21. $x + 5y = 4$
$x + 5y = 4y$ (−1, 1), independent

22. $2x + y = 3x$
$3x - y = 2y$ $\{(x, y) \mid y = x\}$, dependent

Write a system of two equations in two unknowns for each problem. Solve each system by substitution. See Example 5.

23. Mrs. Miller invested $20,000 and received a total of $1,600 in interest. If she invested part of the money at 10% and the remainder at 5%, then how much did she invest at each rate? $12,000 at 10%, $8,000 at 5%

24. Mr. Walker invested $30,000 and received a total of $2,880 in interest. If he invested part of the money at 10%

and the remainder at 9%, then how much did he invest at each rate? $18,000 at 10%, $12,000 at 9%

25. The sum of two numbers is 2, and their difference is 14. Find the numbers. 8 and −6

26. The sum of two numbers is −3, and their difference is 15. Find the numbers. −9 and 6

27. When Mr. Wilson came back from his spring vacation, he paid Frank $50 for mowing his lawn three times and shoveling his sidewalk two times. During Mr. Wilson's spring vacation last year, Frank earned $45 for mowing the lawn two times and shoveling the sidewalk three times. How much does Frank make for mowing the lawn once? How much does he make for shoveling the sidewalk once? Lawn: $12, sidewalk: $7

28. Donna ordered four burgers and one order of fries at the Hamburger Palace. However, the waiter put three burgers and two orders of fries in the bag and charged Donna the correct price for three burgers and two orders of fries, $3.15. When Donna discovered the mistake, she went back to complain. She found out that the price for four burgers and one order of fries is $3.45, and she decided to keep what she had. What is the price of one burger? What is the price of one order of fries? Burger: $.75, fries: $.45

9.3 The Addition Method

IN THIS SECTION:

- **Addition Method**
- **Applications**

In Section 9.2 we solved systems of equations by using substitution. We substituted one equation into the other to eliminate a variable. The addition method of this section is another method for eliminating a variable to solve a system of equations.

Addition Method

In the substitution method we solve for one variable in terms of the other variable. When doing this, we may get an expression involving fractions, which must be substituted into the other equation. The addition method avoids fractions and is easier to use on certain systems.

EXAMPLE 1 Solve the system by the addition method:

$$3x - y = 5$$
$$2x + y = 10$$

Solution The addition property of equality allows us to add the same number to each side of an equation. We can also use the addition property of equality to add the two left-hand sides and add the two right-hand sides:

$$3x - y = 5$$
$$2x + y = 10 \quad \text{Add.}$$
$$\overline{5x \quad\ = 15} \quad -y + y = 0$$
$$x = 3$$

Note that the y term was eliminated when we added the equations, because the coefficients of y in the two equations were opposites. Now use $x = 3$ in either one of the original equations to find y:

$$2x + y = 10$$
$$2(3) + y = 10 \quad \text{Let } x = 3.$$
$$y = 4$$

The solution to the system is $(3, 4)$. Check. ◀

Of course we could subtract the equations, but it is usually best to stick with one method here. Always eliminate variables by addition.

The addition method is based on the addition property of equality. We are adding equal quantities to each side of an equation. The form of the equations does not matter as long as the equal signs and the like terms are in line.

In Example 1, y was eliminated by the addition, because the coefficients of y in the two equations were opposites. If no variable will be eliminated by addition, we can use the multiplication property of equality to change the coefficients of the variables. In the next example the coefficient of x in one equation is a multiple of the coefficient of x in the other equation. We use the multiplication property of equality to get opposite coefficients for x.

EXAMPLE 2 Solve the system by the addition method:

$$-x + 4y = -14$$
$$2x - 3y = 18$$

Solution If we add these equations as they are written, we will not eliminate any variables. However, if we multiply each side of the first equation by 2, then we will be adding $-2x$ and $2x$, and x will be eliminated.

$$2(-x + 4y) = 2(-14) \quad \text{Multiply each side by 2.}$$
$$2x - 3y = 18$$
$$-2x + 8y = -28$$
$$\underline{2x - 3y = \quad 18} \quad \text{Add.}$$
$$5y = -10$$
$$y = -2$$

Now replace y by -2 in one of the original equations:

$$-x + 4(-2) = -14$$
$$-x - 8 = -14$$
$$-x = -6$$
$$x = 6$$

The solution to the system is $(6, -2)$. Check.   ◀

In the next example we need to use a multiple of each equation to eliminate a variable by addition.

EXAMPLE 3 Solve the system by the addition method:

$$2x + 3y = 7$$
$$3x + 4y = 10$$

Solution To eliminate x, we multiply the first equation by -3 and the second equation by 2:

$$-3(2x + 3y) = -3(7)$$
$$2(3x + 4y) = 2(10)$$
$$-6x - 9y = -21$$
$$\underline{6x + 8y = 20} \qquad \textbf{Add.}$$
$$-y - -1$$
$$y - 1$$

Replace y with 1 in one of the original equations:

$$2x + 3y = 7$$
$$2x + 3(1) = 7$$
$$2x + 3 = 7$$
$$2x = 4$$
$$x = 2$$

The solution to the system is $(2, 1)$. Check.   ◀

The strategy used in solving a system by the addition method is summarized in the following box.

STRATEGY
Solving a System by
Addition

1. The equations must be in the same form.
2. If a variable will be eliminated by adding, then add the equations.

3. If necessary, obtain multiples of one or both equations so that a variable will be eliminated by adding the equations.
4. After one variable is eliminated, solve for the remaining variable.
5. Use the value of the remaining variable to find the value of the eliminated variable.
6. Check the solution in the original system.

When the addition method is used, an inconsistent system will be indicated by a false statement. A dependent system will be indicated by an identity.

EXAMPLE 4 Use the addition method to solve each system:

a) $-2x + 3y = 9$
$2x - 3y = 18$

b) $2x - y = 1$
$4x - 2y = 2$

Solution

a) Add the equations:

$$-2x + 3y = 9$$
$$\underline{2x - 3y = 18}$$
$$0 = 27 \quad \text{False.}$$

There is no solution to the system. The system is inconsistent.

b) Multiply the first equation by -2, and then add the equations:

$$-2(2x - y) = -2(1)$$
$$4x - 2y = 2$$
$$-4x + 2y = -2$$
$$\underline{4x - 2y = 2}$$
$$0 = 0 \quad \text{This is an identity.}$$

This system is dependent. The set of points satisfying the system is $\{(x, y) \mid 2x - y = 1\}$. ◄

Applications

In the next example we solve a problem using a system of equations and the addition method.

EXAMPLE 5 Lea purchased two gallons of milk and three loaves of bread for $8.25. Yesterday, she purchased three gallons of milk and two loaves of bread for $9.25. What is the price of a single gallon of milk? What is the price of a single loaf of bread?

Solution Let x represent the price of one gallon of milk. Let y represent the price of one loaf of bread. We can write two equations about the milk and bread:

$$2x + 3y = 8.25 \qquad \text{Today's purchase}$$
$$3x + 2y = 9.25 \qquad \text{Yesterday's purchase}$$

To eliminate x, multiply the first equation by -3 and the second by 2:

$$-3(2x + 3y) = -3(8.25)$$
$$2(3x + 2y) = 2(9.25)$$

$$
\begin{array}{rl}
-6x - 9y = & -24.75 \\
6x + 4y = & 18.50 \qquad \text{Add.} \\
\hline
-5y = & -6.25 \\
y = & 1.25
\end{array}
$$

Replace y by 1.25 in one of the original equations:

$$2x + 3(1.25) = 8.25$$
$$2x + 3.75 = 8.25$$
$$2x = 4.50$$
$$x = 2.25$$

A gallon of milk costs \$2.25, and a loaf of bread costs \$1.25. Check. ◀

Warm-ups

True or false?

1. The addition method is used to eliminate a variable in a system of equations. T

2. Either variable can be eliminated by the addition method. T

3. The point $(1, 2)$ is a solution to the system
$$3x + 2y = 7$$
$$4x - 5y = -6. \quad \text{T}$$

4. Both $(0, 2)$ and $(1, -1)$ satisfy the system
$$y = -3x + 2$$
$$2y + 6x - 4 = 0. \quad \text{T}$$

5. The equations $y = 5x + 4$ and $y = 5x + 6$ are independent. F

6. The equations $x + y = 7$ and $2x + 2y = 14$ are inconsistent. F

7. The solution to the system

$$y = x - 5$$
$$x = y + 5$$

is $\{(x, y) \mid y = x - 5\}$.   T

8. When we use the addition method, we must have the equal signs lined up.   T

9. The equations $y = x$ and $y = -x$ are inconsistent.   F

10. To eliminate x by addition in the system

$$3x - 8y = 9$$
$$4x + 5y = 1$$

we multiply the first equation by 4 and the second equation by 3.   F

9.3 EXERCISES

Solve each system by the addition method. See Examples 1–3.

1. $2x + y = 5$
   $3x - y = 10$ $(3, -1)$

2. $x + 2y = 7$
   $-x + 4y = 5$ $(3, 2)$

3. $x + 2y = 7$
   $-x + 3y = 18$ $(-3, 5)$

4. $3x - y = 3$
   $4x + y = 11$ $(2, 3)$

5. $x + 2y = 2$
   $-4x + 3y = 25$ $(-4, 3)$

6. $2x - 3y = -7$
   $5x + y = -9$ $(-2, 1)$

7. $x + 3y = 4$
   $2x - y = -1$ $\left(\frac{1}{7}, \frac{9}{7}\right)$

8. $x - y = 0$
   $x - 2y = 0$ $(0, 0)$

9. $y = 4x - 1$
   $y = 3x + 7$ $(8, 31)$

10. $2x = y - 9$
   $x = -1 - 3y$ $(-4, 1)$

11. $4x = 3y + 1$
   $2x = y - 1$ $(-2, -3)$

12. $45 = x - 2y$
   $15 = x + 2y$ $\left(30, -\frac{15}{2}\right)$

13. $2x - 5y = -22$
   $-6x + 3y = 18$ $(-1, 4)$

14. $4x - 3y = 7$
   $5x + 6y = -1$ $(1, -1)$

15. $2x + 3y = 4$
   $-3x + 5y = 13$ $(-1, 2)$

16. $-5x + 3y = 1$
   $2x - 7y = 17$ $(-2, -3)$

17. $2x - 5y = 11$
   $3x - 2y = 11$ $(3, -1)$

18. $4x - 3y = 17$
   $3x - 5y = 21$ $(2, -3)$

19. $5x + 4y = 13$
   $2x + 3y = 8$ $(1, 2)$

20. $4x + 3y = 8$
   $6x + 5y = 14$ $(-1, 4)$

Use either the addition method or substitution to solve each system. State whether the system is independent, inconsistent, or dependent. See Example 4.

21. $x + y = 5$
   $x + y = 6$ No solution, inconsistent

22. $x + y = 5$
   $x + 2y = 6$ $(4, 1)$, independent

23. $x + y = 5$
   $2x + 2y = 10$ $\{(x, y) \mid x + y = 5\}$, dependent

24. $2x + 3y = 4$
   $2x - 3y = 4$ $(2, 0)$, independent

25. $2x = y + 3$
   $2y = 4x - 6$ $\{(x, y) \mid 2x = y + 3\}$, dependent

26. $y = 2x - 1$
   $2x - y + 5 = 0$ No solution, inconsistent

27. $x + 3y = 3$
   $\frac{1}{3}x = 1 - y$ $\{(x, y) \mid x + 3y = 3\}$, dependent

28. $y = 3x + 2$
   $y = -3x + 2$ $(0, 2)$, independent

29. $6x - 2y = -2$
$y = 3x + 4$ No solution, inconsistent

30. $x + y = 8$
$\frac{1}{3}x - \frac{1}{2}y = 1$ (6, 2), independent

31. $\frac{1}{2}x - \frac{2}{3}y = -6$
$-\frac{3}{4}x - \frac{1}{2}y = -18$ (12, 18), independent

32. $\frac{1}{2}x - y = 3$
$\frac{1}{5}x + 2y = 6$ (10, 2), independent

33. $.04x + .09y = 7$
$x + y = 100$ (40, 60), independent

34. $.08x - .05y = .2$
$2x + y = 140$ (40, 60), independent

35. $.1x - .2y = -.01$
$.3x + .5y = .08$ (.1, .1), independent

36. $.5y = .2x - .25$
$.1y = .8x - 1.57$ (2, .3), independent

Use two variables and a system of equations to solve each problem. See Example 5.

37. An automobile dealer had 250 vehicles on his lot during the month of June. He must pay a monthly inventory tax of $3 per car and $4 per truck. If his tax bill for June was $850, then how many cars and how many trucks did he have on his lot during June?   150 cars, 100 trucks

38. Kimberly has 30 coins consisting of dimes and nickels. If the value of these coins is $2.30, then how many of each type does she have?   16 dimes, 14 nickels

39. The Audubon Zoo charges $5.50 for each adult admission and $2.75 for each child. The total bill for the 30 people on the Spring Creek Elementary School kindergarten field trip was $99. How many adults and how many children went on the field trip?   6 adults, 24 children

40. Jorge has worked at Dandy Doughnuts for so long that he has memorized the amounts for many of the common or-

ders. For example, six doughnuts and five coffees cost $4.35, while four doughnuts and three coffees cost $2.75. What are the prices of one cup of coffee and one doughnut?   Coffee: $.45, doughnut: $.35

41. The Independent Marketing Research Corporation found 130 smokers among 300 adults surveyed. If one-half of the men and one-third of the women were smokers, then how many men and how many women were in the survey?

42. In one month, Shelly earned $1800 for 210 hours of work. If she earns $8 per hour for regular time and $12 per hour for overtime, then how many hours of each type did she work?   180 hours regular, 30 hours overtime

41. 180 men, 120 women

9.4 Linear Inequalities in Two Variables

IN THIS SECTION:

- Definition
- Determining a Solution
- Graph of a Linear Inequality
- Using a Test Point to Graph an Inequality

We studied linear equations and inequalities in one variable in Chapter 2. In this section we extend the ideas of linear equations in two variables to study linear inequalities in two variables.

Definition

Linear inequalities in two variables have the same form as linear equations in two variables. An inequality symbol is used in place of the equal sign.

DEFINITION
Linear Inequality in
Two Variables

If A, B, and C are real numbers with A and B not both zero, then

$$Ax + By \leq C$$

is called a **linear inequality** in two variables. In place of $\leq$, we can also use $\geq$, $<$, or $>$.

The inequalities

$$3x - 4y \leq 8, \qquad y > 2x - 3, \qquad \text{and} \qquad x - y + 9 < 0$$

are linear inequalities. Not all of these are in the form of the definition, but they could all be rearranged into that form.

Determining a Solution

An ordered pair is a solution to an inequality in two variables if the ordered pair satisfies the inequality.

EXAMPLE 1   Determine whether each of the following points satisfies the inequality $2x - 3y \geq 6$:

a) $(4, 1)$ **b)** $(3, 0)$ **c)** $(3, -2)$

Solution

a) To determine whether $(4, 1)$ is a solution to the inequality, we replace x by 4 and y by 1 in the inequality $2x - 3y \geq 6$:

$$2(4) - 3(1) \geq 6$$
$$8 - 3 \geq 6$$
$$5 \geq 6$$

Because $5 \geq 6$ is incorrect, the point $(4, 1)$ does not satisfy the inequality $2x - 3y \geq 6$.

b) Replace x by 3 and y by 0:

$$2(3) - 3(0) \geq 6$$
$$6 \geq 6$$

Because $6 \geq 6$ is true, the point $(3, 0)$ satisfies the inequality.

c) Replace x by 3 and y by -2:

$$2(3) - 3(-2) \geq 6$$
$$6 + 6 \geq 6$$
$$12 \geq 6$$

Because the inequality $12 \geq 6$ is correct, the point $(3, -2)$ satisfies the inequality $2x - 3y \geq 6$.   ◀

Graph of a Linear Inequality

The graph of a linear inequality in two variables consists of all points in the rectangular coordinate system that satisfy the inequality. For example, the graph of the inequality

$$y > x + 2$$

consists of all points where the y-coordinate is larger than the x-coordinate plus 2. Consider the point $(3, 5)$ on the line

$$y = x + 2.$$

The y-coordinate of $(3, 5)$ is equal to the x-coordinate plus 2. If we move straight up from $(3, 5)$ on the line, say to $(3, 6)$, the y-coordinate gets larger, but the x-coordinate does not change. Thus for points above the line $y = x + 2$ it is true that $y > x + 2$. Likewise, all points below the line $y = x + 2$ satisfy the inequality $y < x + 2$. See Fig. 9.5.

To graph the inequality, we shade all points above the line $y = x + 2$. To indicate that the line is not included in the graph of $y > x + 2$, we use a dashed line.

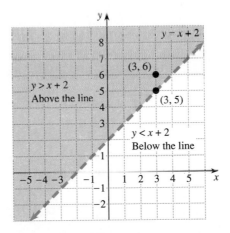

Figure 9.5

A summary of the strategy for graphing linear inequalities is contained in the following box.

STRATEGY
Graphing a Linear Inequality in Two Variables

---

1. Solve the inequality for y, then graph $y = mx + b$.

$y > mx + b$ is the region above the line.

$y = mx + b$ is the line itself.

$y < mx + b$ is the region below the line.

2. If the inequality involves only x, then graph the vertical line $x = k$.

$x > k$ is the region to the right of the line.

$x = k$ is the line itself.

$x < k$ is the region to the left of the line.

EXAMPLE 2   Graph the inequalities:

a) $y < \dfrac{1}{3}x + 1$

b) $y \ge -2x + 3$

c) $2x - 3y < 6$

Solution

a) The set of points satisfying this inequality is the region below the line $y = (1/3)x + 1$. To show this region, we first graph the boundary line. The slope of the line is $1/3$, and the y-intercept is $(0, 1)$. We draw the line dashed because it is not part of the graph of $y < (1/3)x + 1$. See Fig. 9.6, where the graph is the shaded region.

b) Since the inequality symbol is $\ge$, every point on or above the line satisfies this inequality. We use the fact that the slope of this line is -2 and the y-intercept is $(0, 3)$ to draw the graph of the line. To show that the line $y = -2x + 3$ is included in the graph, we make it a solid line and shade the region above. See Fig. 9.7.

c) First solve for y:

$$2x - 3y < 6$$
$$-3y < -2x + 6$$
$$y > \frac{2}{3}x - 2 \qquad \textbf{Divide by } -3 \textbf{, and reverse the inequality.}$$

To graph this inequality, we first graph the line with slope $2/3$ and y-intercept $(0, -2)$. We use a dashed line for the boundary, since it is not included, and we shade the region above the line. Remember that "$<$" *means below the line and*

This method of graphing a linear inequality corresponds to graphing equations from slope-intercept form.

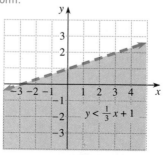

Figure 9.6

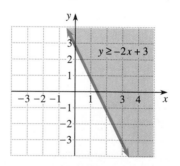

Figure 9.7

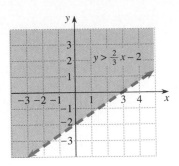

Figure 9.8

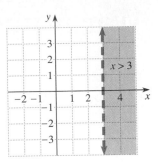

Figure 9.9

Figure 9.10

">" means above the line only when the inequality is solved for y. See Fig. 9.8 for the graph. ◄

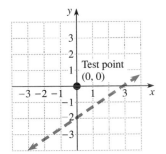

Figure 9.11

EXAMPLE 3 Graph the inequalities:

a) $y \le 4$ **b)** $x > 3$

Solution

a) The line $y = 4$ is the horizontal line with y-intercept $(0, 4)$. We draw a solid horizontal line and shade below it as in Fig. 9.9.

b) This inequality has no y term, but it is easy to graph. First draw a dashed vertical line through $(3, 0)$ on the x-axis. The graph, shown in Fig. 9.10, is the region to the right of the vertical line $x = 3$. ◄

Using a Test Point to Graph an Inequality

The graph of a linear equation such as $2x - 3y = 6$ separates the coordinate plane into two regions. One region satisfies the inequality $2x - 3y > 6$, and the other region satisfies the inequality $2x - 3y < 6$. We can tell which region satisfies which inequality by testing a point in one region. *With this method, it is not necessary to solve the inequality for y.*

This method corresponds to graphing equations using x- and y-intercepts.

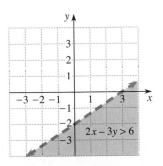

Figure 9.12

EXAMPLE 4 Graph the inequality $2x - 3y > 6$.

Solution First graph the equation $2x - 3y = 6$ using the x-intercept and the y-intercept. If $x = 0$, then $y = -2$. If $y = 0$, then $x = 3$. Use the x-intercept $(3, 0)$ and the y-intercept $(0, -2)$ to graph the line as shown in Fig. 9.11.

Select a point on one side of the line, say $(0, 0)$, to test in the inequality. Since $2(0) - 3(0) > 6$ is false, the region on the other side of the line satisfies the inequality. The graph of $2x - 3y > 6$ is shown in Fig. 9.12. ◄

Warm-ups

True or false?

1. The point $(-1, 4)$ satisfies the inequality $y > 3x + 1$. T
2. The point $(2, -3)$ satisfies the inequality $3x - 2y \geq 12$. T
3. The graph of the inequality $y > x + 9$ is the region above the line $y = x + 9$. T
4. The graph of the inequality $x < y + 2$ is the region below the line $x = y + 2$. F
5. The graph of $x = 3$ is a single point on the x-axis. F
6. The graph of $y \leq 5$ is the region below the horizontal line $y = 5$. F
7. The graph of $x < 3$ is the region to the left of the vertical line $x = 3$. T
8. In graphing the inequality $y \geq x$ we use a dashed boundary line. F
9. The point $(0, 0)$ is on the graph of the inequality $y \geq x$. T
10. The graph of $x \geq 5$ consists of only those points on the x-axis that are to the right of $(5, 0)$. F

9.4 EXERCISES

Determine which of the points following each inequality satisfy that inequality. See Example 1.

1. $x - y > 5$: $(2, 3)$, $(-3, -9)$, $(8, 3)$ $(-3, -9)$
2. $2x + y < 3$: $(-2, 6)$, $(0, 3)$, $(3, 0)$ $(-2, 6)$
3. $y \geq -2x + 5$: $(3, 0)$, $(1, 3)$, $(-2, 5)$ $(3, 0)$, $(1, 3)$
4. $y \leq -x + 6$: $(2, 0)$, $(-3, 9)$, $(-4, 12)$ $(2, 0)$, $(-3, 9)$
5. $x > -3y + 4$: $(2, 3)$, $(7, -1)$, $(0, 5)$ $(2, 3)$, $(0, 5)$
6. $x < -y - 3$: $(1, 2)$, $(-3, -4)$, $(0, -3)$ $(-3, -4)$

Graph each inequality. See Examples 2 *and* 3. Graphs in Answers sections.

7. $y < x + 4$
8. $y < 2x + 2$
9. $y > -x + 3$
10. $y < -2x + 1$

11. $y > \dfrac{2}{3}x - 3$
12. $y < \dfrac{1}{2}x + 1$
13. $y \leq -\dfrac{2}{5}x + 2$
14. $y \geq -\dfrac{1}{2}x + 3$

15. $y - x \geq 0$
16. $x - 2y \leq 0$
17. $x > y - 5$
18. $2x < 3y + 6$

19. $x - 2y + 4 \leq 0$
20. $2x - y + 3 \geq 0$
21. $y \geq 2$
22. $y < 7$

23. $x > 9$
24. $x \leq 1$
25. $x + y \leq 6$
26. $y \leq 6$

27. $x \leq 6$
28. $y \geq 6x$
29. $3x - 4y \leq 8$
30. $2x + 5y \geq 10$

Graph each inequality. Use the test point method of Example 4. Graphs in Answers sections.

31. $2x - 3y < 6$
32. $x - 4y > 4$
33. $\dfrac{1}{2}x - 2y \leq 4$
34. $3y - 5x \geq 15$

35. $2y - 7x \leq 14$
36. $\dfrac{2}{3}x + 3y \leq 12$
37. $x - y < 5$
38. $y - x > -3$

39. $3x - 4y < -12$
40. $4x + 3y > 24$
41. $x - 5y < -10$
42. $-x > 7 - y$

9.5 Systems of Linear Inequalities

IN THIS SECTION:

- **The Solution to a System of Inequalities**
- **Graphing a System of Inequalities**

A point is a solution to a system of equations if it satisfies both equations. A system of inequalities is similar. In this section we will solve systems of linear inequalities.

The Solution to a System of Inequalities

A point is a solution to a system of inequalities if it satisfies both inequalities.

EXAMPLE 1 Determine whether each point is a solution to the system of inequalities:

$$2x + 3y < 6$$
$$y > 2x - 1$$

a) $(-3, 2)$ **b)** $(4, -3)$ **c)** $(5, 1)$

Solution

a) The point $(-3, 2)$ is a solution to the system if it satisfies both inequalities. Let $x = -3$ and $y = 2$ in each inequality:

$$
\begin{array}{c|c}
2x + 3y < 6 & y > 2x - 1 \\
2(-3) + 3(2) < 6 & 2 > 2(-3) - 1 \\
0 < 6 & 2 > -7
\end{array}
$$

Since both inequalities are satisfied, the point $(-3, 2)$ is a solution to the system.

b) Let $x = 4$ and $y = -3$ in each inequality:

$$
\begin{array}{c|c}
2x + 3y < 6 & y > 2x - 1 \\
2(4) + 3(-3) < 6 & -3 > 2(4) - 1 \\
-1 < 6 & -3 > 7
\end{array}
$$

Since only one inequality is satisfied, the point $(4, -3)$ is not a solution to the system.

c) Let $x = 5$ and $y = 1$ in each inequality:

$$2x + 3y < 6 \qquad\qquad y > 2x - 1$$
$$2(5) + 3(1) < 6 \qquad\qquad 1 > 2(5) - 1$$
$$13 < 6 \qquad\qquad\qquad 1 > 9$$

Since neither inequality is satisfied, the point $(5, 1)$ is not a solution to the system. ◀

Graphing a System of Inequalities

There are infinitely many points that satisfy a typical system of inequalities. The best way to describe the solution to a system of inequalities is with a graph showing all points that satisfy the system. When we graph the points that satisfy a system, we say that we are graphing the system.

EXAMPLE 2   Graph the system of inequalities:

$$y > x - 2$$
$$y < -2x + 3$$

Solution   We want a graph showing all points that satisfy both inequalities. The two lines divide the coordinate plane into four regions as shown in Fig. 9.13. To determine which of the four regions contains points that satisfy the system, we

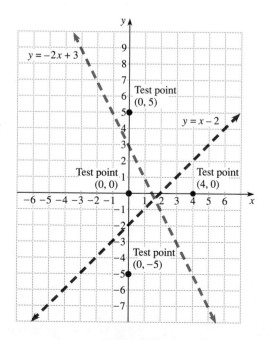

Figure 9.13

check one point in each region to see whether it satisfies both inequalities. The points are shown in Fig. 9.13.

Checking Points in	Check (0, 5)	Check (4, 0)	Check (0, −5)	Check (0, 0)
$y > x - 2$	$5 > 0 - 2$	$0 > 4 - 2$	$-5 > 0 - 2$	$0 > 0 - 2$
$y < -2x + 3$	$5 < -2(0) + 3$	$0 < -2(4) + 3$	$-5 < -2(0) + 3$	$0 < -2(0) + 3$

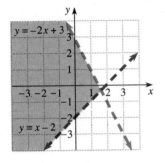

Figure 9.14

The only point that satisfies *both* inequalities of the system is (0, 0). So every point in the region containing (0, 0) also satisfies both inequalities. The points that satisfy the system are graphed in Fig. 9.14. ◀

EXAMPLE 3 Graph the system of inequalities:

$$y > -3x + 4$$
$$2y - x > 2$$

Solution First graph the equations $y = -3x + 4$ and $2y - x = 2$. We leave it to you to select a point in each region and check the point in *both* inequalities. You will find that only points in the region shown in Fig. 9.15 satisfy *both* inequalities. ◀

EXAMPLE 4 Graph the system of inequalities:

$$x > 4$$
$$y < 3$$

Solution We first graph the vertical line $x = 4$ and the horizontal line $y = 3$. The points that satisfy both inequalities are those points that lie to the right of the vertical line $x = 4$ and below the horizontal line $y = 3$. See Fig. 9.16 for the graph of the system. ◀

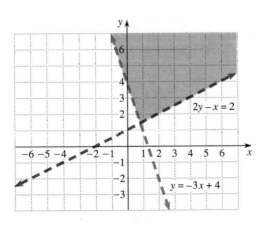

Figure 9.15

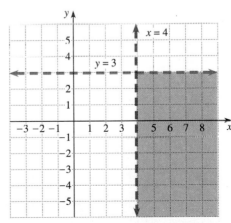

Figure 9.16

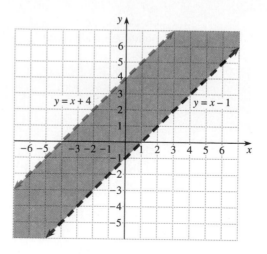

Figure 9.17

EXAMPLE 5 Graph the system of inequalities:

$$y < x + 4$$
$$y > x - 1$$

Solution First graph the lines $y = x + 4$ and $y = x - 1$. Notice that these are parallel lines. These lines divide the plane into three regions. You should check a point in each region to see whether it satisfies both inequalities. Only points in the region between the two parallel lines satisfy both inequalities. See Fig. 9.17. ◄

Warm-ups

True or false?

1. The point $(1, 4)$ is a solution to the system

$$x > 3$$
$$y < 6. \quad \text{F}$$

2. The point $(2, -3)$ is a solution to the system

$$y > -3x + 5$$
$$y < 2x - 3. \quad \text{F}$$

3. The point $(-4, 3)$ is a solution to the system

$$y > 3x - 6$$
$$y < -x + 5. \quad \text{T}$$

4. The inequality $x - y > 4$ is equivalent to the inequality $y < x + 4$. F

5. The graph of the inequality $x + 2 < y$ is the region below the line $y = x + 2$. F

6. The graph of the system

$$y > x - 3$$
$$y < 2x + 5$$

consists of the points that satisfy both inequalities. T

7. The graph of the system

$$y > 2x - 3$$
$$y > 2x + 3$$

is the region between two parallel lines. F

8. The graph of the system

$$x + y > 4$$
$$x - y < 0$$

is the region containing the point $(0, 5)$. T

9. The graph of the inequality $y < 3x + 4$ consists of the points below the line $y = 3x + 4$. T

10. There are no solutions to the system

$$y > 3x + 4$$
$$y < 3x + 4.$$ T

9.5 EXERCISES

Determine which of the points following each system is a solution to the system. See Example 1.

1. $x - y < 5$ $(4, 3), (8, 2), (-3, 0)$ $(4, 3)$
    $2x + y > 3$

2. $x + y < 4$ $(2, -3), (1, 1), (0, -1)$ $(1, 1), (0, -1)$
    $2x - y < 3$

3. $y > -2x + 1$ $(-3, 2), (-1, 5), (3, 6)$ $(3, 6)$
    $y < 3x + 5$

4. $y < -x + 7$ $(-3, 8), (0, 8), (-5, 15)$ $(-3, 8)$
    $y < -x + 9$

5. $x > 3$ $(-5, 4), (9, -5), (6, 0)$ $(9, -5)$
    $y < -2$

6. $y < -5$ $(-2, 4), (0, -7), (6, -9)$ $(0, -7)$
    $x < 1$

Graph each system of inequalities. See Examples 2–5.   Graphs in Answers sections.

7. $y > -x - 1$
    $y > x + 1$

8. $y < x + 3$
    $y < -2x + 4$

9. $y < 2x - 3$
    $y > -x + 2$

10. $y > 2x - 1$
    $y < -x - 4$

11. $x + y > 5$
    $x - y < 3$

12. $2x + y < 3$
    $x - 2y > 2$

13. $2x - 3y < 6$
    $x - y > 3$

14. $3x - 2y > 6$
    $x + y < 4$

15. $x > 5$
    $y > 5$

16. $x < 3$
    $y > 2$

17. $y < -1$
   $x > -3$

18. $y > -2$
   $x < 1$

19. $y > 2x - 4$
   $y < 2x + 1$

20. $y < -2x + 3$
   $y > -2x$

21. $y > x$
   $x > 3$

22. $y < x$
   $y < 1$

23. $y > -x$
   $x < -1$

24. $y < -x$
   $y > -3$

25. $2x - 5y < 5$
   $x + 2y > 4$

26. $3x + 2y < 2$
   $-x - 2y > 4$

27. $x + y > 3$
   $x + y > 1$

28. $x - y < 5$
   $x - y < 3$

29. $y > 3x + 2$
   $y < 3x + 3$

30. $y > x$
   $y < -x$

31. $x + y < 5$
   $x - y > -1$

32. $2x - y > 4$
   $x - 5y < 5$

33. $2x - 3y < 6$
   $3x + 4y < 12$

34. $x - 3y > 3$
   $x + 2y < 4$

35. $3x - 5y < 15$
   $3x + 2y < 12$

36. $x - 4y < 0$
   $x + y > 0$

CHAPTER 9

SUMMARY

Concepts		Examples
Solving systems of linear equations in two variables	Graphing method: Sketch each graph and identify the points they have in common.	
	Substitution method: Solve one equation for one variable in terms of the other, then substitute into the other equation.	$y = x - 4$ $3x + 5y = 1$ Substitution $3x + 5(x - 4) = 1$
	Addition method: Multiply each equation as necessary to eliminate a variable upon addition of the equations.	$5x - 3y = 4$ $\underline{x + 3y = 1}$ $6x \quad\;\; = 5$
Types of linear systems in two variables	Independent: Only one point satisfies both equations. The graphs cross at one point.	$y = x - 4$ $y = 2x + 5$
	Inconsistent: No solution. The graphs are parallel lines.	$y = x - 3$ $y = x + 5$
	Dependent: Infinitely many solutions. One equation is a multiple of the other. The graphs coincide.	$5x + 3y = 2$ $10x + 6y = 4$
Linear inequalities in two variables	If A, B, and C are real numbers with A and B not both zero, then $Ax + By \le C$ is called a linear inequality. We may also use the symbols $\ge$, $<$, and $>$.	$x + y \le 2$ $3x - 5y > 4$ $x < 3$ $y \ge x + 5$

Graphing a
linear inequality

1. Solve the inequality for y, then graph $y = mx + b$.

$y > mx + b$ is the region above the line. $y > x + 3$

$y = mx + b$ is the line itself. $y = x + 3$

$y < mx + b$ is the region below the line. $y < x + 3$

Remember that $<$ means below the line and $>$ means above the line only when the inequality is solved for y.

2. If the inequality involves only x, then graph the vertical line $x = k$. $x > 5$

$x > k$ is the region to the right of the line.

$x = k$ is the line itself.

$x < k$ is the region to the left of the line.

Test points

A linear inequality may also be graphed by graphing the equation and then testing a point to determine which region satisfies the inequality.

Graphing a
system of
inequalities

Graph the equations and use test points to see which regions satisfy both inequalities.

REVIEW EXERCISES

9.1 *Solve by graphing.*

1. $y = 2x + 1$
$x + y = 4$ (1, 3)

2. $y = -x + 1$
$y = -x + 3$ No solution

3. $y = 2x + 3$
$y = -2x - 1$ (-1, 1)

4. $x + y = 6$
$x - y = -10$ (-2, 8)

9.2 *Solve each system by the substitution method.*

5. $y = 3x$
$2x + 3y = 22$ (2, 6)

6. $x + y = 3$
$3x - 2y = -11$ (-1, 4)

7. $x = y - 5$
$2x - 3y = -7$ (-8, -3)

8. $2x + y = 5$
$6x - 9 = 3y$ (2, 1)

9.3 *Solve each system by the addition method. Indicate whether each system is independent, inconsistent, or dependent.*

9. $x - y = 4$
$2x + y = 5$ (3, -1), independent

10. $x + 2y = -5$
$x - 3y = 10$ (1, -3), independent

11. $2x - 4y = 8$ $\{(x, y) \mid x - 2y - 4\}$,
$x - 2y = 4$ dependent

12. $x + 3y = 7$
$2x + 6y = 5$ No solution, inconsistent

13. $y = 3x - 5$
$2y = -x - 3$ (1, -2), independent

14. $3x + 4y = 6$
$4x + 3y = 1$ (-2, 3), independent

15. $2x + 7y = 0$
$7x + 2y = 0$ (0, 0), independent

16. $3x - 5y = 1$
$10y = 6x - 1$ No solution, inconsistent

17. $x - y = 6$ $\{(x, y) \mid x - y = 6\}$,
$2x - 12 = 2y$ dependent

18. $y = 4x$
$y = 3x$ (0, 0), independent

19. $y = 4x$
$y = 4x + 3$ No solution, inconsistent

20. $3x - 5y = 21$
$4x + 7y = -13$ (2, -3), independent

9.4 *Graph each inequality.* Graphs in Answers sections.

21. $y > \dfrac{1}{3}x - 5$

22. $y < \dfrac{1}{2}x + 2$

23. $y \le -2x + 7$

24. $y \ge x - 6$

25. $y \le 8$

26. $x \ge -6$

27. $2x + 3y \le -12$

28. $x - 3y < 9$

9.5 *Graph each system of inequalities.*   Graphs in Answers sections.

29. $x < 5$
$y < 4$

30. $y > -2$
$x < 1$

31. $x + y < 2$
$y > 2x - 3$

32. $x - y > 4$
$2y > x - 4$

33. $y > 5x - 7$
$y < 5x + 1$

34. $y > x - 6$
$y < x - 5$

35. $y < 3x + 5$
$y < 3x$

36. $y > -2x$
$y > -3x$

Miscellaneous

Use a system of equations in two variables to solve each problem. Solve the system by the method of your choice.

37. Two apples and three oranges cost $1.95, and three apples and two oranges cost $2.05. What are the costs of one apple and one orange?   Apple: $.45, orange: $.35

38. Three small drinks and one medium drink cost $2.30, and two small drinks and four medium drinks cost $3.70. What is the cost of one small drink? What is the cost of one medium drink?   Small: $.55, medium: $.65

39. After a long day at the casino, Louis returned home and told his wife Lois that he had won $430 in $5 bills and $10 bills. Upon counting them again he realized that he had mixed up the number of bills of each denomination, and he really had only $380 in his pocket. How many bills of each denomination did Louis have?   32 fives, 22 tens

40. Diane divided her $10,000 bonus between two investments. One paid 8% interest, and the other paid 10% interest. If the total income from these investments for one year was $880, then how much did she invest at each rate?

41. One serving of green beans contains 1 gram of protein and 4 grams of carbohydrates. One serving of chicken soup contains 3 grams of protein and 9 grams of carbohydrates. The Westdale Diet recommends a lunch of 13 grams of protein and 43 grams of carbohydrates. How many servings of green beans and chicken soup are necessary to obtain the recommended amounts?

42. A television station aired four 30-second commercials and three 60-second commercials during the first hour of the midnight movie. During the second hour, it aired six 30-second commercials and five 60-second commercials. The advertising revenue for the first hour was $7,700, and that for the second hour was $12,300. What is the cost of each type of commercial?

40. $6000 at 8%, $4000 at 10%
41. Four servings green beans, three servings chicken soup
42. $1,500 for 60 seconds, $800 for 30 seconds

CHAPTER 9 TEST

Solve the system by graphing.

1. $x + y = 2$
$y = 2x + 5$ $(-1, 3)$

Solve each system by substitution.

2. $y = 2x - 3$
$2x + 3y = 7$ $(2, 1)$

3. $x - y = 4$
$3x - 2y = 11$ $(3, -1)$

Solve each system by the addition method.

4. $2x + 5y = 19$
$4x - 3y = -1$ $(2, 3)$

5. $3x - 2y = 10$
$2x + 5y = 13$ $(4, 1)$

Determine whether each system is independent, inconsistent, or dependent.

6. $y = 4x - 9$
$y = 4x + 8$ Inconsistent

7. $3x - 3y = 12$
$y = x - 4$ Dependent

8. $y = 2x$
$y = 5x$ Independent

Graph each inequality.   Graphs in Answers sections.

9. $y > 3x - 5$

10. $x - y < 3$

11. $x - 2y \geq 4$

Graph each system of inequalities.   Graphs in Answers sections.

12. $x < 6$
$y > -1$

13. $2x + 3y > 6$
$3x - y < 3$

14. $y > 3x - 4$
$3x - y > 3$

For each problem, write a system of equations in two variables. Use the method of your choice to solve each system.

15. The Rest-Is-Easy Motel just outside Amarillo rented five singles and three doubles on Monday night for a total of $188. On Tuesday night it rented three singles and four doubles for a total of $170. On Wednesday night it rented only one single and one double. How much did the motel make on Wednesday night? $48

16. Kathy and Chris studied a total of 54 hours for the CPA exam. If Chris studied only one-half as many hours as Kathy, then how many hours did each of them study?
16. Kathy: 36 hours, Chris: 18 hours

Tying It All Together

CHAPTERS 1–9

Solve each equation.

1. $2(x - 5) + 3x = 25$ 7

2. $(x - 5)^2 = 4$ 3, 7

3. $3x - 5 = 0$ $\dfrac{5}{3}$

4. $3x^2 - 5 = 0$ $-\dfrac{\sqrt{15}}{3}, \dfrac{\sqrt{15}}{3}$

5. $\dfrac{x}{3} - \dfrac{2}{5} = \dfrac{x}{2} - \dfrac{12}{5}$ 12

6. $x^2 + 2x = 24$ −6, 4

7. $\dfrac{x + 5}{x - 1} = \dfrac{2}{3}$ −17

8. $\dfrac{x}{2} - \dfrac{3}{x}$ $\sqrt{6}, \sqrt{6}$

9. $4 - \sqrt{x - 3}$ 19

10. $x - \sqrt{2 - x}$ −1

11. $2x^2 - 6x + 3 = 0$ $\dfrac{3 - \sqrt{3}}{2}, \dfrac{3 + \sqrt{3}}{2}$

12. $4x^2 + 20x + 25 = 0$ $-\dfrac{5}{2}$

Solve each inequality in one variable and sketch the graph on the number line.   Number lines in Answers sections.

13. $3(2 - x) < -6$ $x > 4$

14. $-3 \leq 2x - 4 \leq 6$ $\dfrac{1}{2} \leq x \leq 5$

15. $x > 1$ $x > 1$

Sketch the graph of each equation.   Graphs in Answers sections.

16. $y = 3x - 7$

17. $y = 5 - x$

18. $y = x - 1$

19. $y = x + 1$

20. $y = -2x + 4$

21. $y = -4x - 1$

Graph each inequality in two variables.   Graphs in Answers sections.

22. $y \geq 3x - 7$

23. $x - 2y < 6$

24. $x > 1$

Perform the operations and simplify.

25. $\sqrt{6} \div \sqrt{3}$ $\sqrt{2}$

26. $\sqrt{12} - \sqrt{3}$ $\sqrt{3}$

27. $2\sqrt{3} \cdot 5\sqrt{3}$ 30

28. $\sqrt{5} \cdot 2\sqrt{3}$ $2\sqrt{15}$

29. $(2 - \sqrt{7})(3 + \sqrt{7})$
29. $-1 - \sqrt{7}$

30. $(3 - \sqrt{7})(3 + \sqrt{7})$ 2

31. $(3 - \sqrt{5})^2$ $14 - 6\sqrt{5}$

32. $\dfrac{6}{\sqrt{12}}$ $\sqrt{3}$

APPENDIX A

Table of Squares and Square Roots

n	n^2	$\sqrt{n}$	n	n^2	$\sqrt{n}$	n	n^2	$\sqrt{n}$
1	1	1.0000	41	1681	6.4031	81	6561	9.0000
2	4	1.4142	42	1764	6.4807	82	6724	9.0554
3	9	1.7321	43	1849	6.5574	83	6889	9.1104
4	16	2.0000	44	1936	6.6332	84	7056	9.1652
5	25	2.2361	45	2025	6.7082	85	7225	9.2195
6	36	2.4495	46	2116	6.7823	86	7396	9.2736
7	49	2.6458	47	2209	6.8557	87	7569	9.3274
8	64	2.8284	48	2304	6.9282	88	7744	9.3808
9	81	3.0000	49	2401	7.0000	89	7921	9.4340
10	100	3.1623	50	2500	7.0711	90	8100	9.4868
11	121	3.3166	51	2601	7.1414	91	8281	9.5394
12	144	3.4641	52	2704	7.2111	92	8464	9.5917
13	169	3.6056	53	2809	7.2801	93	8649	9.6437
14	196	3.7417	54	2916	7.3485	94	8836	9.6954
15	225	3.8730	55	3025	7.4162	95	9025	9.7468
16	256	4.0000	56	3136	7.4833	96	9216	9.7980
17	289	4.1231	57	3249	7.5498	97	9409	9.8489
18	324	4.2426	58	3364	7.6158	98	9604	9.8995
19	361	4.3589	59	3481	7.6811	99	9801	9.9499
20	400	4.4721	60	3600	7.7460	100	10000	10.0000
21	441	4.5826	61	3721	7.8102	101	10201	10.0499
22	484	4.6904	62	3844	7.8740	102	10404	10.0995
23	529	4.7958	63	3969	7.9373	103	10609	10.1489
24	576	4.8990	64	4096	8.0000	104	10816	10.1980
25	625	5.0000	65	4225	8.0623	105	11025	10.2470
26	676	5.0990	66	4356	8.1240	106	11236	10.2956
27	729	5.1962	67	4489	8.1854	107	11449	10.3441
28	784	5.2915	68	4624	8.2462	108	11664	10.3923
29	841	5.3852	69	4761	8.3066	109	11881	10.4403
30	900	5.4772	70	4900	8.3666	110	12100	10.4881
31	961	5.5678	71	5041	8.4261	111	12321	10.5357
32	1024	5.6569	72	5184	8.4853	112	12544	10.5830
33	1089	5.7446	73	5329	8.5440	113	12769	10.6301
34	1156	5.8310	74	5476	8.6023	114	12996	10.6771
35	1225	5.9161	75	5625	8.6603	115	13225	10.7238
36	1296	6.0000	76	5776	8.7178	116	13456	10.7703
37	1369	6.0828	77	5929	8.7750	117	13689	10.8167
38	1444	6.1644	78	6084	8.8318	118	13924	10.8628
39	1521	6.2450	79	6241	8.8882	119	14161	10.9087
40	1600	6.3246	80	6400	8.9443	120	14400	10.9545

Answers to Selected Exercises

CHAPTER 1

Section 1.1   Warm-ups TTFTTTTTFT

1. $\frac{1}{2}$ **3.** $\frac{2}{3}$ **5.** 3   **7.** $\frac{1}{2}$ **9.** 2   **11.** $\frac{6}{8}$ **13.** $\frac{32}{12}$

15. $\frac{10}{2}$ **17.** $\frac{75}{100}$ **19.** $\frac{30}{100}$ **21.** $\frac{10}{27}$ **23.** 5

25. $\frac{7}{10}$ **27.** 120   **29.** $\frac{3}{5}$ **31.** 3   **33.** $\frac{1}{15}$ **35.** 20

37. $\frac{6}{5}$ **39.** $\frac{7}{12}$ **41.** $\frac{1}{12}$ **43.** $\frac{19}{24}$ **45.** $\frac{1}{2}$ **47.** $\frac{1}{4}$

49. .6, 60%   **51.** $\frac{2}{25}$, 8%   **53.** $\frac{1}{100}$, 1%

55. $\frac{9}{100}$, .09   **57.** $\frac{1}{50}$, .02   **59.** 1   **61.** $\frac{1}{8}$ **63.** $\frac{13}{12}$

65. $\frac{49}{12}$ **67.** $\frac{3}{8}$ **69.** $\frac{19}{24}$

Section 1.2   Warm-ups TTFFTFTTFF

1. 1, 2, 3, 4, 5

3. 0, 1, 2, 3, 4

5. 0, 1, 2, 3, 4

7. 1, 2, 3, 4, 5, . . .

9. 1, 2, 3, 4, 5, . . .

11. T   **13.** F   **15.** T   **17.** T   **19.** T   **21.** F   **23.** 6

25. 0   **27.** 7   **29.** 30   **31.** 9   **33.** $\frac{3}{4}$ **35.** 5.09

37. T   **39.** T   **41.** T   **43.** F   **45.** F   **47.** T   **49.** T

51. T

Section 1.3   Warm-ups FTTTFTFTFF

1. -13 **3.** -1.15 **5.** $-\frac{1}{2}$ **7.** 0   **9.** 2   **11.** -6

13. 5.6   **15.** $-\frac{1}{4}$ **17.** -2 **19.** -12 **21.** 8

23. 1.5   **25.** -4 **27.** -10 **29.** 11   **31.** -11

33. $-\frac{1}{4}$ **35.** $\frac{3}{4}$ **37.** 7   **39.** .93   **41.** 9.3

43. -5.03 **45.** 3   **47.** -9 **49.** -12 **51.** 23

53. -27 **55.** -7 **57.** -21 **59.** -12 **61.** -15.97

63. -2.92 **65.** -3.73 **67.** $\dfrac{3}{20}$ **69.** $\dfrac{1}{4}$ **71.** -3.49
73. $-.3422$ **75.** -48.84 **77.** -8.85

Section 1.4   Warm-ups TFTFTTTTFF

1. -27 **3.** $-.3$ **5.** 144 **7.** $-\dfrac{1}{3}$ **9.** -1 **11.** -80

13. 3 **15.** $-.25$ **17.** $\dfrac{5}{6}$ **19.** -100 **21.** 27

23. -3 **25.** -4 **27.** -30 **29.** 19 **31.** $-.18$
33. $.3$ **35.** -6 **37.** 1.5 **39.** -164.25 **41.** 1529.41
43. 16 **45.** -8 **47.** 0 **49.** 0 **51.** -3.9 **53.** -40

55. $.4$ **57.** $.4$ **59.** -2 **61.** $-\dfrac{3}{4}$ **63.** $-\dfrac{1}{30}$

65. $-\dfrac{1}{10}$ **67.** 7.5617 **69.** 19.35

Section 1.5   Warm-ups FFTFFFFTFT

1. 4^3 **3.** $(-y)^3$ **5.** $5 \cdot 5 \cdot 5$
7. $(-a)(-a)(-a)(-a)(-a)$ **9.** $(-1)^2$ **11.** 81
13. 125 **15.** 1 **17.** 1 **19.** -128 **21.** 144
23. 100 **25.** 21 **27.** -24 **29.** -7 **31.** 500
33. -6 **35.** 3 **37.** 6 **39.** -216 **41.** 27 **43.** 2
45. 23 **47.** 7 **49.** -1 **51.** -64 **53.** -60
55. -81 **57.** 5 **59.** 1 **61.** 21 **63.** 111 **65.** 12
67. 9 **69.** 16 **71.** 28 **73.** 121 **75.** 17 **77.** 25
79. -1 **81.** -2 **83.** 12 **85.** 1 **87.** -18 **89.** -24

Section 1.6   Warm-ups TFTFTFFFFF

1. Difference   **3.** Cube   **5.** Sum   **7.** Difference
9. Sum   **11.** The difference of x^2 and a^2
13. The square of $x - a$ **15.** The quotient of $x - 4$ and 2
17. The difference of $x/2$ and 4   **19.** The cube of ab
21. $2x + 3x$ **23.** $2 - 3x$ **25.** $(a + b)^2$
27. $(x + 2)(x + 3)$ **29.** $\dfrac{x - 7}{7 - x}$ **31.** 3 **33.** 8

35. -8 **37.** $-\dfrac{2}{3}$ **39.** -3 **41.** 12 **43.** 16

45. -9 **47.** 4 **49.** 1 **51.** -1 **53.** 3 **55.** 3
57. Yes   **59.** Yes   **61.** No   **63.** Yes   **65.** Yes
67. Yes   **69.** 8 **71.** 5 **73.** 10 **75.** 0 **77.** 0
79. 10 **81.** 16 **83.** 2 **85.** -13 **87.** 5 **89.** 14
91. 3 **93.** 17 **95.** 41 **97.** -47 **99.** 9 **101.** 60
103. 14.65 **105.** 37.12

Section 1.7   Warm-ups TFFTFTTTT

1. $r + 9$ **3.** $3(x + 2)$ **5.** $-5x + 4$ **7.** $6x$
9. $-2(x - 4)$ **11.** $4 - 8y$ **13.** $4w^2$ **15.** $3a^2b$
17. $7x^3$ **19.** -3 **21.** -10 **23.** -21 **25.** $.6$

27. -22.4 **29.** $3x - 15$ **31.** $2(m + 6)$ **33.** $2a + at$
35. $-3w + 18$ **37.** $-20 + 4y$ **39.** $4(x - 1)$
41. $-a + 7$ **43.** $-t - 4$ **45.** $4(y - 4)$ **47.** $4(a + 2)$
49. 2 **51.** $-\dfrac{1}{5}$ **53.** $\dfrac{1}{7}$ **55.** 4 **57.** $\dfrac{10}{9}$ **59.** $\dfrac{2}{5}$
61. 1 **63.** Commutative   **65.** Distributive
67. Associative   **69.** Inverse   **71.** Commutative
73. Identity   **75.** Distributive   **77.** Inverse
79. Multiplication property of 0   **81.** Distributive

83. $y + a$ **85.** $(5a)w$ **87.** $\dfrac{1}{2}(x + 1)$ **89.** $3(2x + 5)$

91. 1 **93.** 0 **95.** $\dfrac{100}{33}$

Section 1.8   Warm-ups TFTTFFFFFT

1. 7000 **3.** 1 **5.** 356 **7.** 350 **9.** 36 **11.** $36,000$
13. 0 **15.** 0 **17.** w **19.** $5x$ **21.** $-7mw^2$ **23.** $-a$
25. $9 - 4w$ **27.** $8x^2$ **29.** $-4x + 2x^2$ **31.** $3x$
33. $-2a$ **35.** $12h$ **37.** $12d^2$ **39.** $-15ab$
41. $-6a - 3ab$ **43.** y^2 **45.** $-18b$ **47.** $-k + k^2$
49. $9m^2$ **51.** y **53.** $-6ab$ **55.** y **57.** $4a$
59. $2a - 1$ **61.** $3x - 2$ **63.** $-2x + 1$ **65.** $8 - y$
67. $m - 6$ **69.** $w - 5$ **71.** $8x + 15$ **73.** $5x - 1$
75. $-2a - 1$ **77.** $5a - 2$ **79.** $3m - 18$ **81.** $-3x - 7$
83. $.95x - .5$ **85.** $-3.2x + 12.71$ **87.** $4x - 4$
89. $-20a^2$ **91.** $.15x - .4$ **93.** $2y + 4$
95. $2y + m - 1$ **97.** $-14k + 23$

Chapter 1 Review Exercises

1. $\dfrac{17}{24}$ **3.** 6 **5.** $\dfrac{3}{7}$ **7.** $\dfrac{14}{3}$ **9.** $\dfrac{13}{12}$ **11.** $0, 1, 2, 10$

13. $-2, 0, 1, 2, 10$ **15.** $-\sqrt{5}, \pi$ **17.** T   **19.** F
21. F   **23.** T   **25.** 2 **27.** -13 **29.** -7 **31.** -7

33. 11.95 **35.** $-.05$ **37.** $-\dfrac{1}{6}$ **39.** $-\dfrac{11}{15}$ **41.** -15

43. 4 **45.** 5 **47.** $\dfrac{1}{6}$ **49.** $-.3$ **51.** $-.24$ **53.** 1

55. 66 **57.** 49 **59.** 41 **61.** 1 **63.** 50 **65.** -135
67. 6 **69.** -2 **71.** 16 **73.** 5 **75.** 9 **77.** 7

79. $-\dfrac{1}{3}$ **81.** 1 **83.** -9 **85.** 4 **87.** 6 **89.** 15

91. -4 **93.** Distributive   **95.** Inverse   **97.** Identity
99. Associative   **101.** Commutative   **103.** Inverse
105. Identity   **107.** $-a + 12$ **109.** $6a^2 - 6a$
111. $-12t + 39$ **113.** $-.9a - .57$ **115.** $-.05x - 4$
117. $15x^2 + 6x + 5$ **119.** $-2a$ **121.** $x^2 + 4x - 3$

123. 0 **125.** 8 **127.** -21 **129.** $\dfrac{1}{2}$ **131.** $-.5$

133. -1 **135.** $x + 2$ **137.** $4 + 2x$ **139.** $2x$
141. $-4x + 8$ **143.** $6x$ **145.** x **147.** $8x$

Chapter 1 Test

1. $0, 8$ **2.** $-3, 0, 8$ **3.** $-3, -\dfrac{1}{4}, 0, 8$

4. $-\sqrt{3}, \sqrt{5}, \pi$ **5.** -21 **6.** -4 **7.** 6 **8.** -7
9. $-.95$ **10.** -56 **11.** 978 **12.** 13 **13.** -1

14. 0 **15.** 9740 **16.** $-\dfrac{7}{24}$ **17.** -20 **18.** $-\dfrac{1}{6}$

19. -39 **20.** Distributive   **21.** Commutative
22. Associative   **23.** Inverse   **24.** Identity
25. Multiplication property of 0   **26.** $3(x - 10)$
27. $7(w - 1)$ **28.** $6x + 6$ **29.** $4x - 2$ **30.** $7x - 3$
31. $.9x + 7.5$ **32.** $14a^2 + 5a$ **33.** $x + 2$ **34.** $4t$
35. 41 **36.** 5 **37.** -12 **38.** 2 **39.** 13 **40.** -3

CHAPTER 2

Section 2.1   Warm-ups TTFFFFFFTT

1. 1 **3.** -17 **5.** 1 **7.** 12 **9.** -3 **11.** -12
13. 4 **15.** -6 **17.** -8 **19.** -12 **21.** 1.7 **23.** $.1$
25. 4.6 **27.** 1.03 **29.** -5 **31.** -8 **33.** 5

35. $-\dfrac{2}{3}$ **37.** 1.8 **39.** 20 **41.** $\dfrac{1}{4}$ **43.** $\dfrac{2}{5}$ **45.** $\dfrac{2}{3}$

47. $\dfrac{1}{2}$ **49.** 4.8 **51.** -80 **53.** -12 **55.** 14

57. 12 **59.** 10 **61.** -24 **63.** 12 **65.** $\dfrac{3}{2}$ **67.** -1

69. 8 **71.** -6 **73.** 34 **75.** $\dfrac{9}{4}$ **77.** -6 **79.** 200

81. -10 **83.** 18 **85.** -20 **87.** -1 **89.** -4.3

Section 2.2   Warm-ups TTFFFTTFTT

1. No solution, inconsistent   **3.** All real numbers, identity
5. 1, conditional   **7.** No solution, inconsistent

9. 4, conditional   **11.** $\dfrac{5}{3}$, conditional

13. All real numbers, identity
15. All real numbers, identity   **17.** 7 **19.** 24 **21.** 16
23. -12 **25.** 60 **27.** 24 **29.** 80 **31.** 60 **33.** 6
35. -2 **37.** 200 **39.** 600 **41.** 1000 **43.** 800
45. 90 **47.** 9 **49.** -5 **51.** 3 **53.** -12 **55.** -4

57. 0 **59.** 5 **61.** -12 **63.** 2 **65.** $\dfrac{4}{3}$ **67.** $\dfrac{1}{2}$

69. $\dfrac{1}{4}$ **71.** $\dfrac{1}{6}$ **73.** 6 **75.** 25 **77.** -2 **79.** 5

81. -4 **83.** No solution   **85.** 4 **87.** 100 **89.** $-\dfrac{3}{2}$

91. 30 **93.** $\dfrac{1}{25}$ **95.** $.5$ **97.** $19,608$

Section 2.3   Warm-ups FFFFFTFTFT

1. $R = \dfrac{D}{T}$ **3.** $P = \dfrac{I}{rt}$ **5.** $C = \dfrac{5}{9}(F - 32)$

7. $h = \dfrac{2A}{b}$ **9.** $L = \dfrac{P - 2W}{2}$ **11.** $\pi = \dfrac{C}{D}$

13. $a = 2A - b$ **15.** $r = \dfrac{S - P}{Pt}$ **17.** $a = \dfrac{2A - hb}{h}$

19. $x = \dfrac{b - a}{2}$ **21.** $x = -7a$ **23.** $x = 12 - a$

25. $x = 7ab$ **27.** $y = -x - 9$ **29.** $y = -x + 6$
31. $y = 2x - 2$ **33.** $y = 3x + 4$ **35.** $y = x + 7$

37. $y = -\dfrac{1}{2}x + 2$ **39.** $y = x - \dfrac{1}{2}$ **41.** $y = 3x - 14$

43. $y = \dfrac{3}{2}x + 6$ **45.** $y = -3x + 6$ **47.** $y = -6x - 19$

49. $y = -\dfrac{1}{2}x + 9$ **51.** 2 **53.** 7 **55.** 0 **57.** 1

59. 1.33 **61.** 7 yards   **63.** 225 feet   **65.** 160 feet
67. 24 cubic feet   **69.** 4 inches   **71.** 4% **73.** 4 years
75. 8 feet   **77.** 12 inches   **79.** 20% **81.** $\$300$

Section 2.4   Warm-ups TTTFTFFFFTF

1. $x + 3$ **3.** $x - 3$ **5.** $5x$ **7.** $.1x$ **9.** $x, x + 2$
11. $x, 6 - x$ **13.** $x, x + 15$ **15.** $x, x + 1$ **17.** $3x$

19. $\dfrac{x}{20}$ **21.** $\dfrac{x - 100}{12}$ **23.** $5x$ **25.** $2w + 2(w - 3)$

27. $150 - x$ **29.** $2x + 1$ **31.** $x(x + 5)$
33. $.18(x + 1000)$ **35.** $16.50/x$ **37.** $x + 5 = 13$
39. $x(x + 5) = 8$ **41.** $x + (x + 1) + (x + 2) = 42$
43. $x(x + 1) = 182$ **45.** $.12x = 3000$ **47.** $.05x = 13$
49. $40x = 120$ **51.** $x(x + 5) = 126$
53. $.05x + .10(x + 2) = 3.80$ **55.** $5n + 10(n - 1) = 95$
57. $x - .07x = 84,532$

Section 2.5   Warm-ups FFTFTFFTFT

1. $17, 18, 19$ **3.** $36, 38, 40$ **5.** $75, 77$
7. 26 inches, 13 inches   **9.** 42 inches by 46 inches
11. 13 inches   **13.** $\$320$ **15.** $\$400$ **17.** $\$15,000$
19. 75% **21.** $\$2000$ at 8%, $\$5000$ at 12%
23. $\$15,000$ at 9%, $\$10,000$ at 12% **25.** 30 gallons
27. 20 liters of 5%, 10 liters of 20% **29.** 65 mph
31. 55 mph   **33.** $\$80,000$ **35.** $\$9850$ **37.** 600

39. 42 private, 30 semiprivate
41. 4 nickels, 6 dimes

Section 2.6   Warm-ups TTFTFTFFTF

1. T   **3.** T   **5.** F   **7.** T   **9.** T   **11.** T   **13.** T
15. (number line: −1 0 1 2 3 4 5)   **17.** (number line: −4 −3 −2 −1 0 1 2)
19. (number line: −5 −4 −3 −2 −1 0 1)   **21.** (number line: −4 −3 −2 −1 0 1 2)
23. (number line: −5 −4 −3 −2 −1 0 1)   **25.** (number line: 3 4 5 6 7 8 9)
27. (number line: 0 200 400 600)   **29.** (number line: 5.3; 1 2 3 4 5 6 7)
31. (number line: −4 −3 −2 −1 0 1 2)   **33.** (number line: 2 3 4 5 6 7 8)
35. (number line: −5 −4 −3 −2 −1 0 1)   **37.** (number line: 4 5 6 7 8 9 10)

39. Yes   **41.** No   **43.** Yes   **45.** No   **47.** Yes
49. Yes   **51.** Yes   **53.** No   **55.** Yes   **57.** No

59. $.08x > 1500$ **61.** $\dfrac{44 + 72 + x}{3} \geq 60$

63. $396 < 8x < 453$ **65.** Any number less than 0
67. Any number larger than 5
69. Any number less than or equal to 3

71. Any number less than $\dfrac{5}{2}$

73. Any number less than $\dfrac{5}{2}$

75. Any number less than -3
77. Any number less than 1

79. Any number between $-\dfrac{2}{3}$ and 1

81. Any number between $\dfrac{3}{2}$ and 2

Section 2.7   Warm-ups TFFTFTFTTF

1. $x > -3$ (number line: −5 −4 −3 −2 −1 0 1)

3. $x > 2$ (number line: 0 1 2 3 4 5 6)

5. $x < 3$ (number line: −1 0 1 2 3 4 5)

7. $x \geq -2$ (number line: −4 −3 −2 −1 0 1 2)

9. $x \leq 24$ (number line: 20 22 24 26)

11. $x \leq 12$ (number line: 8 9 10 11 12 13 14)

13. $x < -11$ (number line: −15 −14 −13 −12 −11 −10 −9)

15. $x < 13$ (number line: 9 10 11 12 13 14 15)

17. $x \leq -2$ (number line: −6 −5 −4 −3 −2 −1 0)

19. $x > -10$ (number line: −12 −11 −10 −9 −8 −7 −6)

21. $x < 614.3$ (number line: 614.3)

23. $8 < x < 10$ (number line: 6 7 8 9 10 11 12)

25. $1 < x < \dfrac{9}{2}$ (number line: $\frac{9}{2}$; 0 1 2 3 4 5 6)

27. $4 < x < 10$ (number line: 4 5 6 7 8 9 10)

29. $-5 \leq x < 3$ (number line: −5 −4 −3 −2 −1 0 1 2 3)

31. $-5 < m \leq 5$ (number line: −5 −3 −1 0 1 3 5)

33. $-2 \leq x \leq 9$ (number line: −2 0 1 3 5 7 9)

35. $102.1 < x < 108.3$ (number line: 102.1 108.3)

37. $w \geq 28$ **39.** $x \leq \$550$ **41.** $81 \leq x \leq 94.5$
43. $x \geq \$12,250$ **45.** $x \geq 64$ **47.** $x > 3$ **49.** $x \leq 2$
51. $0 < x < 2$ **53.** $-5 < x \leq 7$ **55.** $x > -4$

Chapter 2 Review Exercises

1. 7   **3.** $\dfrac{7}{3}$ **5.** -2 **7.** 0   **9.** -4 **11.** -30

13. 400   **15.** 100   **17.** $\dfrac{3}{7}$ **19.** No solution

21. All real numbers   **23.** 80   **25.** 1000   **27.** 24
29. All real numbers except 0   **31.** 8   **33.** 9   **35.** -6

37. -20 **39.** $\dfrac{3}{2}$ **41.** -4 **43.** 3   **45.** $-\dfrac{1}{2}$

47. $-\dfrac{2}{3}$ **49.** 0   **51.** All real numbers   **53.** 2

55. No solution   **57.** $x = -\dfrac{b}{a}$ **59.** $x = \dfrac{b + 2}{a}$

61. $x = \dfrac{V}{LW}$ **63.** $x = -\dfrac{b}{3}$ **65.** $y = -\dfrac{5}{2}x + 3$

67. $y = -\dfrac{1}{2}x + 4$ **69.** $y = -2x + 16$ **71.** -13

73. $-\dfrac{2}{5}$ **75.** 17   **77.** $x + 9$ **79.** $x, x + 8$ **81.** $.65x$

83. $x(x + 5) = 98$ **85.** $2(x + 10) = 3x$ **87.** 77, 79, 81

89. 150 feet, 100 feet   **91.** 400   **93.** No   **95.** No

97. $x > 1$ **99.** $x \geq 2$ **101.** $-3 \leq x < 3$ **103.** $x < -1$

105. $x > -4$

107. $x > -1$

109. $x \leq -4$

111. $-1 < x < 5$

113. $0 \leq x \leq 3$

115. $x < 6$

Chapter 2 Test

1. -7 **2.** 2   **3.** -9 **4.** 700   **5.** $y = \dfrac{2}{3}x - 3$

6. $a = \dfrac{m + w}{P}$ **7.** $-3 < x \leq 2$ **8.** $x > 1$

9. $w > 19$

10. $-7 < x < -1$

11. $1 < x < 3$

12. $y > -6$

13. No solution   **14.** All real numbers   **15.** 1   **16.** $\dfrac{7}{6}$

17. 14 meters   **18.** 9 inches   **19.** \$1200   **20.** 150 liters

Tying It All Together   Chapters 1–2

1. $8x$ **2.** $15x^2$ **3.** $2x + 1$ **4.** $4x - 7$ **5.** $-2x + 13$

6. 60   **7.** 72   **8.** -10 **9.** $-2x^3$ **10.** -1 **11.** 1

12. All real numbers   **13.** 0   **14.** 1   **15.** 2   **16.** 2

17. $\dfrac{13}{2}$ **18.** 200

CHAPTER 3

Section 3.1   Warm-ups FFTFTTTFTFT

1. 0   **3.** -5 **5.** $-\dfrac{1}{4}$ **7.** 1   **9.** Binomial, 1

11. Trinomial, 10   **13.** Monomial, 3   **15.** Binomial, 6

17. Monomial, 0   **19.** Trinomial, 2   **21.** $4x - 8$

23. $x^2 + 3x - 2$ **25.** $2x$ **27.** $2x^2 - 3$

29. $3a^2 - 7a - 4$ **31.** $-3w^2 - 8w + 5$

33. $9.66x^2 - 1.93x - 1.49$ **35.** $-x^2 + 2x$ **37.** $2x + 13$

39. $-4x + 6$ **41.** $x^5 - x^4 - x^3 + x^2$ **43.** $2x^2 + 2x + 7$

45. 1   **47.** $-22.85x - 423.2$ **49.** $4a + 2$

51. $-2x + 4$ **53.** $7x^2 - 4x - 14$ **55.** $x + 7$

57. $3x + 1$ **59.** $a^3 - 9a^2 + 2a + 7$ **61.** $-3x + 9$

63. $2a$ **65.** $-2b$ **67.** $2p + 2q$ **69.** $5x + 40$ miles

71. $6x + 3$ meters   **73.** $400x^2 + 350x + 250$

75. $.17x + 74.47$

Section 3.2   Warm-ups FFTFTTTTTF

1. $27x^5$ **3.** $-16x^7$ **5.** $27x^{17}$ **7.** $14a^{11}$ **9.** $24t^7w^8$

11. $-36qs^2$ **13.** $25y^2$ **15.** $4x^6$ **17.** $7x^5 - 35x^4 - 7x^2$

19. $a^3b - ab^3$ **21.** $-3y^2 + 15y - 18$ **23.** $-xy^2 + x^3$

25. $x^2 + 3x + 2$ **27.** $x^2 + 2x - 15$

29. $x^3 + 3x^2 + 4x + 2$ **31.** $2y^3 + 3y^2 + y + 6$

33. $3y^3 + 9y^2 + 3y$ **35.** $t^2 - 13t + 36$

37. $2a^2 + 7a - 15$ **39.** $14x^2 + 95x + 150$

41. $3x^4 + 9x^3 - 15x^2 - 6x$ **43.** $20x^2 - 7x - 6$

45. $2x^2 - 11x + 12$ **47.** $-4a^4 + 6a^3 - 8a$

49. $x^4 - 2x^3 + 2x^2 + 6x - 15$ **51.** $x^2 - y^2$ **53.** f

55. b **57.** c **59.** j **61.** $-6x^2 + 27x$ **63.** $-x - 7$

65. $3x - 2$ **67.** $6x^7 - 8x^4$ **69.** $x^2 - 12x + 36$

71. $4x^2 - 81$ **73.** $-6a^3b^{10}$ **75.** $32s^2tx^2$ **77.** $9a^6b^2$

79. $x^2 + 4x$ square feet   **81.** $x^2 + \dfrac{1}{2}x$ **83.** $x^2 + 5x$

85. $8.05x^2 + 15.93x + 6.12$ square meters

Section 3.3   Warm-ups FTTTFTFFF

1. $x^2 + 6x + 8$ **3.** $a^2 - a - 6$ **5.** $2x^2 + 3x - 2$

7. $2a^2 - a - 3$ **9.** $w^2 + 4w - 5$ **11.** $10m^2 - 9m - 9$

13. $a^2 - 5a - 14$ **15.** $y^2 - ay + 5y - 5a$

17. $5w - w^2 + 5m - mw$ **19.** $x^4 - 7x^2 + 10$

21. $3b^6 + 14b^3 + 8$ **23.** $5n^8 + 14n^4 - 3$

25. $y^3 + 2y^2 - 3y - 6$ **27.** $b^2 + 9b + 20$

29. $x^2 + 6x - 27$ **31.** $a^2 + 10a + 25$

33. $4x^2 - 4x + 1$ **35.** $z^2 - 100$ **37.** $a^2 + 2ab + b^2$

39. $a^2 - 3a + 2$ **41.** $2x^2 + 5x - 3$ **43.** $5t^2 - 7t + 2$

45. $h^2 - 16h + 63$ **47.** $h^2 + 14h + 49$

49. $4h^2 - 4h + 1$ **51.** $2x^2 + 5x - 3$ square feet

53. $5.2555x^2 + .41095x - 1.995$ square meters

Section 3.4   Warm-ups FTTTFTTTFF

1. $x^2 + 2x + 1$ **3.** $y^2 + 8y + 16$ **5.** $x^2 + 16x + 64$

7. $s^2 + 2st + t^2$ **9.** $4x^2 + 2x + .25$ **11.** $4t^2 + 4t + 1$

13. $a^2 - 6a + 9$ **15.** $4t^2 - 4t + 1$ **17.** $t^2 - 4t + 4$

19. $s^2 - 2st + t^2$ **21.** $a^2 - .8a + .16$

23. $9z^2 - 30z + 25$ **25.** $a^2 - 25$ **27.** $y^2 - 1$

29. $r^2 - s^2$ **31.** $9x^2 - 64$ **33.** $36x^2 - 1$ **35.** $x^4 - 1$
37. $x^2 + 15x + 56$ **39.** $t^2 - 25$ **41.** $y^2 - 22y + 121$
43. $a^2 - 400$ **45.** $16x^2 - 1$ **47.** $81y^2 - 18y + 1$
49. $6t^2 - 7t - 20$ **51.** $4t^2 - 20t + 25$ **53.** $4t^2 - 25$
55. $2.25x^2 + 11.4x + 14.44$ **57.** $12.25t^2 - 6.25$
59. $x^2 - 25$ square feet, 25 square feet smaller
61. $3.14b^2 + 6.28b + 3.14$ square meters

Section 3.5   Warm-ups FFTFTFTTTT

1. $4x^2$ **3.** x^6 **5.** $3a^5$ **7.** $-3x$ **9.** $-y$ **11.** $-x$
13. $x - 2$ **15.** $x^3 + 3x^2 - x$ **17.** $4xy - 2x + y$
19. $1 - 3x$ **21.** $x + 2, 7$ **23.** $a^2 + 2a + 8, 13$
25. $x - 4, 4$ **27.** $5, 5$ **29.** $4x^2 - 6x + 9, 0$
31. $2x - 3, 1$ **33.** $x^2 + 1, -1$ **35.** $3 + \dfrac{15}{x - 5}$

37. $x - 1 + \dfrac{1}{x + 1}$ **39.** $x + 1 + \dfrac{2}{x - 1}$

41. $-3 + \dfrac{3}{x + 1}$ **43.** $x^2 + 2x + 4 + \dfrac{8}{x - 2}$

45. $x^2 + \dfrac{3}{x}$ **47.** $1 - \dfrac{1}{x}$ **49.** $2 + \dfrac{1}{y}$ **51.** $-3a$
53. $-a + 4$ **55.** $x - 3$ **57.** $h^2 + 3h + 9$
59. $-6x^2 + 2x - 3$ **61.** $x^4 + 2x^2 + 4$ **63.** $x - 5$ meters

Chapter 3 Review Exercises

1. $5w - 2$ **3.** $-6x + 4$ **5.** $2x^2 - 7x - 4$
7. $-2x^2 + 3x - 1$ **9.** $-50x^{11}$ **11.** $144b^6$
13. $3x^3 - 8x^2 + 16x - 8$ **15.** $x^3 + 8$
17. $15m^5 - 3m^3 + 6m^2$ **19.** $-4x + 15$
21. $3x^2 - 10x + 12$ **23.** $q^2 + 2q - 48$
25. $2t^2 - 21t + 27$ **27.** $20y^2 - 7y - 6$
29. $6x^4 + 13x^2 + 5$ **31.** $x^2 + 10x + 21$ **33.** $z^2 - 49$
35. $y^2 + 14y + 49$ **37.** $t^2 - 7t + 12$
39. $a^2 - 12a + 36$ **41.** $2w^2 - 9w - 18$
43. $9a^2 + 6a + 1$ **45.** $y^2 - 8y + 16$
47. $3x^2 + 2x - 21$ **49.** $5x, 0$ **51.** $-2a^2b^3, 0$
53. $-x + 3, 0$ **55.** $3x - 5, 0$ **57.** $x^2 + 2x - 9, 1$
59. $m^3 + 2m^2 + 4m + 8, 0$ **61.** $m^2 - 3m + 6, 0$

63. $-1, 0$ **65.** $4x^2 - 2x - 9, 0$ **67.** $x - 1 - \dfrac{2}{x + 1}$

69. $2 + \dfrac{6}{x - 3}$ **71.** $-2 + \dfrac{2}{1 - x}$ **73.** $x - 1 + \dfrac{1}{x + 1}$

75. $-3 + \dfrac{12}{x + 4}$ **77.** $2 - \dfrac{3}{x}$

Chapter 3 Test

1. $7x^3 + 4x^2 + 2x - 11$ **2.** $-x^2 - 9x + 2$ **3.** $-35x^8$
4. $9x^6$ **5.** $6x^3$ **6.** 1 **7.** $2ab^4$ **8.** $3a^4b^2$
9. $-2y^2 + 3y$ **10.** $x^3 - 7x^2 + 13x - 6$ **11.** $x^2 + x - 1$

12. $15x^5 - 21x^4 + 12x^3 - 3x^2$ **13.** -1
14. $x^2 + 3x - 10$ **15.** $a^2 - 14a + 49$ **16.** $b^2 - 9$
17. $16x^2 + 24x + 9$ **18.** $4x^4 + 5x^2 - 6$ **19.** $9t^4 - 49$
20. $2 + \dfrac{6}{x - 3}$ **21.** $x - 5 + \dfrac{15}{x + 2}$

Tying It All Together   Chapters 1–3

1. 8 **2.** -9 **3.** 41 **4.** 2^{25} **5.** 2^5 **6.** 992
7. 144 **8.** -1 **9.** 64 **10.** 34 **11.** 899 **12.** 961
13. $x^2 + 8x + 15$ **14.** $x + 3$ **15.** $4x + 15$
16. $x^3 + 13x^2 + 55x + 75$ **17.** $-15t^5v^7$ **18.** $5tv$

19. $3y - 4$ **20.** $4y^2 - 5y - 3$ **21.** $-\dfrac{1}{2}$ **22.** 7

23. $\dfrac{3}{2}$ **24.** 4 **25.** -3 **26.** $-\dfrac{2}{3}$

CHAPTER 4

Section 4.1   Warm-ups FFFTTTTTFFT

1. $2 \cdot 3^2$ **3.** $2^2 \cdot 13$ **5.** $2 \cdot 7^2$ **7.** $2^2 \cdot 5 \cdot 23$
9. $2^2 \cdot 3 \cdot 7 \cdot 11$ **11.** 4 **13.** 12 **15.** 8 **17.** 4
19. 1 **21.** $2x$ **23.** xy **25.** $12ab$ **27.** $2x$ **29.** $6ab$
31. $x(x^2 - 6)$ **33.** $5a(x + y)$ **35.** $2x(x^2 - 3x + 4)$
37. $6x^2(2x^2 + 5x - 4)$ **39.** $h^3(h^2 - 1)$
41. $3h^3t^2(-2h^2 + t^4)$ **43.** $(x - 3)(a + b)$
45. $(y + 1)^2(a + b)$ **47.** $3(x - y), -3(-x + y)$
49. $4x(-1 + 2x), -4x(1 - 2x)$
51. $a^2(-a + 5), -a^2(a - 5)$ **53.** $x(x + 1), -x(-x - 1)$
55. $1(3x - 5), -1(-3x + 5)$ **57.** $1(b + 4), -1(-b - 4)$
59. $1(4 - a), -1(-4 + a)$ **61.** $x + 2$ hours

Section 4.2   Warm-ups FTFFTFFTTT

1. $(a - 2)(a + 2)$ **3.** $(x - 7)(x + 7)$
5. $(2y + 3x)(2y - 3x)$ **7.** $(5a + b)(5a - b)$
9. $(m + 1)(m - 1)$ **11.** $(3w - 5c)(3w + 5c)$
13. Perfect square trinomial   **15.** Neither
17. Perfect square trinomial   **19.** Neither   **21.** Neither
23. Perfect square trinomial   **25.** $(x + 6)^2$ **27.** $(a - 2)^2$
29. $(2w + 1)^2$ **31.** $(4x - 1)^2$ **33.** $(2t + 5)^2$
35. $(n + t)^2$ **37.** $(3w + 7)^2$ **39.** $5(x - 5)(x + 5)$
41. $-2(x - 3)(x + 3)$ **43.** $3(x + 1)^2$ **45.** $x(x - y)^2$
47. $2y(4x - y)(4x + y)$ **49.** $3a(b - 3)^2$
51. $3(2w - 1)(2w + 1)$ **53.** $a(a - b)(a + b)$
55. $-3(x - y)(x + y)$ **57.** $(b + c)(x + y)$
59. $(x - 2)(x + 2)(x + 1)$ **61.** $(3 - x)(a - b)$
63. $(a^2 + 1)(a + 3)$ **65.** $(y^2 + 8)(y - 5)$
67. $(c - 3)(ab + 1)$ **69.** $(a + b)(x - 1)(x + 1)$
71. $(y + b)(y + 1)$

Section 4.3 Warm-ups TTFFTFTFFF

1. $(x + 3)(x + 1)$ **3.** $(x + 3)(x + 6)$ **5.** $(y + 2)(y + 5)$
7. $(b - 6)(b + 1)$ **9.** Prime **11.** Prime **13.** Prime
15. $(w - 4)(w + 2)$ **17.** Prime **19.** $(m + 2)(m + 8)$
21. $(m - 8)(m + 2)$ **23.** $(m - 16)(m - 1)$
25. $(m - 16)(m + 1)$ **27.** Prime **29.** $(t + 8)(t - 3)$
31. $(t - 6)(t + 4)$ **33.** $(t - 20)(t + 10)$
35. $(x - 15)(x + 10)$ **37.** $(y + 6)(y + 4)$
39. $(x - 6y)(x + 2y)$ **41.** $(x - 12y)(x - y)$
43. $(x - 8s)(x + 3s)$ **45.** $2(w - 9)(w + 9)$
47. $(w + 3)(w + 27)$ **49.** $20(w^2 + 5w + 2)$
51. $w(w - 6)(w + 3)$ **53.** Prime **55.** $a^2b(a + b)$
57. $(2 - w)(2 + w)$ **59.** $8v(w + 2)^2$
61. $6xy(x + 3y)(x + 2y)$ **63.** $x + 4$ feet

Section 4.4 Warm-ups TFTFTFFFFT

1. $(2x + 1)(x + 1)$ **3.** $(2x + 1)(x + 4)$
5. $(3t + 1)(t + 2)$ **7.** $(2x - 1)(x + 3)$
9. $(3x - 1)(2x + 3)$ **11.** $(2x - 3)(x - 2)$
13. $(5b - 3)(b - 2)$ **15.** $(4y + 1)(y - 3)$ **17.** Prime
19. $(4x + 1)(2x - 1)$ **21.** $(4x - 1)(2x - 1)$
23. $(3t - 1)(3t - 2)$ **25.** $(5x + 1)(3x + 2)$
27. $(5x - 1)(3x - 2)$ **29.** $(5x - 2)(3x + 1)$
31. $(15x - 1)(x - 2)$ **33.** $2(x^2 + 9x - 45)$
35. $(3x - 5)(x + 2)$ **37.** $2(2w - 5)(w + 3)$
39. $w^2(81w - 1)$ **41.** $y^2(10x - 9)(x + 1)$
43. $y^2(2x^2 + x + 3)$ **45.** $2(x - 7)^2$
47. $3z(x - 3)(x + 2)$ **49.** $(a + 5b)(a - 3b)$
51. $t(3t + 2)(2t - 1)$ **53.** $2t^2(3t - 2)(2t + 1)$
55. $y(2x - y)(2x - 3y)$

Section 4.5 Warm-ups FFTTTFTFTTF

1. $2(x - 3)(x + 3)$ **3.** $4(x + 5)(x - 3)$ **5.** $x(x + 2)^2$
7. $5am(x^2 + 4)$ **9.** $(3x + 1)^2$ **11.** $y(3x + 2)(2x - 1)$
13. Prime **15.** $2(4m + 1)(2m - 1)$ **17.** $(3a + 4)^2$
19. $2(3x - 1)(4x - 3)$ **21.** $3a(a - 9)$
23. $2(2 - x)(2 + x)$ **25.** $x(6x^2 - 5x + 12)$
27. $ab(a - 2)(a + 2)$ **29.** $(x - 2)(x + 2)^2$
31. $3w(a - 3)^2$ **33.** $5(x - 10)(x + 10)$
35. $(2 - w)(m + n)$ **37.** $4(w^2 + w - 1)$
39. $a^2(a + 10)(a - 3)$ **41.** $aw(2w - 3)^2$ **43.** $(t + 3)^2$
45. $(x + 4)(x - 3)(x + 2)$ **47.** $(x - 1)(x + 3)(x + 2)$
49. $(x - 2)(x^2 + 2x + 4)$ **51.** $(x + 5)(x^2 - x + 2)$
53. $(x + 1)(x^2 + x + 1)$

Section 4.6 Warm-ups FFTTTFTTTF

1. $-4, -5$ **3.** $-\dfrac{5}{2}, \dfrac{4}{3}$ **5.** $2, 7$ **7.** $0, -7$ **9.** $-5, 4$

11. $\dfrac{1}{2}, -3$ **13.** $0, -8$ **15.** $-3, -2$ **17.** $-\dfrac{3}{2}, -4$

19. $-4, 4$ **21.** $-3, 3$ **23.** $0, -3, 3$ **25.** $0, -1, 1$
27. $-4, -2, 2$ **29.** $-1, 1, 3$ **31.** $0, 4, 5$
33. 12 feet, 5 feet **35.** 12 feet, 5 feet
37. 2, 3 or $-3, -2$ **39.** 5 and 6 **41.** 7 seconds
43. 6 inches, 13 inches **45.** 20 feet by 20 feet
47. 80 feet **49.** 3 yards by 3 yards, 6 yards by 6 yards.

Chapter 4 Review Exercises

1. $2^4 \cdot 3^2$ **3.** $2 \cdot 29$ **5.** $2 \cdot 3 \cdot 5^2$ **7.** 8 **9.** 6
11. $x + 2$ **13.** $-a + 10$ **15.** $-2 + a$ **17.** 2
19. $-m + n$ **21.** $-x - b$ **23.** $-a + 7$ **25.** $3x^2y$
27. $x^2 - 4x - 3y$ **29.** $(y - 20)(y + 20)$ **31.** $(w - 4)^2$
33. $(2y + 5)^2$ **35.** $(r - 2)^2$ **37.** $2t(2t - 3)^2$
39. $(x + 6y)^2$ **41.** $(x - y)(x + 5)$ **43.** $(b + 8)(b - 3)$
45. $(r - 10)(r + 6)$ **47.** $(y - 11)(y + 5)$
49. $(7t - 3)(2t + 1)$ **51.** $(3x + 1)(2x - 7)$
53. $(3p + 4)(2p - 1)$ **55.** $5x(x^2 + 8)$
57. $(3x - 1)(3x + 2)$ **59.** $(x + 2)(x - 1)(x + 1)$
61. $xy(x - 16y)$ **63.** $(a + 1)^2$ **65.** $(x^2 + 1)(x - 1)$
67. $(a + 2)(a + b)$ **69.** $-2(x - 6)(x - 2)$
71. $(x + 2)(x^2 - 2x + 5)$ **73.** $(x + 4)(x + 5)(x - 3)$

75. $0, 5$ **77.** $0, 5$ **79.** $-\dfrac{1}{2}, 5$ **81.** $-5, -1, 1$

83. $5, 11$ **85.** 6 inches, 8 inches

Chapter 4 Test

1. $2 \cdot 3 \cdot 11$ **2.** $2^4 \cdot 3 \cdot 7$ **3.** 16 **4.** 6 **5.** $5x(x - 2)$
6. $6y^2(x^2 + 2x + 2)$ **7.** $3ab(a - b)(a + b)$
8. $(a + 6)(a - 4)$ **9.** $(2b - 7)^2$ **10.** $3m(m^2 + 9)$
11. $(a + b)(x - y)$ **12.** $(a - 5)(x - 2)$
13. $(3h - 5)(2h + 1)$ **14.** $(m + 2n)^2$
15. $(2a - 3)(a - 5)$ **16.** $z(z + 3)(z + 6)$

17. $(x - 1)(x - 2)(x - 3)$ **18.** $\dfrac{3}{2}, -4$ **19.** $0, -2, 2$

20. 12 feet, 9 feet

Tying It All Together Chapters 1–4

1. -1 **2.** 2 **3.** -3 **4.** 57 **5.** 16 **6.** 7 **7.** $2x^2$
8. $3x$ **9.** $3 + x$ **10.** $6x$ **11.** $24yz$ **12.** $6y + 8z$
13. $4z - 1$ **14.** t^6 **15.** t^{10} **16.** $4t^6$ **17.** 40

18. $\dfrac{15}{7}$ **19.** $-5, \dfrac{3}{2}$ **20.** $0, -\dfrac{7}{2}$ **21.** $-3, 3$ **22.** 0

23. $x < -9$
$\overset{\longleftarrow}{\underset{-13\,-12\,-11\,-10\,-9\;-8\,-7}{+\!+\!+\!+\!+\!\circ\!+\!+}}$

24. $x \geq 3$
$\underset{1\;\;2\;\;3\;\;4\;\;5\;\;6\;\;7}{+\!+\!\bullet\!+\!+\!+\!+\!+}$

25. $x > 12$
$\underset{10\;\,11\;\,12\;\,13\;\,14\;\,15\;\,16}{+\!+\!\circ\!+\!+\!+\!+}$

26. 3 **27.** -5 **28.** $\dfrac{3}{2}$ **29.** $-\dfrac{1}{2}$ **30.** $3, -5$

31. $\dfrac{3}{2}, -\dfrac{1}{2}$ **32.** 0, 3   **33.** 0   **34.** 6   **35.** -4

36. All real numbers   **37.** No solution   **38.** -10

39. $-\dfrac{1}{3}$ **40.** -6

CHAPTER 5

Section 5.1   Warm-ups TFFTTFFTFT

1. -1 **3.** $-1, 1$ **5.** $\dfrac{5}{3}$ **7.** $-2, 3$ **9.** 0   **11.** $\dfrac{2}{9}$

13. $\dfrac{7}{15}$ **15.** $\dfrac{2a}{5}$ **17.** $\dfrac{13}{5w}$ **19.** $\dfrac{3x+1}{3}$ **21.** $\dfrac{2}{3}$

23. $\dfrac{a-2}{a}$ **25.** $\dfrac{a-1}{a+1}$ **27.** $\dfrac{x+1}{2x-2}$ **29.** $\dfrac{x+3}{7}$

31. x^3 **33.** $\dfrac{1}{z^5}$ **35.** $2x^2$ **37.** $\dfrac{-1}{6x^6}$ **39.** $\dfrac{b^5}{a^3}$

41. m^3n^2 **43.** $\dfrac{3}{4c^3}$ **45.** $\dfrac{5c}{3a^4b^{16}}$ **47.** $\dfrac{35}{44}$ **49.** $\dfrac{11}{8}$

51. $\dfrac{21}{10x^4}$ **53.** $\dfrac{33a^4}{16}$ **55.** $-h-t$ **57.** $\dfrac{-2}{3h+g}$

59. -1 **61.** $\dfrac{-x-2}{x+3}$ **63.** -1 **65.** $\dfrac{-2y}{3}$

67. $\dfrac{x+2}{2-x}$ **69.** $\dfrac{-6}{a+3}$ **71.** $\dfrac{x+2}{2x}$ **73.** $\dfrac{-2}{x+2}$

75. $\dfrac{x^4}{2}$ **77.** -1 **79.** $\dfrac{x+2}{x-2}$ **81.** $\dfrac{-2}{x+3}$ **83.** x^2

85. $\dfrac{x+2}{x-8}$ **87.** $\dfrac{300}{x+10}$ **89.** $\dfrac{4.50}{x+4}$ **91.** $\dfrac{1}{x}$

Section 5.2   Warm-ups FTTFTFFTTT

1. $\dfrac{7}{9}$ **3.** $\dfrac{ab}{44}$ **5.** $\dfrac{8a+8}{5a^2+5}$ **7.** $\dfrac{2}{x-w}$ **9.** $\dfrac{a^3+8}{2a-4}$

11. $\dfrac{1}{2}$ **13.** $\dfrac{2}{x+2}$ **15.** -3 **17.** $\dfrac{x+2}{2}$ **19.** $\dfrac{x+7}{15}$

21. $9x+9y$ **23.** $\dfrac{2x}{9}$ **25.** 2   **27.** 12   **29.** 3

31. $\dfrac{1}{27}$ **33.** $\dfrac{1}{4}$ **35.** $\dfrac{4}{3}$ **37.** $\dfrac{x}{3}$ **39.** -1 **41.** -3

43. $\dfrac{a+b}{a}$ **45.** $2x^2h$ **47.** $2x-2y$ **49.** $\dfrac{2}{7}$ **51.** $\dfrac{x}{10b}$

53. $\dfrac{b}{3a}$ **55.** $\dfrac{y}{x}$ **57.** $-\dfrac{1}{2}$ **59.** $\dfrac{x+3}{x-3}$ **61.** $\dfrac{x^2+5x}{3x-1}$

63. $\dfrac{-a^6b^8}{2}$ **65.** $\dfrac{1}{9m^3n}$ **67.** 1   **69.** $\dfrac{(m+3)^2}{(m-3)(m+k)}$

Section 5.3   Warm-ups FFTTFFFFTT

1. $\dfrac{10}{30}$ **3.** $\dfrac{5a^2}{5a^3}$ **5.** $\dfrac{3x+3}{x^2+2x+1}$ **7.** $\dfrac{-20}{-8x-8}$

9. $\dfrac{x^2-x-30}{x^2+x-20}$ **11.** $\dfrac{-8}{1-b}$ **13.** $\dfrac{15b}{3b}$ **15.** 48

17. $30a^2$ **19.** 120   **21.** a^4b^6 **23.** $(x-4)(x+4)^2$

25. $x(x+2)(x-2)$ **27.** $x(x-4)(x+4)$

29. $\dfrac{9}{252}, \dfrac{20}{252}$ **31.** $\dfrac{2}{6x}, \dfrac{9}{6x}$ **33.** $\dfrac{4x^2}{36xyz}, \dfrac{3y^2z}{36xyz}$

35. $\dfrac{2x^2+4x}{(x-3)(x+2)}, \dfrac{5x^2-15x}{(x-3)(x+2)}$

37. $\dfrac{x^2-3x}{(x-3)^2(x+3)}, \dfrac{5x^2+15x}{(x-3)^2(x+3)}$ **39.** $\dfrac{4}{a-6}, \dfrac{-5}{a-6}$

41. 5   **43.** 4   **45.** $6x$ **47.** -1 **49.** x^2+x **51.** 1

53. $3x$ **55.** $2-x$ **57.** $x+3$ **59.** $a+1$ **61.** $x-1$

Section 5.4   Warm-ups FTFTFTFTTT

1. $\dfrac{1}{5}$ **3.** $\dfrac{3}{4}$ **5.** $-\dfrac{2}{3}$ **7.** $-\dfrac{3}{4}$ **9.** $\dfrac{5}{9}$ **11.** $\dfrac{23}{144}$

13. $-\dfrac{31}{40}$ **15.** $\dfrac{5}{24}$ **17.** 3   **19.** $\dfrac{3-a}{3}$ **21.** $\dfrac{3}{x}$

23. $\dfrac{17}{10a}$ **25.** $\dfrac{4a-1}{2}$ **27.** $\dfrac{3m+1}{3}$ **29.** $\dfrac{2-a}{a(a-1)}$

31. $\dfrac{5x-1}{(x+1)(x-1)}$ **33.** $\dfrac{b-5a}{a^2b^2}$ **35.** $\dfrac{b^2-4ac}{4a}$

37. $\dfrac{a^2+5a}{(a-3)(a+3)}$ **39.** $\dfrac{7x+17}{(x+2)(x-1)(x+3)}$

41. $\dfrac{5x^2-7x}{(x-3)(x+3)(x+1)}$ **43.** $\dfrac{2x^2-x-4}{x(x-1)(x+2)}$

45. $\dfrac{15-4x}{5x(x+1)}$ **47.** 0   **49.** $\dfrac{7}{2a-2}$ **51.** 1   **53.** $\dfrac{9}{20}$

55. $\dfrac{7}{6}$ **57.** $\dfrac{a-2}{3}$ **59.** $\dfrac{3a+1}{3}$ **61.** $\dfrac{a+1}{a}$

63. $\dfrac{3-3x}{x}$ **65.** $\dfrac{1}{8}$ **67.** $-\dfrac{13}{7}$ **69.** $\dfrac{19}{14}$

71. $\dfrac{2a-1}{a}$ **73.** $\dfrac{2a-1}{a(a-1)}$ **75.** $\dfrac{2a}{a^2-4}$ **77.** $\dfrac{a+1}{3a}$

Section 5.5   Warm-ups FTFFFFFTFT

1. $-\dfrac{10}{3}$ **3.** $\dfrac{22}{7}$ **5.** $\dfrac{14}{17}$ **7.** $\dfrac{8}{9}$ **9.** $\dfrac{3a+b}{a-3b}$

11. $\dfrac{5a-3}{3a+1}$ **13.** $\dfrac{x^2-4x}{6x^2-2}$ **15.** $\dfrac{10b}{3b^2-4}$

17. $\dfrac{x^2-2x+4}{x^2-3x-1}$ **19.** $\dfrac{y^2-3y+2}{3y^2+y-2}$ **21.** $\dfrac{5x-14}{2x-7}$

23. $\dfrac{-w + 3}{9w - 4}$ **25.** -1 **27.** $\dfrac{a - 6}{3a - 1}$ **29.** $\dfrac{-3m + 12}{4m - 3}$

Section 5.6   Warm-ups FFFTFTFFTF

1. 4   **3.** 4   **5.** 30   **7.** 3   **9.** $-5, 2$ **11.** 2, 3
13. 2   **15.** No solution   **17.** No solution   **19.** 10
21. 3   **23.** 3, 5   **25.** 3   **27.** 1   **29.** 0   **31.** 4
33. $-5, 5$ **35.** -20 **37.** 0

Section 5.7   Warm-ups TFFTTTFFFT

1. $\dfrac{5}{7}$ **3.** $\dfrac{8}{15}$ **5.** $\dfrac{7}{2}$ **7.** $\dfrac{9}{14}$ **9.** $\dfrac{5}{2}$ **11.** $\dfrac{15}{1}$

13. $\dfrac{3}{2}$ **15.** $\dfrac{9}{16}$ **17.** $\dfrac{31}{1}$ **19.** 6   **21.** $-\dfrac{2}{5}$

23. $-\dfrac{27}{5}$ **25.** 5   **27.** $-\dfrac{3}{4}$ **29.** $\dfrac{5}{4}$ **31.** 108

33. 176,000   **35.** 60 sport, 40 luxury   **37.** 84 inches
39. 536.7 miles

Section 5.8   Warm-ups TTTFTTFFFT

1. $y = 2x - 5$ **3.** $y = -\dfrac{1}{2}x - 2$ **5.** $y = mx - mb - a$

7. $y = -\dfrac{1}{3}x - \dfrac{1}{3}$ **9.** $C = \dfrac{B}{A}$ **11.** $p = \dfrac{a}{1 + am}$

13. $m_1 = \dfrac{r^2 F}{km_2}$ **15.** $a = \dfrac{bf}{b - f}$ **17.** $r = \dfrac{S - a}{S}$

19. $P_2 = \dfrac{P_1 V_1 T_2}{T_1 V_2}$ **21.** $h = \dfrac{3V}{4\pi r^2}$ **23.** $C = \dfrac{5}{12}$

25. $a = \dfrac{6}{23}$ **27.** $k = \dfrac{128}{3}$ **29.** $b = -6$ **31.** $a = \dfrac{6}{5}$

33. Marcie: 4 mph, Frank: 3 mph
35. Bob: 25 mph, Pat: 20 mph   **37.** 5 mph   **39.** 6 hours
41. 40 minutes   **43.** 1 hour 36 minutes
45. 8 pounds bananas, 10 pounds apples   **47.** 80 gallons

Chapter 5 Review Exercises

1. $\dfrac{c^2}{a^2}$ **3.** $\dfrac{13x^5}{5}$ **5.** $\dfrac{b^3}{a^3}$ **7.** $\dfrac{1}{2}$ **9.** $\dfrac{2x}{3}$ **11.** $3k$

13. $12x(x - 1)$ **15.** $24a^7 b^3$ **17.** $\dfrac{10x}{15x^2 y}$ **19.** $\dfrac{-10}{12 - 2y}$

21. $\dfrac{3x + 8}{2(x + 2)(x - 2)}$ **23.** $\dfrac{2a - b}{a^2 b^2}$ **25.** $\dfrac{3x + 4}{x}$

27. $\dfrac{a^2 - 1}{a}$ **29.** $\dfrac{3}{a - 8}$ **31.** $-\dfrac{3}{14}$ **33.** $\dfrac{6b + 4a}{3a - 18b}$

35. $-\dfrac{15}{2}$ **37.** 9   **39.** -3 **41.** $\dfrac{21}{2}$ **43.** 5   **45.** 8

47. 56 cups water, 28 cups rice   **49.** $y = mx + b$

51. $m = \dfrac{1}{F - v}$ **53.** $y = 4x - 13$ **55.** 200 hours

57. Bert: 60, Ernie: 50   **59.** $\dfrac{3}{2x}$ **61.** $\dfrac{4 + y}{6xy}$

63. $\dfrac{8}{a - 5}$ **65.** $-1, 2$ **67.** $-\dfrac{5}{3}$ **69.** 6   **71.** $\dfrac{1}{2}$

73. $\dfrac{3x + 7}{(x - 5)(x + 5)(x + 1)}$ **75.** $\dfrac{-5a}{(a - 3)(a + 3)(a + 2)}$

77. $\dfrac{2}{5}$ **79.** 10   **81.** -2 **83.** $3x$ **85.** $2m$ **87.** $\dfrac{1}{6}$

89. $a + 1$ **91.** $\dfrac{5}{14}$ **93.** $\dfrac{2a}{3}$ **95.** $\dfrac{5 - a}{5a}$ **97.** $\dfrac{a - 2}{2}$

99. $\dfrac{a + 1}{a}$ **101.** 1   **103.** $b - a$ **105.** -1 **107.** $\dfrac{1}{10a}$

109. $-\dfrac{1}{5}$

Chapter 5 Test

1. $-1, 1$ **2.** $\dfrac{2}{3}$ **3.** 0   **4.** $-\dfrac{14}{45}$ **5.** $\dfrac{1 + 3y}{y}$

6. $\dfrac{4}{a - 2}$ **7.** $\dfrac{-x + 4}{(x + 2)(x - 2)(x - 1)}$ **8.** $\dfrac{2}{3}$ **9.** $\dfrac{2}{a + b}$

10. $\dfrac{a^6}{18b^4}$ **11.** $-\dfrac{4}{3}$ **12.** $\dfrac{3x - 4}{-2x + 6}$ **13.** $\dfrac{15}{7}$

14. 2, 3   **15.** 12   **16.** $y = -\dfrac{1}{5}x + \dfrac{13}{5}$

17. $c = \dfrac{3M - bd}{b}$ **18.** 7.2 minutes

19. Brenda: 15 mph, Randy: 20 mph;
or Brenda: 10 mph, Randy: 15 mph
20. 72 billion dollars

Tying It All Together   Chapters 1–5

1. $-\dfrac{15}{2}$ **2.** $-6, 6$ **3.** $-\dfrac{6}{5}$ **4.** $-2, 4$ **5.** $-\dfrac{10}{3}$

6. $\dfrac{7}{3}$ **7.** -2 **8.** No solution   **9.** 0   **10.** $-4, -2$

11. $-1, 0, 1$ **12.** $-3, -1$ **13.** $y = \dfrac{c - 2x}{3}$

14. $y = \dfrac{1}{2}x + \dfrac{1}{2}$ **15.** $y = \dfrac{c}{2 - a}$ **16.** $y = \dfrac{AB}{C}$

17. $y = 3B - 3A$ **18.** $y = \dfrac{6A}{5}$ **19.** $y = \dfrac{8}{3 - 5a}$

20. $y = 0$ or $y = B$ **21.** $y = \dfrac{2A - hb}{h}$ **22.** $y = -\dfrac{b}{2}$

23. 64   **24.** 16   **25.** 49   **26.** 121   **27.** $-2x - 2$

28. $2a^2 - 11a + 15$ **29.** x^4 **30.** $\dfrac{2x+1}{5}$ **31.** $\dfrac{1}{2x}$

32. $\dfrac{x+2}{2x}$ **33.** $\dfrac{x}{2}$ **34.** $\dfrac{x-2}{2x}$ **35.** $-\dfrac{7}{5}$ **36.** $\dfrac{3a}{4}$

37. $x^2 - 64$ **38.** $3x^3 - 21x$ **39.** $10a^{14}$ **40.** x^{10}

41. $k^2 - 12k + 36$ **42.** $j^2 + 10j + 25$ **43.** -1

44. $3x^2 - 4x$ **45.** $\dfrac{11}{15}$ **46.** $\dfrac{7}{30}$

CHAPTER 6

Section 6.1   Warm-ups FFFFTFTFFF
1. 16   **3.** 125   **5.** 0   **7.** 1   **9.** 1   **11.** 625   **13.** 64
15. 64   **17.** 300   **19.** 2^{15} **21.** 1   **23.** -16 **25.** 4

27. -1 **29.** $\dfrac{1}{2a^4}$ **31.** $\dfrac{2a^6}{5}$ **33.** x^6 **35.** $2x^{12}$

37. $\dfrac{1}{t^2}$ **39.** $\dfrac{1}{2}$ **41.** $8x^{17}y^{14}$ **43.** $8x^6$ **45.** $-8t^{15}$

47. x^6y^{15} **49.** a^9b^2 **51.** $\dfrac{x^{12}}{64}$ **53.** $\dfrac{a^8}{b^{12}}$ **55.** $\dfrac{8x^6}{y^6}$

57. $\dfrac{x^4y^8}{z^6}$ **59.** 35   **61.** 125   **63.** $\dfrac{8}{27}$ **65.** 200

67. 800   **69.** 1   **71.** 0   **73.** 600,000   **75.** 1   **77.** $\dfrac{8}{x^6}$

79. $x^{11}t^{10}$

Section 6.2   Warm-ups TFTFTTFFTT
1. 36   **3.** 6   **5.** 32   **7.** 2   **9.** 1000   **11.** 10
13. Not a real number   **15.** 0   **17.** -1 **19.** 1
21. Not a real number   **23.** 2   **25.** 5   **27.** -10
29. -6 **31.** m **33.** y^{15} **35.** y^3 **37.** y^5 **39.** m
41. 27   **43.** 32   **45.** 125   **47.** 10^9 **49.** $3\sqrt{y}$
51. $2a$ **53.** x^2y **55.** $3m$ **57.** $9w^5$ **59.** $3 \cdot \sqrt[3]{z^2}$

61. $2z$ **63.** $-5m^2$ **65.** $3 \cdot \sqrt[4]{w}$ **67.** $\dfrac{\sqrt{t}}{2}$ **69.** $\dfrac{25}{4}$

71. $\dfrac{\sqrt{2}}{5}$ **73.** $\dfrac{\sqrt[3]{a}}{3}$ **75.** $\dfrac{a^3}{3}$ **77.** 1.732   **79.** 2.236

81. 3.873   **83.** 3.873   **85.** 4.583

Section 6.3   Warm-ups TFTFTFTFFF
1. $2\sqrt{2}$ **3.** $2\sqrt{6}$ **5.** $2\sqrt{7}$ **7.** $3\sqrt{10}$ **9.** $10\sqrt{5}$

11. $5\sqrt{6}$ **13.** $\dfrac{\sqrt{5}}{5}$ **15.** $\dfrac{3\sqrt{2}}{2}$ **17.** $\dfrac{\sqrt{6}}{2}$

19. $\dfrac{-3\sqrt{10}}{10}$ **21.** $\dfrac{-10\sqrt{17}}{17}$ **23.** $\dfrac{\sqrt{77}}{7}$ **25.** $3\sqrt{7}$

27. $\dfrac{\sqrt{6}}{2}$ **29.** $\dfrac{\sqrt{10}}{4}$ **31.** $\dfrac{\sqrt{15}}{5}$ **33.** 5   **35.** $\dfrac{\sqrt{6}}{2}$

37. a^4 **39.** $a^4 \cdot \sqrt{a}$ **41.** $2a^3 \cdot \sqrt{2}$ **43.** $2a^2b^4 \cdot \sqrt{5b}$

45. $3xy\sqrt{3xy}$ **47.** $\dfrac{\sqrt{x}}{x}$ **49.** $\dfrac{\sqrt{6a}}{3a}$ **51.** $\dfrac{\sqrt{5y}}{5y}$

53. $\dfrac{\sqrt{6xy}}{2y}$ **55.** $\dfrac{\sqrt{6xy}}{3x}$ **57.** $\dfrac{2x\sqrt{2xy}}{y}$ **59.** $4\sqrt{5}$

61. $3y^4 \cdot \sqrt{y}$ **63.** xy **65.** $\dfrac{\sqrt{6}}{3}$ **67.** $\sqrt{a}$ **69.** .707

71. .707   **73.** 2

Section 6.4   Warm-ups FTFFTFFFTT
1. $7\sqrt{5}$ **3.** $2\sqrt{7}$ **5.** $2\sqrt{14}$ **7.** $2 \cdot \sqrt[3]{2}$
9. $4\sqrt{3} - 4\sqrt{2}$ **11.** $-4\sqrt{x} - \sqrt{y}$ **13.** $5x\sqrt{y} + 2\sqrt{a}$
15. $5\sqrt{6}$ **17.** $-14\sqrt{3}$ **19.** $-\sqrt{3}$ **21.** $3x\sqrt{x}$

23. $\dfrac{2\sqrt{3}}{3}$ **25.** $\dfrac{7\sqrt{3}}{3}$ **27.** 36   **29.** $12\sqrt{10}$

31. $30\sqrt{3}$ **33.** $30\sqrt{6}$ **35.** $2a^2 \cdot \sqrt{3}$ **37.** $3x^2 \cdot \sqrt{2x}$
39. $2 + \sqrt{6}$ **41.** $12\sqrt{3} + 6\sqrt{5}$ **43.** $10 - 30\sqrt{2}$
45. 2   **47.** $-7 + \sqrt{5}$ **49.** $28 - \sqrt{5}$ **51.** $-42 - \sqrt{15}$
53. $18 + 28\sqrt{3}$ **55.** $37 + 20\sqrt{3}$ **57.** $5 - 2\sqrt{6}$

59. $\sqrt{2}$ **61.** $\dfrac{\sqrt{15}}{3}$ **63.** $\dfrac{2\sqrt{15}}{5}$ **65.** $\dfrac{2\sqrt{30}}{9}$

67. $\dfrac{5\sqrt{7}}{3}$ **69.** $4\sqrt{2}$ **71.** $1 + \sqrt{2}$ **73.** $-2 + \sqrt{5}$

75. $\dfrac{2 - \sqrt{5}}{3}$ **77.** $\dfrac{2 + \sqrt{6}}{3}$ **79.** $5\sqrt{3} + 5\sqrt{2}$

81. $\dfrac{\sqrt{15} + 3}{2}$ **83.** $\dfrac{13 + 7\sqrt{3}}{22}$ **85.** $\dfrac{4\sqrt{3} - 2}{11}$

87. $3\sqrt{5}$ **89.** 2   **91.** 60   **93.** 395   **95.** $2 - \sqrt{5}$
97. 3   **99.** $2\sqrt{2} + \sqrt{3}$ **101.** 1.866   **103.** .089
105. -1.974 **107.** 1.866

Section 6.5   Warm-ups FTFFTFFTTT
1. $-4, 4$ **3.** $-2\sqrt{3}, 2\sqrt{3}$ **5.** $-\dfrac{\sqrt{6}}{3}, \dfrac{\sqrt{6}}{3}$

7. No solution   **9.** $-1, 3$ **11.** $5 - \sqrt{3}, 5 + \sqrt{3}$
13. -19 **15.** 90   **17.** No solution   **19.** $-5, 5$
21. 3   **23.** 0, 1   **25.** 6   **27.** 3, 5   **29.** 3   **31.** 6, 8

33. $r = \pm\sqrt{\dfrac{V}{\pi h}}$ **35.** $b = \pm\sqrt{c^2 - a^2}$ **37.** $b = \pm2\sqrt{ac}$

39. $t = \dfrac{v^2}{2p}$ **41.** $3\sqrt{2}$ feet or 4.243 feet

43. $3\sqrt{2}$ feet or 4.243 feet   **45.** $\sqrt{2}$ feet or 1.414 feet
47. 5 feet   **49.** 2.5 seconds

51. $\dfrac{200\sqrt{13}}{3}$ feet or 240.37 feet   **53.** $-\sqrt{2}, \sqrt{2}$

55. No solution   **57.** $\dfrac{1 + 2\sqrt{2}}{2}, \dfrac{1 - 2\sqrt{2}}{2}$ **59.** $\dfrac{9}{2}$

61. 3 **63.** $-4, 2$ **65.** $\frac{5}{2}$ **67.** $-1.803, 1.803$
69. .993 **71.** .362, 4.438

Section 6.6   Warm-ups TTTTTTTFFT

1. $\frac{1}{3}$ **3.** $\frac{1}{16}$ **5.** $-\frac{1}{4}$ **7.** $\frac{1}{4}$ **9.** 1 **11.** x **13.** $\frac{1}{x^4}$

15. $\frac{1}{a^5}$ **17.** t^{10} **19.** $\frac{1}{t^8}$ **21.** x **23.** $\frac{1}{x^{10}}$ **25.** a^9

27. $\frac{x^{12}}{16}$ **29.** $\frac{y^6}{x^4}$ **31.** $\frac{x^2}{y^6}$ **33.** $\frac{a^4}{16}$ **35.** 5 **37.** 5

39. 25 **41.** 125 **43.** $\frac{1}{81}$ **45.** $\frac{1}{8}$ **47.** $x^{1/2}$ **49.** x^2

51. $x^{3/2}$ **53.** $t^{1/4}$ **55.** xy^3 **57.** $\frac{x}{y^4}$ **59.** $\frac{y^3}{x^2}$

61. $\frac{x^3}{2}$ **63.** $\frac{1}{6}$ **65.** $\frac{5}{6}$ **67.** $\frac{1}{5}$ **69.** -1 **71.** 4

73. $\frac{1}{2}$ **75.** $\frac{1}{125}$ **77.** $\frac{1}{3}$

Section 6.7   Warm-ups TTFFTTTTFT

1. 9,860,000,000 **3.** $-.0000155$ **5.** 40,070
7. 100,000 **9.** .1 **11.** -555 **13.** 9×10^3
15. 7.8×10^{-4} **17.** -3.45×10^6 **19.** 8.5×10^{-6}
21. -2.55×10^{-5} **23.** 5×10^1 **25.** 6×10^{-10}
27. 2×10^{-38} **29.** 5×10^{27} **31.** 9×10^{24}
33. 1.25×10^{14} **35.** 2.5×10^{-33} **37.** 8.6×10^9
39. 2.1×10^2 **41.** 2.7×10^{-23} **43.** 3×10^{15}
45. 9.135×10^7 **47.** 5.715×10^{-4} **49.** 1.577×10^{182}
51. 1.1×10^{100} **53.** 4.426×10^7
55. 4.910×10^{11} feet **57.** 4.65×10^{-28} hours
59. 9.040×10^8 feet

Chapter 6 Review Exercises

1. 32 **3.** 1000 **5.** x^{13} **7.** a^5 **9.** $\frac{1}{a^4}$ **11.** x^{12}

13. $8x^9$ **15.** $\frac{a^2}{9b^6}$ **17.** 2 **19.** 10,000 **21.** 10

23. x^{12} **25.** x^6 **27.** x^2 **29.** $2x$ **31.** $5x^2$ **33.** $\frac{2x^8}{y^7}$

35. $\frac{w}{4}$ **37.** $6\sqrt{2}$ **39.** $\frac{\sqrt{3}}{3}$ **41.** $\frac{\sqrt{15}}{5}$ **43.** $\sqrt{11}$

45. $\frac{\sqrt{6}}{4}$ **47.** y^3 **49.** $2t^4 \cdot \sqrt{6}$ **51.** $2t\sqrt{3t}$

53. $\frac{\sqrt{2x}}{x}$ **55.** $\frac{\sqrt{6as}}{2s}$ **57.** $10\sqrt{7}$ **59.** $-\sqrt{3}$
61. 30 **63.** $-15\sqrt{3}$ **65.** $-15 - 3\sqrt{3}$
67. $3\sqrt{2} - 3\sqrt{5}$ **69.** -22 **71.** $26 - 4\sqrt{30}$

73. $\frac{\sqrt{10}}{4}$ **75.** $\frac{2 - \sqrt{5}}{5}$ **77.** $-\frac{3 + 3\sqrt{5}}{4}$

79. $-20, 20$ **81.** $-\frac{\sqrt{10}}{5}, \frac{\sqrt{10}}{5}$

83. $4 - 3\sqrt{2}, 4 + 3\sqrt{2}$ **85.** 81 **87.** 4 **89.** 6

91. $t = \pm 2\sqrt{2sw}$ **93.** $t = \frac{9a^2}{b}$ **95.** $\frac{1}{125}$ **97.** 5

99. $\frac{1}{8}$ **101.** $\frac{1}{x^8}$ **103.** $\frac{x^2}{4}$ **105.** w^4 **107.** $\frac{x^6y^9}{8}$

109. $\frac{t^3}{3s^2}$ **111.** $\frac{1}{xy}$ **113.** $u^{3/4}$ **115.** 5×10^3

117. 340,000 **119.** 4.61×10^{-5} **121.** .00000569
123. 7×10^{-4} **125.** 1.6×10^{-15} **127.** 8×10^1
129. 3.2×10^{-34}

Chapter 6 Test

1. 32 **2.** 12 **3.** -3 **4.** $\frac{1}{4}$ **5.** 2 **6.** $2\sqrt{6}$

7. $\frac{\sqrt{6}}{4}$ **8.** 1 **9.** $3\sqrt{2}$ **10.** $7 + 4\sqrt{3}$ **11.** 11

12. $\sqrt{7}$ **13.** $\frac{2\sqrt{15}}{3}$ **14.** $1 + \sqrt{2}$ **15.** 81

16. $3\sqrt{2} - 3$ **17.** $15x^9$ **18.** $8x^{18}$ **19.** $-27x^{15}y^3$

20. $\frac{1}{4y^{14}}$ **21.** $\frac{1}{t^4}$ **22.** xy^3 **23.** $\frac{s^6}{4t^4}$ **24.** $\frac{8w^9}{u^{12}}$

25. $\frac{\sqrt{3t}}{t}$ **26.** $2y^3$ **27.** $2y^4$ **28.** $3t^3 \cdot \sqrt{2t}$

29. $-9, 3$ **30.** 18 **31.** $-\frac{\sqrt{10}}{5}, \frac{\sqrt{10}}{5}$ **32.** $\frac{4}{3}$

33. $\frac{5\sqrt{2}}{2}$ meters **34.** 5.433×10^6 **35.** 6.5×10^{-6}

36. 4.8×10^{-1} **37.** 8.1×10^{-27}

Tying It All Together   Chapters 1–6

1. $-\frac{3}{2}$ **2.** $\frac{3}{2}$ **3.** $x > -\frac{3}{2}$

4. $x < \frac{3}{2}$

5. -3 **6.** $-\frac{\sqrt{6}}{2}, \frac{\sqrt{6}}{2}$ **7.** $-\sqrt{6}, \sqrt{6}$ **8.** $\frac{2}{3}$

9. $-\frac{3}{2}$ **10.** $-\frac{3}{2}, 3$ **11.** No solution **12.** $-2, -1$

13. No solution **14.** 3 **15.** 9 **16.** 72 **17.** 81

18. 9 **19.** 12 **20.** -6 **21.** 3 **22.** $-\frac{3}{2}$

23. $(x-3)^2$ **24.** $(x+5)^2$ **25.** $(x+6)^2$
26. $(x-10)^2$ **27.** $2(x-2)^2$ **28.** $3(x+1)^2$
29. $7x-3$ **30.** $-15t^2-13t+20$
31. $-24j^2+14j+24$ **32.** $6j+6$ **33.** $j+1$
34. j^2+4j+4 **35.** j^2-49 **36.** $-4j^2+9$
37. j^2-2j+1 **38.** $2t-1$ **39.** $-4j^2-j+3$
40. $-18j^2+2$ **41.** $3j-5$ **42.** $-6j+2$ **43.** $2-3x$
44. $-1-3j$ **45.** $-2+3j$ **46.** $-4+j$

CHAPTER 7

Section 7.1 Warm-ups TFFTTFFTFF
1. $-6, 6$ **3.** No solution **5.** $-\sqrt{10}, \sqrt{10}$
7. $-\dfrac{\sqrt{15}}{3}, \dfrac{\sqrt{15}}{3}$ **9.** $-\dfrac{2\sqrt{6}}{3}, \dfrac{2\sqrt{6}}{3}$ **11.** $1, 5$

13. $2-3\sqrt{2}, 2+3\sqrt{2}$ **15.** $-\dfrac{3}{2}, -\dfrac{1}{2}$

17. $\dfrac{2-\sqrt{2}}{2}, \dfrac{2+\sqrt{2}}{2}$ **19.** $\dfrac{-1-\sqrt{2}}{2}, \dfrac{-1+\sqrt{2}}{2}$

21. 11 **23.** $-3, 5$ **25.** -3 **27.** $-1, 2$ **29.** $0, 2$

31. $0, \dfrac{3}{2}$ **33.** $-7, \dfrac{3}{2}$ **35.** 5 **37.** $-\dfrac{5}{2}, 6$

39. $-\sqrt{6}, \sqrt{6}$ **41.** $3, 9$ **43.** $3, 4$ **45.** $\dfrac{5\sqrt{2}}{2}$ meters

47. 9

Section 7.2 Warm-ups FFFFTTTTTT
1. x^2+6x+9 **3.** $x^2+14x+49$ **5.** $x^2-16x+64$

7. $t^2-18t+81$ **9.** $m^2+3m+\dfrac{9}{4}$ **11.** $z^2+z+\dfrac{1}{4}$

13. $x^2-\dfrac{1}{2}x+\dfrac{1}{16}$ **15.** $y^2+\dfrac{1}{4}y+\dfrac{1}{64}$ **17.** $(x+5)^2$

19. $(m-1)^2$ **21.** $(x+1/2)^2$ **23.** $(t+1/6)^2$
25. $(x+1/5)^2$ **27.** $-5, 3$ **29.** $-3, 7$ **31.** -3

33. $\dfrac{1}{2}, 1$ **35.** $\dfrac{3}{2}, 2$ **37.** $-1, \dfrac{1}{3}$

39. $-1-\sqrt{7}, -1+\sqrt{7}$ **41.** $-3-2\sqrt{2}, -3+2\sqrt{2}$
43. $\dfrac{1-\sqrt{13}}{2}, \dfrac{1+\sqrt{13}}{2}$ **45.** $\dfrac{-3-\sqrt{21}}{2}, \dfrac{-3+\sqrt{21}}{2}$

47. $\dfrac{1-\sqrt{33}}{4}, \dfrac{1+\sqrt{33}}{4}$ **49.** $5-\sqrt{7}, 5+\sqrt{7}$

51. $-\dfrac{\sqrt{15}}{3}, \dfrac{\sqrt{15}}{3}$ **53.** No solution

55. $4-2\sqrt{2}, 4+2\sqrt{2}$ **57.** $\dfrac{7}{2}$

59. $-3-2\sqrt{5}, -3+2\sqrt{5}$ **61.** 12
63. $6-\sqrt{2}$ and $6+\sqrt{2}$

Section 7.3 Warm-ups TFFFTFFTTT
1. $-5, 3$ **3.** -5 **5.** $-2, \dfrac{3}{2}$ **7.** $-\dfrac{3}{2}, \dfrac{1}{2}$

9. $\dfrac{3-\sqrt{3}}{2}, \dfrac{3+\sqrt{3}}{2}$ **11.** $\dfrac{-2-\sqrt{2}}{2}, \dfrac{-2+\sqrt{2}}{2}$

13. 0, one **15.** -47, none **17.** 181, two **19.** 0, one

21. -15, none **23.** -59, none **25.** $-2, \dfrac{1}{2}$ **27.** $0, 3$

29. $-\dfrac{6}{5}, 4$ **31.** $-4, -2$ **33.** $0, 9$ **35.** $-8, -2$

37. No solution **39.** $\dfrac{3-\sqrt{33}}{2}, \dfrac{3+\sqrt{33}}{2}$ **41.** $0, \dfrac{3}{2}$

43. $-.791, 3.791$ **45.** $-1.357, 2.357$ **47.** $.304$

Section 7.4 Warm-ups FTFFTTFTTF
1. $1+\sqrt{11}$ meters, $-1+\sqrt{11}$ meters **3.** $4\sqrt{2}$ feet
5. $3+\sqrt{19}$ inches, $-3+\sqrt{19}$ inches
7. $3+\sqrt{5}$ or 5.24 hours **9.** $2\sqrt{19}$ or 8.72 mph
11. $\dfrac{-13+\sqrt{409}}{2}$ or 3.61 feet

13. $\dfrac{15+\sqrt{241}}{8}$ or 3.82 seconds

Section 7.5 Warm-ups FFFFTTTTFT
1. $5+9i$ **3.** 1 **5.** $2-8i$ **7.** $-1-4i$ **9.** -1
11. $6-9i$ **13.** -36 **15.** -36 **17.** $21-i$
19. $21-20i$ **21.** 25 **23.** 2 **25.** $1-3i$

27. $-1+4i$ **29.** $-4i$ **31.** $\dfrac{8}{13}+\dfrac{12}{13}i$ **33.** $\dfrac{3}{5}+\dfrac{4}{5}i$

35. $5+3i$ **37.** $-3-i\sqrt{7}$ **39.** $-1+i\sqrt{3}$

41. $2+\dfrac{1}{2}i\sqrt{5}$ **43.** $-\dfrac{2}{3}+\dfrac{1}{3}i\sqrt{7}$ **45.** $\dfrac{1}{5}-i$

47. $-9i, 9i$ **49.** $-i\sqrt{5}, i\sqrt{5}$ **51.** $-i\dfrac{\sqrt{6}}{3}, i\dfrac{\sqrt{6}}{3}$

53. $2-i, 2+i$ **55.** $3-2i, 3+2i$

57. $2-i\sqrt{3}, 2+i\sqrt{3}$ **59.** $\dfrac{2-i}{3}, \dfrac{2+i}{3}$

61. $\dfrac{1-i\sqrt{3}}{2}, \dfrac{1+i\sqrt{3}}{2}$ **63.** $\dfrac{2-i\sqrt{5}}{2}, \dfrac{2+i\sqrt{5}}{2}$

65. $-6-24i$ **67.** 0 **69.** $x^2-12x+37$

Chapter 7 Review Exercises
1. $-3, 3$ **3.** $-3\sqrt{2}, 3\sqrt{2}$ **5.** $0, 9$ **7.** $-1, 2$

9. $9-\sqrt{10}, 9+\sqrt{10}$ **11.** $\dfrac{3}{2}$ **13.** $4, 5$ **15.** $-7, 2$

17. $-2-\sqrt{11}, -2+\sqrt{11}$ **19.** $-7, 4$

21. $\dfrac{-3 - \sqrt{29}}{2}, \dfrac{-3 + \sqrt{29}}{2}$ **23.** $-5, \dfrac{1}{2}$ **25.** 0, one

27. -44, none **29.** 76, two **31.** $-\dfrac{2}{3}, \dfrac{1}{2}$

33. $\dfrac{1 - \sqrt{17}}{2}, \dfrac{1 + \sqrt{17}}{2}$ **35.** $\dfrac{3 - \sqrt{14}}{5}, \dfrac{3 + \sqrt{14}}{5}$

37. 13 meters

39. $\dfrac{-2 + \sqrt{46}}{2} \approx 2.391$ meters and $\dfrac{2 + \sqrt{46}}{2} \approx 4.391$ meters

41. $-2 + 2\sqrt{11} \approx 4.633$ inches and $2 + 2\sqrt{11} \approx 8.633$ inches
43. $3 + \sqrt{2} \approx 4.414$ and $3 - \sqrt{2} \approx 1.586$
45. $7 + \sqrt{65} \approx 15.062$ hours and $9 + \sqrt{65} \approx 17.062$ hours
47. $3 - 3i$ **49.** $-3 + 7i$ **51.** $15 - 25i$

53. $-55 + 48i$ **55.** $-1 - i\sqrt{2}$ **57.** $\dfrac{3}{37} + \dfrac{19}{37}i$

59. $\dfrac{19}{17} + \dfrac{9}{17}i$ **61.** $-11i, 11i$ **63.** $8 - i, 8 + i$

65. $\dfrac{3 - 3i\sqrt{7}}{4}, \dfrac{3 + 3i\sqrt{7}}{4}$

Chapter 7 Test

1. 0, one **2.** -31, none **3.** 17, two **4.** $-1, \dfrac{3}{5}$

5. $\dfrac{2 - \sqrt{10}}{2}, \dfrac{2 + \sqrt{10}}{2}$ **6.** $-7, 3$

7. $\dfrac{-3 - \sqrt{29}}{2}, \dfrac{-3 + \sqrt{29}}{2}$ **8.** $-5, 4$ **9.** 3, 25

10. $-\dfrac{1}{2}, 3$ **11.** $5 - \sqrt{2}, 5 + \sqrt{2}$ **12.** $10 + 3i$

13. $-6 + 7i$ **14.** -36 **15.** $42 - 2i$ **16.** 68
17. $2 - 3i$ **18.** $-1 + i\sqrt{3}$ **19.** $-2 + i\sqrt{2}$

20. $\dfrac{3}{5} + \dfrac{4}{5}i$ **21.** $-3 - i\sqrt{3}, -3 + i\sqrt{3}$

22. $\dfrac{3}{5} - \dfrac{4}{5}i, \dfrac{3}{5} + \dfrac{4}{5}i$

Tying It All Together Chapters 1–7

1. $\dfrac{1}{2}$ **2.** 1 **3.** $-\dfrac{\sqrt{2}}{2}, \dfrac{\sqrt{2}}{2}$ **4.** $\dfrac{1 - 2\sqrt{2}}{2}, \dfrac{1 + 2\sqrt{2}}{2}$

5. $\dfrac{2 - \sqrt{6}}{2}, \dfrac{2 + \sqrt{6}}{2}$ **6.** 0, 2 **7.** $-1, \dfrac{1}{2}$ **8.** 5

9. $1 - i\sqrt{14}, 1 + i\sqrt{14}$ **10.** $-1, 2$ **11.** $y = \dfrac{5}{4}x - 2$

12. $y = 3x - 9$ **13.** $y = \dfrac{2}{3}x + \dfrac{16}{3}$ **14.** $y = -\dfrac{b}{a}$

15. $y = \dfrac{-b \pm \sqrt{b^2 - 4ac}}{2a}$ **16.** $y = -\dfrac{2}{3}x + 7$

17. $y = -\dfrac{4}{3}x + \dfrac{2}{9}$ **18.** $y = \pm\sqrt{a^2 - x^2}$ **19.** $\dfrac{1}{2}$

20. $\dfrac{1}{4}$ **21.** $\dfrac{1}{7}$ **22.** -2 **23.** -7 **24.** -2

CHAPTER 8

Section 8.1 Warm-ups FFFFTTTTFT

1. $(0, 9), (5, 24), (2, 15)$
3. $(0, -7), (-4, 5), (-2, -1)$
5. $(0, 5), (10, -115), (-1, 17)$
7.–25. (odd)

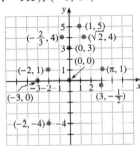

27.

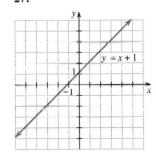

29.

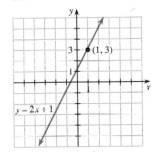

31.

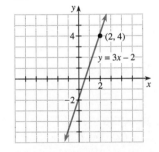

33.

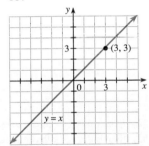

35.

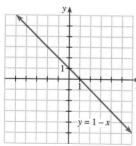

37.

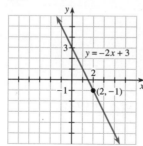

39.

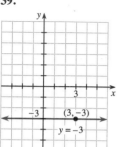

41.

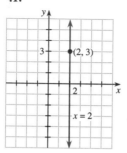

43.

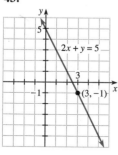

45.

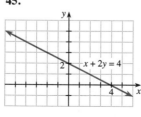

47.

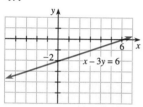

49.

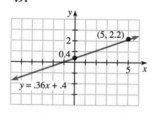

51. Quadrant II   **53.** x-axis   **55.** Quadrant III
57. Quadrant I   **59.** Quadrant II   **61.** y-axis

63.–69. (odd)

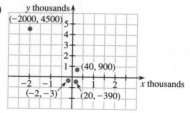

71.

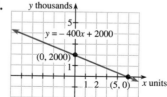

73.

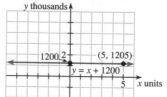

75.

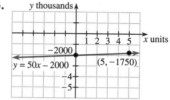

77. $(0, 7)$, $(4, 3)$, $(-2, 9)$ **79.** $(5, 6)$, $(3, 2)$, $(-1, -6)$
81. $(3, 0)$, $(0, -2)$, $(12, 6)$ **83.** $(5, -3)$, $(5, 5)$, $(5, 0)$

Section 8.2   Warm-ups TTTTTFTTTT
1. $2\sqrt{2}$ **3.** $\sqrt{34}$ **5.** $2\sqrt{10}$ **7.** 5   **9.** 10   **11.** $\sqrt{5}$
13. 2.354   **15.** $(5, 7)$ **17.** $(2, -3)$ **19.** $(-3.5, .5)$
21. $(1, 1)$ **23.** $(.5, .5)$ **25.** $(3\sqrt{2}, 2\sqrt{3})$
27. $(-.41, 2.09)$ **29.** Diagonals $5\sqrt{2}$
31. Opposite sides are $2\sqrt{13}$ and $\sqrt{5}$ **33.** Yes   **35.** No
37. Yes   **39.** Yes   **41.** Each side $2\sqrt{5}$

Section 8.3   Warm-ups TTFTFFFFTT
1. $-\dfrac{2}{3}$ **3.** $\dfrac{3}{2}$ **5.** 2   **7.** 0   **9.** $\dfrac{2}{5}$ **11.** $\dfrac{1}{5}$ **13.** 2
15. $-\dfrac{5}{3}$ **17.** $\dfrac{5}{7}$ **19.** $-\dfrac{4}{3}$ **21.** -1 **23.** 1
25. Undefined   **27.** 0   **29.** 3

31.

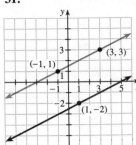

33.

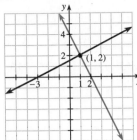

35. $-\dfrac{4}{3}$

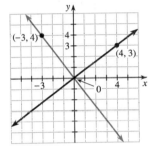

37. $\dfrac{1}{2}$

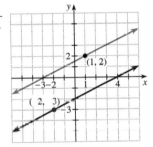

39. 1

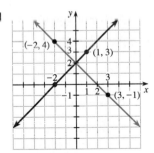

41. Slopes of opposite sides are $\dfrac{1}{5}$ and -3

43. Slopes of opposite sides are $\dfrac{1}{2}$ and -2

45. Two sides have slopes 1 and -1 **47.** No

Section 8.4   Warm-ups TFTTTFFTTF

1. $y = \dfrac{3}{2}x + 1$ **3.** $y = -2x + 2$ **5.** $y = x - 2$

7. $y = -x$ **9.** $y = -1$ **11.** $x = -2$ **13.** 3, $(0, -9)$

15. 0, $(0, 4)$ **17.** -3, $(0, 0)$ **19.** -1, $(0, 5)$

21. $\dfrac{1}{2}$, $(0, -2)$ **23.** $\dfrac{2}{5}$, $(0, -2)$ **25.** 2, $(0, 3)$

27. Undefined, no y-intercept   **29.** $x + y = 2$

31. $x - 2y = -6$ **33.** $9x - 6y = 2$ **35.** $6x + 10y = 7$

37. $x = 6$ **39.** $3y = 20$

41.

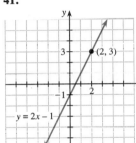

43.

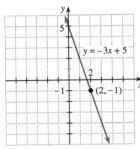

45.

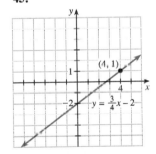

47.

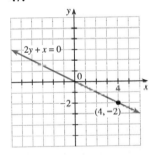

49.

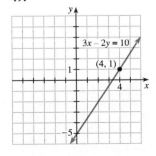

51.

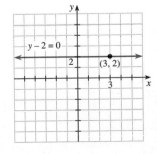

53.

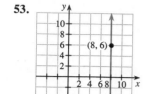

55. $y = -\dfrac{1}{3}x + 6$

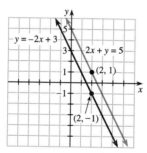

57. $y = -2x + 3$

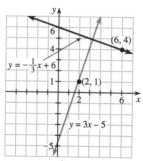

59. $y = 3$

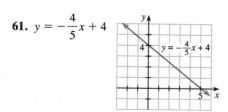

61. $y = -\dfrac{4}{5}x + 4$

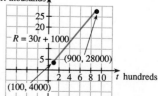

Section 8.5   Warm-ups FFTTFTTTTT

1. $y = \dfrac{1}{3}x + \dfrac{7}{3}$ **3.** $y = -\dfrac{1}{2}x + 4$ **5.** $y = -3x - 4$

7. $3x - 2y = -1$ **9.** $3x + 5y = -11$ **11.** $x - y = -2$

13. $y = -\dfrac{1}{3}x + 5$ **15.** $y = x + 3$ **17.** $y = -\dfrac{2}{3}x + \dfrac{5}{3}$

19. $y = -2x - 5$

21. $y = \dfrac{1}{3}x + \dfrac{7}{3}$

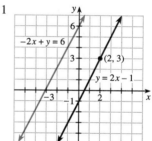

23. $y = 2x - 1$

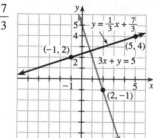

25. $2x - y = 7$ **27.** $x - 2y = 6$ **29.** $2x - 3y = 2$

31. $y = 5x + 11$ **33.** $y = \dfrac{3}{4}x - 20$ **35.** $y = \dfrac{2}{3}x + \dfrac{1}{3}$

Section 8.6   Warm-ups TTFTTFTFTT

1. $F = 3Y$ **3.** $D = 65T$ **5.** $C = \pi D$ **7.** $P = 7.8H$

9.

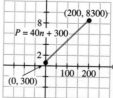

11.

13.

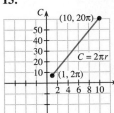

15.

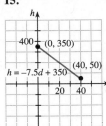

17. $7.05

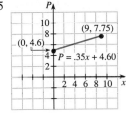

19. $C = 20n + 30$, $170   **21.** $S = 3L - \dfrac{41}{4}$, $8\dfrac{1}{2}$

23. $V = 32t + 10$, 122 feet per second

25. $w = -\dfrac{1}{120}t + \dfrac{3}{2}$, $\dfrac{5}{6}$ inch   **27.** $A = .6w$, 3.6 inches

Section 8.7   Warm-ups FTFTFFFTFT

1. $C = .5t + 5$ **3.** $C = 2\pi r$ **5.** $A = \pi r^2$ **7.** $P = 4s$
9. $s = \sqrt{A}$ **11.** $d = s\sqrt{2}$ **13.** Yes   **15.** Yes
17. No   **19.** Yes   **21.** Yes   **23.** Yes   **25.** No
27. Yes   **29.** Yes   **31.** No   **33.** Yes   **35.** No
37. Yes   **39.** Yes   **41.** Yes   **43.** No
45. {1, 2, 3}, {3, 5, 7}
47. All real numbers, nonnegative real numbers
49. All real numbers, nonnegative real numbers
51. All real numbers, all real numbers
53. $\{s \mid s > 0\}$, $\{A \mid A > 0\}$ **55.** -1 **57.** 13   **59.** 13
61. 1   **63.** 1   **65.** 3   **67.** -2 **69.** -5 or 5
71. The range of g does not include -7 **73.** 2.163
75. .31   **77.** 24.04   **79.** .815 or 3.385

Section 8.8   Warm-ups TFTFFTTTFF

1. $T = kh$ **3.** $y = \dfrac{k}{r}$ **5.** $R = kts$ **7.** $i = kb$

9. $A = kym$ **11.** $y = \dfrac{5}{3}x$ **13.** $A = \dfrac{6}{B}$ **15.** $m = \dfrac{198}{P}$

17. $A = 2tu$ **19.** $T = \dfrac{9}{2}u$ **21.** 25   **23.** 1   **25.** 105

27. 100.3 pounds   **29.** 50 minutes   **31.** $17.40

Chapter 8 Review Exercises

1. Quadrant II   **3.** x-axis   **5.** y-axis   **7.** Quadrant II
9. $(0, -5)$, $(-3, -14)$, $(4, 7)$

11. $\left(0, -\dfrac{8}{3}\right)$, $\left(3, -\dfrac{2}{3}\right)$, $\left(-6, -\dfrac{20}{3}\right)$

13.

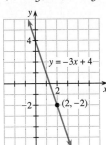

15.

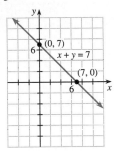

17. $\sqrt{65}$, $(-2, 4.5)$ **19.** 5, $(1.5, 2)$ **21.** 13, $(6, 2.5)$

23. Yes   **25.** Yes   **27.** 1   **29.** $\dfrac{3}{2}$ **31.** $\dfrac{3}{7}$ **33.** Yes

35. Slopes are 1 and -1 **37.** 3, $(0, -18)$
39. 2, $(0, -3)$ **41.** 2, $(0, -4)$

43.

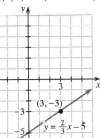

45.

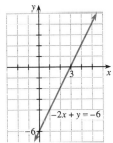

47.

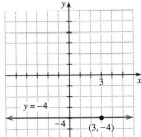

49. $x - 3y = -12$ **51.** $x + 2y = 0$ **53.** $y = 5$

55. $y = \dfrac{2}{3}x + 7$ **57.** $y = \dfrac{3}{7}x - 2$ **59.** $y = -\dfrac{3}{4}x + \dfrac{17}{4}$

61. $y = -2x - 1$ **63.** $y = \dfrac{6}{5}x + \dfrac{17}{5}$ **65.** $y = 3x - 14$

67.

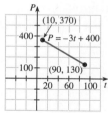

69.

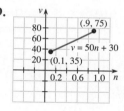

13. $\sqrt{10}$ **14.** $-2\sqrt{2}$ **15.** 30   **16.** -2
17. $17 + 12\sqrt{2}$ **18.** $5\sqrt{5}$ **19.** $-3x + 21$
20. $-5x + 21$ **21.** $2x^2 - 13x + 21$ **22.** $4x^2 - 4x + 1$
23. $z^2 + 10z + 25$ **24.** $w^2 - 49$

25.

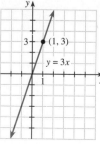

26.

71. $C = 32n + 49$, \$177   **73.** Yes   **75.** No   **77.** Yes
79. Yes   **81.** No   **83.** All real numbers, all real numbers
85. $\{1, 2, 3\}$, $\{0, 2\}$ **87.** All real numbers, $\{3\}$ **89.** 132
91. 2   **93.** 60   **95.** \$15

Chapter 8 Test
1. Quadrant II   **2.** x-axis   **3.** Quadrant IV   **4.** y-axis
5. $\sqrt{2}$ **6.** 10   **7.** $(1.5, 2)$ **8.** $(4.5, 3)$ **9.** 1
10. $-\dfrac{5}{6}$ **11.** $y = \dfrac{1}{2}x + 3$ **12.** $y = \dfrac{3}{7}x - \dfrac{11}{7}$
13. $x - 3y = 11$ **14.** $y = -5x + 19$

27.

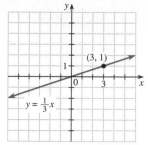

28.

15.

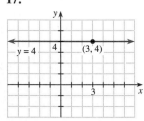

16.

29.

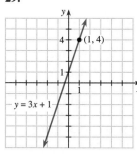

30.

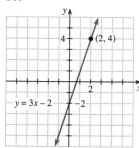

17.

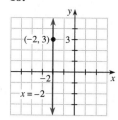

18.

31.

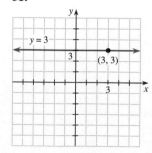

32.

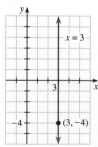

19. All real numbers, $\{y \mid y \geq 1\}$
20. $\{x \mid x \geq 0\}$, $\{y \mid y \geq 0\}$ **21.** 1   **22.** 5   **23.** Yes
24. $S = .75n + 2.50$ **25.** $P = 3v + 20$, 80 cents
26. \$2.80   **27.** 2 hours

Tying It All Together   Chapters 1–8

1. 8   **2.** $2\sqrt{3}$ **3.** 2   **4.** 4   **5.** $\dfrac{1}{4}$ **6.** 4   **7.** 4

8. 1   **9.** $10\sqrt{2}$ **10.** -1 **11.** $2\sqrt{2}$ **12.** $\dfrac{\sqrt{3}}{3}$

33. $y = \dfrac{t - 2}{3\pi}$ **34.** $y = mx + b$ **35.** $y = -1$ or $y = 5$

36. $y = 6$ **37.** $y = 58$ **38.** $y = -5$ or $y = 8$

39. $y = 2 \pm \sqrt{3}$ **40.** $y = \dfrac{4}{5}$ **41.** 3 **42.** $-\sqrt{3}, \sqrt{3}$

43. 28 **44.** $\dfrac{1 - 2\sqrt{2}}{2}, \dfrac{1 + 2\sqrt{2}}{2}$ **45.** $-9, \dfrac{4}{3}$ **46.** $\dfrac{2}{5}$

47. $0, \dfrac{7}{2}$ **48.** $-2, 3$

CHAPTER 9

Section 9.1   Warm-ups TFTTTFTFTT

1. $(3, -2)$ **3.** $(0, -2), (2, 4), (3, 7)$ **5.** None
7. $(2, 4)$ **9.** $(1, 2)$ **11.** $(0, -5)$ **13.** $(-2, 3)$
15. $(0, 0)$ **17.** $(2, 3)$ **19.** No solution, inconsistent
21. $(1, -2)$, independent
23. $\{(x, y) \mid 2x + y = 3\}$, dependent
25. $\{(x, y) \mid y = x\}$, dependent
27. $(-1, -3)$, independent   **29.** $(1, 0)$, independent

Section 9.2   Warm-ups TTTTTTTTTT

1. $(2, 5)$ **3.** $(2, 3)$ **5.** $(-2, 9)$ **7.** $(-5, 5)$
9. $\left(\dfrac{1}{3}, \dfrac{2}{3}\right)$ **11.** $\left(\dfrac{1}{2}, \dfrac{1}{3}\right)$ **13.** $\left(3, \dfrac{5}{2}\right)$, independent
15. No solution, inconsistent
17. No solution, inconsistent   **19.** $\left(\dfrac{7}{3}, 0\right)$, independent
21. $(-1, 1)$, independent
23. \$12,000 at 10%, \$8000 at 5%   **25.** 8 and -6
27. Lawn: \$12, sidewalk: \$7

Section 9.3   Warm-ups TTTTFFTTFF

1. $(3, -1)$ **3.** $(-3, 5)$ **5.** $(-4, 3)$ **7.** $\left(\dfrac{1}{7}, \dfrac{9}{7}\right)$
9. $(8, 31)$ **11.** $(-2, -3)$ **13.** $(-1, 4)$ **15.** $(-1, 2)$
17. $(3, -1)$ **19.** $(1, 2)$ **21.** No solution, inconsistent
23. $\{(x, y) \mid x + y = 5\}$, dependent
25. $\{(x, y) \mid 2x = y + 3\}$, dependent
27. $\{(x, y) \mid x + 3y = 3\}$, dependent
29. No solution, inconsistent
31. $(12, 18)$, independent   **33.** $(40, 60)$, independent
35. $(.1, .1)$, independent   **37.** 150 cars, 100 trucks
39. 6 adults, 24 children   **41.** 180 men, 120 women

Section 9.4   Warm-ups TTTFFFTFTF

1. $(-3, -9)$ **3.** $(3, 0), (1, 3)$ **5.** $(2, 3), (0, 5)$

7.
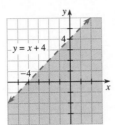

9.
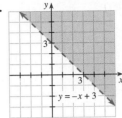

11.
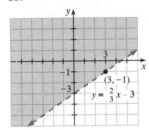

13.
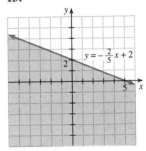

15.
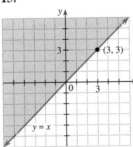

17.
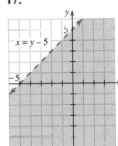

19.
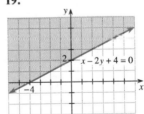

21.
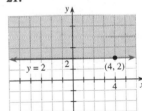

23.
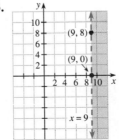

25.
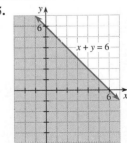

27.

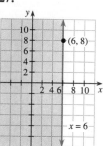

29.

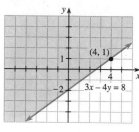

7.

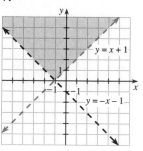

9.

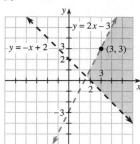

31.

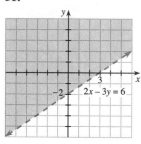

33.

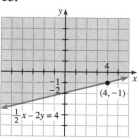

11.

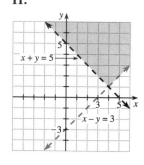

13.

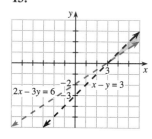

35.

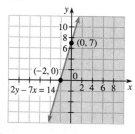

37.

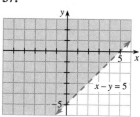

15.

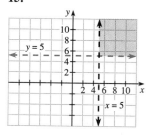

17.

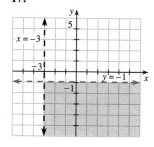

39.

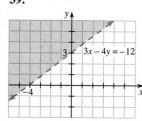

41.

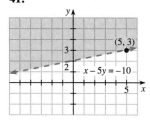

19.

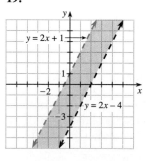

21.

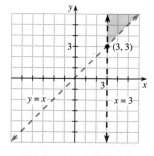

Section 9.5   Warm-ups FFTFFTFTTT

1. $(4, 3)$ **3.** $(3, 6)$ **5.** $(9, -5)$

23.

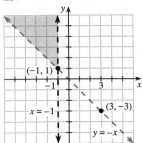

25.

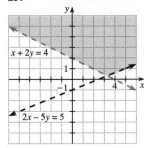

Chapter 9 Review Exercises

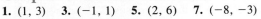

1. $(1, 3)$ **3.** $(-1, 1)$ **5.** $(2, 6)$ **7.** $(-8, -3)$
9. $(3, -1)$, independent
11. $\{(x, y) \mid x - 2y = 4\}$, dependent
13. $(1, -2)$, independent
15. $(0, 0)$, independent
17. $\{(x, y) \mid x - y = 6\}$, dependent
19. No solution, inconsistent

27.

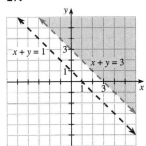

29.

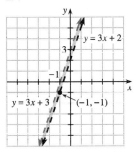

21.

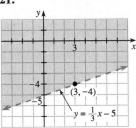

23.

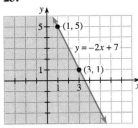

31.

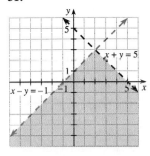

33.

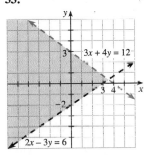

25.

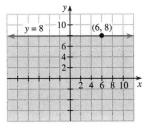

27.

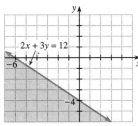

35.

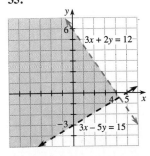

29.

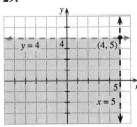

31.

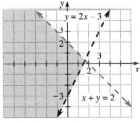

33.

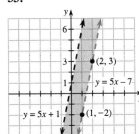

35.

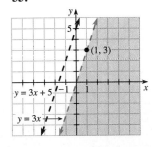

37. Apple: \$.45, orange: \$.35   **39.** 32 fives, 22 tens
41. 4 servings green beans, 3 servings chicken soup

Chapter 9 Test
1. $(-1, 3)$ **2.** $(2, 1)$ **3.** $(3, -1)$ **4.** $(2, 3)$
5. $(4, 1)$ **6.** Inconsistent   **7.** Dependent
8. Independent   **9.**

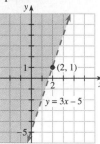

10.

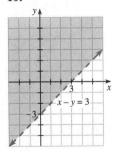

11.

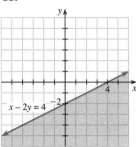

12.

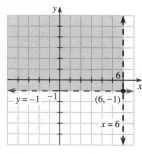

13.

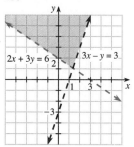

14.

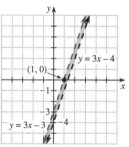

15. \$48   **16.** Kathy: 36 hours, Chris: 18 hours

Tying It All Together   Chapters 1–9
1. 7   **2.** 3, 7   **3.** $\dfrac{5}{3}$ **4.** $-\dfrac{\sqrt{15}}{3}, \dfrac{\sqrt{15}}{3}$ **5.** 12
6. $-6, 4$ **7.** -17 **8.** $-\sqrt{6}, \sqrt{6}$ **9.** 19   **10.** 1
11. $\dfrac{3 - \sqrt{3}}{2}, \dfrac{3 + \sqrt{3}}{2}$ **12.** $-\dfrac{5}{2}$
13. $x > 4$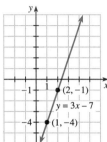
14. $\dfrac{1}{2} \le x \le 5$
15. $x > 1$

16.

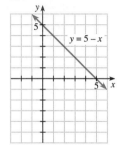

17.

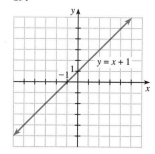

18.

19.

20.

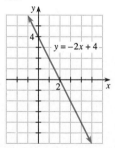

21.

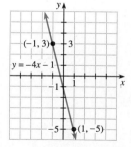

22.

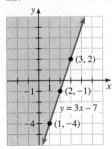

23.

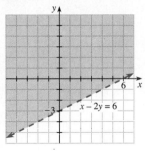

24.

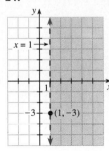

25. $\sqrt{2}$ **26.** $\sqrt{3}$ **27.** 30
28. $2\sqrt{15}$ **29.** $-1 - \sqrt{7}$
30. 2   **31.** $14 - 6\sqrt{5}$
32. $\sqrt{3}$

Index